Springer Climate

Series Editor

John Dodson, Chinese Academy of Sciences, Institute of Earth Environment, Xian, Shaanxi, China

Springer Climate is an interdisciplinary book series dedicated to climate research. This includes climatology, climate change impacts, climate change management, climate change policy, regional climate studies, climate monitoring and modeling, palaeoclimatology etc. The series publishes high quality research for scientists, researchers, students and policy makers. An author/editor questionnaire, instructions for authors and a book proposal form can be obtained from the Publishing Editor. **Now indexed in Scopus® !**

More information about this series at http://www.springer.com/series/11741

Lalit Kumar
Editor

Climate Change and Impacts in the Pacific

 Springer

Editor
Lalit Kumar
School of Environmental and Rural Science
University of New England
Armidale, NSW, Australia

ISSN 2352-0698 ISSN 2352-0701 (electronic)
Springer Climate
ISBN 978-3-030-32880-1 ISBN 978-3-030-32878-8 (eBook)
https://doi.org/10.1007/978-3-030-32878-8

This Springer imprint is published by the registered company Springer Nature Switzerland AG
The registered company address is: Gewerbestrasse 11, 6330 Cham, Switzerland

Preface

Pacific island states are generally small in size with a limited and narrow range of natural resources. Due to their small sizes, generally low elevations and isolation, they are highly vulnerable to natural environmental events. Many of them fall directly in the paths of tropical cyclones. Floods, tropical cyclones, droughts and storm surges have become a part and parcel of these people. However, over many generations, these island people have adapted well to the natural events that are quite regular and over which they have no control. The people of the Pacific have become accustomed to these, and now, such events are inseparable from their lives. As if such challenges in life were not sufficient, we now have a new entrant that is causing havoc in many small islands and is destroying lives and livelihood. This is an event for which humankind is almost totally responsible and over which they have a large degree of control. Yet humankind still refuses to take responsibility for this, and any action to limit its damage is lethargic at best. What we are talking about is climate change.

Climate change and its related issues have become critical for the Pacific islands and its people. While many of the larger countries are modelling what will happen in their surroundings in 50–100 years' time, the people of the Pacific have to deal with it here and now. Communities are being relocated, land is being purchased in other countries to settle entire populations, salt water intrusion and salinity are leading to loss of limited arable land on many islands, crops that have been part of their diet can no longer be cultivated due to rising water tables, and extreme events such as more powerful tropical cyclones are ravaging villages. Yet many of us continue to deny that climate change is a reality and, unfortunately, the leadership of some of the most powerful nations on the planet are providing fuel to such groups. The real-world evidence seems to be having no impact on such thinking. We have pictures of graveyards that are now offshore; who in their right minds would select such locations to have graveyards? We have pictures of houses that have waves crashing into them at every high tide; we have pictures of houses that are almost permanently in the sea now; we have pictures of taro farms regularly inundated by sea water; and the list goes on.

With the people of the Pacific becoming more vocal about this issue in recent years, the issue of climate change impacts in the region has garnered greater attention. Research on likely impacts and best adaptation practices are on the rise. More international support and funds are flowing into the region to support people to deal with the issue of climate change. However, the publicly available information on climate change, its impacts in the Pacific and means of adaptation are few and far in-between and spread over a wide range of scientific publications, web pages and grey literature. One of the objectives of this book is to provide a comprehensive overview of climate change issues in the Pacific, what the people are dealing with now, how susceptible are islands to climate change impacts and how many islanders are adapting to the changes brought about by climate change.

The book starts with a comprehensive overview of climate change and the Pacific, summarising what research has been undertaken and what are the projections for this part of the world over the next 50–100 years. Chapter 2 describes the islands in the Pacific, their settings, distribution and classification. Climate change scenarios and projections for the Pacific are discussed in detail in Chap. 3. This chapter looks at observed climate in the Pacific and compares it with future projections under various climate change scenarios. In Chap. 4, we propose an index that is a relative measure of susceptibility of individual islands in the Pacific to physical change under climate variables. This chapter describes both the physical attributes of islands and environmental variables such as tropical cyclones and significant wave height and how these could be combined to provide information on relative risks of islands. This idea is further refined to a more local (finer) resolution in Chap. 5 where methods are developed for downscaling from whole-island risk assessment to landform susceptibility. A selection of islands from the Pacific is used to demonstrate how this could be incorporated in more local landscape-level risk assessment. Chapter 6 reviews tropical cyclones, its natural variability and potential changes under future climate in the South Pacific. Chapter 7 reports on work undertaken to investigate the distribution of infrastructure in 12 Pacific island countries, with emphasis being on the proportion of built infrastructure in close proximity to the coast and so exposed to coastal climate change impacts. The chapter highlights the very high percentage of infrastructure located very close to the coast and how impacts on such infrastructure could impact on the whole country. Chapter 8 follows the same trend as Chap. 7 but looks at the population distribution across 12 countries in the Pacific. It uses locational data to report on percentage populations in very close proximity to the coastal fringe and how rising sea levels and storm surges related to climate change may impact them. Chapter 9 reports on agriculture under a changing climate in the Pacific. It discusses the significance of agriculture in the Pacific and how climate change and climate extremes may impact on agriculture and sustainability of some agricultural systems. Case studies are used to highlight some of the impacts. Chapter 10 changes from agriculture to marine resources in the Pacific, the importance of such resources to the people and the vulnerability of marine resources to climate change. In Chap. 11, freshwater resources and availability are discussed, including both current issues surrounding freshwater resources and impacts of climate change on water security. Climate change impacts on rainfall and evaporation are also presented. Chapter 12 looks at the impacts of climate change on biodiversity in the Pacific region. It uses the case of terrestrial vertebrate species to show the variety of vulnerable, endangered and critically endangered species that call the Pacific as home and how many of these species occur on one or a few islands only. Many of the species are endemic to the Pacific and so are at an increased risk of extinction due to climate change impacts on the islands they call home. The economic impacts of climate change in the Pacific are explored in Chap. 13. This chapter discusses the economic settings of Pacific islands and how climate change may impact on them. Chapter 14 rounds off the book, looking at the issue of adaptation to climate change. It uses a number of case studies to highlight how different people in different countries of the Pacific are adapting to climate change under different settings. The case studies showcase useful adaptation options and how adaptation could be improved to help people deal with the issues of climate change. So overall, the book covers a wide range of topics very relevant to the climate change debate and to the people of the Pacific and elsewhere.

The travesty is that, quite often, in discourse about climate change and its impacts, the Pacific is overlooked since it is home to only around 12 million people (0.16% of the world's population). What is discounted is that we are talking about 26 countries in this region (13.33% of the 195 countries in the world) having over 30,000 islands, 35 biodiversity hotspots, more than 3200 threatened species of flora and fauna and the world's widest linguistic diversity. The authors contributing to this book hope that it will go some way in highlighting the problems climate change is creating in this part of the world and bringing together a body of literature specifically dealing with the Pacific that will help practitioners make more informed decisions that support them in dealing with climate change.

Finally, I am extremely grateful to all the authors who have volunteered their precious time to share their knowledge with the broader community. I am positive that the knowledge and experience they have willingly shared will have a positive impact on the lives and livelihood of the people of the Pacific. Climate change is now an everyday reality for the Pacific, and their contributions will be appreciated by all.

My heartfelt thanks to the contributors and best wishes to the people of the Pacific in dealing with something they have not contributed to creating but are at the receiving end of probably the greatest impacts. I hope this book serves many researchers and practitioners in this exciting field of climate change.

Armidale, NSW, Australia   Lalit Kumar

Contents

1   **Climate Change and the Pacific Islands**   1
Lalit Kumar, Sadeeka Jayasinghe, Tharani Gopalakrishnan,
and Patrick D. Nunn

2   **Islands in the Pacific: Settings, Distribution and Classification**   33
Patrick D. Nunn, Lalit Kumar, Roger McLean, and Ian Eliot

3   **Climate Change Scenarios and Projections for the Pacific**   171
Savin S. Chand

4   **Comparison of the Physical Susceptibility of Pacific
Islands to Risks Potentially Associated with Variability
in Weather and Climate**   201
Lalit Kumar, Ian Eliot, Patrick D. Nunn, Tanya Stul,
and Roger McLean

5   **Downscaling from Whole-Island to an Island-Coast
Assessment of Coastal Landform Susceptibility
to Metocean Change in the Pacific Ocean**   225
Ian Eliot, Lalit Kumar, Matt Eliot, Tanya Stul, Roger McLean,
and Patrick D. Nunn

6   **A Review of South Pacific Tropical Cyclones:
Impacts of Natural Climate Variability and Climate Change**   251
Savin S. Chand, Andrew Dowdy, Samuel Bell, and Kevin Tory

7   **Impacts of Climate Change on Coastal
Infrastructure in the Pacific**   275
Lalit Kumar, Tharani Gopalakrishnan, and Sadeeka Jayasinghe

8   **Population Distribution in the Pacific Islands,
Proximity to Coastal Areas, and Risks**   295
Lalit Kumar, Tharani Gopalakrishnan, and Sadeeka Jayasinghe

9 Agriculture Under a Changing Climate 323
Viliamu Iese, Siosiua Halavatau, Antoine De Ramon N'Yeurt,
Morgan Wairiu, Elisabeth Holland, Annika Dean, Filipe Veisa,
Soane Patolo, Robin Havea, Sairusi Bosenaqali,
and Otto Navunicagi

**10 Impacts of Climate Change on Marine Resources
in the Pacific Island Region** 359
Johanna E. Johnson, Valerie Allain, Britt Basel, Johann D. Bell,
Andrew Chin, Leo X. C. Dutra, Eryn Hooper, David Loubser,
Janice Lough, Bradley R. Moore, and Simon Nicol

11 Freshwater Availability Under Climate Change 403
Tony Falkland and Ian White

12 Climate Change and Impacts on Biodiversity on Small Islands. 449
Lalit Kumar, Sadeeka Jayasinghe, and Tharani Gopalakrishnan

**13 Economic Impacts and Implications of Climate
Change in the Pacific** 475
Satish Chand

**14 Adaptation to Climate Change: Contemporary
Challenges and Perspectives** 499
Patrick D. Nunn, Roger McLean, Annika Dean, Teddy Fong,
Viliamu Iese, Manasa Katonivualiku, Carola Klöck,
Isoa Korovulavula, Roselyn Kumar, and Tammy Tabe

Index. 525

Abbreviations

ACIAR	Australian Centre for International Agricultural Research
AD	Anno Domini
ADB	Asian Development Bank
ANZ	Australia and New Zealand
AOGCMs	Atmosphere-Ocean General Circulation Models
APCC	APEC Climate Centre
APEC	Asia-Pacific Economic Cooperation
APSIM	Agricultural Production Systems Simulator
AR4	Fourth Assessment Report
AR5	Fifth Assessment Report
BISP	British Solomon Islands Protectorate
BMI	Body Mass Index
BOM	Bureau of Meteorology
$CaCO_3$	Calcium Carbonate
CBA	Community-Based Adaptation
CDF-t	Cumulative Distribution Function Transform
CePaCT	Centre for Pacific Crops and Trees
CH_4	Methane
CMIP3	Coupled Model Intercomparison Project Phase 3
CMIP5	Coupled Model Intercomparison Project Phase 5
CO_2	Carbon Dioxide
CO_3^{2-}	Carbonate Ions
CR	Critically Endangered
CSIRO	Commonwealth Scientific and Industrial Research Organisation
Cv	Coefficient of Variation
DACCRS	Development Assistance Committee Creditor Reporting System
DSAP	Development of Sustainable Agriculture in the Pacific
DSSAT	Decision Support System for Agrotechnology Transfer
EEZ	Exclusive Economic Zone
EN	Endangered
ENSO	El Niño-Southern Oscillation

ESMs	Earth System Models
ESRI	Environmental Systems Research Institute
EU	European Union
FADs	Fish Aggregation Devices
FAO	Food Agriculture Organizations
FRDP	Framework for Resilient Development in the Pacific
FSM	Federated States of Micronesia
GCCA	Global Climate Change Alliance
GCF	Green Climate Fund
GCMs	Global Climate Models
GDP	Gross Domestic Product
GEIC	Gilbert and Ellice Islands Colony
GERD	Gross Domestic Expenditure
GHGs	Greenhouse Gases
GIS	Geographical Information System
GIZ	German Agency for International Cooperation
GSHHG	Global Self-Consistent, Hierarchical, High-Resolution Geography Database
HIV/AIDS	Human Immunodeficiency Virus/Acquired Immunodeficiency Syndrome
H_s	Mean Significant Wave Height
ICM	Integrated Coastal Management
ICT	Information and Communication Technology
IFAD	International Fund for Agricultural Development
IPCC	Intergovernmental Panel on Climate Change
IPO	Interdecadal Pacific Oscillation
IUCN	International Union for Conservation of Nature
LAT to HAT	Lowest Astronomical Tide to Highest Astronomical Tide
LC	Least Concern
LEAP	Local Early Action Planning and Management Planning
LGM	Last Glacial Maximum
LiDAR	Light Detection and Ranging
LMI	Lifetime Maximum Intensity
MEIDECC	Ministry of Meteorology, Energy, Information, Disaster Management, Environment, Climate Change and Communications
MJO	Madden-Julian Oscillation
MORDI	Mainstreaming of Rural Development Innovation
MORDI TT	Mainstreaming of Rural Development and Innovation Tonga Trust
MPAs	Marine Protected Areas
MSL	Mean Sea Level
Mt	Metric Tonnes
N_2O	Nitrous Oxide
NASA	National Aeronautics and Space Administration
NDCs	Nationally Determined Contributions
NESP	National Environmental Science Programme

NGO	Non-government Organisations
NIWA	National Institute of Water and Atmospheric Research (New Zealand)
NT	Near-Threatened
OECD-DAC	Organisation for Economic Co-operation and Development's Development Assistance Committee
PACC	Pacific Adaptation to Climate Change
PACCSAP	Pacific-Australia Climate Change Science and Adaptation Planning Programme
PacRIS	Pacific Risk Information System
PCCSP	Pacific Climate Change Science Program
PCRAFI	Pacific Catastrophe Risk Assessment and Financing Initiative
PDO	Pacific Decadal Oscillation
PICs	Pacific Island Countries
PICTs	Pacific Island Countries and Territories
PLA	Participatory Learning and Action
PNA	Parties to the Nauru Agreement
PNG	Papua New Guinea
PPP	Purchasing Power Parity
PRRP	Pacific Risk Resilience Programme
PSIDS	Pacific Small Island Developing States
PWWA	Pacific Water and Wastes Association
RCMs	Regional Climate Models
RCPs	Representative Concentration Pathways
REDD+	Reducing Emissions from Deforestation and Forest Degradation
RMI	Republic of the Marshall Islands
RO	Reverse Osmosis
SAM	Southern Annular Mode
SEAPODYM	Spatial Ecosystem and Populations Dynamics Model
SLP	Sea-Level Pressure
SLR	Sea-Level Rise
SOI	Southern Oscillation Index
SOPAC	South Pacific Applied Geoscience Commission
SPC	Secretariat of the Pacific Community
SPCZ	South Pacific Convergence Zone
SPCZI	South Pacific Convergence Zone Index
SPEArTC	Southwest Pacific Enhanced Archive for Tropical Cyclones
SST	Sea Surface Temperature
t	Tonnes
TC	Tropical Cyclone
TPI	Tripole Index
UK	United Kingdom
UNDP	United Nations Development Programme
UNESCO	United Nations Educational, Scientific and Cultural Organization
UNFCCC	United Nations Framework Convention on Climate Change

US	United States
US$	US Dollars
USA	United States of America
USAID	United States Agency for International Development
USD	US Dollars
USGS	United States Geological Survey
USP	The University of the South Pacific
VDS	Vessel Day Scheme
VU	Vulnerable
WCPFC	Western and Central Pacific Fisheries Commission
WCPO	Western and Central Pacific Ocean
WDI	World Development Indicators Database
WGS84	World Geodetic System 1984
WHO	World Health Organization
WMO	World Meteorological Organization
WRI	World Resources Institute
WVS	World Vector Shorelines

About the Editor

Lalit Kumar is a Professor in Spatial Modelling, specialising in GIS and remote sensing applications in agriculture, environment and climate change-related impacts. He has an MSc in Environmental Physics from the University of the South Pacific in Fiji and a PhD in Remote Sensing from the University of New South Wales in Sydney, Australia. He worked as a Lecturer at the University of the South Pacific and the University of New South Wales for a number of years before moving to Europe and taking a position as an Associate Professor at the International Institute for Aerospace Survey and Earth Sciences (ITC) where he worked for 5 years. He then moved to the University of New England in Australia where he is currently a Full Professor in Environmental Science.

He has over 25 years' experience in the application of satellite and environmental data layers for broad-scale environmental monitoring, change detection, land-use change and its impacts, above-ground biomass estimation, pasture quality assessment and impacts of climate change on invasive species and biodiversity. He has research and consultancy experience in a number of countries, including Kenya, South Africa, Tanzania, Burkina Faso, Indonesia, India, Nepal, Bhutan, Sri Lanka, Bangladesh, Fiji and Australia, to name a few.

He has published extensively in international peer-reviewed journals, with over 220 journal articles and more than 150 conference papers and technical reports. His work has been cited over 6500 times. He is an Editor on a number of international journals, such as *Remote Sensing*; *ISPRS Journal of Photogrammetry and Remote Sensing*; *Geomatics, Natural Hazards and Risk*; *PLOS One*; *Sustainability*; and *Remote Sensing Applications: Society and Environment*. He has successfully supervised over 25 PhD students to completion and currently has 14 PhD students in his lab group. Some of his work is showcased at lalit-kumar.com, and he can be reached at lkumar@une.edu.au.

Contributors

Valerie Allain Pacific Community (SPC), Noumea, New Caledonia

Britt Basel C2O Pacific, California, USA

Ecothropic, San Cristóbal de las Casas, Chiapas, Mexico

Johann D. Bell Australian National Centre for Ocean Resources and Security, University of Wollongong, Wollongong, NSW, Australia

Conservation International, Arlington, VA, USA

Samuel Bell Center for Informatics and Applied Optimization, Federation University Australia, Mt Helen, VIC, Australia

Sairusi Bosenaqali Pacific Centre for Environment and Sustainable Development, University of the South Pacific, Suva, Fiji

Satish Chand School of Business, Australian Defence Force Academy, UNSW Canberra, Canberra, ACT, Australia

Savin S. Chand Center for Informatics and Applied Optimization, Federation University Australia, Mt Helen, VIC, Australia

Andrew Chin College of Science and Engineering, James Cook University, Townsville, QLD, Australia

Annika Dean University of New South Wales, Sydney, NSW, Australia

Andrew Dowdy Bureau of Meteorology, Melbourne, VIC, Australia

Leo X. C. Dutra CSIRO Oceans and Atmosphere Business Unit, Queensland BioSciences Precinct, St Lucia, Brisbane, QLD, Australia

School of Marine Studies, Faculty of Science, Technology and Environment, The University of the South Pacific, Suva, Fiji

Ian Eliot University of Western Australia and Damara WA Pty Ltd., Innaloo, WA, Australia

Matt Eliot Damara WA Pty Ltd., Innaloo, WA, Australia

Tony Falkland Island Hydrology Services, Canberra, ACT, Australia

Teddy Fong University of the South Pacific, Suva, Fiji

Tharani Gopalakrishnan School of Environmental and Rural Science, University of New England, Armidale, NSW, Australia

Siosiua Halavatau Land Resource Division—Pacific Community (SPC), Suva, Fiji

Robin Havea School of Computing, Information and Mathematical Sciences, The University of the South Pacific, Suva, Fiji

Elisabeth Holland Pacific Centre for Environment and Sustainable Development, University of the South Pacific, Suva, Fiji

Eryn Hooper C2O Pacific, Port Vila, Vanuatu

Viliamu Iese Pacific Centre for Environment and Sustainable Development, University of the South Pacific, Suva, Fiji

Sadeeka Jayasinghe Department of Export Agriculture, Faculty of Animal Science and Export Agriculture, Uva Wellassa University, Badulla, Sri Lanka

Johanna E. Johnson C2O Pacific, Port Vila, Vanuatu

College of Science and Engineering, James Cook University, Townsville, QLD, Australia

Manasa Katonivualiku United Nations Economic and Social Commission for Asia and the Pacific, Bangkok, Thailand

Carola Klöck SciencePo, Paris, France

Isoa Korovulavula University of the South Pacific, Suva, Fiji

Lalit Kumar School of Environmental and Rural Science, University of New England, Armidale, NSW, Australia

Roselyn Kumar University of the Sunshine Coast, Maroochydore, QLD, Australia

David Loubser Ecosystem Services Ltd, Wellington, New Zealand

Janice Lough Australian Institute of Marine Science (AIMS), Townsville, QLD, Australia

Roger McLean School of Physical, Environmental and Mathematical Sciences, University of New South Wales at the Australian Defence Force Academy, Canberra, ACT, Australia

Bradley R. Moore Institute for Marine and Antarctic Studies (IMAS), University of Tasmania, Hobart, TAS, Australia

National Institute of Water and Atmospheric Research (NIWA), Nelson, New Zealand

Otto Navunicagi Pacific Centre for Environment and Sustainable Development, University of the South Pacific, Suva, Fiji

Simon Nicol Pacific Community (SPC), Noumea, New Caledonia

Patrick D. Nunn University of the Sunshine Coast, Maroochydore, QLD, Australia

Antoine De Ramon N'Yeurt Pacific Centre for Environment and Sustainable Development, The University of the South Pacific, Suva, Fiji

Soane Patolo Mainstreaming of Rural Development Innovations, Tonga Trust (MORDI TT), Nuku'alofa, Tonga

Tanya Stul Damara WA Pty Ltd., Innaloo, WA, Australia

Tammy Tabe University of the South Pacific, Suva, Fiji

Kevin Tory Bureau of Meteorology, Melbourne, VIC, Australia

Filipe Veisa Pacific Centre for Environment and Sustainable Development, University of the South Pacific, Suva, Fiji

Morgan Wairiu Pacific Centre for Environment and Sustainable Development, University of the South Pacific, Suva, Fiji

Ian White Fenner School of Environment and Society, Australian National University, Canberra, ACT, Australia

Chapter 1
Climate Change and the Pacific Islands

Lalit Kumar, Sadeeka Jayasinghe, Tharani Gopalakrishnan, and Patrick D. Nunn

1.1 Introduction

Since the late twentieth century, climate change has undeniably been the world's most prominent environmental issue. When it first emerged, climate change was discussed exclusively by scientists. However, in recent years, the general public has become much more involved in the concept, with the subject also creating major political repercussions in several countries. The likely consequences of global climate change have reached an alarming state in view of environmental, physical, and socio-economic aspects and pose a critical threat on a global scale. Increased public involvement in climate change discourse, ensuring subsequent awareness of the potential threats and uncertainties associated with the issue, is crucial.

The term 'climate change' is used with different implications and perspectives. In its broadest sense, climate change refers to any significant change in the statistical properties of the climate system that persists for an extended period, typically 30 years (IPCC 2014). In order to understand climate change, one has to have an understanding of all of the system's components (i.e. atmosphere, ocean, land surface processes, cryosphere, and biosphere), climate variables (temperature and precipitation), and climate descriptors (such as the Earth's surface temperature, ocean

L. Kumar (✉) · T. Gopalakrishnan
School of Environmental and Rural Science, University of New England,
Armidale, NSW, Australia
e-mail: lkumar@une.edu.au; tgopalak@myune.edu.au

S. Jayasinghe
Department of Export Agriculture, Faculty of Animal Science and Export Agriculture,
Uva Wellassa University, Badulla, Sri Lanka
e-mail: ljayasi2@myune.edu.au

P. D. Nunn
University of the Sunshine Coast, Maroochydore, QLD, Australia
e-mail: pnunn@usc.edu.au

© Springer Nature Switzerland AG 2020
L. Kumar (ed.), *Climate Change and Impacts in the Pacific*, Springer Climate,
https://doi.org/10.1007/978-3-030-32878-8_1

temperatures, and snow cover) (IPCC 2001; Weber 2010). This global phenomenon has been created from a combination of natural (such as changes in the sun's radiation and volcanoes) and anthropogenic (such as burning fossil fuels and inappropriate land use changes) activities (Fröhlich and Lean 1998).

Palaeoclimatologists have been investigating how the climate system, including increasing atmospheric temperature trends, rising sea levels, and increasing atmospheric greenhouse gases, has changed on a global scale over many decades (Easterling et al. 2010). An overwhelming majority in the scientific community conclude that future human-induced climate change is inevitable and will have far-reaching environmental impacts that will affect the ways people live in many parts of the world. It is widely agreed that observed global warming is rooted in climate change. Global warming disturbs natural cycles and causes several irreversible changes over the long term. The main cause of the warming trend is the emission of greenhouse gases (GHGs) from human activity which enhances the 'greenhouse effect'. The consequences of a continued enhancement of the natural greenhouse effect is likely to result in warming greater than what has been experienced on average over the past century. Warmer conditions will result in more evaporation and precipitation, but different regions will experience these changes at different scales; some will be wetter and others drier (Van Aalst 2006). Moreover, a stronger greenhouse effect increases sea levels, increases ocean heat content, and promotes the loss of ice mass in Greenland, Antarctica, and the Arctic and mountain glaciers worldwide; it generates more intense and longer droughts in many regions, relatively lower mountain glaciers and snow cover in both hemispheres, higher atmospheric water vapour, ocean acidification, and changes in the historical pattern of extreme weather events (Meinshausen et al. 2009; Nerem et al. 2018).

Since the industrial revolution, the average temperature of the Earth has increased; average global surface temperature rose by 0.9 °C between 1880 and 2015 (Rahmstorf et al. 2017). Much of this heat has been absorbed by the oceans, with the top 700 meters of ocean warming over 0.2 °C since 1969 (Levitus et al. 2017). This warming has been driven mainly by increases in all the major GHGs, particularly carbon dioxide (CO_2), methane (CH_4), and nitrous oxide (N_2O). Emissions of these GHGs continue to increase. For example, concentrations of atmospheric CO_2 rose from approximately 290 ppm to 430 ppm between 1880 and 2014 (IPCC 2014). The IPCC (2014) report states that CO_2 concentrations are likely to rise to around 450 ppm by 2030, and if they continue to increase and reach around 750 ppm to 1300 ppm, the Earth may experience global mean temperature rises of 3.7 °C to 7.8 °C (compared to the 1986–2005 average) by 2100 (Rahmstorf et al. 2017). Net greenhouse gas emissions from anthropogenic activities worldwide increased by 35% from 1990 to 2010. Burning of fossil fuels is still on the rise and is the primary cause of observed growth in GHGs, which accounts for 80% of the overall emissions. Greenhouse gas emissions from agriculture are in the range of 10–15% of the total emissions, and 5–10% of emissions are created from changes in land use patterns. Increased levels of GHGs cause radiative energy to rise and then increase the temperature on Earth's surface. Higher GHG concentrations increase the amount of heat that the atmosphere absorbs and redirects back to the surface. It has been reported that the Earth currently retains approximately 816 terawatts of excess heat per year, which further increases the surface temperature (Henderson et al. 2015).

Scientific evidence of global warming is unambiguous, and many research organizations have built a comprehensive basis of evidence to understand how our climate is already changing (IPCC 2014). Each of the last three decades has been warmer than any previous decade. Changes have been observed since 1950 in many extreme weather and climate events (Gutowski et al. 2008). Greenland and Antarctica's ice sheets have declined in volume and area. Data from NASA's Gravity Recovery and Climate Experiment (NASA 2019) show that, between 1993 and 2016, an average of 286 billion tonnes of ice per year was lost by Greenland, while Antarctica has lost about 127 billion tonnes of ice per year over the same period. Over the past decade, the Antarctic ice mass loss rate has tripled (NASA 2019). Greenland lost 150 km^3 to 250 km^3 of ice annually between 2002 and 2006, while Antarctica lost about 152 km^3 of ice between 2002 and 2005. Glaciers have retreated throughout the world, particularly in the Alps, the Himalayas, the Andes, the Rockies, and Alaska. Declining Arctic sea ice has also been observed over the past several decades (Church et al. 2013). Satellite images show that the extent of snow cover in spring in the northern hemisphere has fallen in the last five decades and that winter snow is now melting earlier than normal (Du Plessis 2018). Over the last century, global sea level rose about 20.3 cm, yet the rate over the past two decades is almost double that of the last century and is slightly accelerating each year (Nerem et al. 2018).

The acidity of ocean waters, particularly surface ocean waters, has increased by about 30% since the beginning of the industrial revolution. This is due to more CO_2 being emitted into the atmosphere with concomitant increases in its absorption by the oceans. The amount of CO_2 absorbed by the upper ocean layer has been increasing by approximately 2 billion tonnes per year (Sabine et al. 2004; Schmutter et al. 2017). The scientific community generally agrees that global warming needs to be limited to 2 °C above pre-industrial levels by the end of the twenty-first century in order to avoid potentially dangerous impacts. This requires concentrations of atmospheric CO_2, estimated at around 430 ppm in 2016, to remain below 450 ppm. Therefore, keeping the Earth within the 2 °C limit requires urgent action. Climate change is a systemic transboundary problem with far-reaching health, security, and prosperity implications for the world. However, despite ongoing efforts to mitigate climate change, global emissions continue to rise. Appropriate approaches will require systematic global efforts to implement systemic changes, and many questions remain as to what form such an effort should take (First 2018).

Many scientists are concerned that the impacts of global warming have developed much more rapidly than expected. Hence the scientific community, the government bodies, and the media have paid considerable attention to climate change and related issues. Signatories to the UNFCCC, the Kyoto Protocol, and the Paris Agreement are discussing how best to tackle this problem, in particular by developing mitigation and adaptation strategies to prevent excessively negative impacts for future generations and to reduce the world's vulnerability to these changes (Saxena et al. 2018; Schelling 2002).

The world is addressing climate change in two ways: mitigation and adaptation. Mitigation involves a reduction in greenhouse gas emissions to alleviate the acceleration of climate change, whereas adaptation involves learning how to live with existing climate change and protecting ourselves against unavoidable future climate change effects (IPCC 2014). The growing body of scientific evidence has led to a clear global consensus on the need for action. UNFCCC commits parties to address climate change by 'preventing dangerous anthropogenic interference with the climate system' by stabilizing GHG levels. Yet the implementation of strategies to mitigate or survive under turbulent climatic conditions requires a broad acceptance/awareness of climate change. A broadened perspective on adaptation and mitigation strategies could help all nations understand the adjustments or actions that can ultimately increase resilience or reduce vulnerability to expected climate and weather changes (IPCC 2014, 2018).

1.2 Impacts of Climate Change

1.2.1 Global Warming of 1.5 °C

In 2018, the IPCC published a special report on the impacts of exceeding 1.5 °C global warming. The report prescribed that limiting global warming to 1.5 °C would need rapid, far-reaching, and unprecedented changes in all aspects of society (First 2018). By limiting global warming to 1.5 °C compared to 2 °C, for example, the negative impacts of climate change would be significantly reduced. While previous estimates focused on estimating the damage where average temperatures were to rise by 2 °C or more (New et al. 2011), this report shows that there will still be many adverse effects of climate change at 1.5 °C. For example, by 2100, global sea-level rise would be 10 cm lower with global warming of 1.5 °C compared to 2 °C. With global warming of 1.5 °C, coral reefs would decline by 70–90%, while almost all would be lost with a 2 °C increase (Hoegh-Guldberg 2014). Global net human-induced CO_2 emissions would have to fall by approximately 45% from 2010 levels by 2030, reaching 'net zero' by 2050, in order to limit global warming to 1.5 °C (First 2018).

1.2.2 Global Warming and Sea-Level Rise

Given the current concentrations and ongoing greenhouse gas emissions, the global mean temperature is likely to continue to rise above pre-industrial levels by the end of this century. This has resulted in extensive melting of ice sheets, both in the Arctic and Antarctic, resulting in rising sea levels regionally and globally. The Arctic Ocean is anticipated to become essentially devoid of summer ice before the middle of the twenty-first century as a result of the warming. Rates of sea-level rise have accelerated since 1870 and now average around 3.5 mm per year (Chen et al. 2017). The average sea-level rise is projected to be 24–30 cm by 2065 and 40–63 cm by 2100 under various scenarios compared to the reference period of 1986–2005 (Allen et al. 2014; Pachauri et al. 2014).

Accelerated sea-level rise will result in higher inundation levels, rising water tables, higher and more extreme flood frequency and levels, greater erosion, increased salt water intrusion, and ecological changes in coastal flora and fauna. These will lead to significant socio-economic impacts, such as loss of coastal resources, infrastructure, and agricultural land and associated declines in economic, ecological, and cultural values (Church et al. 2013). An important issue concerning rising sea levels is that it could submerge parts of low-lying coastal lands which are the habitat of an estimated 470–760 million people (Dasgupta et al. 2007). A number of islands are already submerged, including 11 in Solomon Islands and several in Pohnpei (Federated States of Micronesia (Albert et al. 2016; Nunn et al. 2017). It is predicted that between 665,000 and 1.7 million people in the Pacific will be forced to migrate owing to rising sea levels by 2050, including from atoll islands in the Marshall Islands, Tuvalu, and Kiribati (Church et al. 2013). Very large proportions of the population of Bangladesh (46%) and the Netherlands (70%) are likely to be forced to relocate. By 2100, coastal properties worth $238 billion to $507 billion in the United States alone are likely to be below sea level, with particular risk of inundation and flooding in major cities including Miami, Florida, and Norfolk, Virginia (United Nations 2017).

1.2.3   Changing Weather Patterns and Extreme Events

Climate change will also lead to more frequent and/or severe extreme weather events (Trenberth et al. 2007) and possibly even large-scale, abrupt climate change (Alley et al. 2003). Extreme weather events occur when an individual climate variable (such as temperature or rainfall) exceeds a specific threshold and forces significant divergence from mean climate conditions. The world has already witnessed direct and indirect impacts of climate forcing on extreme events such as storms, hurricanes, tornadoes, severe thunderstorms, floods, and hail, and this trend is expected to continue (Walsh et al. 2016).

Climate change is an urgent threat to the entire human population, contributing to a range of increases in natural disasters. Global rainfall patterns are shifting with rising temperatures. Since the late 1990s, Somalia, Kenya, and other East African countries have experienced lower than average rainfall, contributing to a 30% drop in crop yields and famines in 2010, 2011, and 2016 (Henderson et al. 2015). Hurricanes and other destructive weather events have also increased in prevalence. For instance, the worst typhoons (tropical cyclones) recorded in the Philippines occurred in 2013, resulting in more than 6000 deaths and a displacement of almost four million people (Acosta et al. 2016). Since the early 1980s, the intensity, frequency, and duration of North Atlantic hurricanes and the frequency of the most severe hurricanes have increased (Kossin et al. 2013). Hurricane-related storm intensity and rainfall rates are projected to rise as the climate keeps warming. Storm surges, flooding, and coastal erosion threaten coastal settlements and associated infrastructure, transportation, water, and sanitation (IPCC 2007).

1.2.4 Pressure on Water and Food

Food production is closely related to water availability. In 2014, 16% of the Earth's croplands were irrigated as opposed to rain-fed farming, yet the irrigated land accounted for 36% of global harvest (Pimentel 2012). It is estimated that by 2020, approximately 75–250 million people could be affected by increased water stress in Africa, while rain-fed agriculture-related yields could decrease by up to 50% in some regions (Moriondo et al. 2006). In Pakistan and India, the warming Earth combined with water shortages has been blamed for threatening the viability of the region's agriculture (Henderson et al. 2015). Without significant GHG emission reductions, the proportion of the world's land surface in extreme drought could rise by 2090 to 30%, compared to the current 1–3%.

Warmer temperatures, increased CO_2 levels, and extreme weather events also affect global food production. Agriculture and fisheries depend on specific climatic configurations. Increased CO_2 or warmer weather has the potential to accelerate crop growth or increase yields in some crops; however, crop yield starts to decrease above an optimal temperature that varies from crop to crop (Pimentel 2012). On the other hand, some plant species can respond favourably to increased atmospheric CO_2 and grow more vigorously and more efficiently using less water (Bowes 1993). Higher temperatures and changing climate trends can affect the composition of natural plant cover and change the areas where crops grow best (Rahmstorf et al. 2017). Warmer weather facilitates for the spread of pests, weeds, and parasites, while extreme weather has the potential to harm farmlands, crops, and livestock. Climate change could have a direct and indirect impact on livestock production (Thornton 2010). The warmer climate, particularly heatwaves, has a negative impact on livestock. Drought will impact pasture and feed supplies, posing a risk to livestock retention, while increased prevalence of pests and diseases will affect livestock negatively. Temperature changes could affect fisheries by changing the natural habitat and migration ranges of many aquatic creatures (Brierley and Kingsford 2009).

1.2.5 Human Health Risks

Higher temperatures increase the possibility of injury and death related to heat. In the 2003 European heatwave, as many as 70,000 people died, and in 2010, more than 50,000 died in a heatwave in Russia (Parry 2011). Thousands more have been affected by severe heatwaves in India in 2015, in Europe in 2006, and in other regions around the world (Parry 2011). Water and vector-borne diseases are also projected to increase in a warmer world as insects and other carriers move into higher latitudes and altitudes (Benitez 2009; Conn 2014). Mosquito-breeding regions will also change, leading to potentially greater threats from mosquito-borne diseases (Khormi and Kumar 2014, 2016). A warmer climate also tends to increase lung-related health risk, while fossil fuel burning can lead to premature deaths. The World Health Organization found that, in 2012, seven million people died from air pollution worldwide (Lee and Dong 2012).

1.2.6 Impact on Wildlife and Ecosystems

Climate change also harms many natural habitats and increases many species' risk of extinction (IPCC 2014; Van Aalst 2006). The current extinction rate is 100 times the normal rate, and some scientists predict that the Earth is heading for the sixth mass extinction event in its history (Barnosky et al. 2011). By 2100, 30–50% of the world's terrestrial and marine species may be extinct. Climate change also has significant ocean-related effects (IPCC 2014). Oceans absorb about 25% of CO_2 emitted from the atmosphere, leading to the acidification of seawater. Over the past 100 years, warming has raised near-surface ocean temperatures by about 0.74 °C and has made the sea considerably more acidic, likely affecting marine animal reproduction and survival. As a result of various factors, coral coverage is only half of what it was in the 1960s in some places, and scientists predict that the world's coral reefs could become completely extinct by 2050 (Henderson et al. 2015). Projected future increases in sea surface temperatures of around 1–3 °C are very likely to result in more frequent coral bleaching events and widespread coral mortality if corals are unable to acclimatize or adapt (First 2018).

Ecosystems will continue to change with climate, with some species moving further poleward or becoming more successful at adapting to changes, while some species may be unable to adapt and could become extinct (Parmesan 2006). Changes in temperature and rainfall and extreme events may affect the timing of reproduction in animals and plants, animal migration, length of cropping season, distribution of species and population sizes, and availability of food species. Increased acidification and catastrophic flooding could reduce marine biodiversity and mangrove wealth (Hoegh-Guldberg 2014; Pearson et al. 2019; Schmutter et al. 2017).

1.3 The Pacific Ocean: Location, Size, and Distribution

The Pacific Ocean is the world's largest ocean, with an areal extent of 165 million km^2 and average depth of 4000 m, covering more than 30% of the Earth and bordering 50 countries or territories' coastlines (NOAA 2018). The equator divides the

Pacific Ocean into the North Pacific Ocean and the South Pacific Ocean. The South Pacific Ocean is generally taken to be located between 0° and 60°S latitude and 130°E and 120°W longitude. The Pacific Ocean plays host to a wide range of habitats, such as coral reefs, mangroves, seagrass, and seamounts, and accounts for much of the world's marine biodiversity (Cheung et al. 2010) while also playing a key role in regulating global climate and biogeochemical cycles (Cheung and Sumaila 2013).

The islands in this region cover nearly 528,090 km² of land (0.39%) spread throughout the ocean, with a combined exclusive economic zone (EEZ) of approximately 30 million km² (Carlos et al. 2008) and a total coastline of 135,663 km. Islands are distributed unevenly across the Pacific basin, most being located in the western, especially in the south and western tropical regions, and the fewest in the northeastern quadrant (Fig. 1.1) (Nunn et al. 2016b). The islands belong to a mixture of independent states, semi-independent states, parts of non-Pacific Island countries, and dependent states. The massive realm of islands of the tropical Pacific Ocean includes approximately 30,000 islands of various sizes and topography. In general, the size of the islands in the Pacific decreases from west to east. New Guinea, the largest island, accounts for 83% of the total land area, while Nauru, Tuvalu, and Tokelau have an area less than 30 km². Most Pacific Island nations are comparatively small with total areas less than 1000 km².

The ocean and its resources play a significant role in the livelihoods of the people of the Pacific Islands. Oceania's terrestrial diversity and endemism per unit area are

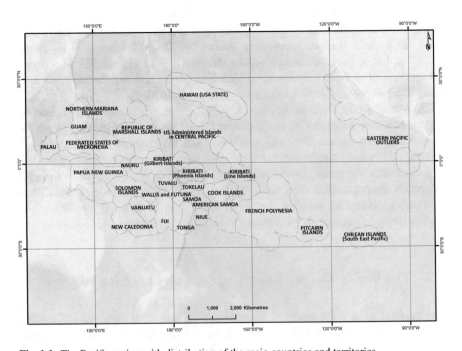

Fig. 1.1 The Pacific region with distribution of the main countries and territories

among the highest on the planet (Keppel et al. 2012; Kier et al. 2009). The region encompasses three global biodiversity hotspots with more than 30,000 plants and 3000 vertebrate species.

Pacific Island countries have been traditionally grouped along the lines of ethno-geographic and cultural lines as Melanesia, Micronesia, and Polynesia. This grouping excludes the adjoining continent of Australia, the Asian-linked Indonesian, Philippine, and Japanese archipelagos as well as those comprising the Ryukyu, Bonin, Volcano, and Kuril arcs which project seaward from Japan.

Melanesia is a subregion of Oceania located in the southwestern region of the Pacific basin, north of Australia, and bordering Indonesia to its east. The region includes the four independent countries of Fiji, Vanuatu, Solomon Islands, and Papua New Guinea and New Caledonia which is a French overseas territory. The dominant feature of Melanesia is relatively large high islands; it includes 98% of the total land area of the Pacific Islands and approximately 82% of the total population. Papua New Guinea is the largest among Melanesian countries as well as the largest country in the Pacific realm with total land area of 67,754 km^2 followed by Solomon Islands (29,675 km^2), New Caledonia (21,613 km^2), Fiji (20,857 km^2), and Vanuatu (13,526 km^2).

Micronesia consists of some 2500 islands spanning more than seven million square kilometres of the Pacific Ocean north of the equator. Micronesia comprises only 0.3% of the total land area of the Pacific Islands and about 5% of the Pacific population. It includes Kiribati, Guam, Nauru, Marshall Islands, Northern Mariana Islands, Palau, and the Federated States of Micronesia (FSM). Kiribati is the largest country in Micronesia with an area of 995 km^2, followed by the Federated States of Micronesia (799 km^2), Guam (588 km^2), Northern Mariana Islands (537 km^2), Palau (495 km^2), Marshall Islands (286 km^2), and Nauru, the smallest single island country of Micronesia with 23 km^2.

Polynesia is the largest region of the Pacific, made up of around 1000 islands scattered over 8000 km^2 in the Pacific Ocean. It is defined as the islands enclosed within a huge triangle connecting Hawaii to the north, New Zealand to the southwest, and Easter Island to the east. It encompasses more than a dozen of the main island groups of central and southern Pacific groups with large distances between them. Polynesia includes Tuvalu, Tokelau, Wallis and Futuna, Samoa (formerly Western Samoa), American Samoa, Tonga, Niue, the Cook Islands, French Polynesia, Easter Islands, and Pitcairn Islands. Polynesia comprises only about 1% of the total Pacific land area but more than 13% of the total population, excluding Hawaii. French Polynesia is the largest country with 3939 km^2 followed by Samoa (3046 km^2), Tonga (847 km^2), Cook Islands (297 km^2), Niue (298 km^2), American Samoa (222 km^2), Easter Island (164 km^2), Tuvalu (44 km^2), Pitcairn Island (54 km^2), and Tokelau with 16 km^2 area.

In terms of geological origin, the islands can be divided into reef islands, volcanic islands, limestone islands, and islands of mixed geological type. The reef islands are generally composed of unconsolidated sediments and commonly form linear groups where a reef has grown above a line of submerged volcanic islands. Examples include most islands in Kiribati, Marshall Islands and Tuvalu, and reef-island groups in the Federated States of Micronesia, French Polynesia, and the western islands of the Hawaii group. They are commonly characterized by their tendency to develop on wide reef surfaces in lower latitudes of the Pacific Ocean (Nunn et al. 2016a).

Volcanic islands are formed when volcanoes erupt (Nunn 1994) and produce islands often with high altitudes in the centre and extremely rugged inner cores. The high island terrain of volcanic islands is characterized by often abrupt changes in elevation (mountains, sheer cliffs, steep ridges, and valleys), with these characteristics varying in altitude and size depending on the island's age (Keener 2013). High islands receive more rainfall than the surrounding ocean from orographic precipitation. This occurs because of the height of the interior of the island, with the warm ocean air being forced up to the higher altitudes, cooling down and falling as rain. The high island landscape is favourable to the formation and persistence of freshwater streams and soil development capable of supporting large and diverse populations of plants and animals (Keener 2013).

The mixed geology-type islands are formed in various ways, principally as a combination of volcanic and coral reef formation. This commonly occurs when the volcanic island forms a high island and a coral reef forms a doughnut-shaped island around it above the water, serving as a barrier from erosion (these are the makatea island types described by Nunn (1994)). Table 1.1 gives some pertinent details, such as population, land area, political status, colonial connections, and dominant lithology of the main Pacific Island countries.

Sea-level rise will directly impact people living in coastal areas of Pacific Island countries. Population distribution is increasingly skewed and concentrated along or near coasts. This is a worldwide phenomenon that is much more pronounced in the Pacific. Kumar et al. (see Chap. 12) analysed the distribution of populations for 12 countries (Cook Islands, Kiribati, Marshall Islands, Nauru, Niue, Palau, Samoa, Solomon Islands, Tonga, Tuvalu, and Vanuatu) in the Pacific and found that around 55% of the population in these countries live within 500 m of the coast, with 20% residing within 100 m. For some of Pacific Island countries, almost the entire population resides in very close proximity to the shoreline. For example, in Kiribati, Marshall Islands, and Tuvalu, the percentage of people living within 500 m of the coast are 98%, 98%, and 99%, respectively.

1.4   Emissions by Pacific Island Countries

Greenhouse gas emissions are spread very unevenly across the world, with the top ten countries generating more than 73.01% of total GHG emissions, and three countries, China (26.83%), the United States (14.36%), and European Union (9.66%), are by far the largest contributors (IPCC 2014). The world's poorest countries have made the least per capita contribution to carbon emissions in the world. These countries burn trivial amounts of fossil fuel compared to countries like China, the United States, Russia, and Australia, and yet they have to bear the greatest impact of climate change (Padilla and Serrano 2006).

Table 1.1 Some key characteristics of the main Pacific Island countries

Country or territory	Population (2014)	Land area (km²)	Political status	Colonial connections[a]	Dominant lithology
Melanesia					
Fiji	903,207	20,857	Independent	UK	Volcanic
New Caledonia	267,840	21,613	Territory	France	Limestone
Papua New Guinea	6,552,730	67,754	Independent	Australia	Volcanic
Solomon Islands	547,540	29,675	Independent	UK	Volcanic
Vanuatu	245,860	13,526	Independent	UK/France	Volcanic
Micronesia					
Fed. States of Micronesia	111,560	799	Free Association	USA	Reef
Guam	161,001	588	Territory	USA	Composite
Kiribati	104,488	995	Independent	UK	Reef
Marshall Islands	54,820	286	Free Association	USA	Reef
Nauru	10,800	23	Independent	UK	Limestone
Northern Mariana Islands	51,483	537	Territory	USA	Volcanic
Palau	20,500	495	Free Association	USA	Limestone
Polynesia					
American Samoa	54,517	222	Territory	USA	Volcanic
Cook Islands	19,800	297	Free Association	New Zealand	Reef
French Polynesia	280,026	3939	Territory	France	Reef
Niue	1480	298	Free Association	New Zealand	Limestone
Samoa	182,900	3046	Independent	New Zealand	Volcanic
Tokelau	1337	16	Territory	New Zealand	Reef
Tonga	103,350	847	Independent	UK	Limestone
Tuvalu	9561	44	Independent	UK	Reef
Wallis and Futuna	15,561	190	Territory	France	Reef/volcanic

[a]The current colonial government or prior to attaining independence status
Based on information from Campbell and Barnett (2010), Kumar and Taylor (2015), and Nunn et al. (2016a)

The Pacific Island region accounts for only 0.03% of the world's total greenhouse gas emissions but is one of the regions that is facing the greatest impacts of climate change from rising sea levels, warming oceans, drought, coral ecosystem destruction, ocean acidification, and extreme weather (Rogers and Evans 2011). For example, CO_2 emissions from Kiribati and Tuvalu are among the lowest of all nations, both in total and per capita terms, yet these are the two countries currently suffering the most from rising sea levels. From Table 1.2, large differences between

Table 1.2 Total CO_2 emissions per country per year and emissions per capita per year measured in 2017 for representative countries in the Pacific, together with selected larger emitters for comparison

Country	Total CO_2 emissions (Mt CO_2/year)	CO_2 emissions per capita (t CO_2/person/year)
Cook Islands	0.07	3.70
Federated States of Micronesia	0.20	1.70
Fiji	1.37	1.55
Kiribati	0.07	0.45
Marshall Islands	0.10	2.30
Nauru	0.10	4.90
New Caledonia	5.76	20.70
Palau	0.86	12.34
Papua New Guinea	5.88	0.70
Samoa	0.17	0.95
Solomon Islands	0.17	0.30
Tonga	0.12	1.30
Tuvalu	0.01	1.10
Vanuatu	0.15	0.50
USA	5188.69	15.85
China	10358.10	7.35
Australia	407.62	16.75
New Zealand	36.39	7.75
India	2460.88	1.80

Notes: (1) Values are fossil fuel-related emissions. They do not consider land use changes or forestry. (2) Presented numbers are averages taken from various sources, including https://en.wikipedia.org/wiki/List_of_countries_by_carbon_dioxide_emissions and http://www.globalcarbonatlas.org/en/CO2-emissions

emissions by the Pacific Island countries and some of the industrialized nations are evident. For comparison, it is more logical to look at CO_2 emissions on a per capita basis. For most of the Pacific Island countries, the per capita emissions are below 2.0 t CO_2 per year, yet for countries such as Australia and the United States, these figures are 16.75 and 15.85 t CO_2 per year, respectively. Australia is one of the world's highest polluters on a per capita basis.

1.5 Projected Climate Change and Impacts

The IPCC report on the impact of global warming states that, if warming continues to increase at the current rate, it is likely to reach 1.5 °C between 2030 and 2052 (high confidence) and small islands are projected to experience higher risks as a consequence (IPCC 2018). In the Pacific, under the RCP4.5 scenario, sea level is likely to increase 0.5 to 0.6 m by 2100 compared to 1986 to 2005 (Church et al. 2013).

The frequency of occurrence of tropical cyclones is likely to remain unchanged or decrease according to the IPCC AR5. On the other hand, the intensity of tropical cyclones is likely to increase with increasing temperatures and precipitation (Christensen et al. 2013; CSIRO 2015). An increase of even 32 cm sea-level rise is projected to have serious consequences for the continued sustainability of ecological and social systems on low coral atolls (Pearce 2000). Wave actions, storm surges, sea-level rise, and river flooding can damage the freshwater supply and in turn have adverse effects on various sectors such as agriculture, tourism, public health, and hydro-electricity production (Campbell and Barnett 2010).

Projected data for Suva, Fiji, show trends of temperature (Fig. 1.2) and rainfall (Fig. 1.3) over the next 80 years to 2100, with the GCMs used in the ensemble modelling shown in Table 1.3.

Figure 1.2 compares temperatures for two RCP scenarios and different time periods. Based on historical data, we can see that the temperature in the Pacific Island region increased slowly from 1951 to 1975, followed by a steady increase until 2010. Observed temperature (1979–2010) was also consistent with this trend. The mean historical temperature data derived from GCMs shows a warming of 0.58 °C within the period 1950 to 2010. In the period from 1979 to 2010, the observed average surface temperature increased by 0.14 °C. Observed data confirms that the average temperature of Suva, Fiji, rose by 0.05 °C per decade since 1979. The projected mean surface temperature change for 2050 relative to 2010 under RCP4.5 is 0.7 °C, while it is 0.84 °C under RCP8.5. The temperature change for 2100 relative to 2010 is projected to be 1.19 °C and 2.9 °C for RCP4.5 and RCP8.5, respectively. Temperature increase for the projected period becomes quite prominent under both

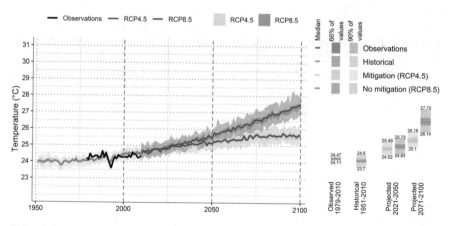

Fig. 1.2 Observed (1979–2010) and projected (until 2100) temperature for Suva, Fiji, under an ensemble of 30 GCMs (Table 1.3). Data for the projections of temperature and rainfall was obtained from the Climate Data Factory website (The Climate Data Factory 2019) <https://theclimatedatafactory.com/> for the period of 1951 to 2100. Different numbers of Global Climate Models (GCMs) obtained from the official IPCC data portal (ESGF 2009) (ESGF <https://esgf.llnl.gov/>) were used to project climate data

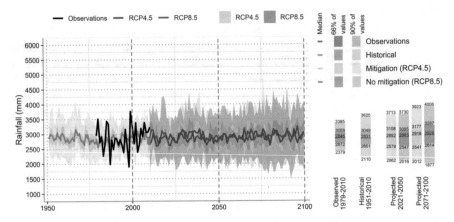

Fig. 1.3 Observed (1979–2010) and projected (until 2100) rainfall for Suva, Fiji, under an ensemble of 31 GCMs (Table 1.3)

RCP4.5 and RCP8.5 towards the end of the century. The difference in terms of temperature values between the RCPs will begin to expand after 2025 (Fig. 1.2).

Suva is already experiencing an increased temperature regime that is evident from the observed temperature which is 0.12 °C (median value) higher than the historical period (1951–2010) (Fig. 1.4). By the end of the twenty-first century, under the business-as-usual scenario (no mitigation, RCP8.5), the temperature will increase by 2.59 °C. Even if the mitigation strategies are implemented successfully (RCP4.5), a 1.33 °C increase in temperature will take place compared to the median value of the observed period. Not only does the temperature increase, but also the inter-annual variance increases in the latter half of the century under both the RCPs (see the confidence intervals on the right of Fig. 1.2). This implies that many hot spells will dominate in the future and, in extreme cases, the annual mean temperature can go even higher than 28 °C, while it was below 24.5 °C during the observed period. However, if mitigation policies are properly implemented as assumed by the RCP4.5 scenario, the temperature is likely to stabilize after 2071, with a median value of 25.6 °C.

Suva receives an annual rainfall of around 2800 mm (median = 2846 mm for 1979–2010 period), and the projections show that rainfall will generally remain similar by 2100 under both selected RCPs (Fig. 1.3). The difference between the radiative forcing of RCP8.5 and RCP4.5 (IPCC 2014) will cause only about 10 mm difference in median values of rainfall during 2021–2050 and 2071–2100 for Suva. In the projected period, the average rainfall under RCP8.5 will be slightly higher than that for RCP4.5; rainfall anomalies (inter-annual variability) will also be considerably higher. This may result in more pronounced wetter and drier seasons in the future, which will have implications for flooding and drought.

Over recent decades, the El Niño-Southern Oscillation (*ENSO*) characteristics have changed quite sharply, even in the absence of obvious external forcing

Table 1.3 Models used for projection of rainfall and temperature data

S. No.	Rainfall RCP8.5	Temperature RCP8.5	Rainfall RCP4.5	Temperature RCP4.5
1	ACCESS1.0	ACCESS1.0	ACCESS1.0	ACCESS1.0
2	ACCESS1.3	ACCESS1.3	ACCESS1.3	ACCESS1.3
3	bcc.csm1.1.m	bcc.csm1.1.m	bcc.csm1.1.m	bcc.csm1.1.m
4	BNU.ESM	BNU.ESM	BNU.ESM	BNU.ESM
5	CanESM2	CanESM2	CESM1.BGC	CCSM4
6	CCSM4	CCSM4	CESM1.CAM5	CESM1.BGC
7	CESM1.BGC	CESM1.BGC	CMCC.CM	CESM1.CAM5
8	CESM1.CAM5	CESM1.CAM5	CNRM.CM5	CMCC.CM
9	CMCC.CESM	CMCC.CESM	EC.EARTH	CNRM.CM5
10	CMCC.CM	CMCC.CM	GFDL.CM3	EC.EARTH
11	CMCC.CMS	CMCC.CMS	GFDL.ESM2G	GFDL.CM3
12	CNRM.CM5	CNRM.CM5	GFDL.ESM2M	GFDL.ESM2G
13	EC.EARTH	EC.EARTH	HadGEM2.CC	GFDL.ESM2M
14	FGOALS.g2	GFDL.CM3	HadGEM2.ES	HadGEM2.CC
15	GFDL.CM3	GFDL.ESM2G	inmcm4	inmcm4
16	GFDL.ESM2G	GFDL.ESM2M	IPSL.CM5A.MR	IPSL.CM5A.MR
17	GFDL.ESM2M	HadGEM2.CC	IPSL.CM5B.LR	IPSL.CM5B.LR
18	HadGEM2.CC	HadGEM2.ES	MIROC.ESM	MIROC.ESM
19	HadGEM2.ES	inmcm4	MIROC.ESM. CHEM	MIROC.ESM. CHEM
20	inmcm4	IPSL.CM5A.MR	MIROC5	MIROC5
21	IPSL.CM5A.MR	IPSL.CM5B.LR	MPI.ESM.LR	MPI.ESM.LR
22	IPSL.CM5B.LR	MIROC.ESM	MPI.ESM.LR.1	MPI.ESM.LR.1
23	MIROC.ESM	MIROC.ESM. CHEM	MPI.ESM.MR	MPI.ESM.MR
24	MIROC.ESM. CHEM	MIROC5	MRI.CGCM3	MRI.CGCM3
25	MIROC5	MPI.ESM.LR		
26	MPI.ESM.LR	MPI.ESM.LR.1		
27	MPI.ESM.LR.1	MPI.ESM.MR		
28	MPI.ESM.MR	MRI.CGCM3		
29	MRI.CGCM3	MRI.ESM1		
30	MRI.ESM1	NorESM1.M		
31	NorESM1.M			

The GCM data were downscaled and bias-corrected using cumulative distribution function transform (CDF-t) method embedded in the CDFt() function of R (Michelangeli et al. 2009). Two packages of R, namely, tidyverse (Wickham 2018) and grid (R Core Team 2019), were used for processing and visualization of the data

(Cobb et al. 2003). Therefore, it is also appropriate to expect similar abrupt changes in climate variability of the tropical Pacific region in the future, with or without a trigger from ongoing greenhouse forcing (Kleypas et al. 2015). However, under the RCP8.5 scenario, the equatorial Pacific is likely to experience an increase in mean

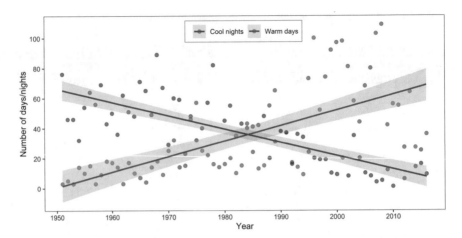

Fig. 1.4 Observed time series of annual total number of warm days (red) and cool nights (blue) for Suva, Fiji, indicating a general warming trend. Grey bands around the linear regression line show one standard error of the estimate (Data: Fiji Meteorological Service)

annual precipitation by 2100 (IPCC 2018). The South Pacific is projected to experience changes in precipitation, relative to 1961–1990, ranging from −3.9% to 3.4% by 2020, −8.23% to 6.7% by 2050, and −14% to 14.6% by 2080 (Barnett 2011).

The changing climate will have impacts across the landscape that will be variable. For example, the rising sea levels and changes in currents will result in significant wave height changes that will affect different regions differently (Fig. 1.5). Mean significant wave height (H_s) data obtained from the South Pacific Applied Geoscience Commission (SOPAC) was modelled using two concentration pathways, RCP 4.5 and RCP 8.5, under the Coupled Model Intercomparison Project Phase 5 (CMIP5) model (http://wacop.gsd.spc.int/) (WACOP 2016). The GCMs used were CNRM-CM5, HadGEM2-ES, INMCM4, and ACCESS1.0. An average value was obtained for 2081–2100 by using the above models, and the difference between the projected and the historical scenario (1986–2005) was derived for projected changes in H_s. From Fig. 1.5, it can be observed that there is likely to be considerable variability in changes in H_s across the Pacific, with H_s differences of up to 0.4 m seen by 2081–2100. The highest increase in H_s will be experienced in the north-west Pacific around Palau and Northern Mariana Islands as well as in the south around Tonga and Niue. Several regions in the Central Pacific are projected to experience no changes in H_s. This projected data for H_s shows that the impacts of climate change will be highly variable across the Pacific region, with some areas being impacted considerably more than others.

Anthropogenic CO_2 has caused a decrease of 0.06 pH units in the tropical Pacific since the beginning of the industrial era (Howes et al. 2018). Currently, the pH of the tropical Pacific Ocean is decreasing at a rate of 0.02 units per decade, and it is projected to decrease by 0.15 units relative to 1986–2005 by 2050 (Hoegh-Guldberg et al. 2014). In addition, the CMIP5 ensemble model projects a further decrease of

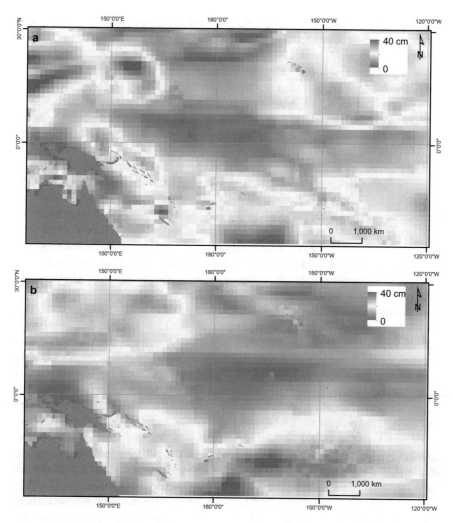

Fig. 1.5 Projected differences in H_s values under RCP4.5 (**a**) and RCP8.5 (**b**) for 2081–2100 compared to historical scenario of 1986–2005 under CMIP5 and an ensemble of GCMs (CNRM-CM5, HadGEM2-ES, INMCM4, and ACCESS1.0). Maximum H_s values were calculated from modelled monthly data supplied by SOPAC (http://wacop.gsd.spc.int/)

0.23–0.28 pH units relative to 1986 to 2005 by 2100 (Howes et al. 2018). This declining seawater pH level corresponds to a decrease in concentration of dissolved carbonate ions (CO_3^{2-}) which may lead to a 'saturation state', lowering the potential of $CaCO_3$ precipitation. According to IPCC AR5, under the RCP8.5 scenario, the aragonite saturation states in the subtropical gyre region will continuously decrease to around 800 ppm by 2100, which will intensify the calcification process with detrimental effects for many shallow-water organisms (Hoegh-Guldberg 2014).

This phenomenon is anticipated to affect the biological and physical complexity of corals; coral cover is projected to decline from the current maximum of 40% to 15 to 30% by 2035 and 10% to 20% by 2050, primarily due to the acidification of the ocean and increasing sea surface temperature (Bruno and Selig 2007; Hoegh-Guldberg 2014). This will also negatively affect the ability of corals to compete with microalgae for space; hence, microalgae are likely to smother a significant proportion of corals by 2035. This pressure on coral reefs will also affect the reproduction of coral reef fish species, numbers of which are projected to decrease 20% by 2050 (Bell et al. 2013).

Climate change will have detrimental impacts on human health directly and indirectly in almost all the regions of the world. Pacific Island countries are particularly vulnerable to health impacts from changing climate due to their unique geologic, social, and economic characteristics (Hanna and McIver 2014; Woodward et al. 2000). Comparatively small size and isolation, their tropical locations, often stagnant economies, and limited health infrastructure are some of the reasons. The direct impacts include damages to health infrastructure, deaths, and traumatic injuries occurring during extreme hydro-meteorological events and physiological effects from heatwaves. For example, in 2015, Cyclone Pam caused severe damages to the health-care system of Vanuatu, destroying 21 of 24 health facilities (hospitals, health centres, and dispensaries) across 22 affected islands in the most affected province (Esler 2015). Indirect impacts occur from the disruption of existing ecosystems, including increased geographic ranges of vectors and increased pathogen loads in food and water (McIver et al. 2012). For example, with the prevailing severe water shortage issue, the changing climate is likely to worsen the diarrheal disease in many Pacific Island countries (Singh et al. 2001). A strong positive correlation was identified between the extreme weather events and outbreaks of dengue fever and diarrhoeal disease in Fiji (McIver et al. 2012). Another foodborne disease of concern is ciguatera, a toxidrome believed to be caused by a toxic dinoflagellate-contaminated reef fish (WHO 2015). Increased incidents of ciguatera in the Pacific Island countries have been reported over the past two decades (Skinner et al. 2011). The ciguatera incidence was linked with marine surface temperatures and ENSO cycles (Llewellyn 2010; Skinner et al. 2011). In addition, the sensitive zones of vectors transmitting pathogens may expand with increases in temperature and alterations in precipitation and humidity (Hanna and McIver 2014).

The biodiversity of Pacific Island regions is also facing pressure from global climate change. Three of 35 global biodiversity hotspots are located in the Pacific Island region, enriched with large numbers of endemic species. The limited amount of suitable habitat and limited capacity for rapid adaptation of small islands make the consequences of accelerating climate change likely to be severe for the region's biodiversity (Taylor and Kumar 2016). Sea-level rise poses a major threat to the restricted species ranges on smaller and atoll islands. In addition, high-elevation ecosystems such as cloud montane forests are projected to disappear by the end of this century (Taylor and Kumar 2016). In an assessment of 23 countries in the Pacific, Kumar and Tehrany (2017) showed that 674 of the islands hosted at least 1 terrestrial vertebrate species that was either vulnerable, endangered, or critically endangered. A total of 84 terrestrial vertebrate species are endemic to this region, and many of them occupy one island only, increasing their chances of extinction.

Climate change is one of the major threats to the culture and traditions of indigenous communities of Pacific Island countries (Keener 2013). A community's response to every dimension of climate change including understanding the causes and responses is mediated by culture (Adger 2006). Nowhere has culture already been threatened by climate change than in the small island states of the Pacific Island region, a trend likely to continue for some time (Ede 2003; Funk 2009; Hunter 2002; Patel 2006). Indigenous people of such islands whose culture is intricately connected to their ancestral lands will experience significant cultural disruption (Farbotko and McGregor 2010). For example, in Samoan culture, the place where families and forebears lived plays an important role in their culture and personal identity; yet increasing numbers of islanders are moving inland or to other countries in search of a more secure future, while some are determined to hold their ground (Piggott-McKellar et al. 2019). In this context, relocations and resettlements have been significantly affecting the state of Samoan culture in terms of loss of heritage and sense of being cut off from the ancestral communities left behind. For instance, the personal connection to the sea has subsequently been lost by those who moved inland or offshore where fishing is no longer their primary source of food (Wing 2017). Such impacts on culture and traditions will be more likely in the future with the accelerating pace of climate change.

1.6   Economic Impacts in the Pacific

Island economies face significant costs due to climate change. According to a recent study by the Asian Development Bank (2013), it is estimated that under the 'business-as-usual scenario', climate change could cost 2.2 to 3.5% of the annual GDP of Pacific Island countries by 2050 and 12.7% by 2100. The agriculture sector was identified as one of the most vulnerable sectors, contributing 5.4% of annual GDP loss by 2100 under the high emission scenario. Agriculture is likely to be affected in various ways, including loss of arable land and contamination of freshwater. For example, in Fiji in 2003, Cyclone Ami caused damage to crops to the value of US$ 35 million (McKenzie et al. 2005), while severe flooding occurred in the Wainbuka and Rewa Rivers in 2004, destroying 50–70% of crops (Connell and Lowitt 2019). The World Bank estimates that climate change may cost Tarawa atoll in Kiribati USD 8–16 million, equivalent to 17–34% of current GDP, by 2050 (World Bank 2017).

Regardless of their size and population, the major socio-economic reality regarding small island countries of the Pacific is that their cost of adapting to climate change is significantly higher in terms of GDP than for larger countries, a phenomenon referred to as 'indivisibility' in economics. For example, for the construction of a similar coastal protection structure, the unit cost per capita in small island countries is substantially higher than for bigger countries with larger populations. In addition, compared to larger or continental territories, the relative impact of a coastal hazard or extreme event has a disproportionate impact on small island countries' GDP compared to continental or larger territories where it only affects a small portion of its total land mass (Pachauri et al. 2014). According to the World Bank Climate Vulnerability Assessment Report of Fiji (World Bank 2017), the country's economic growth has been relatively slow in the last couple of decades because of the impacts of climate change. Fiji is particularly vulnerable to floods and tropical cyclones which have already made a significant impact on the economy. Tropical Cyclone Winston in 2016, with the strongest winds ever recorded in the southern hemisphere, caused damages costing F$2 billion (USD 0.95 billion), equivalent to 20% of Fiji's GDP. During this event, the average losses of assets due to the tropical cyclones and floods alone are estimated at more than F$500 million (USD 230 million).

Tourism is one of the fastest growing sectors in the world. The tourism sector is a common industry in almost all Pacific Island countries and a major source of employment and foreign exchange, contributing an average 20% of GDP and 15% of total jobs (ESCAP 2010). It is also considered as crucial to poverty alleviation and a pathway for achieving economic security coupled with broader development goals around employment and infrastructure (Everett et al. 2018). Climate change has a profound and negative impact on tourism by reducing the value of attractiveness of the tourism destinations (Becken and Hay 2012). Sea-level rise and storm surges pose threats to coastal assets and infrastructure. Kumar and Taylor (2015) have shown that 57% of all infrastructure in 12 Pacific Island countries are within 500 m of the coast, with 20% being within 100 m. This exposes a very large proportion of national infrastructure in these island countries to coastal climate change impacts.

Oceans are intrinsically linked with the atmosphere as they absorb more than 90% of the surplus heat produced by global warming and about two-thirds of CO_2 emitted through anthropogenic activities (Rhein et al. 2013). This affects both the ocean dynamics and ecosystems and consequently has a major impact on the resources they provide to the community (Pörtner et al. 2014). In the Pacific Island countries, fishing and aquaculture contribute substantially to economic development, government revenues, food security, and livelihoods. Climate change impacts on oceans are expected to have major effects on the distribution of fish habitats, the food webs, the fish stocks they support, and, as a consequence, the productivity of fisheries. For example, the combined impacts of increasing temperature, sea-level rise, and alteration of mixing the ocean layer thickness will affect the nutrient supply, lagoon flushing, and ocean acidity and will ultimately affect plankton productivity and survival of corals (FAO 2008; Lal 2004). Stormy weather and more intense cyclones can also make fishing trips unsafe and less productive. This will most likely affect the fish supply, deprive fishermen of income, and potentially threaten the economic security of some island communities (FAO 2008).

1.7 Migration and Displacement Due to Climate Change

Change in the climate system will significantly affect small islands, with severe impacts projected for local economies and livelihoods of people, resulting in human mobility and cross-border displacement and migration (Perch-Nielsen et al. 2008). In certain contexts, particularly in low-lying coastal areas, climate change can be a driving factor in human mobility. Significant migrations from rural atolls to coastal towns and cities or to larger islands have taken place over the past decades in the Pacific Island region (Campbell and Warrick 2014). This has a negative impact on resources in urban coastal areas, and climate change is expected to exacerbate these pressures. In this context, one adaptive strategy for climate change is international migration, especially for the island population who lose livelihood opportunities or whose land disappears or who have limited land. As opportunities and resources diminish, freedom and attraction of movement to other countries or larger islands increase. This, in turn, encourages international migration for those with sufficient resources to move abroad. Therefore, essentially, climate change and rural hardships may encourage people to seek economic opportunities in other countries. Many Pacific Island countries currently have large proportions of their population living abroad; Table 1.4 shows the percentage of population abroad and the main destinations for some Pacific Island countries. Fifty-six percent of the Pacific Islanders who live abroad are settled in New Zealand and Australia, with almost 20,000 more Pacific migrants in the former. North America is the second most popular destination region, with 25% of Pacific immigrants, with the United States having a much larger share than Canada. The special visa schemes for Pacific Islanders in the United States, New Zealand, and Australia provide opportunities for

Table 1.4 Pacific Island countries and territories by share of the total population and major destinations domiciled abroad (2015)

No.	Pacific Island	% total population abroad	Main destinations in order of importance
1	Guam	44.8	Philippines, Northern Mariana Islands, Palau
2	American Samoa	41.8	Samoa, Australia
3	Northern Mariana Islands	39.3	Guam, Palau
4	Tokelau	39.0	New Zealand, Australia
5	Niue	34.6	New Zealand, Australia
6	Nauru	31.1	Kiribati, Australia
7	Palau	26.6	Guam, Northern Mariana Islands, Federated States of Micronesia
8	New Caledonia	24.4	French Polynesia, Australia, Wallis and Futuna Islands
9	Wallis and Futuna Islands	21.7	New Caledonia
10	Cook Islands	19.9	New Zealand, Australia

Adapted from DESA (2015)

temporary and sometimes permanent migration for people living in climatically vulnerable areas (DESA 2015).

Pacific Islanders have been described as one of the world's most mobile groups (Ash and Campbell 2016). Global estimates of migrants relocating as a result of rising sea levels vary. In particular, 'disappearing' or 'sinking' islands force islanders to relocate either within their country or beyond its borders. In fear of future climate change and natural disasters, countries such as Tuvalu, Kiribati, Fiji, Solomon Islands, Vanuatu, and Papua New Guinea have considered new plans for relocations. The move is less challenging when relocation takes place within existing customary land boundaries. However, if relocations occur outside of land boundaries, then the relevant government bodies need to be consulted in order to avoid any conflicts (Ash and Campbell 2016). Kiribati's government has purchased land in Vanua Levu, Fiji, with speculation that ultimately this land will be used to relocate Kiribati to Fiji. However, the Government of Kiribati's statements have tended to focus on the potential of the land for agriculture (Hermann and Kempf 2017). Forced displacement from climate change is highly disruptive to livelihoods, culture, and society unless proper and well-planned interventions support people to adapt to the challenges (Gharbaoui and Blocher 2016; Piggott-McKellar et al. 2019).

Some Pacific Island countries have agreements with Australia, New Zealand, and the United States which already host large groups of immigrants from these countries. Yet, many of those countries with the greatest migration pressures, including Tuvalu, Kiribati, and Nauru, have the fewest available international destinations (Doherty and Roy 2017). Relocation due to climate change has many economic, social, cultural, and psychological costs, although economic and social reasons may be the primary reasons for migration.

1.8   Adaptation, Adaptive Capacity, and Lack of Information and Information Communication Infrastructure

Improving the adaptive capacity of communities in the Pacific Islands is one way to reduce vulnerability. Adaptive capacity is conventionally assumed to be based on the extent to which people can access, understand, and use new knowledge to inform their decision-making processes. This is true in some sense – the pace and nature of current/future climate change is unprecedented – yet much of this knowledge was generated outside the Pacific Island region and is therefore perceived by many people within the region as 'alien', even reflecting a foreign preoccupation that applies to others not to 'us' (Nunn 2009). This is one of the reasons for the widespread and conspicuous failure of most external interventions for climate change adaptation in the Pacific Islands over the past 30–40 years (Piggott-McKellar et al. 2019). It is not that the adaptive capacity of people in the Pacific Island region is low; it is rather that the adaptation pathways they are being offered are unfamiliar and underpinned by unfamiliar reasoning.

Yet to have survived on often quite remote islands in the Pacific for three millennia or more, it is clear that Pacific Island people must have evolved effective ways of coping with climate extremes, be these short-onset events or longer-term periods of changed climate (McNeill 1994; Nunn 2007). Evidence for the former abounds. In several Pacific Island societies, it has been demonstrated that there were methods for ensuring food security in the aftermath of tropical cyclones as well as ways of identifying their precursors (Johnston 2015; Lee and Dong 2012). It is also clear that Pacific Island people survived longer-term climate changes such as the AD 1300 event by changing livelihood strategies (Nunn 2007). In today's globalized world, it is easy for people, especially those outside the Pacific region, to make assumptions about vulnerability and need in an era of rapidly changing climate and to overlook traditional coping strategies. Recently there have been many calls for the renewal and revitalization of such strategies, at least in combination with global knowledge, to help Pacific people cope with the future (Mercer et al. 2007; Nunn and Kumar 2018).

Another reason for adaptation failure that comes as a surprise to many outsiders is that the adaptive solutions being offered to Pacific Island people are invariably secular in nature. These are in conflict with the deeply held religious beliefs through which many decisions, especially around environmental governance, are filtered in Pacific Island communities (Nunn et al. 2016b). Unless adaptation pathways are developed that acknowledge people's spiritual beliefs, it seems unlikely that external interventions for climate change adaptation can become either effective or sustainable in most instances.

In terms of raising awareness about climate change, education is key; yet, public media reports, which often focus on extreme scenarios, are often more persuasive in a Pacific Island context. Many Pacific Island school students are gaining education regarding climate change through school curricula and are experiencing anxiety and frustration at their elders' lack of awareness and foresight (Scott-Parker and Kumar 2018). It seems clear that the localization of climate change awareness and knowledge is key to effective anticipatory adaptation in many Pacific Island contexts.

Telecommunications can help ease the isolation experienced by many of the more remote islands and provide significant access to health care, education, and government services. Unfortunately, due to the remoteness and isolation of the islands in the Pacific, these regions face problems such as lack of access to transport, communications, basic services, and economic opportunities (Dornan and Newton Cain 2014). Pacific Island countries have some of the lowest ICT penetration rates in the world in terms of Internet and mobile phone connectivity. Bandwidth is therefore limited and prices for broadband are high (Cave 2012). Significant progress has been made in recent years in improving telecommunications services in the Pacific Islands. Mobile technology has flourished in this environment. By 2013, one in three residents in Fiji, Tonga, and Tuvalu had access to the Internet (Firth 2018). Mobile phone technology advances were clearly a factor in providing remote areas with Internet access. Fiji has shown significant growth in Internet access and mobile telephone services. The geographic location, service culture, pro-business policies, English-speaking population, and well-connected e-society have supported this trend. Fiji has a relatively reliable and efficient telecommunications system with access to the Southern Cross submarine cable linking New Zealand, Australia, and North America relative to many other South Pacific islands.

Without timely and relevant information, developing Pacific Island states will find it difficult to monitor their progress towards sustainable development. A mature ICT infrastructure is critical for enhancing scientific research, upgrading the technological capabilities of industrial sectors, and encouraging innovation. Research and development expenditure as a proportion of GDP and researchers (in full-time equivalent) per million inhabitants are the two indicators chosen by the United Nations to measure progress (UNESCO 2015). Fiji is the only developing country in the South Pacific with recent data on research and development gross domestic expenditure (GERD). In 2012, the National Statistics Bureau cites a GERD/GDP ratio of 0.15%. Research and development in the private sector are insignificant, while government investment between 2007 and 2012 tended to favour agriculture.

1.9   Conclusions

Climate change has been identified as one of this century's critical challenges for the Pacific region as a whole. The unique vulnerability of the Pacific Island countries to climate change is determined by their geography and environment, frailty of their economic structures, and demographics as well as the interactions between these factors. The vulnerability to climate change in the Pacific Islands is multidimensional and inextricably linked to broader challenges of development. Key impacts include damage to coastal systems, settlements, and infrastructure, undermining recent economic developments, ameliorating existing challenges to water and food security, increasing human health threats, and degrading regional biodiversity (Barnett 2001; Keener 2013). Climate change threatens prosperity and the viability of Pacific Island countries. If the world does not respond effectively to rising greenhouse gas emissions, significant additional stress will be placed on coastal communities, natural ecosystems, water and food security, and the health of islanders in the Pacific. In the face of often menacing climatic conditions, the people of the Pacific have a long history of resilience, and the nations and communities of the Pacific are now actively responding to the new challenges of climate change. With Pacific Island leaders already implementing adaptation measures and looking at relocation options for their climate refugees, islanders will have a better chance of survival if the global warming is limited to a 1.5 °C temperature rise (McNamara and Gibson 2009). The Paris Agreement of the United Nations has committed the world to 'net zero' global greenhouse gas emissions, and it is imperative that this is followed through for the long-term survival of many Pacific Island nations.

References

Acosta LA, Eugenio EA, Macandog PBM, Magcale-Macandog DB, Lin EKH, Abucay ER, Primavera MG (2016) Loss and damage from typhoon-induced floods and landslides in the Philippines: community perceptions on climate impacts and adaptation options. Int J Global Warming 9(1):33–65

Adger W (2006) Vulnerability. Glob Environ Chang 16(3):268–281. https://doi.org/10.1016/j.gloenvcha.2006.02.006

Albert S, Leon JX, Grinham AR, Church JA, Gibbes BR, Woodroffe CD (2016) Interactions between sea-level rise and wave exposure on reef island dynamics in the Solomon Islands. Environ Res Lett 11(5):054011. https://doi.org/10.1088/1748-9326/11/5/054011

Allen MR, Barros VR, Broome J, Cramer W, Christ R, Church JA et al (eds) (2014) IPCC Fifth Assessment Synthesis Report—Climate change 2014 synthesis report. IPCC, Geneva

Alley R, Marotzke J, Nordhaus W, Overpeck J, Peteet D, Pielke RA, Wallace JM (2003) Abrupt climate change. Science 299(5615):2005–2010. https://doi.org/10.1126/science.1081056

Ash J, Campbell J (2016) Climate change and migration: the case of the Pacific Islands and Australia. J Pacific Stud 36(1):53–72

Asian Development Bank (2013) The economics of climate change in the Pacific. Asian Development Bank, Philippines

Earth System Grid Federation (ESGF) (2009) Climate Model Data Service, 1998–2012. NASA, Goddard Space Flight Center. Accessed on 17 May 2019. https://cds.nccs.nasa.gov/data/by-project/esgf/

Barnett J (2001) Adapting to climate change in Pacific Island countries: the problem of uncertainty. World Dev 29(6):977–993

Barnett J (2011) Dangerous climate change in the Pacific Islands: food production and food security. Reg Environ Chang 11(1):229–237

Barnosky A, Matzke N, Tomiya S, Wogan G, Swartz B, Quental T, Mersey B (2011) Has the Earth's sixth mass extinction already arrived? Nature 471(7336):51. https://doi.org/10.1038/nature09678

Becken S, Hay J (2012) Climate change and tourism: from policy to practice. Taylor and Francis Group, Routledge, London

Bell J, Ganachaud A, Gehrke P, Griffiths S, Hobday A, Hoegh-Guldberg O, Matear R (2013) Mixed responses of tropical Pacific fisheries and aquaculture to climate change. Nat Clim Chang 3(6):591. https://doi.org/10.1038/NCLIMATE1838

Benitez M (2009) Climate change could affect mosquito-borne diseases in Asia. Lancet 373(9669):1070

Bowes G (1993) Facing the inevitable: plants and increasing atmospheric CO2. Annu Rev Plant Biol 44(1):309–332

Brierley A, Kingsford M (2009) Impacts of climate change on marine organisms and ecosystems. Curr Biol 19(14):R602–R614. https://doi.org/10.1016/j.cub.2009.05.046

Bruno J, Selig E (2007) Regional decline of coral cover in the Indo-Pacific: timing, extent, and subregional comparisons. PLoS One 2(8):711. https://doi.org/10.1371/journal.pone.0000711

Campbell J, Barnett J (2010) Climate change and small island states: power, knowledge and the South Pacific. Routledge, London

Campbell J, Warrick O (2014) Climate change and migration issues in the Pacific Economic and Social Commission for Asia and the Pacific. United Nations, New York, p 34. https://www.ilo.org/dyn/migpractice/docs/261/Pacific.pdf

Carlos G, Velmurugan A, Jerard B, Karthick R, Jaisankar I (2008) Biodiversity of Polynesian Islands: distribution and threat from climate change. In: Biodiversity and climate change adaptation in tropical islands. Academic Press, Cambridge, MA, pp 105–125

Cave D (2012) Digital islands: how the Pacific's ICT revolution is transforming the region. Lowy Institute for International Policy, Sydney

Chen X, Zhang X, Church J, Watson C, King M, Monselesan D et al (2017) The increasing rate of global mean sea-level rise during 1993–2014. Nat Clim Chang 7(7):492. https://www.nature.com/articles/nclimate3325

Cheung W, Sumaila R (2013) Managing multiple human stressors in the ocean: a case study in the Pacific Ocean, vol 11. Elsevier, Burlington, MA

Cheung W, Lam VW, Sarmiento J, Kearney K, Watson R, Zeller D, Pauly D (2010) Large-scale redistribution of maximum fisheries catch potential in the global ocean under climate change. Glob Chang Biol 16(1):24–35. https://doi.org/10.1111/j.1365-2486.2009.01995.x

Christensen J, Kanikicharla K, Marshall G, Turner J (2013) Climate phenomena and their relevance for future regional climate change. In: IPCC WGI Fifth Assessment Report. Cambridge University Press, Cambridge

Church J, Clark P, Cazenave A, Gregory JM, Jevrejeva S, Levermann A, Payne AJ (2013) Sea-level and ocean heat-content change. Int Geophys 103:697–725. https://doi.org/10.1016/B978-0-12-391851-2.00027-1

Cobb KM, Charles CD, Cheng H, Edwards RL (2003) El Niño/Southern Oscillation and tropical Pacific climate during the last millennium. Nature 424(6946):271

Conn D (2014) Aquatic invasive species and emerging infectious disease threats: a one health perspective. Paper presented at the Aquatic Invasions, Niagara Falls, Canada

Connell J, Lowitt K (2019) Food security in small island states. Springer, Singapore

CSIRO (2015) Tidal Dataset – CAMRIS – Lowest Astronomical Tide. v1. CSIRO. Data Collection. https://doi.org/10.4225/08/55148535DD183. Accessed 25 May 2019

Dasgupta S, Laplante B, Meisner C, Wheeler D, Yan J (2007) The impact of sea level rise on developing countries: a comparative analysis. World Bank Policy Research Working Paper 4136. Paper presented at the Velichko (2007). Ecosystems, their properties, goods, and services. Climate change 2007: impacts, adaptation and vulnerability; Contribution of Working Group II to the Fourth Assessment Report of the Intergovernmental Panel on Climate, Canada. http://hdl.handle.net/10986/7174

DESA (2015) World population prospects: the 2015 revision, key findings and advance tables. United Nations, New York

Doherty, B., Roy, E. A. (2017). World Bank: let climate-threatened Pacific islanders migrate to Australia or NZ. The Guardian. Accessed 12 June 2019. https://www.theguardian.com/environment/2017/may/08/australia-and-nz-should-allow-open-migration-for-pacific-islanders-threatened-by-climate-says-report

Dornan M, Newton Cain T (2014) Regional service delivery among Pacific Island countries: an assessment. Asia Pacific Policy Stud 1(3):541–560

Du Plessis A (2018) Current and Future water scarcity and stress. In: Water as an inescapable risk. Springer, Cham. https://doi.org/10.1007/978-3-030-03186-2_2

Easterling D, Meehl G, Parmesan C, Changnon S, Karl T, Mearns L (2010) Climate extremes: observations, modeling, and impacts. Science's Compass 289(5487):2068–2074. https://doi.org/10.1126/science.289.5487.2068

Ede P (2003) Come hell or high water: rising sea levels and extreme flooding threaten to make the South Pacific's Tuvalu the first victim of global warming. Alternatives J Waterloo 29(1):8–10

ESCAP (2010) Statistical yearbook For Asia and the Pacific 2009. Economic and Social Commission for Asia and the Pacific, Bangkok, Thailand, p 260. https://www.unisdr.org/files/13373_ESCAPSYB2009.pdf

Esler S (2015) Vanuatu post-disaster needs assessment tropical cyclone Pam. Government of Vanuatu, Vanuatu. https://dfat.gov.au/about-us/publications/Documents/post-disaster-needs-assessment-cyclone-pam.pdf

Everett H, Simpson D, Wayne S (2018) Tourism as a river of growth in the Pacific. In: A pathway to growth and prosperity for Pacific Island countries, vol 2. Asian Development Bank, Manila, p 26

FAO (2008) Climate change and food security in Pacific Island Countries: issues and requirements. In: Climate change and food security in Pacific Island Countries. United Nations Food and Agriculture Organization, Rome

Farbotko C, McGregor H (2010) Copenhagen, climate science and the emotional geographies of climate change. Aust Geogr 41(2):159–166. https://doi.org/10.1080/00049181003742286

First P (2018) Global warming of 1.5°C. An IPCC Special Report on the impacts of global warming of 1.5°C above pre-industrial levels and related global greenhouse gas emission pathways, in the context of strengthening the global response to the threat of climate change, sustainable development, and efforts to eradicate poverty. World Meteorological Organization, Geneva

Firth S (2018) Instability in the Pacific Islands. A status report. Lowy Institute. https://www.lowy-institute.org/publications/instability-pacific-islands-status-report

Fröhlich C, Lean J (1998) The Sun's total irradiance: cycles, trends and related climate change uncertainties since 1976. Geophys Res Lett 25(23):4377–4380. https://agupubs.onlinelibrary.wiley.com/doi/pdf/10.1029/1998GL900157

Funk M (2009) Come hell or high water. World Policy J 26(2):93–101. https://doi.org/10.1162/wopj.2009.26.2.93

Gharbaoui D, Blocher J (2016) The reason land matters: relocation as adaptation to climate change in Fiji Islands. In: Migration, risk management and climate change: evidence and policy responses, vol 8. Springer, Switzerland, pp 149–173

Gutowski J, Hegerl C, Holland J, Knutson R, Mearns O, Stouffer J, Zwiers W (2008) Causes of observed changes in extremes and projections of future changes. In: Weather and climate extremes in a changing climate; Regions of Focus: North America, Hawaii, Caribbean, and U.S. Pacific Islands, vol 3. The US Climate Change Science Program, pp 81–116. https://www.gfdl.noaa.gov/bibliography/related_files/wjg0801.pdf

Hanna E, McIver L (2014) 19 small island states–canaries in the coal mine of climate change and health. Clim Change Global Health:181–192

Henderson R, Reinert S, Dekhtyar P, Migdal A (2015) Climate change in 2018: implications for business. vol. 1, pp. 39. https://www.hbs.edu/environment/Documents/climate-change-2018.pdf

Hermann E, Kempf W (2017) Climate change and the imagining of migration: emerging discourses on Kiribati's land purchase in Fiji. Contemp Pac 29(2):231–263

Hoegh-Guldberg O (2014) Coral reef sustainability through adaptation: glimmer of hope or persistent mirage? Curr Opin Environ Sustain 7:127–133. https://doi.org/10.1071/MF99078

Hoegh-Guldberg O, Cai R, Poloczanska ES, Brewer PG, Sundby S, Hilmi K et al (2014) The Ocean. In: Climate change 2014: impacts, adaptation, and vulnerability. Part B: Regional aspects. Contribution of Working Group II to the Fifth Assessment Report of the Intergovernmental Panel on Climate Change. Cambridge University Press, Cambridge/New York, NY

Howes E, Birchenough S, Lincoln S (2018) Effects of climate change relevant to the Pacific islands. Pacific Marine Climate Change Report Card Science Review:1–19

Hunter J (2002) A note on relative sea level change at Funafuti, Tuvalu. Antarctic Cooperative Research Center, Hobart, Austrilia, p 125

IPCC (2001) Climate change 2001: impacts, adaptation, and vulnerability. Intergovernmental Panel on Climate Change Cambridge University Press, Cambridge

IPCC (2007) Climate change 2007: impacts, adaptation and vulnerability. In: Parry ML, Canziani OF, Palutikof JP, Van Der Linden PJ, Hanson CE (eds) Contribution of Working Group II to the Fourth Assessment Report of the Intergovernmental Panel on Climate Change, Cambridge. https://www.ipcc.ch/site/assets/uploads/2018/03/ar4_wg2_full_report.pdf

IPCC (2014) Summary for policymakers in climate change 2014: impacts, adaptation, and vulnerability. Part A: Global and sectoral aspects. Contribution of Working Group II to the Fifth Assessment Report of the Intergovernmental Panel on Climate Change. In: Field CB, Barros VR, Dokken D, Mach K, MAS-TRANDREA M, Bilir T, Chatterjee M, Ebi K, ES-TRADA Y, Genova R (eds) Contribution of Working Group II to the Fifth Assessment Report of the Intergovernmental Panel on Climate Change, pp 1–32. https://www.ipcc.ch/site/assets/uploads/2018/03/ar5_wgII_spm_en-1.pdf: Intergovernmental Panel on Climate Change

IPCC (2018) Global Warming of 1.5° C: An IPCC Special Report on the Impacts of Global Warming of 1.5° C Above Pre-industrial Levels and Related Global Greenhouse Gas Emission

Pathways, in the Context of Strengthening the Global Response to the Threat of Climate Change, Sustainable Development, and Efforts to Eradicate Poverty. Geneva, Switzerland: Intergovernmental Panel on Climate Change

Johnston I (2015) Traditional warning signs of cyclones on remote islands in Fiji and Tonga. Environ Hazard 14(3):210–223. https://doi.org/10.1080/17477891.2015.1046156

Keener V (2013) Climate change and pacific islands: indicators and impacts. In: Report for the 2012 pacific islands regional climate assessment. Island Press, Washington, DC

Keppel G, Morrison C, Watling D, Tuiwawa MV, Rounds I (2012) Conservation in tropical Pacific Island countries: why most current approaches are failing. Conserv Lett 5(4):256–265. https://doi.org/10.1111/j.1755-263X.2012.00243.x

Khormi H, Kumar L (2014) Climate change and the potential global distribution of Aedes aegypti: spatial modelling using geographical information system and CLIMEX. Geospat Health 8(2):405–415

Khormi HM, Kumar L (2016) Future malaria spatial pattern based on the potential global warming impact in South and Southeast Asia. Geospat Health 11(416):290–298. https://doi.org/10.4081/gh.2016.416

Kier G, Kreft H, Lee T, Jetz W, Ibisch P, Nowicki C et al (2009) A global assessment of endemism and species richness across island and mainland regions. Proc Natl Acad Sci 106(23):9322–9327. https://doi.org/10.1073/pnas.0810306106

Kleypas J, Castruccio F, Curchitser E, Mcleod E (2015) The impact of ENSO on coral heat stress in the western equatorial Pacific. Glob Chang Biol 21(7):2525–2539. https://doi.org/10.1111/gcb.12881

Kossin JP, Olander TL, Knapp KR (2013) Trend analysis with a new global record of tropical cyclone intensity. J Climate 26(24):9960–9976. https://doi.org/10.1175/JCLI-D-13-00262.1

Kumar L, Taylor S (2015) Exposure of coastal built assets in the South Pacific to climate risks. Nat Clim Chang 5(11):992. https://www.nature.com/articles/nclimate2702

Kumar L, Tehrany M (2017) Climate change impacts on the threatened terrestrial vertebrates of the Pacific Islands. Sci Rep 7(1):5030. https://www.nature.com/articles/s41598-017-05034-4

Lal M (2004) Implications of climate change in small Island developing countries of the South Pacific. Fiji Stud 2(1):15–35. https://search.informit.com.au/documentSummary;dn=7088923 73132239;res=IELNZC.

Lee Y, Dong G (2012) Air pollution and health effects in children. In: Khare M (ed) Air pollution-monitoring, modelling and health, vol 15. InTech, China, p 386

Levitus S, Antonov J, Boyer T, Baranova O, Garcia H, Locarnini R, et al. (2017) NCEI ocean heat content, temperature anomalies, salinity anomalies, thermosteric sea level anomalies, halosteric sea level anomalies, and total steric sea level anomalies from 1955 to present calculated from in situ oceanographic subsurface profile data (NCEI Accession 0164586). Accessed 14 April 2019, from NOAA, National Centers for Environmental Information. https://doi.org/10.7289/V53F4MVP

Llewellyn L (2010) Revisiting the association between sea surface temperature and the epidemiology of fish poisoning in the South Pacific: reassessing the link between ciguatera and climate change. Toxicon 56(5):691–697. https://doi.org/10.1016/j.toxicon.2009.08.011

McIver L, Naicker J, Hales S, Singh S, Dawainavesi A (2012) Climate change and health in Fiji: environmental epidemiology of infectious diseases and potential for climate-based early warning systems. Fiji J Public Health 1:7–13. https://researchonline.jcu.edu.au/41240/

McKenzie E, Prasad B, Kaloumaira A (2005) Economic impacts of natural disasters on development in the Pacific. Tool one: guidelines for estimating the economic impact of natural disasters on development in the Pacific, University of the South Pacific (USP) and the South Pacific Applied Geoscience Commission (SOPAC): Australian Agency for International Development (AusAID), p. 102

McNamara KE, Gibson C (2009) We do not want to leave our land': Pacific ambassadors at the United Nations resist the category of 'climate refugees. Geoforum 40(3):475–483

McNeill J (1994) Of rats and men: a synoptic environmental history of the island Pacific. J World Hist 5(2):299–349. https://www.jstor.org/stable/20078602

Meinshausen M, Meinshausen N, Hare W, Raper S, Frieler K, Knutti R, Allen M (2009) Greenhouse-gas emission targets for limiting global warming to 2 C. Nat Lett 458(7242):1158. https://doi.org/10.1038/nature08017

Mercer J, Dominey-Howes D, Kelman I, Lloyd K (2007) The potential for combining indigenous and western knowledge in reducing vulnerability to environmental hazards in small island developing states. Environ Hazard 7(4):245–256. https://doi.org/10.1016/j.envhaz.2006.11.001

Michelangeli, P. A., Vrac, M., & Loukos, H. (2009). Probabilistic downscaling approaches: Application to wind cumulative distribution functions. Geophysical Research Letters, 36(11).

Moriondo M, Good P, Durao R, Bindi M, Giannakopoulos C, Corte-Real J (2006) Potential impact of climate change on fire risk in the Mediterranean area. Climate Res 31(1):85–95. https://www.int-res.com/articles/cr2006/31/c031p085.pdf

NASA (2019). Ice Sheets. Global climate change; vital signs of the planet. Accessed 18 May 2019. https://climate.nasa.gov/vital-signs/ice-sheets/

Nerem R, Beckley B, Fasullo J, Hamlington B, Masters D, Mitchum G (2018) Climate-change–driven accelerated sea-level rise detected in the altimeter era. Proc Natl Acad Sci 115(9):2022–2025

New M, Liverman D, Schroder H, Anderson K (2011) Four degrees and beyond: the potential for a global temperature increase of four degrees and its implications. Philos Trans A Math Phys Eng Sci 369(1934):6–19. https://doi.org/10.1098/rsta.2010.0303

NOAA (2018) Ocean Explorer. https://oceanexplorer.noaa.gov/facts/pacific-size.html. Accessed 13 May 2019

Nunn P (1994) Oceanic Islands (natural environment), 1st edn. Wiley-Blackwell, Oxford

Nunn P (2007) The AD 1300 event in the Pacific Basin. Geogr Rev 97(1):1–23. https://doi.org/10.1111/j.1931-0846.2007.tb00277.x

Nunn P (2009) Responding to the challenges of climate change in the Pacific Islands: management and technological imperatives. Clim Res Vlim Res 40(2–3):211–231

Nunn P, Kumar R (2018) Understanding climate-human interactions in Small Island Developing States (SIDS) Implications for future livelihood sustainability. Int J Clim Change Strategies Manage 10(2):245–271. https://doi.org/10.1108/IJCCSM-01-2017-0012

Nunn P, Kumar L, Eliot I, McLean R (2016a) Classifying pacific islands. Geosci Lett 3(1):7. https://doi.org/10.1186/s40562-016-0041-8

Nunn P, Mulgrew K, Scott-Parker B, Hine D, Marks AD, Mahar D, Maebuta J (2016b) Spirituality and attitudes towards Nature in the Pacific Islands: insights for enabling climate-change adaptation. Clim Change 136(3–4):477–493. https://doi.org/10.1007/s10584-016-1646-9

Nunn P, Kohler A, Kumar R (2017) Identifying and assessing evidence for recent shoreline change attributable to uncommonly rapid sea-level rise in Pohnpei, Federated States of Micronesia, Northwest Pacific Ocean. J Coast Conserv 21(6):719–730. https://doi.org/10.1007/s11852-017-0531-7

Pachauri R, Allen M, Barros V, Broome J, Cramer W, Christ R, et al. (2014) Climate change 2014: synthesis report. Contribution of Working Groups I, II and III to the Fifth Assessment Report of the Intergovernmental Panel on Climate Change. In Pachauri R, Meyer L (Eds.), Vol. 151, Intergovernmental Panel on Climate Change, Geneva, Switzerland

Padilla E, Serrano A (2006) Inequality in CO2 emissions across countries and its relationship with income inequality: a distributive approach. Energy Policy 34(14):1762–1772

Parmesan C (2006) Ecological and evolutionary responses to recent climate change. Annu Rev Ecol Evol Syst 37:637–669. https://doi.org/10.1146/annurev.ecolsys.37.091305.110100

Parry, W. (2011). Recent heat waves likely warmest since 1500 in Europe. Live Science. Accessed 15 May 2019. https://www.livescience.com/13296-europeanrussia-heat-waves-climate-change.html

Patel S (2006) A sinking feeling. Nature 440(7085):734–736

Pearce, F. (2000). Turning back the tide. New Scientist, 165(2225), 44–7. Accessed 16 May 2019. https://www.turnbackthetide.ca/

Pearson J, McNamara E, Nunn P (2019) Gender-specific perspectives of mangrove ecosystem services: case study from Bua Province, Fiji Islands. Ecosyst Serv 38:100970. https://doi.org/10.1016/j.ecoser.2019.100970

Perch-Nielsen S, Bättig M, Imboden D (2008) Exploring the link between climate change and migration. Clim Change 91(3–4):375–393. https://doi.org/10.1007/s10584-008-9416-y

Piggott-McKellar A, McNamara K, Nunn P, Watson J (2019) What are the barriers to successful community-based climate change adaptation? A review of grey literature. Local Environ 24(4):374–390. https://doi.org/10.1080/13549839.2019.1580688

Pimentel D (2012) Food and natural resources. Elsevier Science, Amsterdam

Pörtner H, Karl D, Boyd P, Cheung W, Lluch-Cota S, Nojiri Y, Armstrong C (2014) Ocean systems. In Climate change 2014: impacts, adaptation, and vulnerability. In: Field CB, Barros V, Dokken D, Mach K, Mastrandrea M, Bilir T, Chatterjee M, Ebi K, Estrada Y, Genova R, Girma B, Kissel E, Levy A, MacCracken S, Mastrandrea P, White L (eds) Part A: Global and sectoral aspects. Contribution of Working Group Ii TO THE Fifth Assessment Report of the Intergovernmental Panel on Climate Change. Cambridge University Press, Cambridge, pp 411–484

Rahmstorf S, Foster G, Cahill N (2017) Global temperature evolution: recent trends and some pitfalls. Environ Res Lett 12(5):054001. https://doi.org/10.1088/1748-9326/aa685

Rhein M, Rintoul S, Aoki S, Campos E, Chambers D, Feely R et al (2013) Observations: ocean. In: Freeland H, Garzoli S, Nojiri Y (eds) Climate change 2013: the physical science basis. Contribution of Working Group I to the Fifth Assessment Report of the Intergovernmental Panel on Climate Change. Cambridge University Press, Cambridge, p 62

Rogers S, Evans L (2011) World carbon dioxide emissions data by country: China speeds ahead of the rest. The guardian

Sabine C, Feely R, Gruber N, Key R, Lee K, Bullister J et al (2004) The oceanic sink for anthropogenic CO_2. Science 305(5682):367–371. https://doi.org/10.1126/science.1097403

Saxena A, Qui K, Robinson S (2018) Knowledge, attitudes and practices of climate adaptation actors towards resilience and transformation in a 1.5 C world. Environ Sci Policy 80:152–159. https://doi.org/10.1016/j.envsci.2017.11.001

Schelling TC (2002) What makes Greenhouse sense?-Time to rethink the Kyoto protocol

Schmutter K, Nash M, Dovey L (2017) Ocean acidification: assessing the vulnerability of socio-economic systems in Small Island Developing States. Reg Environ Chang 17(4):973–987. https://doi.org/10.1007/s10113-016-0949-8

Scott-Parker B, Kumar R (2018) Fijian adolescents' understanding and evaluation of climate change: implications for enabling effective future adaptation. Asia Pac Viewp 59(1):47–59. https://doi.org/10.1111/apv.12184

Singh R, Hales S, De Wet N, Raj R, Hearnden M, Weinstein P (2001) The influence of climate variation and change on diarrheal disease in the Pacific Islands. Environ Health Perspect 109(2):155–159. https://doi.org/10.1289/ehp.01109155.

Skinner M, Brewer T, Johnstone R, Fleming L, Lewis R (2011) Ciguatera fish poisoning in the Pacific Islands (1998 to 2008). PLoS Negl Trop Dis 5(12):1416. https://doi.org/10.1371/journal.pntd.0001416

Taylor S, Kumar L (2016) Global climate change impacts on pacific islands terrestrial biodiversity: a review. Trop Conserv Sci 9(1):203–223. https://doi.org/10.1177/194008291600900111

Team, R. C. (2019). A language and environment for statistical computing. Vienna, Austria: R Foundation for Statistical Computing; 2012. URL https://www. Rproject.org

The climate data factory, (2019) Climate Projections, 1951 to 2100. 12 rue de Belzunce, Paris, France. Accessed 18 May 2019. https://theclimatedatafactory.com/search-results/?q=suva

Thornton PK (2010) Livestock production: recent trends, future prospects. Philos Trans Royal Soc Biol Sci 365(1554):2853–2867. https://doi.org/10.1098/rstb.2010.0134

Trenberth K, Jones P, Ambenje P, Bojariu R, Easterling D, Klein Tank A., Soden B (2007) Observations: surface and atmospheric climate change Climate change. IPCC Working Group I: National Oceanic and Atmospheric Administration, NOAA, United Kingdom, pp. 235–336

UNESCO (2015) In: Bokova IG (ed) UNESCO Science Report: towards 2030. United Nations, France

United Nations (2017) International Migration Report 2017: Highlights (ST/ESA/SER.A/404). Department of Economic and Social Affairs, United Nations, New York, p 38

Van Aalst M (2006) The impacts of climate change on the risk of natural disasters. Disasters 30(1):5–18. https://doi.org/10.1111/j.1467-9523.2006.00303.x

WACOP (2016) Changing Waves and Coasts in the Pacific. https://www.pacificclimatechange.net/project/wacop-changing-waves-coasts-pacific. Accessed 10 May 2019

Walsh K, McBride J, Klotzbach J, Balachandran S, Camargo S, Holland G et al (2016) Tropical cyclones and climate change. Clim Change 7(1):65–89. https://doi.org/10.1002/wcc.371

Weber E (2010) What shapes perceptions of climate change? Clim Change 1(3):332–342. https://doi.org/10.1002/wcc.41

Wickham, H. (2018). tidyverse: Easily Install and Load 'Tidyverse'Packages (2017). r package version 1.1. 1.

WHO (2015) World report on ageing and health. World Health Organization, Geneva, Switzerland

Wing T (2017) Submerging paradise: climate change in the Pacific Islands. http://climate.org/submerging-paradise-climate-change-in-the-pacific-islands/. Accessed 13 April 2019

Woodward A, Hales S, Litidamu N, Phillips D, Martin J (2000) Protecting human health in a changing world: the role of social and economic development. Bull World Health Organ 78:1148–1155

World Bank (2017) Climate vulnerability assessment—making Fiji climate resilient. The Government of the Republic of Fiji, pp. 172. http://documents.worldbank.org/curated/en/163081509454340771/Climate-vulnerability-assessment-making-Fiji-climate-resilient

Chapter 2
Islands in the Pacific: Settings, Distribution and Classification

Patrick D. Nunn, Lalit Kumar, Roger McLean, and Ian Eliot

2.1 Introduction

In most geographies of the world, accounts of continents are extensive, and accounts of islands—especially those in the middle of oceans—are generally quite short. The reasons for this are obvious. Models of the world, its formation and changing configuration, are underpinned by global science that had its origins on continents, mostly in Europe. People from European centres of learning spread out across the world, observing and analysing what they saw in order to contribute data to nascent models of the world. Yet these people were not objective detached observers but rather burdened by their own prejudices and beliefs, their own intellectual baggage. Inevitably this informed their observations; for example, the tendency to regard continental landmasses as 'normal' and in hemispheric balance led to an impression of vast ocean basins as anomalies, probably therefore places where continents had once 'disappeared' (Nunn 2009b).

P. D. Nunn (✉)
University of the Sunshine Coast, Maroochydore, QLD, Australia
e-mail: pnunn@usc.edu.au

L. Kumar
School of Environmental and Rural Science, University of New England,
Armidale, NSW, Australia
e-mail: lkumar@une.edu.au

R. McLean
School of Physical, Environmental and Mathematical Sciences, University of New South
Wales at the Australian Defence Force Academy, Canberra, ACT, Australia
e-mail: r.mclean@adfa.edu.au

I. Eliot
University of Western Australia and Damara WA Pty Ltd., Innaloo, WA, Australia
e-mail: ian.eliot@bigpond.com

© Springer Nature Switzerland AG 2020
L. Kumar (ed.), *Climate Change and Impacts in the Pacific*, Springer Climate,
https://doi.org/10.1007/978-3-030-32878-8_2

Many of the earliest continental observers of oceanic islands hardly knew what to make of them. Combined with the difficulties of reaching such islands, their generally huge distances from continental centres of learning, and their 'small' areas, set the scene for a history of marginalization of islands in the natural sciences that is still not redressed, despite an upsurge of interest in islands over the last few decades (Nunn 1994; Menard 1986; Mueller-Dombois and Fosberg 1998).

Marginalization of this kind inevitably leads to generalization, the overlooking of diversity, an enforced homogenization that is far from an objective appraisal of the actual situation. Today, far from regarding oceanic islands as anomalous and of only peripheral importance to the understanding of our planet, we now acknowledge them as special places, the study of which is able to inform global issues. For example, oceanic-island genesis can inform us about that of the ocean basins, which occupy >70% of the Earth's surface, far better than can studies of most parts of the continents (Neall and Trewick 2008); oceanic islands have long been recognized as 'dipsticks' that record their own changes in level and those of the surrounding ocean far more easily than many continental shores (Bloom 1970). The often-singular nature of island biotas can inform us about the nature of organic evolution, dispersal and even issues like adaptation and speciation (Whittaker and Fernandez-Palacios 2007). Within the last decade or so, the global community has recognized the special vulnerabilities of islands to climate change (Mimura et al. 2007; Nurse et al. 2014).

The problems of marginalization of oceanic islands, especially in an age of globalization, become especially acute when applied to the contemporary situation of island peoples challenged by issues like economic development in the face of inequitable access to world markets on the one hand and climate change—one of a range of environmental stressors to which islanders are disproportionately exposed (Nunn and Kumar 2017; Shope et al. 2016; Connell 2013)—on the other. Together with many other external interventions intended to remove such inequities, most attempts at climate change adaptation in oceanic-island contexts over the past 30 years have failed to be either effective or sustainable (Nunn 2009a; McNamara 2013; Betzold 2015). Among the most common reason for this failure is that islands are commonly treated by the international community as 'continents in miniature', which they are not, so that the continental solutions imposed on them are inappropriate, both environmentally and culturally (Gillis 2014).

In the various fields that the authors have worked over the past few decades, especially in the Pacific Ocean, the need for a simple method of explaining oceanic-island diversity has become increasingly pressing. In response, this paper presents an earth-science-based classification of Pacific oceanic islands (not continental outliers) that captures their diversity and is intended to become the basis for more focused study (Nunn et al. 2016).

2.2 Island Settings: The Pacific Basin and Its Oceanic Islands

Comprising almost one-third of the Earth's surface, the Pacific Basin is bounded along its western side by East Asia and Australasia and on its eastern side by the western parts of the Americas; Antarctica forms a southern boundary. The oceanic-island groups within this region are listed in Table 2.1 and their locations shown in

Table 2.1 Key data from the island database in the Appendix

Country/group of islands	Number of islands	Total area of islands (km²)	Average island area (km²)	Average island maximum elevation (m)
Cook Islands	15	297	20	73
East Pacific outliers[a]	24	8236	343	509
Federated States of Micronesia	127	799	6	45
Fiji	211	20,857	99	134
French Polynesia	126	3940	31	154
Guam	1	588	588	400
Hawaii	16	19,121	1195	869
Kiribati	33	995	30[c]	6
Marshall Islands	34	286	8[c]	3
Nauru	1	23	23	71
New Caledonia	29	21,613	745	121
Niue	1	298	298	60
Northern Mariana Islands	16	537	34	444
Palau	33	495	15	58
Papua New Guinea (+ Irian Jaya)[b]	439	67,757	154	134
Pitcairn Islands	4	54	13	97
Samoa	7	3046	435	504
Solomon Islands	413	29,672	72	88
Tokelau	3	16	5[c]	5
Tonga	124	847	7	56
Tuvalu	10	44	4[c]	4
US-administered islands (central Pacific)	8	37	5	5
Vanuatu	81	13,526	167	330
Wallis and Futuna	14	190	14	94
Total	**1779**	**193,713**	**169**	**190**

[a]This group is comprised mostly of the Galapagos Islands, politically part of Ecuador
[b]The island of New Guinea which is included in the database is divided politically between Papua New Guinea and Indonesia. Although the latter country is not otherwise included in the database, that part of New Guinea island (named Irian Jaya) it controls is included
[c]Average island areas for these atoll countries are overestimates as they are based on polygons that subsume multiple islands (see text)

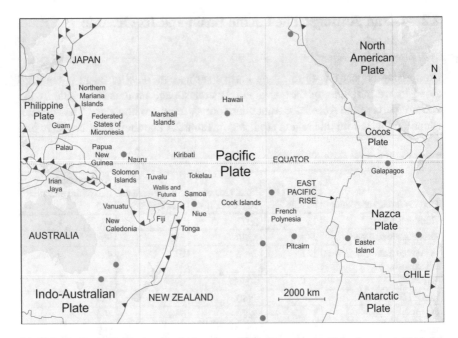

Fig. 2.1 The Pacific Basin showing the locations of island countries and island groups in Table 2.1 for which data about islands were obtained. The geotectonic context is also outlined; plate boundaries are shown in red and hotspots by orange circles. Convergent plate boundaries are those with filled triangles pointing in the direction of downthrusting. Other plate boundaries are mostly transform except for the East Pacific Rise where divergence is occurring. Locations of hotspots (active since 43 Ma) are from King and Adam (2014). Figure based on Nunn et al. (2016)

Fig. 2.1. Island groups (like Indonesia, Japan, Philippines) that are close to the continental rim of the Pacific Basin are not included, while those comprising Papua New Guinea, largely oceanic (not continental) in origin, are included. The continental outliers of the New Caledonia group are included because they are surrounded by oceanic crust.

The question of how to define an island remains important but is not considered here; for the purpose of this classification, all discrete ocean-bounded landmasses within the region of interest (the Pacific Basin) that are ≥ 1 ha (0.01 km^2) in area when measured above high water level are deemed islands. Reefs that lack islands are not included. Transient islands that may alternately appear and disappear are excluded, as are those that existed in the remembered past but do so no longer. After all these filters are applied, 1779 islands are found in the study region: a total land area of almost 200,000 km^2 after excluding the massive island of New Guinea, an average island area of 171 km^2 and an average maximum island elevation of 190 m above sea level (Table 2.1).

2.3   Classifying Pacific Islands: Data Sources

Previous classifications of oceanic islands developed categories based on size, shape or location that largely failed to provide a platform suited to a range of more detailed classification. Given that this classification seeks to capture physical and natural attributes of islands, it is based at its highest level on *elevation* and *lithology* (rock type). Together these two variables allow information about erodibility and resistance, drainage, landscape and landscape-changing processes. While climate is implicit in elevation for such a large population of islands, other possible classificatory parameters like exposure to natural hazards and climate change are deemed more appropriate to a focused use of this baseline classification.

The database of islands ($n = 1779$) is available in the Appendix. Reef islands strung out (short distances apart) along a single linear reef, as is common on barrier and atoll reefs, are treated as a single island because they often exhibit changeable forms and are periodically joined or bisected. Data about the locations and shapes of islands were obtained from the WVS (World Vector Shorelines) database which shows all the world's shorelines at a scale of 1:250,000, a resolution adjudged adequate for classifying Pacific islands. As a preliminary, information was sought for every island about its location, area, name, (maximum) elevation and lithology; these are discussed in the following subsections.

2.3.1   Island Locations and Shapes

Latitude and longitude for an island were obtained from Google Earth by placing its cursor at the island's central point, a method that could readily be applied consistently and which avoids having any islands overlap in location. Coordinates obtained by this method were cross-checked with published sources where these could be found. The disadvantage of this approach is that nothing can be said from a single point about island shape although coordinates were then converted to a GIS shapefile using the WGS84 (World Geodetic System 1984) before being overlain on a WVS polygon file to allow names to be assigned to each island (see Fig. 2.3c). Minor inaccuracies in island locations and shapes were rectified by comparing those derived to those in a world base map provided through ESRI, a process that allowed individual polygons to be moved slightly as needed. In those cases where WVS polygons were not a good match with actual island shape or were in fact missing, new polygons were digitized from the 1:20,000 base map.

2.3.2   Island Areas

Across published and Internet sources, areas given for the same island often vary greatly, so for this purpose, island areas were calculated directly from the polygon shapefile layer in the GIS described above. While the latter areas did not always coincide with published areas for the same island, largely because polygons were generated from coarse-scale satellite imagery, they were generally within the range of these, so their level of inaccuracy was considered admissible. Use of the same polygon shapefile ensured that errors in the calculation of island areas were consistent.

2.3.3 Island Names

While incomplete, the most comprehensive list of island names in the Pacific is that by Motteler (2006) which was supplemented for the purpose of assigning the correct name to particular islands by other sources, notably Langdon (1978). This is a less straightforward issue than it may sound, especially with smaller, more remote islands, where various sources may give more than one name or even no name at all. That said, only a handful of islands in the dataset (see Appendix) appear to have no detectable name. Information about other island names came from a great diversity of sources, published or not; in several troublesome cases, neither the Internet nor Google Earth proved especially helpful.

2.3.4 Island Elevations

Capturing both island-building (tectonic) and denudation processes, island elevation also reflects lithology and is therefore a first-order classifier. After various trials, it was found that a simple distinction between high (≥30 m above mean sea level) and low (<30 m above mean sea level) *maximum* elevation best separates higher (more resistant lithologies, less denuded, more uplifted or younger) from lower (less resistant lithologies, more denuded, less uplifted or older) islands: a conclusion similar to that reached in other studies (Menard 1983; Ramalho et al. 2013).

While needing to know only whether maximum island elevation is ≥30 m above mean sea level or not, the use of Google Earth did not prove adequate for many islands, so expert knowledge and published sources (notably Dahl 1980; Karolle 1993; Lobban and Schefter 1997; Mueller-Dombois and Fosberg 1998; Rapaport 2013) were also used.

2.3.5 Island Lithologies

Lithology is the other first-order classifier. It can be used to infer something about an island's habitability through soil type and surface water availability and is an expression of particular island-forming processes (Walsh 1982; Herzberg 2011).

The lithology classification is kept intentionally simple, not least because of the risk of losing sight of the broad classificatory aims and becoming mired in controversy over minutiae. This classification therefore uses five types of lithology: volcanic (synonymous with igneous for this purpose); limestone (synonymous with calcareous and non-volcanic sedimentary); composite (<80% volcanic and <80% limestone); reef (including all islands made from unconsolidated sediments); and continental (of non-oceanic origin).

All data about lithology came from archival sources; some sources were especially helpful at a Pacific-wide scale (Menard 1986; Nunn 1994; Nunn 1998b; Neall and Trewick 2008; Gillespie and Clague 2009; Vacher and Quinn 1997; Nunn 1999; Wiens 1962) and others of more use at a subregional level (including Bonvallot et al. 1993; Keating and Bolton 1992; Greene and Wong 1988; Scholl and Vallier 1985; Brocher 1985; Derrick 1957; Jost 1998; Wood 1967; Coleman 1970; Dow 1977; Anthony 2004; Tracey et al. 1964; Macdonald et al. 1983; Mcbirney et al. 1969; Bonatti et al. 1977; Dana 1875; Davis 1920; Duncan and McDougall 1976). For islands about which information was not found in such sources, island-specific studies were sought, typically through Google Scholar. Where these were inapplicable, recourse was sometimes made to Google Earth photographs that were examined for diagnostic signs of volcanic or limestone landforms, for instance.

2.4 Island Types: Outcomes of Classification

Using elevation and lithology data obtained as described above, each of the 1779 islands in the database was assigned to one of eight island 'types'.

Volcanic high islands are those composed of at least 80% igneous rock that reach a maximum elevation of at least 30 m above mean sea level. These island types are commonest in the Pacific in places that are within 500 km of sites of active (often undersea) volcanism, typically either along volcanic island arcs that run parallel to lines to plate convergence or at the younger ends of intraplate island hotspot chains (Nunn 1994).

Volcanic low islands are composed of at least 80% igneous rock but do not rise 30 m or more above mean sea level. These island types often are found farther away from lines of active/recent plate convergence and the younger ends of active hotspot chains, a spatial difference that reflects the fact of island subsidence (on cooler lithosphere) and/or an increased degree of post-volcanism denudation (Scott and Rotondo 1983).

Limestone high islands are composed of at least 80% calcareous rock types and reach at least 30 m above mean sea level. In the Pacific, such islands are commonest in forearc areas close to convergent plate boundaries where one (oceanic) plate is thrust above another but are also found where plates are colliding or being compressed without subduction. Many such places have been experiencing tectonic uplift for hundreds of millennia, resulting in uplifted (reef) islands sometimes hundreds of metres high (Ferry et al. 2004; Neef and McCulloch 2001).

Limestone low islands are composed of at least 80% calcareous rock types and have a maximum elevation of less than 30 m above sea level. While common in places where higher limestone islands are found, perhaps in locations where uplift rates have been lower or have occurred over shorter time periods, limestone low islands are also found in tectonically quiet parts of the Pacific. In such places, the Last Interglacial sea-level maximum (about 125,000 years ago) may have facilitated reef growth as much as 6 m above present mean sea level; the Holocene maximum sea level (about 5000 years ago) may have allowed reefs to grow as much as 2.1 m above present mean sea level. In both situations, subsequent net sea-level fall demonstrably led to the emergence of fossil reef, forming limestone low islands (Kayanne et al. 2002; Pirazzoli and Montaggioni 1986; Furness 2004).

Reef islands are those composed of at least 80% unconsolidated sediments, derived from adjacent (offshore/terrestrial) areas that have accumulated on shallow sea floor, often in the tropical Pacific on (biogenic) reef flats. Sometimes difficult to distinguish clearly from limestone low islands (see example from Majuro, Marshall Islands—Yasukochi et al. 2014), such islands are defined as those that form when sediment is supplied at rates greater than those at which it is removed, leading to net accumulation, whether at the mouth of a river or—as is common in the database (see Appendix)—on barrier and atoll reefs rising from the submerged flanks of a formerly (more) emergent volcanic island.

Composite high islands are defined as those composed of *both* less than 80% volcanic rock types and less than 80% limestone rock types that reach a maximum elevation of at least 30 m above sea level. While these islands are common around convergent plate boundaries where large islands have formed as a result of both uplift and volcanism, there is a special type of composite (usually high) island found elsewhere. These are the *makatea* islands (named by Nunn 1994) formed when the subsidence of a reef-fringed volcanic island is interrupted by uplift, commonly associated with lithospheric flexure, that gives rise to a volcanic island fringed by uplifted reef; examples come from French Polynesia and Solomon Islands (Stoddart and Spencer 1987; Taylor 1973).

Composite low islands are composed of *both* less than 80% volcanic rock types and less than 80% limestone rock types that reach a maximum elevation of less than 30 m above sea level. Often found in places where composite high islands occur, such islands may simply have been subject to less uplift and/or more denudation. In many cases, large composite high islands are surrounded by a number of smaller limestone low islands, representing fringing/coastal fragments of an original contiguous landmass.

Continental islands are those formed of at least 80% continental rocks, not any that originated within the ocean basins. Since these island types are few, mostly rise above 30 m and are found exclusively in the New Caledonia group, they are not subdivided by elevation. The presence of this continental sliver in the southwest Pacific is well understood (Cluzell et al. 2012).

2.5   Characteristics of Islands in the Pacific

Following an account of where the 1779 islands in the dataset are located within the Pacific and why, this section plots key data (area, maximum elevation) for each island by location and explains how each pair of variables is linked.

2.5.1   Distribution of Islands

Islands are not uniformly distributed across the Pacific Basin (Fig. 2.2). The southwest quadrant has most islands (as defined for the database), while the others have less; much of the northeast Pacific Ocean lacks any islands. Aside from continental islands, all other islands in the database originated as ocean-floor volcanoes, meaning that their original locations are determined by places where this volcanism takes/took place. In the Pacific Ocean Basin, there are three geotectonic contexts in which ocean-floor volcanism that can form (above-sea) islands occurs (Nunn 1994; Neall and Trewick 2008; Nunn 1999).

The first is along mid-ocean ridges (divergent plate boundaries), which are comparatively few in the Pacific and involve voluminous ridge crest/flank volcanism of the kind that may include Easter Island (off the East Pacific Rise) as well as islands

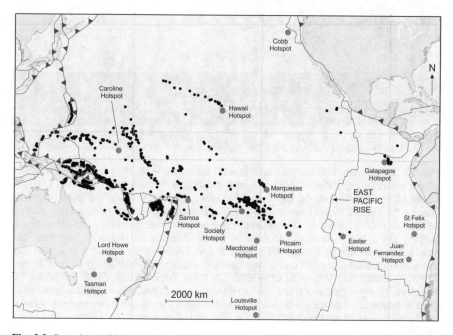

Fig. 2.2 Locations of islands in the Pacific Basin showing their relationship with places (plate boundary and hotspot) where island formation is commonest. Figure based on Nunn et al. (2016)

in back-arc basins of the western Pacific like Niuafo'ou (Tonga) and Mota Lava (Vanuatu) (Sorbadere et al. 2013; Tian et al. 2011).

The second context is along convergent plate boundaries where one lithospheric plate is being thrust (subducted) beneath another. There are more convergent plate boundaries in the Pacific than any other ocean basin, and most are concentrated in its southwest part where the island groups of Papua New Guinea, Solomon Islands, Vanuatu, Fiji, Tonga and New Zealand track the boundary between the Pacific Plate and the Indo-Australian Plate. In such situations, the downthrust plate commonly melts at depths of around 100 km below the ocean floor releasing magma that moves upwards, sometimes erupting on the ocean floor and starting to build islands. Lines of active volcanoes (island arcs) formed in this way include those in Tonga and Vanuatu (Fig. 2.3a).

Fig. 2.3 Illustrations of archetypal Pacific islands. (**a**) Mount Garet, the principal volcano comprising (volcanic high) Gaua Island (Vanuatu), erupting in September 2010. Photo: T. Boyer, Creative Commons licenced. (**b**) The coast of (volcanic high) Bora Bora island (French Polynesia) showing the peak of Mount Otemanu. Such photos illustrate the 'peakiness' of many (younger) volcanic islands that needs to be considered when analysing their maximum elevations. Photo: Sergio Calleja, Creative Commons licenced. (**c**) The reef island of Kehpara on the barrier reef surrounding the (volcanic high) island of Pohnpei (Federated States of Micronesia) is built from sand and gravel deposited on the reef during large wave events and partly stabilized by the development of indurated rocks (like beachrock) along its shores. Photo: Petra Nunn, used with permission. (**d**) Terraces of emerged coral reef above Togo on the east coast of (limestone high) Niue Island show signs of weathering, but their original surfaces are still clearly visible. The main terrace is likely to be of Last Interglacial age (about 125,000 years old), the higher one (top right) perhaps dating from the previous interglaciation (about 200,000 years ago). Photo: Susan and Ken FitzGerald, used with permission. The third situation is in intraplate (mid-plate) areas, far from plate boundaries, where a plate may move across a fixed mantle plume (hotspot) that 'leaks' magma to form—over long time periods—a line of volcanoes. Well-studied hotspot island groups include the Hawaiian-Emperor island-seamount chain that has existed for at least 80 million years (O'Connor et al. 2013)

Most uplift in ocean basins occurs close to convergent plate boundaries, which further explains the concentration of islands there, especially composite and limestone islands, although flexures in the intraplate lithosphere, perhaps in association with nearby subduction (as with Niue Island—Nunn and Britton 2004) or localized swells (like the South Pacific Superswell—McNutt and Fischer 1987), also cause uplift of limestone islands.

Not every Pacific island fits these categories. Islands like Manihiki (Cook Islands) and Ontong Java (Solomon Islands) and Pohnpei (Federated States of Micronesia [FSM]) may all have an origin associated with the presence of locally thickened crust that forms oceanic plateaux (Taylor 2006). The Galapagos Islands appear to have formed at a plate triple junction (Smith et al. 2013).

The absence of islands in vast tracts of the Pacific is explainable largely by an absence of island-forming and island-preserving processes. Outside the southwest quadrant of the Pacific, there are few convergent plate boundaries, so most islands originated at intraplate swells or hotspots, which generally produce smaller islands within comparatively small areas. There are simply no island-forming processes in operation elsewhere in such places. Added to this is the conspicuous absence of islands in higher latitudes, both in the north and south Pacific, something attributable in large measure to an absence of island-forming processes but also to the cooler water which means that reef-supported islands sink (and disappear from view) once they pass into these areas.

2.5.2 Areas of Islands

Most (77%) islands in the database are <10 km^2 in area (Fig. 2.4), something for which there are two explanations. First, given that since the Last Glacial Maximum (LGM) about 20,000 years ago, sea level in the Pacific has risen about 120 m, many islands have become partly drowned, fragmented into ever smaller pieces as sea level rose. It should be noted that while this explanation seems instinctively correct, we cannot be certain of it given that we do not know how many islands existed in the Pacific during the LGM and how many of these were submerged completely by rising Postglacial sea level. Second, once sea level in the Pacific began to stabilize about 6000 years ago (Grossman et al. 1998), broad reef platforms began to develop around many tropical islands creating a substrate on which (comparatively small) reef islands might form; the subsequent ~2-m fall of sea level in the Pacific caused many of these substrates to emerge, making them suitable foci for reef-island growth (McLean and Kench 2015). In addition, it seems clear that reef islands depend more on sediment supply and accommodation space (on a reef platform) than simply on sea level, meaning that some reef islands developed in the Pacific more than 6000 years ago and persist still (Kench et al. 2014).

Larger islands are fewer in the database; only 6% of islands are larger than 100 km^2 in area (see Fig. 2.4). This is because of the intrinsic greater difficulty of forming and sustaining a large island rather than a small one—by any of the mecha-

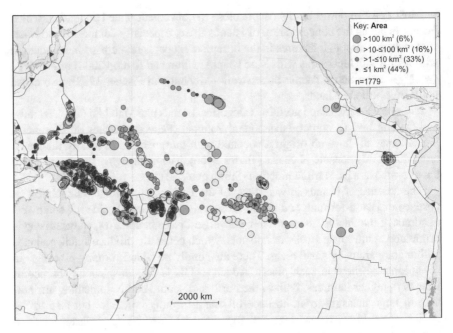

Fig. 2.4 Areas of islands in the Pacific Basin showing how their locations relate to those where island formation is commonest. Figure based on Nunn et al. (2016)

nisms outlined above. Unless the supply of island-forming materials, be they hard or soft, and/or the process of uplift is anomalously great and enduring compared to other situations, such large islands will not generally form. And even if they do, sea-level rise, flank collapse or even denudation is liable to reduce them in size.

Conversely, many of the larger islands are amalgams of smaller islands, perhaps on emergent reef flats or perhaps because of uplift associated with plate convergence. A good example is the island of Viti Levu, the largest in the Fiji group (10,388 km^2), which is essentially a result of two volcanic island arcs raised and bent as a result of forming part of a microplate that has been twisted and crushed by the oblique convergence of the Pacific and Indo-Australian Plates (Stratford and Rodda 2000).

It is easier to explain the distribution of larger islands in the Pacific than smaller ones. Larger islands are almost all concentrated in places where island-forming processes are unusually effective and have been so for some time, as is the case along the convergent boundaries in the southwest Pacific. Their comparative absence elsewhere in the region means that smaller islands dominate in such places, although the distribution of smaller islands resists straightforward analysis.

2.5.3   Maximum Elevations of Islands

The distinction used to separate high from low islands (discussed in Sect. 2.4) informs the classification of island types, yet does not allow for the independent analysis of maximum elevation as a characteristic of islands in the Pacific. To this end, Fig. 2.5 shows five categories for maximum elevation obtained from analysis of the database (Appendix).

In general terms, the greater the maximum elevation of an island, the more likely it is to have experienced island-forming processes at a greater pace than lower islands. Further, the chances are that, if those processes have ceased (or become comparatively subdued), they did so only recently; older islands invariably subside and become reduced in elevation as a result of denudation (Menard 1986; Menard 1983).

That said, some islands where island-forming processes have ceased can periodically experience rejuvenation, perhaps from passing close to a site of volcanic activity (as with Savai'i in Samoa—Keating 1992) or from moving across a lithospheric swell that causes them to be uplifted (as in southeast Fiji—Nunn 1995).

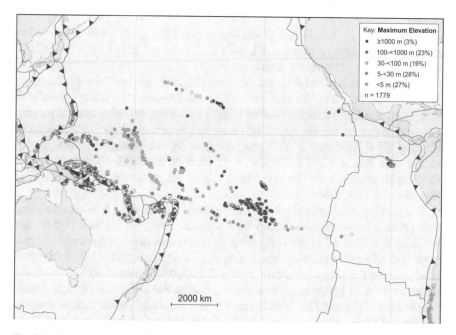

Fig. 2.5 Maximum elevations of islands in the Pacific Basin showing how their locations relate to those where island formation is commonest. Figure based on Nunn et al. (2016)

Elevation is also a function of lithology as well as structure, climate and vegetation. Some rock types are more resistant to denudation and surface lowering than others. Reef limestones, for example, weather only very slowly, not least because most drainage is underground. In contrast, many volcanic rocks weather comparatively fast, especially on islands where orographic rainfall dominates. Climatic controls are also important; islands like Easter Island in uncommonly dry parts of the Pacific often exhibit comparatively low rates of surface lowering than those in wetter parts (Li 1988). While vegetation often increases resistance to ground-surface erosion, its removal during droughts or following storms may cause more rapid than expected erosion (Terry 1999).

It can be seen in Fig. 2.5 that islands with higher maximum elevations are often found in places where island-forming processes are most active or have been so during an island's recent history. Much also depends on the topography of the island; volcanic islands are more likely to be 'peaky' so that their maximum elevations are much greater than their mean elevations (Fig. 2.3b); limestone islands are likely to be far less so, especially in the case of emerged reef islands, so that their maximum and mean elevations are likely to be closer.

2.6   Distribution of Island Types

The distribution of island types (identified in Sect. 2.4) is shown in Fig. 2.6. The most common island types are reef islands (36%) and volcanic high islands (31%) which have quite different distributions.

Most reef islands in the database are those that form (part-cemented) accumulations of sand and gravel on shallow reefs in the low-latitude Pacific, typically in intraplate locations where the process of slow uninterrupted subsidence is particularly conducive to the formation of broad barrier and atoll reefs (McLean and Hosking 1991; Dickinson 2004; Yamaguchi et al. 2009). Most such islands tend to be low and elongate and are common in parts of the FSM, French Polynesia, Kiribati, Marshall Islands, Tokelau and Tuvalu as well as the western outliers of the Hawaii group (Fig. 2.3c).

The prevalence of volcanic high islands reflects the fact that every oceanic island in the Pacific began life as an ocean-floor volcano. That most volcanic high islands are today concentrated in parts of the Pacific close either to convergent plate boundaries, typically along volcanic island arcs, or to intraplate hotspots shows that most must be comparatively young, expressions of the fast island-building processes that characterize such geotectonic locations. Concentrations of volcanic high islands near convergent boundaries include many in the southwest Pacific such as those in the Central Chain of Vanuatu and in the Tonga Volcanic Arc (Greene and Wong 1988; Scholl and Vallier 1985). Numerous volcanic high islands are found in

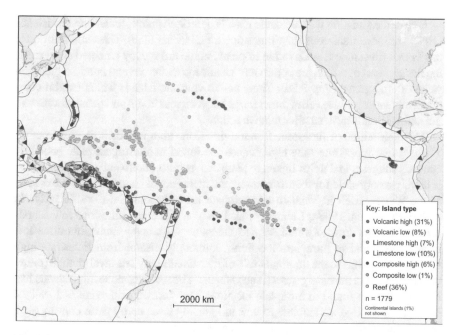

Fig. 2.6 Types of islands in the Pacific Basin showing how their locations relate to those where island formation is commonest. Figure based on Nunn et al. (2016)

association with the Hawaii hotspot, currently below Hawai'i Island and Lo'ihi Seamount, and the Samoa hotspot, currently near Rose Atoll (Keating 1992; Ballmer et al. 2011).

Volcanic low islands (8%) evolve from/into volcanic high islands, so it is no surprise they have similar distributions. Their comparative overall paucity is more surprising, being perhaps a result of the tendency of island-forming processes in such situations to readily form high islands that endure a long time. It is also important to consider that the 'peakiness' of volcanic islands (see Fig. 2.3b) means that higher types will persist longer than they might with limestone or composite islands.

Composite high islands (6%) and composite low islands (1%) have similar distributions. Most composite islands are associated with convergent plate boundaries, reflecting the prevalence there of both volcanism and tectonic uplift that both contribute to the formation of composite islands. The composite *makatea* islands in intraplate locations are volcanic islands with a fringe of raised reef, uplifted as a result of local lithospheric flexure (Spencer et al. 1987).

Unlike the situation with volcanic islands described above, limestone low islands (10%) are more numerous than limestone high islands (7%). The main reason for this is that limestone islands in the tropical Pacific are mostly emerged coral reefs, the flat surfaces of which are commonly visible even after several hundred thousand years of emergence (Fig. 2.3d). Thus, unlike volcanic islands which exhibit comparatively large relative relief (large range between maximum and mean elevations), limestone islands have smaller relative relief.

The distribution of limestone low and limestone high islands is also instructive. The fact that most limestone high islands are found in convergent plate boundary contexts suggests that uplift linked to plate convergence, collision and compression is largely responsible for their formation; examples come from island forearcs such as the Western Belt in Vanuatu and remnant arcs like the Lau-Colville Ridge in eastern Fiji (Nunn 1998a; Calmant et al. 1999). While many limestone low islands occur in similar locations and formed the same way, many more limestone low islands are found in intraplate locations, commonly distant from limestone high islands. This points to the dominant role of Late Quaternary sea-level change (rather than tectonics) in the emergence of such islands. The few limestone high islands like Nauru and Niue found in intraplate locations have formed as a result of localized lithospheric flexure attributable to volcano loading and nearby plate convergence, respectively (Hill and Jacobson 1989; Nunn and Britton 2004).

2.7   Conclusions

The island classification presented in this chapter is geoscience-based and descriptive and adequately captures the diversity of island types in the Pacific Basin. Based on an analysis of 1779 islands, it is clear that this classification can be used as an explanatory tool for understanding why islands are located where they are; why islands are the size they are; why islands have the maximum elevations they have; and why particular types of island are found in particular places and sometimes not in others. In this sense, this classification represents the first that is able to answer fundamental questions about individual islands, but its value also lies in explaining the spatial diversity of island groups at subregional as well as regional scales.

This classification could be used as the basis for more detailed (second-order) classifications addressing particular issues. These include (non-partisan) areal assessments of environmental risk exposure as well as climate change vulnerability. It also has a potential role in the understanding of airborne or seaborne pollution and in national or regional planning for issues like infrastructure development or settlement relocation.

Appendix: The Island Database

Region/country	Island group	Island name	Island type	Latitude	Longitude	Area (ha)	Maximum elevation
American Samoa	Manu'a Islands	Ofu (1)	Volcanic high island	−14.1725	−169.667	866.99	491
American Samoa	Manu'a Islands	Olosega	Volcanic high island	−14.1811	−169.621	627.02	629
American Samoa	Manu'a Islands	Rose Atoll	Reef island	−14.5467	−168.152	10.74	4
American Samoa	Manu'a Islands	Swains Island (Olohega)	Reef island	−11.055	−171.078	427.61	4
American Samoa	Manu'a Islands	Ta'u	Volcanic high island	−14.2256	−169.436	4982.83	931
American Samoa	Outliers	Aunu'u	Volcanic high island	−14.2856	−170.559	154.33	61
American Samoa	Outliers	Tutuila	Volcanic high island	−14.3025	−170.733	15095.21	653
Chilean Islands (South East Pacific)	Valparaiso Province Islands	Easter Island (Isla de Pascua)	Volcanic high island	−27.1211	−109.366	21350.09	507
Chilean Islands (South East Pacific)	Valparaiso Province Islands	Sala y Gomez Island (Isla Sala y Gomez)	Volcanic high island	−26.4719	−105.362	191.95	30
Cook Islands	Northern	Manihiki	Reef island	−10.4272	−160.99	854.4	3
Cook Islands	Northern	Manuae	Reef island	−19.2525	−158.945	892.7	5
Cook Islands	Northern	Mauke	Composite low island	−20.1606	−157.338	2261.08	30
Cook Islands	Northern	Mitiaro	Composite low island	−19.8775	−157.702	2528.8	9
Cook Islands	Northern	Nassau	Reef island	−11.5594	−165.416	133.37	9

(continued)

(continued)

Region/country	Island group	Island name	Island type	Latitude	Longitude	Area (ha)	Maximum elevation
Cook Islands	Northern	Palmerston	Reef island	−18.0417	−163.147	261.29	4
Cook Islands	Northern	Penrhyn	Reef island	−9.005	−157.976	1540.22	4
Cook Islands	Northern	Pukapuka	Reef island	−10.85	−165.839	174.78	4
Cook Islands	Northern	Rakahanga	Reef island	−10.0386	−161.091	990.84	4
Cook Islands	Northern	Rarotonga	Composite high island	−21.2292	−159.776	8148.73	658
Cook Islands	Northern	Suwarrow	Reef island	−13.2692	−163.124	179.34	3
Cook Islands	Northern	Takutea	Reef island	−19.8108	−158.288	178.26	3
Cook Islands	Southern	Aitutaki	Composite high island	−18.8489	−159.782	2200.59	123
Cook Islands	Southern	Atiu	Composite high island	−19.9939	−158.118	3335.16	71
Cook Islands	Southern	Mangaia	Composite high island	−21.935	−157.916	5983.52	169
Eastern Pacific Outliers	Galapagos Islands	Baltra (South Seymour)	Volcanic high island	−0.45944	−90.2714	2619.16	100
Eastern Pacific Outliers	Galapagos Islands	Bartolome (Bartholomew)	Volcanic high island	−0.28389	−90.5472	132.08	114
Eastern Pacific Outliers	Galapagos Islands	Crossman Islands (Isla Los Hermanos)	Volcanic high island	−0.85639	−90.7717	70.07	168
Eastern Pacific Outliers	Galapagos Islands	Culpepper (Darwin (of Ecuador)	Volcanic high island	1.653056	−92.0003	66.76	168
Eastern Pacific Outliers	Galapagos Islands	Espanola (Hood)	Volcanic high island	−1.37583	−89.6719	6163.83	206
Eastern Pacific Outliers	Galapagos Islands	Fernandina (Narborough)	Volcanic high island	−0.41222	−91.4819	64935.9	1494

Eastern Pacific Outliers	Galapagos Islands	Floreana (Charles, Santa Maria)	Volcanic high island	-1.30806	-90.4311	19996.89	640
Eastern Pacific Outliers	Galapagos Islands	Isabela (Albemarle)	Volcanic high island	-0.82917	-91.1353	472272.2	1710
Eastern Pacific Outliers	Galapagos Islands	Jervis (Robida)	Volcanic high island	-0.41694	-90.7103	515.09	367
Eastern Pacific Outliers	Galapagos Islands	Marchena (Bindloe)	Volcanic high island	0.318333	-90.4689	13633.83	343
Eastern Pacific Outliers	Galapagos Islands	Pinta (Abingdon)	Volcanic high island	0.591944	-90.7628	6007.24	780
Eastern Pacific Outliers	Galapagos Islands	Pinzen (Duncan)	Volcanic high island	-0.61083	-90.6678	1832.24	458
Eastern Pacific Outliers	Galapagos Islands	San Cristobal (Chatham)	Volcanic high island	-0.64722	-89.39	55821.93	759
Eastern Pacific Outliers	Galapagos Islands	Santa Cruz (Indefatigable)	Volcanic high island	-0.59278	-90.3397	99853.33	864
Eastern Pacific Outliers	Galapagos Islands	Santa Fe (Barrington)	Volcanic high island	-0.81972	-90.0653	2446.07	259
Eastern Pacific Outliers	Galapagos Islands	Santiago (James, San Salvador)	Volcanic high island	-0.26194	-90.7153	56439.29	920
Eastern Pacific Outliers	Galapagos Islands	Seymour (North Seymour)	Volcanic low island	-0.39583	-90.2878	182.4	28
Eastern Pacific Outliers	Galapagos Islands	Tortuga (Brattle)	Volcanic high island	-1.00917	-90.8708	128.24	128
Eastern Pacific Outliers	Galapagos Islands	Wenman (Wolf)	Volcanic high island	1.383611	-91.8158	129.36	253
Eastern Pacific Outliers	Revillagige do Islands	Clarion Island (Isla Clarion)	Volcanic high island	18.35722	-114.718	2091.13	335

(continued)

(continued)

Region/country	Island group	Island name	Island type	Latitude	Longitude	Area (ha)	Maximum elevation
Eastern Pacific Outliers	Revillagige do Islands	Clipperton Island (Ile Clipperton) (Colony of France, administered from French Polynesia)	Composite low island	10.29694	−109.216	1000.43	29
Eastern Pacific Outliers	Revillagige do Islands	Cocos Island (Isla del Coco) (part of Costa Rica territory)	Volcanic high island	5.523056	−87.0722	2316.81	634
Eastern Pacific Outliers	Revillagige do Islands	San Benedicto Island (Isla San Benedicto)	Volcanic high island	19.31278	−110.805	582.96	332
Eastern Pacific Outliers	Revillagige do Islands	Socorro Island (Isla Socorro)	Volcanic high island	18.79472	−110.972	14385.83	1130
Federated States of Micronesia	Chuuk	(Satowan) Moch	Reef island	7.514167	151.9667	11.27	3
Federated States of Micronesia	Chuuk	Anangenimon	Reef island	7.425	151.9964	4.06	1
Federated States of Micronesia	Chuuk	Engenenimo	Reef island	7.314444	152.0231	3.26	1
Federated States of Micronesia	Chuuk	Eor	Reef island	8.689444	152.3408	17.2	8
Federated States of Micronesia	Chuuk	Eot	Volcanic high island	7.386667	151.7397	52.34	61
Federated States of Micronesia	Chuuk	Ettal	Reef island	5.565556	153.5808	111.71	5
Federated States of Micronesia	Chuuk	Etten	Volcanic low island	7.355556	151.8853	59.55	13
Federated States of Micronesia	Chuuk	Fanadik	Reef island	7.586111	149.3939	12.82	2
Federated States of Micronesia	Chuuk	Fanan	Reef island	7.204167	151.9878	22.05	5

Federated States of Micronesia	Chuuk	Fananu	Reef island	8.558611	151.9061	34.6	2
Federated States of Micronesia	Chuuk	Fanapanges	Volcanic high island	7.353611	151.6672	193.4	61
Federated States of Micronesia	Chuuk	Fayu/Fayo	Reef island	8.550278	151.3408	48.75	2
Federated States of Micronesia	Chuuk	Fefan	Volcanic high island	7.347778	151.8414	1311.64	298
Federated States of Micronesia	Chuuk	Ferit	Reef island	7.415556	151.9336	5.02	1
Federated States of Micronesia	Chuuk	Fono	Volcanic high island	7.486389	151.8811	40.55	52
Federated States of Micronesia	Chuuk	Houk (Pulusuk)	Reef island	6.686667	149.3014	314.5	3
Federated States of Micronesia	Chuuk	Igup	Reef island	8.563889	151.9189	42.59	2
Federated States of Micronesia	Chuuk	Losap	Reef island	6.893889	152.74	86.7	5
Federated States of Micronesia	Chuuk	Lukunor	Reef island	5.505278	153.8197	162.82	8
Federated States of Micronesia	Chuuk	Magererik	Reef island	8.946389	150.0664	32.44	4
Federated States of Micronesia	Chuuk	Makur (Namonuito)	Reef island	8.983333	150.1275	50.2	8
Federated States of Micronesia	Chuuk	Meseong	Reef island	7.145278	151.9072	84.18	4
Federated States of Micronesia	Chuuk	Murilo	Reef island	8.692222	152.3408	29.63	5

(continued)

(continued)

Region/country	Island group	Island name	Island type	Latitude	Longitude	Area (ha)	Maximum elevation
Federated States of Micronesia	Chuuk	Nachu/Pisar	Reef island	7.220278	151.9906	4.86	1
Federated States of Micronesia	Chuuk	Nama	Reef island	6.9925	152.5747	117.51	14
Federated States of Micronesia	Chuuk	Namoluk	Reef island	5.923611	153.1164	178.6	5
Federated States of Micronesia	Chuuk	Neoch (Feneppi/Fenebe)	Reef island	7.113056	151.8775	28.45	1
Federated States of Micronesia	Chuuk	Nomwin	Reef island	8.431389	151.7464	23.19	2
Federated States of Micronesia	Chuuk	Numurus	Reef island	8.592778	152.0364	42.38	5
Federated States of Micronesia	Chuuk	Ocha	Reef island	7.150556	151.89	26.64	4
Federated States of Micronesia	Chuuk	Onari (Unanu)	Reef island	8.751944	150.3397	44.11	2
Federated States of Micronesia	Chuuk	Oneop	Reef island	5.506944	153.7111	60.77	3
Federated States of Micronesia	Chuuk	Onnang/Ollan	Reef island	7.224444	151.6356	26.06	3
Federated States of Micronesia	Chuuk	Onou (Ono)	Reef island	8.798056	150.2919	55.53	2
Federated States of Micronesia	Chuuk	Onoun/Ulul	Reef island	8.581667	149.685	288.17	7
Federated States of Micronesia	Chuuk	Pakuru	Reef island	7.225278	151.7133	2.24	2
Federated States of Micronesia	Chuuk	Parem	Volcanic high island	7.377778	151.7903	152.1	61

Federated States of Micronesia	Chuuk	Pata	Volcanic high island	7.375833	151.5819	447.1	185
Federated States of Micronesia	Chuuk	Piherarh (Pisaras)	Reef island	8.569722	150.4192	29.09	2
Federated States of Micronesia	Chuuk	Piis-emmwar	Limestone low island	6.835556	152.7008	33.29	8
Federated States of Micronesia	Chuuk	Pis (Piis-Paneu)	Reef island	7.6775	151.7644	33.46	8
Federated States of Micronesia	Chuuk	Pollap/Pulap	Reef island	7.638889	149.4297	80.64	2
Federated States of Micronesia	Chuuk	Polle	Volcanic high island	7.344167	151.5994	930.24	171
Federated States of Micronesia	Chuuk	Poluwat/Puluwat	Reef island	7.3625	149.1897	230.1	2
Federated States of Micronesia	Chuuk	Pones	Reef island	7.223333	151.9958	9.08	1
Federated States of Micronesia	Chuuk	Pones (Fanan)	Reef island	7.191667	151.9822	53.23	3
Federated States of Micronesia	Chuuk	Romonum/Ramanum	Volcanic low island	7.411389	151.6692	97.64	4
Federated States of Micronesia	Chuuk	Ruo	Reef island	8.609167	152.2408	66.38	3
Federated States of Micronesia	Chuuk	Salat/Sanat	Reef island	7.233889	152.005	34.38	1
Federated States of Micronesia	Chuuk	Satawan (Kuttu)	Reef island	5.328611	153.7342	141.51	5
Federated States of Micronesia	Chuuk	Siis (Tsis)	Volcanic high island	7.2975	151.8244	67.5	91

(continued)

(continued)

Region/country	Island group	Island name	Island type	Latitude	Longitude	Area (ha)	Maximum elevation
Federated States of Micronesia	Chuuk	Ta	Reef island	5.294444	153.6822	209.41	5
Federated States of Micronesia	Chuuk	Tamatam	Reef island	7.539167	149.4128	72.83	2
Federated States of Micronesia	Chuuk	Tol	Volcanic high island	7.366667	151.6222	1742.11	439
Federated States of Micronesia	Chuuk	Tonowas	Volcanic high island	7.375833	151.8783	969.52	349
Federated States of Micronesia	Chuuk	Totiu	Volcanic high island	7.34	151.78	55.87	34
Federated States of Micronesia	Chuuk	Udot	Volcanic high island	7.383611	151.7122	547.1	152
Federated States of Micronesia	Chuuk	Uijec/Oror en Wirong	Reef island	7.164444	151.9383	42.07	2
Federated States of Micronesia	Chuuk	Uman	Volcanic high island	7.300278	151.8811	437.16	244
Federated States of Micronesia	Chuuk	Weltot	Reef island	8.608889	150.3825	9.38	2
Federated States of Micronesia	Chuuk	Weno (Moen)	Volcanic high island	7.442222	151.8564	2106.74	370
Federated States of Micronesia	Chuuk	Wonei	Volcanic high island	7.386389	151.6039	368.75	180
Federated States of Micronesia	Kosrae	Kosrae	Volcanic high island	5.309167	162.9722	11428.13	628
Federated States of Micronesia	Pohnpei	Ahnd (Ant)	Reef island	6.724374	157.933	358.28	5
Federated States of Micronesia	Pohnpei	Dehpehk (Takaieu)	Volcanic high island	6.944167	158.2961	391.71	105

Federated States of Micronesia	Pohnpei	Dekehtik (Sokehs)	Volcanic high island	6.983056	158.2083	173.07	202
Federated States of Micronesia	Pohnpei	Doletik	Volcanic low island	6.814722	158.1522	12.11	1
Federated States of Micronesia	Pohnpei	Kapingamarangi	Reef island	1.083056	154.8069	23.28	3
Federated States of Micronesia	Pohnpei	Kepara	Reef island	6.794722	158.1164	5.01	3
Federated States of Micronesia	Pohnpei	Laiap	Reef island	6.777222	158.2072	5.55	1
Federated States of Micronesia	Pohnpei	Lenger	Volcanic high island	6.995278	158.2311	40.93	77
Federated States of Micronesia	Pohnpei	Minto Reef	Reef island	8.109167	154.2817	4751.59	1
Federated States of Micronesia	Pohnpei	Mudok (Dauen Pahn)	Volcanic low island	6.765833	158.2697	15.47	1
Federated States of Micronesia	Pohnpei	Mwahnd Peidak	Volcanic high island	6.991944	158.2597	109.08	95
Federated States of Micronesia	Pohnpei	Mwahnd Peidi	Volcanic high island	6.980833	158.2839	69.86	85
Federated States of Micronesia	Pohnpei	Mwoakilloa (Mokil)/Manton	Reef island	6.693056	159.7658	92.43	2
Federated States of Micronesia	Pohnpei	Na	Reef island	6.848333	158.3481	88.19	1
Federated States of Micronesia	Pohnpei	Nahlap	Reef island	6.783611	158.1547	22.08	3
Federated States of Micronesia	Pohnpei	Nukuoro	Reef island	3.84	154.9736	48.15	5

(continued)

(continued)

Region/country	Island group	Island name	Island type	Latitude	Longitude	Area (ha)	Maximum elevation
Federated States of Micronesia	Pohnpei	Oroluk	Reef island	7.621111	155.1589	21.75	3
Federated States of Micronesia	Pohnpei	Pakin	Reef island	7.078611	157.7733	181.25	3
Federated States of Micronesia	Pohnpei	Parempei	Volcanic high island	7.002222	158.2519	306.7	93
Federated States of Micronesia	Pohnpei	Pingelap	Reef island	6.206111	160.7058	224.58	6
Federated States of Micronesia	Pohnpei	Pohnpei	Volcanic high island	6.876389	158.2206	35637.89	791
Federated States of Micronesia	Pohnpei	Sapwtik	Volcanic high island	7.004722	158.2253	15.95	35
Federated States of Micronesia	Pohnpei	Sapwuahfik (Ngatik)	Reef island	5.7875	157.1558	133.48	5
Federated States of Micronesia	Pohnpei	Temwen	Volcanic high island	6.847778	158.3267	364.33	65
Federated States of Micronesia	Yap	Asor	Reef island	10.02972	139.7664	62.47	3
Federated States of Micronesia	Yap	Bulubul/Bulbul	Reef island	9.927778	139.8228	5.86	3
Federated States of Micronesia	Yap	Eau/Yaaw	Reef island	9.864722	139.6097	4.25	7
Federated States of Micronesia	Yap	Eauripik	Reef island	6.686389	143.0425	31.79	3
Federated States of Micronesia	Yap	Elato	Reef island	7.511944	146.1719	47.18	3
Federated States of Micronesia	Yap	Fais	Limestone low island	9.762222	140.5169	331.76	18

Federated States of Micronesia	Yap	Falalop (Ulithi)	Reef island	10.01778	139.7906	105.16	4
Federated States of Micronesia	Yap	Faraulep	Reef island	8.585833	144.5269	30.48	7
Federated States of Micronesia	Yap	Fassarai/Fedarai	Reef island	9.908889	139.6583	54.31	8
Federated States of Micronesia	Yap	Feitabul/Feeatabol	Reef island	9.863889	139.6975	18.33	3
Federated States of Micronesia	Yap	Gaferut	Reef island	9.228056	145.3844	22.43	5
Federated States of Micronesia	Yap	Gagil–Tomil	Composite high island	9.543333	138.1669	3060.84	81
Federated States of Micronesia	Yap	Gielap/Giil'ab	Reef island	9.9525	139.9083	6.56	3
Federated States of Micronesia	Yap	Iar/Yaaor	Reef island	9.928889	139.9017	6.62	3
Federated States of Micronesia	Yap	Ifalik	Reef island	7.2475	144.4553	172.95	3
Federated States of Micronesia	Yap	L'am (Sorenleng)/L'aamw	Reef island	10.04694	139.5892	14.75	3
Federated States of Micronesia	Yap	Lamotrek	Reef island	7.455556	146.3803	86.81	3
Federated States of Micronesia	Yap	Losiep/L'oosiyep	Reef island	9.906667	139.8481	29.26	6
Federated States of Micronesia	Yap	Lossau	Reef island	9.888056	139.6792	26.94	3
Federated States of Micronesia	Yap	Maap	Volcanic high island	9.596667	138.1647	1077.34	61

(continued)

(continued)

Region/country	Island group	Island name	Island type	Latitude	Longitude	Area (ha)	Maximum elevation
Federated States of Micronesia	Yap	Mangejang	Reef island	9.945833	139.6867	19.25	3
Federated States of Micronesia	Yap	Mogmog	Reef island	10.08806	139.7081	32.01	3
Federated States of Micronesia	Yap	Ngulu	Reef island	8.45	137.4833	118.02	2
Federated States of Micronesia	Yap	Olimarao	Reef island	7.703611	145.8811	37.31	5
Federated States of Micronesia	Yap	Pig/Piig	Reef island	9.780833	139.6736	7.53	4
Federated States of Micronesia	Yap	Pigelelel (Song)/ Pigl'eeal'ey	Reef island	9.989444	139.6256	18.01	3
Federated States of Micronesia	Yap	Pikelot	Reef island	8.104722	147.6467	23.75	3
Federated States of Micronesia	Yap	Potoangroas/Potangeras	Reef island	10.07583	139.6728	30.08	3
Federated States of Micronesia	Yap	Rumung	Volcanic high island	9.621667	138.1547	469.92	84
Federated States of Micronesia	Yap	Satawal	Limestone low island	7.381111	147.0306	163	5
Federated States of Micronesia	Yap	Sogoroi (Sorlen)	Reef island	10.07972	139.7361	58.27	3
Federated States of Micronesia	Yap	Song/Soong	Reef island	10.00611	139.6189	14.8	2
Federated States of Micronesia	Yap	Songetigech/ Songoachigchi−g	Reef island	9.973056	139.6122	5.32	7
Federated States of Micronesia	Yap	Sorenleng/Dorooleng	Reef island	10.04056	139.5761	12.86	3

Federated States of Micronesia	Yap	Sorol	Reef island	8.146111	140.3847	2.46	3
Federated States of Micronesia	Yap	West Fayu	Reef island	8.083611	146.73	16.64	1
Federated States of Micronesia	Yap	Woleai	Reef island	7.373611	143.9075	576.67	3
Federated States of Micronesia	Yap	Yal'eepiig/Fusshimaro	Reef island	9.982222	139.6258	6.61	3
Federated States of Micronesia	Yap	Yap	Composite high island	9.511389	138.0936	6000.77	176
Federated States of Micronesia	Yap	Yew/Yeew	Reef island	9.929444	139.8164	19.79	3
Fiji	Beqa	Moturiki (2)	Volcanic high island	−18.408	178.1525	24.66	98
Fiji	Beqa Group	Beqa	Volcanic high island	−18.3931	178.1253	4221.91	439
Fiji	Beqa Group	Bird	Reef island	−18.3458	178.0003	3.18	1
Fiji	Beqa Group	Unganga (Royal Davui)	Volcanic low island	−18.4336	178.0739	6.78	20
Fiji	Kadavu Group	Bulia	Volcanic high island	−18.8397	178.5325	184.68	140
Fiji	Kadavu Group	Denham (Nagigia in Fiji file)	Limestone low island	−19.1275	177.9453	22.67	4
Fiji	Kadavu Group	Dravuni	Volcanic high island	−18.7636	178.5219	90.29	40
Fiji	Kadavu Group	Galoa (2)	Volcanic high island	−19.0814	178.1833	450.97	114
Fiji	Kadavu Group	Kadavu	Volcanic high island	−19.0111	178.2047	50626.38	610

(continued)

(continued)

Region/country	Island group	Island name	Island type	Latitude	Longitude	Area (ha)	Maximum elevation
Fiji	Kadavu Group	Matanuku	Volcanic high island	−19.1697	178.105	345.38	174
Fiji	Kadavu Group	Matava/Tawadromu	Volcanic low island	−19.0228	178.2533	30.27	21
Fiji	Kadavu Group	Ono	Volcanic high island	−18.9028	178.4856	3478.9	344
Fiji	Kadavu Group	Qasibale/Nggasimbale	Volcanic low island	18.8	178.5	5.77	18
Fiji	Kadavu Group	Vanuakula	Volcanic high island	−18.7364	178.5072	21.27	76
Fiji	Kadavu Group	Yabu	Volcanic high island	18.85	178.5	8.74	52
Fiji	Kadavu Group	Yandatavaya	Volcanic low island	−19.0594	178.2156	4.04	20
Fiji	Kadavu Group	Yanuyanu-i-loma	Volcanic low island	18.78306	18.78306	4.42	43
Fiji	Kadavu Group	Yanuyanu-i-sau	Volcanic low island	−18.7833	178.5	19.2	28
Fiji	Kadavu Group	Yaukuvelailai	Volcanic high island	−18.8136	178.5222	20.94	64
Fiji	Kadavu Group	Yaukuvelevu	Volcanic high island	−18.8036	178.5283	72	122
Fiji	Lau Group	Aiwa	Limestone high island	−18.3319	−178.683	77.56	51
Fiji	Lau Group	Avea	Composite high island	−17.1892	−178.903	283.29	183
Fiji	Lau Group	Booby Rock	Limestone low island	−17.8814	−178.306	8.68	4

Fiji	Lau Group	Cicia	Composite high island	−17.7492	−179.311	3877.01	165
Fiji	Lau Group	Cikobia-i-Lau (Thikombia)	Composite high island	−17.2822	−178.786	309.19	168
Fiji	Lau Group	Davura	Volcanic high island	−20.6533	−178.704	12.63	34
Fiji	Lau Group	Doi	Volcanic high island	−20.6378	−178.715	95.17	64
Fiji	Lau Group	Evuevu	Limestone high island	−17.1689	−179.04	143.05	136
Fiji	Lau Group	Fulaga	Limestone high island	−19.1553	−178.578	2136.98	79
Fiji	Lau Group	Kabara	Composite high island	−18.9575	−178.954	3720.58	143
Fiji	Lau Group	Kaibu	Composite high island	−17.2567	−179.487	196.48	388
Fiji	Lau Group	Kanacea	Volcanic high island	−17.2667	−179.151	1389.59	259
Fiji	Lau Group	Karoni	Limestone high island	−18.7019	−178.478	30.89	37
Fiji	Lau Group	Kimbombo Islet/Kibobo?	Limestone high island	−17.0478	−179.028	17.36	58
Fiji	Lau Group	Komo	Volcanic high island	−18.6844	−178.627	165.22	82
Fiji	Lau Group	Komo Ndriti/Komo Driki?	Limestone low island	−18.6819	−178.649	9.07	13
Fiji	Lau Group	Lakeba	Composite high island	−18.2142	−178.793	6253.73	210

(continued)

(continued)

Region/country	Island group	Island name	Island type	Latitude	Longitude	Area (ha)	Maximum elevation
Fiji	Lau Group	Late-i-Tonga	Limestone low island	−17.9333	−178.283	12.48	15
Fiji	Lau Group	Loa	Volcanic low island	−18.4303	−178.443	3.02	13
Fiji	Lau Group	Mago	Composite high island	−17.4423	−179.161	2556.31	204
Fiji	Lau Group	Malima	Volcanic high island	−17.1178	−179.193	26.58	40
Fiji	Lau Group	Mana (2)	Limestone low island	−20.6739	−178.696	9.46	6
Fiji	Lau Group	Marabo	Limestone high island	−18.9919	−178.824	165.07	49
Fiji	Lau Group	Moce	Volcanic high island	−18.6533	−178.506	1267.43	180
Fiji	Lau Group	Munia	Volcanic high island	−17.3686	−178.879	493.13	290
Fiji	Lau Group	Naitaba/Naitauba	Composite high island	−17.0181	−179.28	911.09	186
Fiji	Lau Group	Namalata	Composite high island	−17.3364	−178.984	205.47	128
Fiji	Lau Group	Namuka-i-Lau	Limestone high island	−18.8483	−178.642	1515.57	79
Fiji	Lau Group	Navutu-i-Loma	Limestone high island	−18.9656	−178.485	165.95	52
Fiji	Lau Group	Navutu-i-Ra	Limestone high island	−18.9435	−178.509	36.7	82
Fiji	Lau Group	Nayau	Limestone high island	−17.9794	−179.05	2079.16	179

Fiji	Lau Group	Ogea Levu	Limestone high island	−19.1478	−178.407	1626.68	82
Fiji	Lau Group	Olorua	Volcanic high island	−18.6058	−178.753	22.59	76
Fiji	Lau Group	Oneata (1)	Composite high island	−18.44	−178.475	512.56	49
Fiji	Lau Group	Ono-i-Lau	Volcanic high island	−20.6608	−178.739	471.85	113
Fiji	Lau Group	Oqea Driki	Limestone high island	−19.2075	−178.416	552.62	91
Fiji	Lau Group	Qilaqila	Limestone high island	−17.1783	−179.027	68.73	60
Fiji	Lau Group	Sovu	Limestone high island	−17.1625	−178.828	22.13	70
Fiji	Lau Group	Susui	Composite high island	−17.3472	−178.957	436.72	131
Fiji	Lau Group	Tavunasici	Limestone high island	−18.7267	−179.088	54.9	61
Fiji	Lau Group	Tuvana-i-Ra	Reef island	−21.0364	−178.847	11.81	3
Fiji	Lau Group	Tuvuca	Limestone high island	−17.6733	−178.82	1426.84	244
Fiji	Lau Group	Undui	Reef island	−20.6322	−178.75	26.23	5
Fiji	Lau Group	Vanua Balavu	Composite high island	−17.2206	−178.947	5984.34	283
Fiji	Lau Group	Vanua Vatu	Limestone high island	−18.3714	−179.271	483.17	102
Fiji	Lau Group	Vatoa	Limestone high island	−19.8183	−178.242	498.99	64

(continued)

(continued)

Region/country	Island group	Island name	Island type	Latitude	Longitude	Area (ha)	Maximum elevation
Fiji	Lau Group	Vatu Vara	Limestone high island	−17.4283	−179.525	461.73	314
Fiji	Lau Group	Vekai	Limestone low island	−17.5608	−178.814	6.65	6
Fiji	Lau Group	Vuaqava	Limestone high island	−18.8753	−178.903	952.02	107
Fiji	Lau Group	Yacata	Composite high island	−17.2567	−179.525	943.95	256
Fiji	Lau Group	Yanuca (2)	Limestone low island	−20.6081	−178.682	8.61	9
Fiji	Lau Group	Yanuyanu	Volcanic high island	−17.2978	−178.978	32.86	44
Fiji	Lomaiviti Group	Batiki	Volcanic high island	−17.7869	179.1583	1182.79	173
Fiji	Lomaiviti Group	Gau	Volcanic high island	−18.0186	179.3108	15734.47	715
Fiji	Lomaiviti Group	Koro	Volcanic high island	−17.3114	179.4069	11743.41	522
Fiji	Lomaiviti Group	Makodroga	Volcanic high island	−17.4228	178.9489	132.11	138
Fiji	Lomaiviti Group	Makogai	Volcanic high island	−17.4489	178.9653	1095.55	259
Fiji	Lomaiviti Group	Nairai	Volcanic high island	−17.8061	179.4081	2859.43	336
Fiji	Lomaiviti Group	Wakaya	Volcanic high island	−17.6169	179.0042	1251.29	183
Fiji	Lomaiviti Group	Yaciwa (Yathiwa)	Limestone low island	−18.1375	179.3428	5.85	4

Fiji	Mamanuca Group	Eori	Volcanic high island	−17.4381	177.0636	24.5	64
Fiji	Mamanuca Group	Kadavulailai (Bounty)	Reef island	−17.6733	177.3061	27.52	3
Fiji	Mamanuca Group	Kadomo	Volcanic high island	−17.4933	177.0519	52.96	100
Fiji	Mamanuca Group	Luvuka/Eluvuka?	Reef island	−17.6553	177.2672	18.17	3
Fiji	Mamanuca Group	Mala mala	Reef island	−17.7239	177.275	10.29	4
Fiji	Mamanuca Group	Malolo	Volcanic high island	−17.7486	177.1697	1120.03	229
Fiji	Mamanuca Group	Malolo Lailai (Malolo Sewa in Fiji file)	Volcanic high island	−17.7756	177.1981	277.22	70
Fiji	Mamanuca Group	Mana (1)	Volcanic high island	−17.6739	177.0986	175.01	70
Fiji	Mamanuca Group	Matamanoa	Volcanic high island	−17.6378	177.0661	19.85	32
Fiji	Mamanuca Group	Monu	Volcanic high island	−17.5969	177.0422	96.19	222
Fiji	Mamanuca Group	Monuriki	Volcanic high island	−17.61	177.0339	54.88	180
Fiji	Mamanuca Group	Namotu (1)	Reef island	−17.8444	177.1836	7.38	2
Fiji	Mamanuca Group	Nautanivono	Volcanic high island	−17.6278	177.1275	9.56	73
Fiji	Mamanuca Group	Navadra	Volcanic high island	−17.4556	177.0522	44.67	128
Fiji	Mamanuca Group	Navini	Reef island	−17.705	177.2253	8.67	3
Fiji	Mamanuca Group	Qalito	Volcanic high island	−17.7336	177.1342	78.74	119
Fiji	Mamanuca Group	Tai	Reef island	−17.6544	177.2547	8.48	4

(continued)

(continued)

Region/country	Island group	Island name	Island type	Latitude	Longitude	Area (ha)	Maximum elevation
Fiji	Mamanuca Group	Tavarua	Reef island	−17.8581	177.2019	10.27	4
Fiji	Mamanuca Group	Tavua	Volcanic high island	−17.6178	177.0919	230.04	174
Fiji	Mamanuca Group	Tokoriki	Volcanic high island	−17.5758	177.0917	98.61	95
Fiji	Mamanuca Group	Vomo	Volcanic high island	−17.4931	177.2686	125.24	116
Fiji	Mamanuca Group	Vomo Lailai	Volcanic high island	−17.4856	177.2506	13.55	61
Fiji	Mamanuca Group	Yakuilau	Reef island	−17.7597	177.3636	15.8	2
Fiji	Mamanuca Group	Yanuya	Volcanic high island	−17.5931	177.0594	164.52	104
Fiji	Mamanuca Group	Yavurimba	Volcanic high island	−17.5061	177.0978	1.3	30
Fiji	Moala Group	Matuku (1)	Volcanic high island	−19.1592	179.7661	3591.71	385
Fiji	Moala Group	Moala	Volcanic high island	−18.6028	179.8714	7068.98	468
Fiji	Moala Group	Totoya	Volcanic high island	−18.9336	−179.841	3802.55	361
Fiji	Ovalau Group	Cagalai (Thangalai)	Reef island	−17.7881	178.7303	5.23	6
Fiji	Ovalau Group	Leleuvia-Suva	Reef island	−17.8081	178.7217	4.23	6
Fiji	Ovalau Group	Moturiki (1)	Volcanic high island	−17.7633	178.745	1694.19	133
Fiji	Ovalau Group	Ovalau	Volcanic high island	−17.6853	178.7911	11992.84	626
Fiji	Ovalau Group	Yanuca Lailai	Volcanic high island	−17.7525	178.7828	59.37	574

Fiji	Ovalau Group	Yanuca Levu	Volcanic high island	-17.7636	178.7794	172.96	61
Fiji	Ringgold Isles	Cobia (Thombia)	Volcanic high island	-16.4661	-179.645	93.44	177
Fiji	Ringgold Isles	Maqewa (Manggewa)	Volcanic high island	-16.4989	-179.682	45.77	85
Fiji	Ringgold Isles	Nanuku Lailai	Reef island	-16.7097	-179.449	3.82	2
Fiji	Ringgold Isles	Nanuku Levu	Reef island	-16.72	-179.454	4.15	3
Fiji	Ringgold Isles	Nukusemanu	Reef island	-16.3375	-179.415	10.88	2
Fiji	Ringgold Isles	Qelelevu	Limestone low island	-16.0906	-179.161	175.88	12
Fiji	Ringgold Isles	Tauraria	Limestone low island	-16.0835	-179.168	6.51	7
Fiji	Ringgold Isles	Vetauua	Limestone low island	-15.9561	-179.383	89.06	3
Fiji	Ringgold Isles	Wailagilala	Reef island	-16.7489	-179.104	98.17	4
Fiji	Ringgold Isles	Yavu (Yanuca)	Volcanic high island	-16.5017	-179.695	141.88	113
Fiji	Rotuma Group	Hatana	Volcanic low island	-12.4742	176.9658	20.03	18
Fiji	Rotuma Group	Hofliua	Volcanic high island	-12.4961	176.9372	13.55	58
Fiji	Rotuma Group	Rotuma	Volcanic high island	-12.5031	177.0775	4882.28	256
Fiji	Rotuma Group	Uea	Volcanic high island	-12.4664	176.9917	98.65	262
Fiji	Vanua Levu Group	Bekana (1)	Volcanic low island	-16.1997	179.8358	23.5	6

(continued)

(continued)

Region/country	Island group	Island name	Island type	Latitude	Longitude	Area (ha)	Maximum elevation
Fiji	Vanua Levu Group	Bekana (2)	Volcanic low island	−17.5888	177.4393	86.74	6
Fiji	Vanua Levu Group	Cikobia (Thikombia)	Limestone high island	−15.7375	−179.95	1746.11	192
Fiji	Vanua Levu Group	Cukini	Reef island	−16.385	179.2044	292.91	8
Fiji	Vanua Levu Group	Galoa (1)	Volcanic high island	−16.6161	178.6864	132.15	90
Fiji	Vanua Levu Group	Gevo	Volcanic high island	−16.2192	179.6019	106.65	141
Fiji	Vanua Levu Group	Kavewa	Volcanic high island	−16.19	179.5758	125.57	49
Fiji	Vanua Levu Group	Kia	Volcanic high island	−16.2303	179.0928	234	238
Fiji	Vanua Levu Group	Kioa	Volcanic high island	−16.6542	179.9131	2199.16	305
Fiji	Vanua Levu Group	Laucala	Volcanic high island	−16.7597	−179.683	1465.61	134
Fiji	Vanua Levu Group	Macuata-i-wai	Volcanic high island	−16.4275	179.0639	398.47	152
Fiji	Vanua Levu Group	Mali	Volcanic high island	−16.3433	179.3522	1095.04	171
Fiji	Vanua Levu Group	Matagi	Volcanic high island	−16.7317	−179.734	123.83	100
Fiji	Vanua Levu Group	Nadogo	Reef island	−16.5194	178.8094	1366	3
Fiji	Vanua Levu Group	Nagano	Reef island	−16.3956	179.1608	211.39	8
Fiji	Vanua Levu Group	Namenalala (Namena)	Volcanic high island	−17.1103	179.0994	91.74	105

Fiji	Vanua Levu Group	NdruaNdrua	Volcanic high island	−16.1975	179.6167	492.68	134
Fiji	Vanua Levu Group	Nukubati	Reef island	−16.4631	179.0247	62.72	2
Fiji	Vanua Levu Group	Nukuira	Reef island	−16.5694	178.72	492.16	3
Fiji	Vanua Levu Group	Nukunuku	Reef island	−16.3711	179.1958	55.22	3
Fiji	Vanua Levu Group	Qamea	Volcanic high island	−16.77	−179.752	4101.33	304
Fiji	Vanua Levu Group	Rabi	Volcanic high island	−16.4995	−179.953	8832	466
Fiji	Vanua Levu Group	Talailau	Reef island	−16.3875	179.1297	234.99	9
Fiji	Vanua Levu Group	Tavea	Volcanic high island	−16.6258	178.7267	118.66	79
Fiji	Vanua Levu Group	Taveuni	Volcanic high island	−16.8617	−179.968	52728.56	1241
Fiji	Vanua Levu Group	Tilagica (Tilangitha)	Reef island	−16.1875	179.7792	77.19	4
Fiji	Vanua Levu Group	Tivi (Titi)	Volcanic high island	−16.2739	179.4686	175.52	105
Fiji	Vanua Levu Group	Tutu	Volcanic high island	−16.2392	179.5714	438.21	193
Fiji	Vanua Levu Group	Vanua Levu	Volcanic high island	−16.5783	179.1911	625900.1	1032
Fiji	Vanua Levu Group	Vatu-i-Cake (Vati-i-Thake)	Reef island	−17.3811	178.7675	12.98	8
Fiji	Vanua Levu Group	Vorovoro	Volcanic high island	−16.3425	179.3136	73.43	89
Fiji	Vanua Levu Group	Yadua	Volcanic high island	−16.8167	178.2996	1549.3	195
Fiji	Vanua Levu Group	Yadua Taba	Volcanic high island	−16.8336	178.2783	79.51	100

(continued)

(continued)

Region/country	Island group	Island name	Island type	Latitude	Longitude	Area (ha)	Maximum elevation
Fiji	Vanua Levu Group	Yalewa Kalou	Volcanic high island	−16.6633	177.7647	57.25	180
Fiji	Vanua Levu Group	Yanucagi (Yanuthangi)	Reef island	−16.6142	178.6636	28.2	3
Fiji	Vanua Levu Group	Yaqaga (Yangganga)	Volcanic high island	−16.5914	178.5878	1194.39	270
Fiji	Vatulele Group	Vatulele	Limestone high island	−18.5292	177.6269	3641.74	34
Fiji	Viti Levu Group	Bau	Limestone low island	−17.9725	178.6156	10.5	24
Fiji	Viti Levu Group	Labiko (Valolo)	Reef island	−18.1311	178.6019	22.21	1
Fiji	Viti Levu Group	Macuata	Volcanic high island	−17.355	178.0347	45.81	122
Fiji	Viti Levu Group	Makuluva	Reef island	−18.1881	178.5181	8.57	2
Fiji	Viti Levu Group	Malake–Rakiraki	Volcanic high island	−17.3211	178.1433	549.9	236
Fiji	Viti Levu Group	Naigani	Composite high island	−17.58	178.675	250.96	186
Fiji	Viti Levu Group	Namuka	Volcanic low island	−18.1489	178.3356	27.6	12
Fiji	Viti Levu Group	Nananu-i-Cake	Volcanic high island	−17.3211	178.2314	279.26	167
Fiji	Viti Levu Group	Nananu-i-Ra	Volcanic high island	−17.2856	178.2178	349.88	180
Fiji	Viti Levu Group	Naqara	Volcanic high island	−18.1764	178.2683	105.9	53
Fiji	Viti Levu Group	Nasoata	Reef island	−18.1461	178.5875	88.83	1
Fiji	Viti Levu Group	Nukulau	Reef island	−18.1739	178.5158	4.09	2

Fiji	Viti Levu Group	Omini	Limestone low island	−17.7558	178.5778	19	24
Fiji	Viti Levu Group	Qata	Limestone low island	−17.7589	178.6086	41.72	9
Fiji	Viti Levu Group	Qoma (2)	Volcanic low island	−17.6367	178.5825	26.48	30
Fiji	Viti Levu Group	Senia	Reef island	−18.135	178.5897	2.22	8
Fiji	Viti Levu Group	Tawainave	Limestone low island	−17.7741	178.6097	20.37	8
Fiji	Viti Levu Group	Telau	Limestone low island	−17.908	178.6196	14.95	30
Fiji	Viti Levu Group	Tovu	Volcanic high island	−17.3464	178.0581	52.91	62
Fiji	Viti Levu Group	Tovulailai	Volcanic low island	−17.3433	178.0667	22.98	14
Fiji	Viti Levu Group	Vatia Lailai	Volcanic high island	−17.3703	177.8253	80.27	35
Fiji	Viti Levu Group	Vatu-i-Ra	Volcanic low island	−17.3158	178.4667	5.75	7
Fiji	Viti Levu Group	Viti Levu	Composite high island	−17.8053	178.0053	1182706	1323
Fiji	Viti Levu Group	Viwa (2)	Limestone high island	−17.9417	178.6158	90.4	49
Fiji	Yasawa Group	Drawaqa	Volcanic high island	−17.1728	177.1917	43.88	84
Fiji	Yasawa Group	Kuata	Volcanic high island	−17.3706	177.1372	190.69	174
Fiji	Yasawa Group	Matacawa Levu	Volcanic high island	−16.9547	177.3372	1145.39	299

(continued)

(continued)

Region/country	Island group	Island name	Island type	Latitude	Longitude	Area (ha)	Maximum elevation
Fiji	Yasawa Group	Nacula	Volcanic high island	−16.8972	177.4194	2652.58	270
Fiji	Yasawa Group	Nanuya Balavu	Volcanic high island	−17.1889	177.1797	120.21	105
Fiji	Yasawa Group	Nanuya Lailai	Volcanic high island	−16.9469	177.3744	166.36	39
Fiji	Yasawa Group	Nanuya Levu (Turtle)	Volcanic high island	−16.9636	177.3756	199.79	68
Fiji	Yasawa Group	Nanuya-i-Ra	Reef island	−16.6533	177.5781	23.62	9
Fiji	Yasawa Group	Nanuya-i-Yata	Reef island	−16.6669	177.5917	16.93	4
Fiji	Yasawa Group	Narara	Volcanic high island	−17.2078	177.1744	43.72	102
Fiji	Yasawa Group	Naukacuvu	Volcanic high island	−17.195	177.1722	39.55	64
Fiji	Yasawa Group	Naviti	Volcanic high island	−17.1125	177.2419	3926.26	388
Fiji	Yasawa Group	Sawa-i-Lau	Limestone high island	−16.8514	177.4697	114.34	210
Fiji	Yasawa Group	Tavewa	Volcanic high island	−16.925	177.3606	194.65	180
Fiji	Yasawa Group	Vawa	Volcanic high island	−16.7975	177.4567	43.57	86
Fiji	Yasawa Group	Viwa (1)	Limestone low island	−17.1408	176.9272	734.79	18
Fiji	Yasawa Group	Waya	Volcanic high island	−17.2908	177.125	2332.61	571

Fiji	Yasawa Group	Wayasewa	Volcanic high island	−17.3414	177.1389	715.21	354
Fiji	Yasawa Group	Yanuca (1)	Volcanic low island	−17.3	178.2333	4.22	18
Fiji	Yasawa Group	Yaqeta (Yanggeta)	Volcanic high island	−17.0078	177.3264	788.79	220
Fiji	Yasawa Group	Yasawa	Volcanic high island	−16.7822	177.5144	3784.55	233
Fiji	Yasawa Group	Yawini	Volcanic high island	−16.6881	177.5606	52.38	50
French Polynesia	Austral Islands	Iles Maria/Maria/Hull Island	Reef island	−21.8022	−154.696	79.65	7
French Polynesia	Austral Islands	Raivavae (Raevavae)	Volcanic high island	−23.8647	−147.661	1871.83	437
French Polynesia	Austral Islands	Rimatara	Composite high island	−22.6369	−152.844	1024.08	95
French Polynesia	Austral Islands	Rurutu	Composite high island	−22.48	−151.338	4044.6	389
French Polynesia	Austral Islands	Tubuai	Volcanic high island	−23.3567	−149.449	5475.62	422
French Polynesia	Bass Islands	Rapa Iti/Rapa	Volcanic high island	−27.5922	−144.352	5413.68	650
French Polynesia	Gambier Group; Acteon Group	Tenarunga	Reef island	−21.3483	−136.54	358.85	9
French Polynesia	Gambier Islands/Mangareva	Agakauitai	Volcanic high island	−23.1653	−135.035	104.06	146
French Polynesia	Gambier Islands/Mangareva	Akamaru	Volcanic high island	−23.1875	−134.914	298.17	246

(continued)

(continued)

Region/country	Island group	Island name	Island type	Latitude	Longitude	Area (ha)	Maximum elevation
French Polynesia	Gambier Islands/ Mangareva	Aukena	Volcanic high island	−23.1331	−134.905	241.1	198
French Polynesia	Gambier Islands/ Mangareva	Kamaka	Volcanic high island	−23.2433	−134.957	72.79	176
French Polynesia	Gambier Islands/ Mangareva	Makaroa	Volcanic high island	−23.2192	−134.971	39.66	136
French Polynesia	Gambier Islands/ Mangareva	Mangareva	Volcanic high island	−23.1094	−134.974	1941.06	445
French Polynesia	Gambier Islands/ Mangareva	Manui	Volcanic high island	−23.2319	−134.945	17.21	54
French Polynesia	Gambier Islands/ Mangareva	Taravai	Volcanic high island	−23.1478	−135.038	683.13	256
French Polynesia	Leeward Islands	Bora Bora	Volcanic high island	−16.5003	−151.741	2236.54	727
French Polynesia	Leeward Islands	Huahine, Fare	Volcanic high island	−16.7403	−151.024	8617.08	669
French Polynesia	Leeward Islands	Manuae (Fenuaura, Scily)	Reef island	−16.5483	−154.686	536.07	8
French Polynesia	Leeward Islands	Maupiti	Volcanic high island	−16.4439	−152.26	453.16	380
French Polynesia	Leeward Islands	Mopelia (Maupihaa, Mopihaa)	Reef island	−16.8131	−153.958	459.88	6
French Polynesia	Leeward Islands	Motu One (Bellingshausen	Reef island	−15.8203	−154.521	341.58	6
French Polynesia	Leeward Islands	Raiatea	Volcanic high island	−16.8039	−151.443	18778.93	1017
French Polynesia	Leeward Islands	Tahaa	Volcanic high island	−16.6094	−151.502	9865.47	590
French Polynesia	Leeward Islands	Tupai (Motu Iti)	Reef island	−16.2647	−151.819	1270.21	4

French Polynesia	Marquesas Islands (Northern)	Eiao, french Polynesia	Volcanic high island	-7.97889	-140.669	4379.26	576
French Polynesia	Marquesas Islands (Northern)	Hatutu	Volcanic high island	-7.91694	-140.568	810.02	428
French Polynesia	Marquesas Islands (Northern)	Motu Iti	Volcanic high island	-8.68222	-140.608	34.7	232
French Polynesia	Marquesas Islands (Northern)	Motu Oa	Volcanic high island	-9.47972	-140.048	50.74	112
French Polynesia	Marquesas Islands (Northern)	Nuku Hiva	Volcanic high island	-8.86028	-140.142	36953.47	1224
French Polynesia	Marquesas Islands (Northern)	Ua Huka	Volcanic high island	-8.9075	-139.548	8696.57	884
French Polynesia	Marquesas Islands (Northern)	Ua Pou (Ua Pu)	Volcanic high island	-9.40417	-140.08	12010.21	1203
French Polynesia	Marquesas Islands (Southern)	Fatu Hiva	Volcanic high island	-10.4825	-138.625	9020	960
French Polynesia	Marquesas Islands (Southern)	Fatu Huku	Volcanic high island	-9.43611	-138.924	104.76	361
French Polynesia	Marquesas Islands (Southern)	Hiva Oa	Volcanic high island	-9.75444	-139.021	33663.25	1213
French Polynesia	Marquesas Islands (Southern)	Moho Tani (Motane)	Volcanic high island	-9.98	-138.834	1509.59	520
French Polynesia	Marquesas Islands (Southern)	Tahuata	Volcanic high island	-9.95167	-139.092	7710.99	1050
French Polynesia	Marquesas Islands (Southern)	Terihi	Volcanic high island	-10.0189	-138.801	16.92	152
French Polynesia	Other Islands of the Gambier Group	Gaioio	Reef island	-23.1122	-134.84	24.49	2

(continued)

(continued)

Region/country	Island group	Island name	Island type	Latitude	Longitude	Area (ha)	Maximum elevation
French Polynesia	Other Islands of the Gambier Group	Kouaku	Reef island	−23.2058	−134.859	12.56	2
French Polynesia	Other Islands of the Gambier Group	Makapu	Reef island	−23.1996	−134.924	13.23	6
French Polynesia	Other Islands of the Gambier Group	Mekiro	Volcanic high island	−23.1711	−134.922	13.54	38
French Polynesia	Other Islands of the Gambier Group	Motu Teiku	Reef island	−23.2083	−134.976	4.14	5
French Polynesia	Other Islands of the Gambier Group	Motu-O-Ari	Reef island	−23.1478	−135.023	5.34	1
French Polynesia	Other Islands of the Gambier Group	Puaumu	Reef island	−23.0136	−134.919	29.25	1
French Polynesia	Other Islands of the Gambier Group	Rumarei	Reef island	−23.1106	−134.991	2.29	1
French Polynesia	Other Islands of the Gambier Group	Tarauru Roa	Reef island	−23.1061	−134.856	70.91	2
French Polynesia	Other Islands of the Gambier Group	Teauaone	Reef island	−23.0033	−134.954	10.4	1
French Polynesia	Other Islands of the Gambier Group	Tekava	Reef island	−23.1617	−134.853	23.93	2
French Polynesia	Other Islands of the Gambier Group	Tepapuri	Reef island	−23.0064	−134.945	30.63	1
French Polynesia	Other Islands of the Gambier Group	Totegegie	Reef island	−23.0778	−134.89	163.58	2
French Polynesia	Other Islands of the Gambier Group	Tuaeu	Reef island	−23.0622	−134.904	4.32	1
French Polynesia	Society Islands	Ma'iao	Volcanic high island	−17.6586	−150.627	410.21	154

French Polynesia	Society Islands	Meheti'a	Volcanic high island	-17.8775	-148.067	224.21	433
French Polynesia	Society Islands	Mo'orea	Volcanic high island	-17.5386	-149.829	14850.92	1207
French Polynesia	Society Islands	Tahiti	Volcanic high island	-17.6508	-149.426	114755.8	1042
French Polynesia	Society Islands	Tetiaroa	Reef island	-17.0019	-149.562	305.89	4
French Polynesia	Toamotu Group	Reitoru (Bird's)	Reef island	-17.8539	143.0725	178.43	6
French Polynesia	Tuamotu Group	Ahe (Peacock)	Reef island	-14.515	-146.32	1956.56	4
French Polynesia	Tuamotu Group	Ahunui (Byam Martin)	Reef island	-19.6356	-140.404	406.99	8
French Polynesia	Tuamotu Group	Akiaki (Lancier, Thrum Cap)	Reef island	-18.5581	-139.212	104.9	9
French Polynesia	Tuamotu Group	Amanu (Moller)	Reef island	-17.7928	-140.82	3020.64	9
French Polynesia	Tuamotu Group	Anaa (Chain)	Reef island	-17.3956	-145.469	852.32	7
French Polynesia	Tuamotu Group	Apataki (Hegemeister)	Reef island	-15.3019	-146.35	2677.12	4
French Polynesia	Tuamotu Group	Aratika (Karlshoff)	Reef island	-15.4883	-145.482	1506.23	4
French Polynesia	Tuamotu Group	Arutua (Rurick)	Reef island	-15.2897	-146.786	1239.05	4
French Polynesia	Tuamotu Group	Faaite (Miloradovitch)	Reef island	-16.7431	-145.238	1043.19	9
French Polynesia	Tuamotu Group	Fakahina (Enterprise)	Reef island	-15.9903	-140.137	1482.18	7
French Polynesia	Tuamotu Group	Fakarava (Wittgenstein)	Reef island	-16.2583	-145.606	4911.15	7
French Polynesia	Tuamotu Group	Fangatau (Arakchev)	Reef island	-15.8253	-140.861	534.92	8
French Polynesia	Tuamotu Group	Fangataufa (Cockburn)	Reef island	-22.2333	-138.75	1150.82	8
French Polynesia	Tuamotu Group	Haorangi (Bow Harp)	Reef island	-18.2092	-140.854	3596.01	8
French Polynesia	Tuamotu Group	Haraiki (Croker, San Quentin)	Reef island	-17.4631	-143.447	369.3	9
French Polynesia	Tuamotu Group	Hereheretue (St. Paul)	Reef island	-19.8733	-144.938	75.12	8
French Polynesia	Tuamotu Group	Hikueru (Melville)	Reef island	-17.5928	-142.621	215.29	8

(continued)

(continued)

Region/country	Island group	Island name	Island type	Latitude	Longitude	Area (ha)	Maximum elevation
French Polynesia	Tuamotu Group	Hiti (Ohiti, Clute)	Reef island	−16.7403	−144.055	443.76	9
French Polynesia	Tuamotu Group	Katiu (Saken)	Reef island	−16.4203	−144.36	1379.43	7
French Polynesia	Tuamotu Group	Kauehi (Vincennes)	Reef island	−15.7939	−145.173	2391.24	5
French Polynesia	Tuamotu Group	Kaukura (Auura, Oura)	Reef island	−15.7672	−146.687	1373.75	5
French Polynesia	Tuamotu Group	Makatea (Aurora)	Limestone high island	−15.845	−148.252	3117.13	111
French Polynesia	Tuamotu Group	Makemo (Koutousof Smolenski)	Reef island	−16.5394	−143.867	2548.22	6
French Polynesia	Tuamotu Group	Manihi (Wilsons)	Reef island	−14.3761	−145.95	2524.48	9
French Polynesia	Tuamotu Group	Manuhangi (Cumberland)	Reef island	−19.1972	−141.245	366.08	5
French Polynesia	Tuamotu Group	Maria Est	Reef island	−22.0144	−136.186	602.16	8
French Polynesia	Tuamotu Group	Marokau (Manaka)	Reef island	−18.0492	−142.329	440.87	5
French Polynesia	Tuamotu Group	Marutea North (Furneaux)	Reef island	−21.5281	−135.572	1152.75	6
French Polynesia	Tuamotu Group	Mataiva (Matahiva, Lazareff)	Reef island	−14.8814	−148.686	1594.1	8
French Polynesia	Tuamotu Group	Matureivavao (Melbourne)	Reef island	−21.4678	−136.393	304.48	6
French Polynesia	Tuamotu Group	Morane	Reef island	−23.1544	−137.131	79.02	9
French Polynesia	Tuamotu Group	Moruroa (Mururoa, Matilda)	Reef island	−21.8331	−138.917	1852.64	7
French Polynesia	Tuamotu Group	Motutunga (Adventure)	Reef island	−17.0986	−144.367	143.71	8
French Polynesia	Tuamotu Group	Napuka (Isle of Disappointment)	Reef island	−14.1675	−141.232	877.81	7
French Polynesia	Tuamotu Group	Niau (Greig)	Reef island	−16.1692	−146.331	6023.81	9
French Polynesia	Tuamotu Group	Nihiru (Nigeri)	Reef island	−16.7003	−142.842	345.6	5

French Polynesia	Tuamotu Group	Nukutavake (Queen Charlotte)	Reef island	−19.2808	−138.782	459.79	5
French Polynesia	Tuamotu Group	Paraoa (Gloucester)	Reef island	−19.1336	−140.677	454.16	6
French Polynesia	Tuamotu Group	Pinaki (Whitsunday)	Reef island	−19.3969	−138.678	209.59	8
French Polynesia	Tuamotu Group	Puka–puka (Dog)	Reef island	−14.8217	−138.814	354.18	5
French Polynesia	Tuamotu Group	Pukaruha (Serle)	Reef island	−18.3025	−137.014	1809.92	5
French Polynesia	Tuamotu Group	Rangiroa (Deans)	Reef island	−15.0847	−147.656	4425.79	6
French Polynesia	Tuamotu Group	Raraka	Reef island	−16.17	−144.975	1091.69	4
French Polynesia	Tuamotu Group	Raroia (Barclay de Tolley)	Reef island	−16.015	−142.387	1306.94	2
French Polynesia	Tuamotu Group	Ravahere (Dawhaida)	Reef island	−18.2503	−142.139	148.95	4
French Polynesia	Tuamotu Group	Reao (Clermont-Tonnere)	Reef island	−18.5547	−136.334	2798.01	8
French Polynesia	Tuamotu Group	Rekareka (Tehuata, Good Hope)	Reef island	−16.8339	−141.917	264.49	8
French Polynesia	Tuamotu Group	Taenga (Holt, Yermalov)	Reef island	−16.3364	−143.136	309.38	3
French Polynesia	Tuamotu Group	Tahanea (Tchitchagoff)	Reef island	−16.9375	−144.653	1265.38	4
French Polynesia	Tuamotu Group	Taiaro (king)	Reef island	−15.7569	−144.564	1150.29	5
French Polynesia	Tuamotu Group	Takapoto (Spiridof)	Reef island	−14.6247	−145.198	2686.86	6
French Polynesia	Tuamotu Group	Takaroa (King George's Islet)	Reef island	−14.4311	−144.929	2795.14	5
French Polynesia	Tuamotu Group	Takume (Volkonosky)	Reef island	−15.7811	−142.201	410.17	9
French Polynesia	Tuamotu Group	Tatakoto (Clerke, Narcissus)	Reef island	−17.3378	−138.391	759	9
French Polynesia	Tuamotu Group	Tauere (St. Simeon)	Reef island	−17.37	−141.511	243.65	7
French Polynesia	Tuamotu Group	Tekokota (Doubtful)	Reef island	−17.3242	−142.566	23.47	1
French Polynesia	Tuamotu Group	Tematangi (Bligh's)	Reef island	−21.6814	−140.625	415.04	9
French Polynesia	Tuamotu Group	Tepoto North (Otooho)	Reef island	−14.1006	−141.43	116.63	7

(continued)

(continued)

Region/country	Island group	Island name	Island type	Latitude	Longitude	Area (ha)	Maximum elevation
French Polynesia	Tuamotu Group	Tepoto South (Eliza)	Reef island	−16.8189	−144.266	170.86	6
French Polynesia	Tuamotu Group	Tikehau (Krusenstern)	Reef island	−15.0222	−148.148	1892.75	8
French Polynesia	Tuamotu Group	Tikei (Romanzoff)	Reef island	−14.9497	−144.55	341.18	3
French Polynesia	Tuamotu Group	Toau (Elizabeth)	Reef island	−15.8131	−145.995	1845.29	6
French Polynesia	Tuamotu Group	Tuanake (Reid)	Reef island	−16.6467	−144.212	444.75	8
French Polynesia	Tuamotu Group	Tureia (Carysfort)	Reef island	−20.8283	−138.539	1168.58	8
French Polynesia	Tuamotu Group	Vahanga (Bedford)	Reef island	−21.3328	−136.652	294.53	6
French Polynesia	Tuamotu Group	Vahitahi (Cook's lagoon)	Reef island	−18.76	−138.817	368.68	6
French Polynesia	Tuamotu Group	Vairaatea (Egmont)	Reef island	−19.3422	−139.236	566.31	6
French Polynesia	Tuamotu Group	Vanavana (Barrow)	Reef island	−20.7786	−139.137	248.85	5
Guam		Guam	Composite high island	13.455	144.7794	58818.78	400
Hawaii (USA State)	Main Hawaiian Islands	Hawai'i	Volcanic high island	19.89667	−155.583	1182220	4205
Hawaii (USA State)	Main Hawaiian Islands	Kaho'olawe	Volcanic high island	20.55778	−156.606	13987.59	452
Hawaii (USA State)	Main Hawaiian Islands	Kaua'i	Volcanic high island	22.09639	−159.526	169047.6	1598
Hawaii (USA State)	Main Hawaiian Islands	Lana'i	Volcanic high island	20.81639	−156.927	42828.43	1026
Hawaii (USA State)	Main Hawaiian Islands	Maui	Volcanic high island	20.79833	−156.332	218792	3055
Hawaii (USA State)	Main Hawaiian Islands	Moloka'i	Volcanic high island	21.14417	−157.023	79981.73	1512
Hawaii (USA State)	Main Hawaiian Islands	Ni'ihau	Volcanic high island	21.89194	−160.157	22691.49	381

Hawaii (USA State)	Main Hawaiian Islands	O'ahu	Volcanic high island	21.43889	−158	181361.9	1220
Hawaii (USA State)	Northwestern Hawaiian Islands	French Frigate Shoals	Volcanic high island	23.74889	−166.146	193.7	37
Hawaii (USA State)	Northwestern Hawaiian Islands	Gardner Pinnacles	Volcanic high island	25.01639	−167.983	11.03	52
Hawaii (USA State)	Northwestern Hawaiian Islands	Kure	Reef island	28.3925	−178.293	169.04	3
Hawaii (USA State)	Northwestern Hawaiian Islands	Laysan	Reef island	25.76778	−171.732	553.09	2
Hawaii (USA State)	Northwestern Hawaiian Islands	Lisianski	Reef island	26.06623	−173.966	223.18	6
Hawaii (USA State)	Northwestern Hawaiian Islands	Necker	Volcanic high island	23.58333	−164.703	68.54	84
Hawaii (USA State)	Northwestern Hawaiian Islands	Nihoa	Volcanic high island	23.06028	−161.922	112.72	272
Hawaii (USA State)	Northwestern Hawaiian Islands	Pearl and Hermes Atoll	Reef island	27.8375	−175.813	93.16	3
Kiribati	Gilbert Islands	Abaiang	Reef island	1.850278	173.0036	2483.15	3
Kiribati	Gilbert Islands	Abemama	Reef island	0.429167	173.9117	3369.34	3
Kiribati	Gilbert Islands	Aranuka	Reef island	0.170278	173.5967	1761.81	3
Kiribati	Gilbert Islands	Arorae	Reef island	−2.63861	176.8256	761.97	2
Kiribati	Gilbert Islands	Banaba (Ocean)	Limestone high island	−0.85611	169.5358	800.53	81
Kiribati	Gilbert Islands	Beru	Reef island	−1.32917	176.0097	1623.42	3
Kiribati	Gilbert Islands	Butaritari	Reef island	3.096667	172.8458	1222.95	3
Kiribati	Gilbert Islands	Kuria	Reef island	0.22	173.4214	1185.53	3
Kiribati	Gilbert Islands	Maiana	Reef island	0.913889	173.0342	1988.64	3

(continued)

(continued)

Region/country	Island group	Island name	Island type	Latitude	Longitude	Area (ha)	Maximum elevation
Kiribati	Gilbert Islands	Makin_Meang	Reef island	3.380833	172.9928	716.47	3
Kiribati	Gilbert Islands	Marakei	Reef island	2.010278	173.2933	1813.24	3
Kiribati	Gilbert Islands	Nikunau	Reef island	-1.36333	176.4597	2093.56	3
Kiribati	Gilbert Islands	Nonouti	Reef island	-0.67861	174.4472	2455.47	2
Kiribati	Gilbert Islands	Onotoa	Reef island	-1.89333	175.6036	1356.48	2
Kiribati	Gilbert Islands	Tabiteuea	Reef island	-1.32806	174.8433	5507.07	3
Kiribati	Gilbert Islands	Tamana	Reef island	-2.49861	175.9833	428.53	2
Kiribati	Gilbert Islands	Tarawa (Bairiki)	Reef island	1.434167	173.0858	3332.2	3
Kiribati	Line Islands	Flint	Reef island	-11.4314	-151.819	265.89	4
Kiribati	Line Islands	Kiritimati (Christmas)	Reef island	1.848889	-157.39	47844.43	3
Kiribati	Line Islands	Malden	Limestone low island	-4.01667	-154.938	3827.95	10
Kiribati	Line Islands	Millennium	Limestone low island	-9.96056	-150.207	604.1	6
Kiribati	Line Islands	Starbuck	Limestone low island	-5.64194	-155.878	2228.38	8
Kiribati	Line Islands	Tabuaeran (Fanning)	Reef island	3.863333	-159.346	5168.68	3
Kiribati	Line Islands	Teraina (Washington)	Reef island	4.683889	-160.382	1226.23	5
Kiribati	Line Islands	Vostok	Reef island	-10.1	-152.383	36.09	5
Kiribati	Phoenix Islands	Birnie	Reef island	-3.58278	-171.52	60.39	4
Kiribati	Phoenix Islands	Enderbury	Limestone low island	-3.12139	-171.086	723.93	7
Kiribati	Phoenix Islands	Kanton	Limestone low island	-2.80917	-171.715	1621.27	7
Kiribati	Phoenix Islands	Manra	Reef island	-4.45333	-171.246	1183.3	4
Kiribati	Phoenix Islands	McKean	Reef island	-3.59583	-174.123	86.03	2

Kiribati	Phoenix Islands	Nikumaroro	Reef island	−4.67472	−174.523	527.98	2
Kiribati	Phoenix Islands	Orona (Hull)	Reef island	−4.51472	−172.177	1110.06	3
Kiribati	Phoenix Islands	Rawaki (Phoenix)	Reef island	−3.71417	−170.728	78.19	3
Marshall Islands	Ralik Chain	Ailinginae (Ailiginae)	Reef island	11.14056	166.4092	456.24	2
Marshall Islands	Ralik Chain	Ailinglaplap	Reef island	7.425278	168.7928	2196.56	3
Marshall Islands	Ralik Chain	Bikini (Pikinni)	Reef island	11.55917	165.3878	971.07	5
Marshall Islands	Ralik Chain	Ebon (Epoon)	Reef island	4.616389	168.7247	825.61	3
Marshall Islands	Ralik Chain	Enewetak (+newetak)	Reef island	11.46528	162.1889	1071.2	4
Marshall Islands	Ralik Chain	Jabat (Jebat)	Reef island	7.751667	168.9767	88.04	3
Marshall Islands	Ralik Chain	Jaluit (Jalwoj)	Reef island	6.0075	169.4875	1878.57	3
Marshall Islands	Ralik Chain	Kili (K)le)	Reef island	5.640278	169.1181	80.72	3
Marshall Islands	Ralik Chain	Kwajalein (Kuwajleen)	Reef island	9.052778	167.435	2949.64	6
Marshall Islands	Ralik Chain	Lae	Reef island	8.940833	166.2361	230.22	3
Marshall Islands	Ralik Chain	Lib (Ellep)	Reef island	8.311667	167.3811	111.35	3
Marshall Islands	Ralik Chain	Namorik (Namdik)	Reef island	5.613056	168.1164	533.13	3
Marshall Islands	Ralik Chain	Namu	Reef island	7.900556	168.2247	804.22	3
Marshall Islands	Ralik Chain	Rongelap	Reef island	11.3625	166.8167	1165.67	3
Marshall Islands	Ralik Chain	Rongerik	Reef island	11.3675	167.4472	342.8	3
Marshall Islands	Ralik Chain	Ujae	Reef island	9.058056	165.6439	236.15	3
Marshall Islands	Ralik Chain	Ujelang (Ujla~)	Reef island	9.822778	160.895	268.19	3
Marshall Islands	Ralik Chain	Wotho	Reef island	10.10222	165.9811	635.4	3
Marshall Islands	Ratak Chain	a) Mili–Nadikdik (Knox Atoll)	Reef island	5.891111	172.1658	135.08	3
Marshall Islands	Ratak Chain	Ailuk (Aelok)	Reef island	10.30472	169.9414	625.26	3
Marshall Islands	Ratak Chain	Arno	Reef island	7.060833	171.6472	2405.71	2
Marshall Islands	Ratak Chain	Aur	Reef island	8.275	171.0975	867.32	3
Marshall Islands	Ratak Chain	Bikar (Pikaar)	Reef island	12.2475	170.1092	97.68	6

(continued)

(continued)

Region/country	Island group	Island name	Island type	Latitude	Longitude	Area (ha)	Maximum elevation
Marshall Islands	Ratak Chain	Bokak (Taongi Atoll)	Reef island	14.65778	168.945	593.91	3
Marshall Islands	Ratak Chain	Erikub (Adkup)	Reef island	9.118889	170.0311	332.91	3
Marshall Islands	Ratak Chain	Jemo	Reef island	10.08	169.525	114.12	3
Marshall Islands	Ratak Chain	Likiep	Reef island	9.925556	169.1322	947.25	4
Marshall Islands	Ratak Chain	Majuro (Mjiro)	Reef island	7.142778	171.0394	2355.09	3
Marshall Islands	Ratak Chain	Maloelap	Reef island	8.773333	171.0311	1373.23	3
Marshall Islands	Ratak Chain	Mejit (Mpjej)	Reef island	10.29361	170.8708	252.62	3
Marshall Islands	Ratak Chain	Mili (Mile)	Reef island	6.133889	171.9097	1904.44	3
Marshall Islands	Ratak Chain	Toke (Taka)	Reef island	11.14	169.6275	91.94	6
Marshall Islands	Ratak Chain	Utirik (Utrik)	Reef island	11.26889	169.7958	369	3
Marshall Islands	Ratak Chain	Wotje (Wojja)	Reef island	9.448056	170.0533	1309.16	6
Nauru		Nauru	Limestone high island	-0.52278	166.9314	2264.32	71
New Caledonia	Geographically part of Loyalty Islands but administered by Isle of Pines	Ile Le Leizour	Reef island	-18.2846	163.043	81.62	4
New Caledonia	Geographically part of Loyalty Islands but administered by Isle of Pines	Baaba/Paaba	Continental island	-20.0531	163.9731	3107.81	7
New Caledonia	Geographically part of Loyalty Islands but administered by Isle of Pines	Balabio	Continental island	-20.1042	164.1925	3732.57	282

New Caledonia	Geographically part of Loyalty Islands but administered by Isle of Pines	Boh	Continental island	−20.3222	164.0881	320.62	21
New Caledonia	Geographically part of Loyalty Islands but administered by Isle of Pines	Chesterfield isles	Reef island	−19.3692	158.9425	1515.49	9
New Caledonia	Geographically part of Loyalty Islands but administered by Isle of Pines	Daos/Iles Daos du nord	Continental island	−19.8203	163.6811	179.98	103
New Caledonia	Geographically part of Loyalty Islands but administered by Isle of Pines	Ile Art (Belep)	Continental island	−19.7403	163.6678	6004	283
New Caledonia	Geographically part of Loyalty Islands but administered by Isle of Pines	Ile des Pins (Kunie)	Continental island	−22.6119	167.4753	17235.83	266
New Caledonia	Geographically part of Loyalty Islands but administered by Isle of Pines	Ile Fabre	Reef island	−18.2953	163.01	84.35	3
New Caledonia	Geographically part of Loyalty Islands but administered by Isle of Pines	Ile Ouen	Continental island	−22.4281	166.8017	4562.08	332

(continued)

(continued)

Region/country	Island group	Island name	Island type	Latitude	Longitude	Area (ha)	Maximum elevation
New Caledonia	Geographically part of Loyalty Islands but administered by Isle of Pines	Ile Pott (Belep)	Continental island	−19.5811	163.5953	1462.05	157
New Caledonia	Geographically part of Loyalty Islands but administered by Isle of Pines	Kotomo (Koutoumo)	Continental island	−22.6619	167.5308	1603.25	36
New Caledonia	Geographically part of Loyalty Islands but administered by Isle of Pines	Neba	Continental island	−20.162	163.933	461.01	82
New Caledonia	Geographically part of Loyalty Islands but administered by Isle of Pines	New Caledonia (La Grande Terre)	Continental island	−21.4339	165.5417	1891479	450
New Caledonia	Geographically part of Loyalty Islands but administered by Isle of Pines	Pam	Continental island	−20.2417	164.2875	716.58	136
New Caledonia	Geographically part of Loyalty Islands but administered by Isle of Pines	Pia (Taanlo)	Continental island	−20.013	163.9316	105.5	17
New Caledonia	Geographically part of Loyalty Islands but administered by Isle of Pines	Walpole	Limestone high island	−22.5972	168.9528	183.64	90

	Geographically part of Loyalty Islands but administered by Isle of Pines						
New Caledonia		Yande	Continental island	−20.0458	163.8097	1602.08	326
New Caledonia	Loyalty Islands	Beautemps-Beaupre	Limestone low island	−20.4094	166.1389	113.22	10
New Caledonia	Loyalty Islands	Ile Dudune	Limestone low island	−21.3528	167.7303	387.91	20
New Caledonia	Loyalty Islands	Ile Leliogat	Limestone high island	−21.3094	167.5711	93.91	38
New Caledonia	Loyalty Islands	Ile Uoa (Oua)	Limestone low island	−21.2558	167.5869	118.13	7
New Caledonia	Loyalty Islands	Lifou	Limestone high island	−20.9731	167.2081	132041	90
New Caledonia	Loyalty Islands	Mare	Limestone high island	−21.5297	167.9789	76657.47	129
New Caledonia	Loyalty Islands	Ouvea	Limestone high island	−20.6131	166.5806	16040.42	39
New Caledonia	Loyalty Islands	Tiga	Limestone high island	−21.1044	167.8169	1306.8	76
New Caledonia	Loyalty Islands	Vauvilliers	Limestone low island	−21.1547	167.5642	250.59	24
New Caledonia	Outliers	Hunter (Fern Island)	Volcanic high island	−22.3953	172.0864	218.82	297
New Caledonia	Outliers	Matthew	Volcanic high island	−22.3431	171.3558	136.52	177
Niue	Outliers	Niue	Limestone high island	−19.0544	−169.867	29789.96	60

(continued)

(continued)

Region/country	Island group	Island name	Island type	Latitude	Longitude	Area (ha)	Maximum elevation
Northern Mariana Islands		Agrihan	Volcanic high island	18.77056	145.6669	5238.21	965
Northern Mariana Islands		Aguiguan	Limestone high island	14.85194	145.5583	826.41	150
Northern Mariana Islands		Alamagan	Volcanic high island	17.59806	145.8336	1338.03	744
Northern Mariana Islands		Anatahan	Volcanic high island	16.35056	145.6789	3584.67	788
Northern Mariana Islands		Asuncion	Volcanic high island	19.69472	145.4056	882.79	884
Northern Mariana Islands		Farallon de Medinilla	Limestone high island	16.01806	146.0586	130.5	81
Northern Mariana Islands		Farallon de Pajaros	Volcanic high island	20.53611	144.8975	262.69	319
Northern Mariana Islands		Guguan	Volcanic high island	17.30861	145.8419	517.21	300
Northern Mariana Islands		Managaha	Reef island	15.24139	145.7125	7.53	1
Northern Mariana Islands		Maug	Volcanic high island	20.01444	145.2314	144.44	218
Northern Mariana Islands		Naftan Rock	Limestone high island	14.83472	145.5331	5.75	102
Northern Mariana Islands		Pagan	Volcanic high island	18.11222	145.7672	6253.38	570
Northern Mariana Islands		Rota	Composite high island	14.15833	145.2122	9399.36	491
Northern Mariana Islands		Saipan	Composite high island	15.18944	145.7514	13240.6	474

Northern Mariana Islands		Sarigan	Volcanic high island	16.705	145.7789	657.14	549
Northern Mariana Islands		Tinian	Limestone high island	15.00722	145.6269	11250.18	474
Palau	Main Group	Babeldaob	Composite high island	7.488333	134.5575	38341.46	124
Palau	Main Group	Bablomekang	Limestone high island	7.142222	134.3186	41.74	81
Palau	Main Group	Beliliou (Peleliu)	Limestone high island	7.002778	134.2389	1976.22	30
Palau	Main Group	Bukrrairong	Limestone high island	7.328056	134.49	32.2	68
Palau	Main Group	Butottoribo	Limestone high island	7.269444	134.3881	201.39	108
Palau	Main Group	Mecherchar	Limestone high island	7.151389	134.3619	1297.25	82
Palau	Main Group	Ngcheangel (Kayangel)	Reef island	8.0825	134.7192	146.81	1
Palau	Main Group	Ngeaur (Angaur)	Limestone high island	6.905833	134.1383	895.79	61
Palau	Main Group	Ngebad	Limestone low island	7.021111	134.2731	93.67	6
Palau	Main Group	Ngebedangel	Limestone high island	7.269167	134.3342	58.68	100
Palau	Main Group	Ngedbus	Limestone low island	7.056667	134.2603	148.33	8
Palau	Main Group	Ngemelachel (Malakal)	Volcanic high island	7.331667	134.4531	117.92	122
Palau	Main Group	Ngemlis	Limestone low island	7.103889	134.2481	68.91	25

(continued)

(continued)

Region/country	Island group	Island name	Island type	Latitude	Longitude	Area (ha)	Maximum elevation
Palau	Main Group	Ngerchaol	Limestone high island	7.342778	134.4425	201.03	90
Palau	Main Group	Ngercheu (Carp)	Limestone high island	7.092778	134.2783	119.45	52
Palau	Main Group	Ngerebelas	Reef island	8.042778	134.6994	13.27	2
Palau	Main Group	Ngerechong	Limestone low island	7.113333	134.3625	70.57	6
Palau	Main Group	Ngerekebesang	Volcanic high island	7.361481	134.4296	279.27	103
Palau	Main Group	Ngeriungs	Reef island	8.058889	134.7111	54.89	2
Palau	Main Group	Ngerukewid	Limestone high island	7.1775	134.2628	100.05	30
Palau	Main Group	Ngeruktabel	Limestone high island	7.260833	134.4069	2572.81	207
Palau	Main Group	Ongael	Limestone high island	7.253889	134.3817	160.75	130
Palau	Main Group	Orak	Reef island	8.04	134.6942	4.17	2
Palau	Main Group	Oreor (Koror)	Composite high island	7.34	134.4764	1223.38	130
Palau	Main Group	Tlutkaraguis	Limestone high island	7.308611	134.4097	8.65	40
Palau	Main Group	Ulebsechel	Limestone high island	7.308611	134.4782	589.11	185
Palau	Main Group	Ulong	Limestone high island	7.278333	134.2961	192.13	81
Palau	Outliers	Fanna	Reef island	5.353889	132.2272	70.62	6
Palau	Outliers	Helen	Reef island	2.973746	131.8121	1.43	3

Palau	Outliers	Merir	Reef island	4.313333	132.3111	135.84	6
Palau	Outliers	Pulo Anna	Limestone low island	4.660833	131.96	52.82	12
Palau	Outliers	Sonsorol	Limestone low island	5.326667	132.2228	169.38	20
Palau	Outliers	Tobi	Reef island	3.006944	131.1239	89.51	1
Papua New Guinea	Bismarck Archipelago	Abau	Volcanic high island	−10.1808	148.7031	30.96	45
Papua New Guinea	Bismarck Archipelago	Abaurai	Reef island	−8.49358	143.6285	5042.98	3
Papua New Guinea	Bismarck Archipelago	Abavi	Composite low island	−10.1806	148.7214	379.03	5
Papua New Guinea	Bismarck Archipelago	Ablingi	Limestone high island	−6.29972	150.0756	352.68	86
Papua New Guinea	Bismarck Archipelago	Agur (1)/Akur (1)	Reef island	−6.2925	150.2867	130.18	1
Papua New Guinea	Bismarck Archipelago	Agur (2)/Akur (2)	Volcanic low island	−6.28333	150.2753	63.36	17
Papua New Guinea	Bismarck Archipelago	Ahet	Reef island	−1.93194	146.6656	29	2
Papua New Guinea	Bismarck Archipelago	Ahu (1)	Reef island	−1.32639	144.1594	125.83	2
Papua New Guinea	Bismarck Archipelago	Ahu (2)	Volcanic low island	−2.33639	146.8611	19.24	2
Papua New Guinea	Bismarck Archipelago	Aivet	Limestone high island	−6.29306	149.9572	250.94	86
Papua New Guinea	Bismarck Archipelago	Akib/Unknown	Volcanic high island	−1.52556	145.0014	38.76	31

(continued)

(continued)

Region/country	Island group	Island name	Island type	Latitude	Longitude	Area (ha)	Maximum elevation
Papua New Guinea	Bismarck Archipelago	Alage/Amge	Limestone high island	−6.3175	149.8097	47.64	93
Papua New Guinea	Bismarck Archipelago	Ali	Reef island	−3.12722	142.4694	105.37	8
Papua New Guinea	Bismarck Archipelago	Alim	Volcanic low island	−2.89361	147.0594	68.32	18
Papua New Guinea	Bismarck Archipelago	Ambitle	Volcanic high island	−4.08389	153.6147	8979.99	450
Papua New Guinea	Bismarck Archipelago	Amerer	Volcanic low island	−6.27556	150.3986	12.53	14
Papua New Guinea	Bismarck Archipelago	Ampul (1)	Limestone high island	−6.32083	149.8733	70.87	77
Papua New Guinea	Bismarck Archipelago	Ampul (2)	Limestone high island	−6.31222	149.7919	163.71	127
Papua New Guinea	Bismarck Archipelago	Analtin	Reef island	−1.15333	144.4144	320.91	3
Papua New Guinea	Bismarck Archipelago	Andra/Unknown	Reef island	−1.93889	147.0028	22.81	2
Papua New Guinea	Bismarck Archipelago	Anum	Volcanic low island	−2.26806	146.8286	35.28	30
Papua New Guinea	Bismarck Archipelago	Aramia	Reef island	−8.045	143.4819	6994.69	3
Papua New Guinea	Bismarck Archipelago	Aua (Durour)	Reef island	−1.46333	143.0608	609.52	2
Papua New Guinea	Bismarck Archipelago	Avahain	Volcanic low island	−6.27389	150.3692	200.62	15
Papua New Guinea	Bismarck Archipelago	Awin	Reef island	−1.6475	144.0297	98.06	2

Papua New Guinea	Bismarck Archipelago	Babagutu	Composite high island	−7.72194	147.6203	49.73	38
Papua New Guinea	Bismarck Archipelago	Babase	Volcanic high island	−4.015	153.7069	2393.71	200
Papua New Guinea	Bismarck Archipelago	Bagabag	Volcanic high island	−4.80417	146.2283	3702.7	600
Papua New Guinea	Bismarck Archipelago	Baluan	Volcanic high island	−2.57028	147.2664	1508.85	254
Papua New Guinea	Bismarck Archipelago	Balumara/Unknown	Reef island	−1.95028	146.4928	2.14	2
Papua New Guinea	Bismarck Archipelago	Bam	Volcanic high island	−3.6125	144.8189	708.36	685
Papua New Guinea	Bismarck Archipelago	Banban (Lolobau)	Volcanic high island	−4.9225	151.1675	6935.37	875
Papua New Guinea	Bismarck Archipelago	Bangatang	Limestone high island	−2.63278	150.5878	1417.7	37
Papua New Guinea	Bismarck Archipelago	Bat (1)/Unknown	Reef island	−2.84167	146.2356	7.43	2
Papua New Guinea	Bismarck Archipelago	Bat (2)/Unknown	Reef island	−2.84917	146.2319	23.56	2
Papua New Guinea	Bismarck Archipelago	Batteru	Composite high island	−7.35083	147.2164	80.03	99
Papua New Guinea	Bismarck Archipelago	Bauddisson	Limestone high island	−2.68861	150.5656	2686.79	97
Papua New Guinea	Bismarck Archipelago	Beligila	Reef island	−2.63583	149.6636	72.06	2
Papua New Guinea	Bismarck Archipelago	Big Ndrova	Volcanic low island	−2.20556	147.1806	132.46	11

(continued)

(continued)

Region/country	Island group	Island name	Island type	Latitude	Longitude	Area (ha)	Maximum elevation
Papua New Guinea	Bismarck Archipelago	Big Pigeon/Pisin	Reef island	−4.26944	152.3436	45.68	4
Papua New Guinea	Bismarck Archipelago	Blup Blup	Volcanic high island	−3.50833	144.6031	585.77	402
Papua New Guinea	Bismarck Archipelago	Boang	Volcanic high island	−3.37389	153.2947	2961.14	145
Papua New Guinea	Bismarck Archipelago	Boigu	Reef island	−9.26639	142.2183	7599.88	4
Papua New Guinea	Bismarck Archipelago	Boisa	Volcanic high island	−4.00056	144.9631	144.88	240
Papua New Guinea	Bismarck Archipelago	Bona Bona	Volcanic high island	−10.4942	149.8453	1409.75	383
Papua New Guinea	Bismarck Archipelago	Bonarua (Brumer)	Volcanic high island	−10.7586	150.3825	261.44	120
Papua New Guinea	Bismarck Archipelago	Bristow (Bobo)	Reef island	−9.12528	143.2369	3562.25	2
Papua New Guinea	Bismarck Archipelago	Browne	Volcanic low island	−1.93722	146.5994	39.9	19
Papua New Guinea	Bismarck Archipelago	Bulangelu/Unknown	Volcanic low island	−2.3875	146.8431	9	27
Papua New Guinea	Bismarck Archipelago	Bundro	Volcanic low island	−2.20694	147.7764	321.81	19
Papua New Guinea	Bismarck Archipelago	Busseau/Pwiseu	Reef island	−2.20167	146.3347	4.79	2
Papua New Guinea	Bismarck Archipelago	Crown	Volcanic high island	−5.12083	146.9456	1460.49	270
Papua New Guinea	Bismarck Archipelago	Damera	Volcanic low island	−8.0525	143.6153	10026.96	5

Papua New Guinea	Bismarck Archipelago	Daru	Composite low island	−9.08417	143.2067	1619.48	27
Papua New Guinea	Bismarck Archipelago	Dauan	Volcanic high island	−9.4225	142.5353	427.09	273
Papua New Guinea	Bismarck Archipelago	Daugo	Volcanic low island	−9.51806	147.0597	420.48	8
Papua New Guinea	Bismarck Archipelago	Dawari	Reef island	−8.69861	143.3806	959.37	3
Papua New Guinea	Bismarck Archipelago	Deirina/Leocadie/Delina	Volcanic high island	−10.6939	150.4367	59.71	37
Papua New Guinea	Bismarck Archipelago	Dibiri	Reef island	−8.21528	143.6636	6682.6	18
Papua New Guinea	Bismarck Archipelago	Dubuwaro	Reef island	−8.60028	143.3414	588.7	3
Papua New Guinea	Bismarck Archipelago	Duke of York	Limestone high island	−4.175	152.4614	5376.87	92
Papua New Guinea	Bismarck Archipelago	Dunung	Reef island	−2.36667	150.1197	97.3	3
Papua New Guinea	Bismarck Archipelago	Dyaul (Djaul)	Volcanic high island	−2.94361	150.8764	11706.1	170
Papua New Guinea	Bismarck Archipelago	Eloaua	Composite high island	−1.57167	149.6381	844.47	58
Papua New Guinea	Bismarck Archipelago	Emananus	Composite high island	−1.575	149.5811	499.1	50
Papua New Guinea	Bismarck Archipelago	Emhoro	Composite high island	−10.3928	149.4561	58.48	67
Papua New Guinea	Bismarck Archipelago	Emirau	Composite high island	−1.66	149.965	3620.4	77

(continued)

(continued)

Region/country	Island group	Island name	Island type	Latitude	Longitude	Area (ha)	Maximum elevation
Papua New Guinea	Bismarck Archipelago	Enuk	Composite low island	−2.64889	150.7258	219.79	25
Papua New Guinea	Bismarck Archipelago	Fly (1)	Composite high island	−7.43556	147.3444	101.29	130
Papua New Guinea	Bismarck Archipelago	Fly (2)	Reef island	−7.47056	147.3636	50.64	1
Papua New Guinea	Bismarck Archipelago	Fly (3)	Reef island	−7.5	147.3264	36.58	1
Papua New Guinea	Bismarck Archipelago	Fly (4)	Composite low island	−7.43222	147.3017	13.42	6
Papua New Guinea	Bismarck Archipelago	Fly (5)	Composite low island	−7.44111	147.3236	15.94	25
Papua New Guinea	Bismarck Archipelago	Garove	Volcanic high island	−4.67167	149.4944	5692.79	368
Papua New Guinea	Bismarck Archipelago	Garua	Volcanic high island	−5.30306	150.0711	742.27	565
Papua New Guinea	Bismarck Archipelago	Gebaro	Reef island	−8.34806	143.2047	5731.85	3
Papua New Guinea	Bismarck Archipelago	Gemo	Volcanic high island	−9.48278	147.1161	81.81	73
Papua New Guinea	Bismarck Archipelago	Haidana	Reef island	−9.44472	147.0358	288.9	2
Papua New Guinea	Bismarck Archipelago	Harengan	Volcanic high island	−1.95833	146.5817	51.56	46
Papua New Guinea	Bismarck Archipelago	Harikoia (Brumer)	Volcanic high island	−10.7636	150.4181	122.89	165
Papua New Guinea	Bismarck Archipelago	Hauwei/Hawei	Reef island	−1.96139	147.2856	33.95	6

Papua New Guinea	Bismarck Archipelago	Hosken	Composite high island	−7.69639	147.5314	86.47	54
Papua New Guinea	Bismarck Archipelago	Hus	Reef island	−1.94139	147.1019	38.91	2
Papua New Guinea	Bismarck Archipelago	Ifo	Volcanic high island	−4.83111	152.9094	29.27	114
Papua New Guinea	Bismarck Archipelago	Igoigoli	Volcanic high island	−10.7258	150.2922	232.34	157
Papua New Guinea	Bismarck Archipelago	Jalun (1)/Unknown	Volcanic low island	−1.5425	145.0064	73.56	30
Papua New Guinea	Bismarck Archipelago	Jalun (Djalun)	Volcanic high island	−1.54972	145.0192	142.39	92
Papua New Guinea	Bismarck Archipelago	Jawani	Composite high island	−7.34167	147.2081	38.98	94
Papua New Guinea	Bismarck Archipelago	Johnston (1)	Reef island	−2.43278	147.0886	29.32	3
Papua New Guinea	Bismarck Archipelago	Johnston (2)	Reef island	−2.43444	147.1081	8.39	2
Papua New Guinea	Bismarck Archipelago	Johnston (3)	Reef island	−2.46222	147.0394	6.29	2
Papua New Guinea	Bismarck Archipelago	Kabakon	Limestone low island	−4.23722	152.3956	100.72	24
Papua New Guinea	Bismarck Archipelago	Kabitong	Reef island	−2.42361	150.0053	15.66	2
Papua New Guinea	Bismarck Archipelago	Kabotteron	Reef island	−2.65944	150.6825	205.59	3
Papua New Guinea	Bismarck Archipelago	Kairiru	Volcanic high island	−3.35139	143.5558	5524.23	722

(continued)

(continued)

Region/country	Island group	Island name	Island type	Latitude	Longitude	Area (ha)	Maximum elevation
Papua New Guinea	Bismarck Archipelago	Kakolan (Heath)/Wulai	Volcanic high island	−4.90889	151.3661	114.04	132
Papua New Guinea	Bismarck Archipelago	Kalipo	Volcanic low island	−2.19806	147.7564	80.73	20
Papua New Guinea	Bismarck Archipelago	Karkar	Volcanic high island	−4.6325	145.9622	40808.61	1839
Papua New Guinea	Bismarck Archipelago	Kat	Reef island	−1.11056	144.4922	109.89	3
Papua New Guinea	Bismarck Archipelago	Kaumag	Reef island	−9.36861	142.6953	898.27	8
Papua New Guinea	Bismarck Archipelago	Kauptimente/Kauptimete	Limestone low island	−6.18222	148.9392	155.98	23
Papua New Guinea	Bismarck Archipelago	Kawa (1)	Volcanic low island	−9.18917	141.9892	1076.07	23
Papua New Guinea	Bismarck Archipelago	Kehoi (Ninigo Atolls)/ Kehoi	Reef island	−1.3225	144.3367	24.35	2
Papua New Guinea	Bismarck Archipelago	Kerawara	Limestone low island	−4.24389	152.4158	74.46	23
Papua New Guinea	Bismarck Archipelago	Keresau/Karasau	Reef island	−3.3925	143.45	128.26	8
Papua New Guinea	Bismarck Archipelago	Killenge	Volcanic low island	−2.27278	146.7983	14.76	27
Papua New Guinea	Bismarck Archipelago	Kiton	Limestone low island	−2.67917	150.635	120.87	19
Papua New Guinea	Bismarck Archipelago	Kiwai	Reef island	−8.56333	143.4222	34660.84	3

Papua New Guinea	Bismarck Archipelago	Koil	Volcanic high island	-3.35667	144.2069	335.37	93
Papua New Guinea	Bismarck Archipelago	Kolenusa/Unknown	Reef island	-2.62833	149.6761	73.84	2
Papua New Guinea	Bismarck Archipelago	Koruniat	Volcanic low island	-1.97361	147.35	101.01	21
Papua New Guinea	Bismarck Archipelago	Krangket (Kranket)	Limestone low island	-5.19778	145.825	152.24	12
Papua New Guinea	Bismarck Archipelago	Kulik	Reef island	-2.39306	150.0639	5.62	2
Papua New Guinea	Bismarck Archipelago	Kung	Reef island	-2.38111	150.0994	92.68	3
Papua New Guinea	Bismarck Archipelago	Laluoro	Limestone low island	-10.3433	149.3447	31.62	10
Papua New Guinea	Bismarck Archipelago	Lamassa	Volcanic low island	-4.69667	152.7656	206.61	3
Papua New Guinea	Bismarck Archipelago	Lambom	Volcanic high island	-4.79806	152.8439	449.5	144
Papua New Guinea	Bismarck Archipelago	Lasanga	Composite high island	-7.41444	147.2531	1282.39	450
Papua New Guinea	Bismarck Archipelago	Lau	Reef island	-1.36139	144.1753	33.02	2
Papua New Guinea	Bismarck Archipelago	Laualau (Heina)	Reef island	-1.13222	144.5058	163.16	3
Papua New Guinea	Bismarck Archipelago	Lavongai (New Hanover)	Volcanic high island	-2.54778	150.2533	123221.9	900
Papua New Guinea	Bismarck Archipelago	Lemus	Limestone low island	-2.62861	150.6264	135.48	14

(continued)

(continued)

Region/country	Island group	Island name	Island type	Latitude	Longitude	Area (ha)	Maximum elevation
Papua New Guinea	Bismarck Archipelago	Lepa (Paeowa)/Lepa	Limestone low island	−5.17694	145.8303	23.67	16
Papua New Guinea	Bismarck Archipelago	Lif	Volcanic high island	−3.50667	153.1769	221.91	283
Papua New Guinea	Bismarck Archipelago	Lihir	Volcanic high island	−3.12889	152.5894	20595.96	700
Papua New Guinea	Bismarck Archipelago	Limellon	Reef island	−2.67389	150.7714	18.87	2
Papua New Guinea	Bismarck Archipelago	Liot	Reef island	−1.40944	144.5131	242.87	3
Papua New Guinea	Bismarck Archipelago	Lissenung	Reef island	−2.66417	150.7336	12.58	3
Papua New Guinea	Bismarck Archipelago	Long (Arop)	Volcanic high island	−5.32528	147.1003	42091.88	1280
Papua New Guinea	Bismarck Archipelago	Longan	Reef island	−1.22167	144.29	72.02	2
Papua New Guinea	Bismarck Archipelago	Lou	Volcanic high island	−2.39	147.3444	3204.17	226
Papua New Guinea	Bismarck Archipelago	Loupomu	Limestone low island	−10.3256	149.3228	43.06	8
Papua New Guinea	Bismarck Archipelago	Luf	Volcanic high island	−1.53472	145.0581	581.69	260
Papua New Guinea	Bismarck Archipelago	Mahur	Limestone high island	−2.78333	152.6542	842.82	240
Papua New Guinea	Bismarck Archipelago	Mailu	Composite high island	−10.3875	149.3564	133.84	90

Papua New Guinea	Bismarck Archipelago	Makada	Limestone high island	−4.12444	152.4231	387.64	119
Papua New Guinea	Bismarck Archipelago	Maklo/Makio	Limestone low island	−6.16139	148.9611	404.59	23
Papua New Guinea	Bismarck Archipelago	Mal	Reef island	−1.41	144.1844	355.55	2
Papua New Guinea	Bismarck Archipelago	Malai	Volcanic low island	−5.89667	147.9403	104.25	20
Papua New Guinea	Bismarck Archipelago	Malendok	Volcanic high island	−3.465	153.2083	3496.94	472
Papua New Guinea	Bismarck Archipelago	Mali (Malie)	Limestone low island	−3.03111	152.6597	64.21	2
Papua New Guinea	Bismarck Archipelago	Mali/Sinambiet	Reef island	−3.03194	152.6828	207.44	3
Papua New Guinea	Bismarck Archipelago	Manam	Volcanic high island	−4.075	145.0339	8334.93	1807
Papua New Guinea	Bismarck Archipelago	Manne	Limestone high island	−2.73556	150.6633	2443.22	34
Papua New Guinea	Bismarck Archipelago	Manu (Allison)	Reef island	−1.31028	143.5814	39.59	2
Papua New Guinea	Bismarck Archipelago	Manubada	Volcanic high island	−9.51417	147.1781	40.22	45
Papua New Guinea	Bismarck Archipelago	Manus	Composite high island	−2.08222	146.9475	192132.1	718
Papua New Guinea	Bismarck Archipelago	Marengan	Reef island	−1.905	146.5822	21.92	2
Papua New Guinea	Bismarck Archipelago	Masahet	Limestone high island	−2.95222	152.6578	859.09	197

(continued)

(continued)

Region/country	Island group	Island name	Island type	Latitude	Longitude	Area (ha)	Maximum elevation
Papua New Guinea	Bismarck Archipelago	Mata Kawa	Reef island	−9.1925	142.0517	1047.12	12
Papua New Guinea	Bismarck Archipelago	Matalik	Reef island	−2.41417	150.0239	24.48	2
Papua New Guinea	Bismarck Archipelago	Mbatmanda?	Reef island	−1.97333	148.0681	26.1	1
Papua New Guinea	Bismarck Archipelago	Mbuke	Volcanic high island	−2.38694	146.8278	104.39	183
Papua New Guinea	Bismarck Archipelago	Meit	Limestone high island	−2.99444	150.7297	285.77	66
Papua New Guinea	Bismarck Archipelago	Meman	Reef island	−1.22139	144.2236	105.25	2
Papua New Guinea	Bismarck Archipelago	Mibu	Reef island	−8.74361	143.4433	2355.36	3
Papua New Guinea	Bismarck Archipelago	Mindregutu	Limestone low island	−7.71361	147.6339	120.35	25
Papua New Guinea	Bismarck Archipelago	Mioko	Limestone low island	−4.23167	152.4594	132.56	29
Papua New Guinea	Bismarck Archipelago	Mole Island/Unknown	Reef island	−2.87111	146.4586	33.83	2
Papua New Guinea	Bismarck Archipelago	Morigio	Reef island	−7.78472	143.9336	7838.13	2
Papua New Guinea	Bismarck Archipelago	Morobegutu	Composite high island	−7.735	147.6078	57.34	44
Papua New Guinea	Bismarck Archipelago	Moroge	Reef island	−8.66639	143.3947	1434.76	3
Papua New Guinea	Bismarck Archipelago	Mouse/Unknown	Reef island	−2.89722	146.4031	21.82	2

Papua New Guinea	Bismarck Archipelago	Mualim	Limestone low island	-4.21889	152.4647	21	27
Papua New Guinea	Bismarck Archipelago	Muschu	Volcanic high island	-3.41389	143.5858	4229.89	126
Papua New Guinea	Bismarck Archipelago	Musik	Composite high island	-7.37556	147.2228	24.37	55
Papua New Guinea	Bismarck Archipelago	Mussau	Volcanic high island	-1.42222	149.6097	34485.59	590
Papua New Guinea	Bismarck Archipelago	Narega (Narage)	Volcanic high island	-4.54611	149.1056	166.13	307
Papua New Guinea	Bismarck Archipelago	Nauna	Composite high island	-2.21444	148.2011	256.46	126
Papua New Guinea	Bismarck Archipelago	Naviu	Volcanic low island	-8.13333	143.64	8270.27	6
Papua New Guinea	Bismarck Archipelago	Ndrilo	Volcanic low island	-1.965	147.3269	157.47	15
Papua New Guinea	Bismarck Archipelago	Neirtri	Reef island	-2.43139	150.0031	57.05	2
Papua New Guinea	Bismarck Archipelago	Neitab	Reef island	-2.35	150.1444	39.82	3
Papua New Guinea	Bismarck Archipelago	Nemto	Reef island	-2.34972	150.3086	190.36	3
Papua New Guinea	Bismarck Archipelago	New Britain	Composite high island	-5.76583	150.7322	3578123	2438
Papua New Guinea	Bismarck Archipelago	New Guinea	Composite high island	-5.50111	141.4292	78787181	4884
Papua New Guinea	Bismarck Archipelago	New Ireland	Composite high island	-3.41611	152.09	710273.4	2379

(continued)

(continued)

Region/country	Island group	Island name	Island type	Latitude	Longitude	Area (ha)	Maximum elevation
Papua New Guinea	Bismarck Archipelago	Ningau	Volcanic high island	−4.62417	149.3517	710.19	176
Papua New Guinea	Bismarck Archipelago	Noru	Volcanic low island	−1.95583	146.6192	127.57	19
Papua New Guinea	Bismarck Archipelago	Nusa	Reef island	−2.57361	150.7789	155.74	2
Papua New Guinea	Bismarck Archipelago	Nusailas	Reef island	−2.67833	150.7344	153.4	3
Papua New Guinea	Bismarck Archipelago	Nusaum	Reef island	−2.63806	150.6414	10.63	2
Papua New Guinea	Bismarck Archipelago	Nuvarege	Volcanic low island	−5.51806	149.4375	142.79	15
Papua New Guinea	Bismarck Archipelago	Okuru	Reef island	−1.90806	146.6042	20.68	2
Papua New Guinea	Bismarck Archipelago	Onneta	Reef island	−1.94444	147.1333	16.26	2
Papua New Guinea	Bismarck Archipelago	Pahi	Reef island	−2.095	146.4256	178.23	2
Papua New Guinea	Bismarck Archipelago	Pahi (2)/Paii	Reef island	−2.05722	146.4608	14.96	2
Papua New Guinea	Bismarck Archipelago	Pak	Volcanic low island	−2.0825	147.6253	970.48	28
Papua New Guinea	Bismarck Archipelago	Palitolla	Reef island	−1.08778	144.3994	187.91	2
Papua New Guinea	Bismarck Archipelago	Pam (1)/Unknown	Reef island	−2.49556	147.3394	33.76	2
Papua New Guinea	Bismarck Archipelago	Pam (2)/Unknown	Reef island	−2.50611	147.3308	39.14	2

Papua New Guinea	Bismarck Archipelago	Papialou (1)/Unknown	Reef island	−2.73111	147.3481	7.08	2
Papua New Guinea	Bismarck Archipelago	Papialou (2)/Unknown	Reef island	−2.73833	147.3375	3.81	2
Papua New Guinea	Bismarck Archipelago	Parama (1)	Reef island	−9.0075	143.4181	3932.15	3
Papua New Guinea	Bismarck Archipelago	Parinte/Unknown	Reef island	−2.20194	146.2344	9.35	2
Papua New Guinea	Bismarck Archipelago	Patio	Limestone low island	−2.59556	150.5381	683.87	18
Papua New Guinea	Bismarck Archipelago	Patuam	Volcanic low island	−2.19333	147.7283	243.6	22
Papua New Guinea	Bismarck Archipelago	Pemei	Volcanic high island	−1.5125	145.0836	120.05	72
Papua New Guinea	Bismarck Archipelago	Pihun	Reef island	−1.30167	144.345	131.76	2
Papua New Guinea	Bismarck Archipelago	Pileo	Limestone high island	−6.17972	149.0439	177.96	51
Papua New Guinea	Bismarck Archipelago	Pityilu	Limestone low island	−1.96333	147.2256	246.86	10
Papua New Guinea	Bismarck Archipelago	Ponam	Reef island	−1.91389	146.8858	95.61	2
Papua New Guinea	Bismarck Archipelago	Purutu	Reef island	−8.39718	143.4122	20511.88	3
Papua New Guinea	Bismarck Archipelago	Rambutyo	Volcanic high island	−2.30972	147.8225	10098.5	306
Papua New Guinea	Bismarck Archipelago	Rat/Unknown	Reef island	−2.9275	146.3586	9.25	2

(continued)

(continued)

Region/country	Island group	Island name	Island type	Latitude	Longitude	Area (ha)	Maximum elevation
Papua New Guinea	Bismarck Archipelago	Ritter	Volcanic high island	−5.52361	148.115	50.17	140
Papua New Guinea	Bismarck Archipelago	Sabben/Paindreh	Reef island	−2.20944	146.2736	10.99	2
Papua New Guinea	Bismarck Archipelago	Sae	Reef island	−0.75806	145.3064	103.83	2
Papua New Guinea	Bismarck Archipelago	Saibai	Reef island	−9.39917	142.6769	10976.03	2
Papua New Guinea	Bismarck Archipelago	Sakar	Volcanic high island	−5.40972	148.08	4620.73	992
Papua New Guinea	Bismarck Archipelago	Salihau/Unknown	Volcanic low island	−2.08833	146.5536	56.22	10
Papua New Guinea	Bismarck Archipelago	Sama	Reef island	−1.40972	144.0864	5.79	2
Papua New Guinea	Bismarck Archipelago	Samasuma/Sumasuma	Reef island	−1.47056	144.0514	271.98	2
Papua New Guinea	Bismarck Archipelago	Sek	Volcanic low island	−5.08278	145.8206	155.92	18
Papua New Guinea	Bismarck Archipelago	Selapiu	Composite high island	−2.66472	150.4489	423.25	125
Papua New Guinea	Bismarck Archipelago	Seleo	Composite low island	−3.14639	142.4872	96.28	18
Papua New Guinea	Bismarck Archipelago	Siar	Volcanic low island	−5.18611	145.8058	15.01	22
Papua New Guinea	Bismarck Archipelago	Simberi	Volcanic high island	−2.63167	151.9825	5656.05	340
Papua New Guinea	Bismarck Archipelago	Similam (1)	Reef island	−1.39444	144.0861	56.6	2

Papua New Guinea	Bismarck Archipelago	Similam (2)	Reef island	−1.39472	144.0789	45.44	2
Papua New Guinea	Bismarck Archipelago	Sisi	Reef island	−2.11139	146.3967	267.78	2
Papua New Guinea	Bismarck Archipelago	Sivisa (1)	Volcanic high island	−2.38333	147.4714	76.75	45
Papua New Guinea	Bismarck Archipelago	Sivisa (2)	Volcanic high island	−2.40028	147.4603	22	42
Papua New Guinea	Bismarck Archipelago	Skok/Unknown	Reef island	−2.16389	146.2914	8.27	2
Papua New Guinea	Bismarck Archipelago	Sosson	Reef island	−2.425	150.0347	16.63	2
Papua New Guinea	Bismarck Archipelago	Sowe	Composite low island	−2.19833	146.6075	17.47	24
Papua New Guinea	Bismarck Archipelago	Suau	Volcanic high island	−10.7017	150.2542	419.78	225
Papua New Guinea	Bismarck Archipelago	Suf	Reef island	−0.88167	145.5569	244.08	2
Papua New Guinea	Bismarck Archipelago	Tabar	Volcanic high island	−2.92222	152.0108	11294.89	622
Papua New Guinea	Bismarck Archipelago	Takuman/Unknown	Reef island	−2.53444	147.3053	11.36	2
Papua New Guinea	Bismarck Archipelago	Tami	Volcanic low island	−6.76083	147.9011	29.59	4
Papua New Guinea	Bismarck Archipelago	Tarawai	Limestone low island	−3.20556	143.2578	465.04	10
Papua New Guinea	Bismarck Archipelago	Tatana	Reef island	−9.43778	147.1261	56.75	2

(continued)

(continued)

Region/country	Island group	Island name	Island type	Latitude	Longitude	Area (ha)	Maximum elevation
Papua New Guinea	Bismarck Archipelago	Tatau	Volcanic high island	−2.78639	151.9528	12538.02	354
Papua New Guinea	Bismarck Archipelago	Tawi	Volcanic high island	−2.22806	146.9822	51.28	46
Papua New Guinea	Bismarck Archipelago	Tefa	Volcanic high island	−3.53583	153.1911	234.77	155
Papua New Guinea	Bismarck Archipelago	Tench (Enus)	Volcanic low island	−1.65306	150.6717	41.32	12
Papua New Guinea	Bismarck Archipelago	Tiiianu (1)	Volcanic low island	−2.27583	147.5508	82.87	3
Papua New Guinea	Bismarck Archipelago	Tiiianu (2)	Volcanic low island	−2.30778	147.5367	32.84	27
Papua New Guinea	Bismarck Archipelago	Tingwon	Reef island	−2.6125	149.7078	296.02	2
Papua New Guinea	Bismarck Archipelago	Tolokiwa (Lokep)	Volcanic high island	−5.315	147.5869	4570.94	1372
Papua New Guinea	Bismarck Archipelago	Tomboi	Volcanic low island	−2.31417	146.8611	11.77	22
Papua New Guinea	Bismarck Archipelago	Tong	Volcanic high island	−2.05306	147.7697	1961.04	34
Papua New Guinea	Bismarck Archipelago	Tsalui	Limestone low island	−2.45444	150.47	576.69	6
Papua New Guinea	Bismarck Archipelago	Tsoilaunung	Limestone low island	−2.54361	150.53	895.79	25
Papua New Guinea	Bismarck Archipelago	Tuam	Volcanic low island	−5.95639	148.0272	116.36	10
Papua New Guinea	Bismarck Archipelago	Tumleo (Tamara)	Composite high island	−3.12167	142.3975	116.99	38

Papua New Guinea	Bismarck Archipelago	Turtle Back	Composite low island	−10.4058	149.3614	17.07	13
Papua New Guinea	Bismarck Archipelago	Ulu	Limestone high island	−4.21472	152.4136	777.46	40
Papua New Guinea	Bismarck Archipelago	Umboi (Rooke)	Volcanic high island	−5.63278	147.9233	88635.74	1548
Papua New Guinea	Bismarck Archipelago	Umunda	Reef island	−8.46972	143.7475	4322.93	3
Papua New Guinea	Bismarck Archipelago	Unea (Ball)	Volcanic high island	−4.89111	149.1433	3497.68	553
Papua New Guinea	Bismarck Archipelago	Ungalabu	Reef island	−2.40278	150.0411	63.12	2
Papua New Guinea	Bismarck Archipelago	Ungalik	Reef island	−2.37556	150.2394	41.35	3
Papua New Guinea	Bismarck Archipelago	Ungan	Reef island	−2.65444	150.6647	51.55	2
Papua New Guinea	Bismarck Archipelago	Urara	Volcanic low island	−4.18194	151.9411	93.28	23
Papua New Guinea	Bismarck Archipelago	Usienlik	Reef island	−2.63722	150.7664	80.87	3
Papua New Guinea	Bismarck Archipelago	Utan (Utuan)	Limestone low island	−4.23083	152.44	59.59	23
Papua New Guinea	Bismarck Archipelago	Vogali	Volcanic low island	−2.37194	146.8281	87.49	30
Papua New Guinea	Bismarck Archipelago	Vokeo	Volcanic high island	−3.2225	144.0967	2921.03	589
Papua New Guinea	Bismarck Archipelago	Wabuda	Reef island	−8.37342	143.6177	9497.58	3

(continued)

(continued)

Region/country	Island group	Island name	Island type	Latitude	Longitude	Area (ha)	Maximum elevation
Papua New Guinea	Bismarck Archipelago	Wadei	Limestone high island	−2.65944	150.6369	448.97	32
Papua New Guinea	Bismarck Archipelago	Waikatu/Unknown	Reef island	−2.45472	147.3972	40.53	2
Papua New Guinea	Bismarck Archipelago	Walis (Valif)	Limestone high island	−3.22972	143.3008	1116.2	32
Papua New Guinea	Bismarck Archipelago	Wamuk	Reef island	−2.205	146.6053	4.56	2
Papua New Guinea	Bismarck Archipelago	Wariuro	Volcanic low island	−8.31806	143.4092	4538.95	25
Papua New Guinea	Bismarck Archipelago	Watom	Volcanic high island	−4.11222	152.0739	1412.05	320
Papua New Guinea	Bismarck Archipelago	Wei (Vial)	Volcanic high island	−3.39083	144.4053	429.05	59
Papua New Guinea	Bismarck Archipelago	Wulai	Composite high island	−5.35139	150.4897	63.55	32
Papua New Guinea	Bismarck Archipelago	Wuvulu (Maty)	Reef island	−1.73417	142.8519	1556.81	2
Papua New Guinea	Bismarck Archipelago	Yule	Reef island	−8.81528	146.5294	1286.76	2
Papua New Guinea	Bismarck Archipelago	Yuo	Volcanic low island	−3.40444	143.4856	83.2	12
Papua New Guinea	Bismarck Archipelago	Zumbale	Composite high island	−7.3875	147.2494	45.37	82
Papua New Guinea	D'Entrecasteaux Islands	Dobu	Volcanic high island	−9.75306	150.8678	847.72	253

Papua New Guinea	D'Entrecasteaux Islands	Dum Dum (Wamera)	Volcanic high island	−9.25194	150.9036	469.85	320
Papua New Guinea	D'Entrecasteaux Islands	Fergusson	Volcanic high island	−9.53889	150.6825	137867.2	2073
Papua New Guinea	D'Entrecasteaux Islands	Goodenough	Volcanic high island	−9.34	150.24	73262.84	2536
Papua New Guinea	D'Entrecasteaux Islands	Gumawana (Urasi)	Volcanic high island	−9.21889	150.8739	350.33	300
Papua New Guinea	D'Entrecasteaux Islands	Nabwageta (Tuboa)	Volcanic high island	−9.21361	150.7853	83.72	95
Papua New Guinea	D'Entrecasteaux Islands	Normanby	Volcanic high island	−10.0633	151.0694	93295.34	1158
Papua New Guinea	D'Entrecasteaux Islands	Nuamata	Volcanic high island	−9.17667	150.2033	87.89	75
Papua New Guinea	D'Entrecasteaux Islands	Sanaroa	Volcanic high island	−9.60639	151.0008	5117.81	138
Papua New Guinea	D'Entrecasteaux Islands	Wagifa	Volcanic high island	−9.49778	150.3694	182.98	165
Papua New Guinea	D'Entrecasteaux Islands	Watota	Volcanic high island	−9.30083	150.7044	354.17	165
Papua New Guinea	D'Entrecasteaux Islands	Wawiwa	Volcanic high island	−9.28139	150.7536	988.54	365
Papua New Guinea	D'Entrecasteaux Islands	Yabwaia	Volcanic high island	−9.29222	150.7897	537.44	586
Papua New Guinea	Green Islands	Barahun	Reef island	−4.49472	154.1669	56.31	3
Papua New Guinea	Green Islands	Nissan	Reef island	−4.50472	154.2281	3225.44	3

(continued)

(continued)

Region/country	Island group	Island name	Island type	Latitude	Longitude	Area (ha)	Maximum elevation
Papua New Guinea	Green Islands	Pinipel	Reef island	−4.39639	154.1325	753.69	4
Papua New Guinea	Green Islands	Sau	Reef island	−4.37917	154.1069	8.94	1
Papua New Guinea	Green Islands	Sirot	Reef island	−4.46694	154.1622	131.88	3
Papua New Guinea	Louisiade Archipelago	Adele (Loa Boloba)	Volcanic high island	−11.4486	154.3961	20.59	43
Papua New Guinea	Louisiade Archipelago	Anaqusa (Bentley)	Volcanic high island	−10.7139	151.2408	174.43	68
Papua New Guinea	Louisiade Archipelago	Aurobu Islet/Aurobu	Volcanic low island	−11.1497	152.6922	7.3	27
Papua New Guinea	Louisiade Archipelago	Babagarai Islet/Babagarai	Volcanic low island	−10.6706	151.1219	18.78	13
Papua New Guinea	Louisiade Archipelago	Bagaman	Volcanic high island	−11.1336	152.6886	765.81	186
Papua New Guinea	Louisiade Archipelago	Basilaki	Volcanic high island	−10.6183	150.9889	11339.31	480
Papua New Guinea	Louisiade Archipelago	Bobo Eina	Volcanic high island	−11.1286	152.7339	273.91	197
Papua New Guinea	Louisiade Archipelago	Boirama	Volcanic high island	−10.285	151.0494	61.03	64
Papua New Guinea	Louisiade Archipelago	Bonna Bonnawan	Reef island	−10.815	152.1842	83.88	3
Papua New Guinea	Louisiade Archipelago	Bonna Wan Islet/Bonna Wan	Volcanic high island	−11.1303	152.6594	118.9	58

Papua New Guinea	Louisiade Archipelago	Bonvouloir	Limestone low island	−10.2567	151.8597	123.33	12
Papua New Guinea	Louisiade Archipelago	Bright/Unknown	Reef island	−10.5339	151.2128	11.93	1
Papua New Guinea	Louisiade Archipelago	Buiari	Volcanic high island	−10.6689	150.9225	383.22	170
Papua New Guinea	Louisiade Archipelago	Butchart	Volcanic low island	−10.5875	151.1986	309.94	3
Papua New Guinea	Louisiade Archipelago	Daiwari/Daiwan	Volcanic high island	−10.3056	151.0481	50.38	128
Papua New Guinea	Louisiade Archipelago	Dawson (1)	Volcanic low island	−10.3928	151.4164	154.42	21
Papua New Guinea	Louisiade Archipelago	Dawson (2)	Volcanic low island	−10.4022	151.4347	35.14	5
Papua New Guinea	Louisiade Archipelago	Dawson (3)	Reef island	−10.4094	151.4411	10.85	1
Papua New Guinea	Louisiade Archipelago	Deedes Islet/Unknown	Reef island	−10.5286	151.2697	26.69	1
Papua New Guinea	Louisiade Archipelago	Doini (Castori)	Volcanic high island	−10.6972	150.7142	450.37	119
Papua New Guinea	Louisiade Archipelago	East	Limestone high island	−10.3956	152.1039	301.29	146
Papua New Guinea	Louisiade Archipelago	Einamu Islet/Einamu	Volcanic high island	−11.1583	152.8656	188.54	91
Papua New Guinea	Louisiade Archipelago	Flat Islet/Flat	Reef island	−10.6078	151.3767	68.9	3
Papua New Guinea	Louisiade Archipelago	Gigila	Volcanic high island	−11.1708	152.94	152.68	112

(continued)

(continued)

Region/country	Island group	Island name	Island type	Latitude	Longitude	Area (ha)	Maximum elevation
Papua New Guinea	Louisiade Archipelago	Gilia Islet/Gilia	Volcanic low island	−11.1353	152.7181	35.47	21
Papua New Guinea	Louisiade Archipelago	Good Island/Unknown	Reef island	−10.5342	151.2453	42.55	1
Papua New Guinea	Louisiade Archipelago	Gulowa	Volcanic high island	−11.0669	152.5336	78.94	130
Papua New Guinea	Louisiade Archipelago	Haines	Volcanic high island	−10.6719	151.0597	89.84	45
Papua New Guinea	Louisiade Archipelago	Hardman/Unknown	Reef island	−10.4247	151.3122	42.74	8
Papua New Guinea	Louisiade Archipelago	Hastings	Volcanic high island	−10.3364	151.8772	151.18	158
Papua New Guinea	Louisiade Archipelago	Haszard	Reef island	−10.5881	151.3678	97.79	6
Papua New Guinea	Louisiade Archipelago	Hemenahei	Volcanic high island	−11.1581	153.0714	1128.82	59
Papua New Guinea	Louisiade Archipelago	Heva Isi	Volcanic high island	−11.2103	153.0925	27.35	58
Papua New Guinea	Louisiade Archipelago	High	Volcanic high island	−11.3053	154.0222	119.09	78
Papua New Guinea	Louisiade Archipelago	Iabama (Lelei−Gana)	Volcanic high island	−10.3047	150.9219	30.5	68
Papua New Guinea	Louisiade Archipelago	Irai	Reef island	−10.7681	151.6981	141.46	3
Papua New Guinea	Louisiade Archipelago	Itamarina	Reef island	−10.7394	151.7242	82.94	3
Papua New Guinea	Louisiade Archipelago	Ito	Volcanic high island	−10.5653	150.7603	335.76	109

Papua New Guinea	Louisiade Archipelago	Kaiti	Volcanic high island	−10.6719	151.1117	71.67	79
Papua New Guinea	Louisiade Archipelago	Kitai Bona Bona	Volcanic low island	−10.6472	151.1094	63.96	25
Papua New Guinea	Louisiade Archipelago	Kuwanak	Volcanic high island	−11.1728	152.9233	417.71	164
Papua New Guinea	Louisiade Archipelago	Kwalaiwa (Watts)	Composite high island	−10.6211	151.2883	293.29	92
Papua New Guinea	Louisiade Archipelago	Kwato	Volcanic high island	−10.6136	150.6321	34.27	35
Papua New Guinea	Louisiade Archipelago	Kwaui Islet/Kwaui	Volcanic high island	−10.5619	150.7219	168.56	91
Papua New Guinea	Louisiade Archipelago	Laiwan Islet/Laiwan	Volcanic low island	−11.1069	152.6392	22.6	25
Papua New Guinea	Louisiade Archipelago	Logeia (Rogeia)	Volcanic high island	−10.6361	150.6475	1005.2	324
Papua New Guinea	Louisiade Archipelago	Lunn/unknown	Reef island	−10.7869	152.0017	65.91	7
Papua New Guinea	Louisiade Archipelago	Mabneian Islet/Mabneian	Volcanic high island	−11.1231	152.76	48.14	47
Papua New Guinea	Louisiade Archipelago	Managun Islet/Managun	Reef island	−10.6883	152.8856	36.46	8
Papua New Guinea	Louisiade Archipelago	Misima	Volcanic high island	−10.6722	152.7511	22465.53	1036
Papua New Guinea	Louisiade Archipelago	Motorina	Volcanic high island	−11.0844	152.5664	904.16	263
Papua New Guinea	Louisiade Archipelago	Nare	Reef island	−10.7514	151.315	138.42	1

(continued)

(continued)

Region/country	Island group	Island name	Island type	Latitude	Longitude	Area (ha)	Maximum elevation
Papua New Guinea	Louisiade Archipelago	Nasakoli	Reef island	−11.1119	152.3156	3.06	1
Papua New Guinea	Louisiade Archipelago	Naunalualua	Volcanic high island	−10.6622	151.0883	338.52	145
Papua New Guinea	Louisiade Archipelago	Nimoa	Volcanic high island	−11.3069	153.2517	413.4	118
Papua New Guinea	Louisiade Archipelago	Nivani/Nivarni	Volcanic high island	−10.7919	152.3919	40.25	58
Papua New Guinea	Louisiade Archipelago	Nuakata	Volcanic high island	−10.2842	151.0178	1125.14	285
Papua New Guinea	Louisiade Archipelago	Obstruction	Volcanic high island	−10.2822	150.9306	44.46	46
Papua New Guinea	Louisiade Archipelago	Oreia Islet/Oreia	Reef island	−10.8419	152.9742	299.13	3
Papua New Guinea	Louisiade Archipelago	Pana Numara (Panangaribu)	Volcanic high island	−11.135	152.8114	46.08	64
Papua New Guinea	Louisiade Archipelago	Pana Rora	Volcanic high island	−11.1083	152.4992	92.68	103
Papua New Guinea	Louisiade Archipelago	Pana Udu Udi	Volcanic high island	−11.0439	152.4892	83.28	78
Papua New Guinea	Louisiade Archipelago	Pana Vara Vara	Volcanic high island	−11.1214	152.3031	67.52	71
Papua New Guinea	Louisiade Archipelago	Pana Waipona Islet (Panawina)	Reef island	−11.1889	152.03	50.5	7
Papua New Guinea	Louisiade Archipelago	Panaeati	Volcanic high island	−10.6836	152.3689	3269.38	175
Papua New Guinea	Louisiade Archipelago	Panantanian	Volcanic high island	−11.1542	152.8269	89.16	96

Papua New Guinea	Louisiade Archipelago	Panapompom	Volcanic high island	−10.7714	152.3975	905.49	125
Papua New Guinea	Louisiade Archipelago	Panarairai	Volcanic low island	−11.2525	152.1378	78.96	16
Papua New Guinea	Louisiade Archipelago	Panarakuum	Reef island	−10.8067	151.9222	96.28	5
Papua New Guinea	Louisiade Archipelago	Panasesa	Reef island	−10.7383	151.8261	74.45	4
Papua New Guinea	Louisiade Archipelago	Panasia	Limestone high island	−11.1356	152.3374	261.95	152
Papua New Guinea	Louisiade Archipelago	Panatinane/Panatinane	Volcanic high island	−11.2386	153.1692	8343.48	296
Papua New Guinea	Louisiade Archipelago	Panaumala	Volcanic high island	−11.1578	152.7803	228.66	118
Papua New Guinea	Louisiade Archipelago	Panawina	Volcanic high island	−11.1736	153.0108	3262.95	260
Papua New Guinea	Louisiade Archipelago	Panua Keikeisa Islet/Panua Keikeisa	Volcanic low island	−11.0992	152.6067	14.04	23
Papua New Guinea	Louisiade Archipelago	Passage	Reef island	−10.7928	152.4889	13.52	1
Papua New Guinea	Louisiade Archipelago	Pender/unknown	Reef island	−10.5619	151.2581	25.03	4
Papua New Guinea	Louisiade Archipelago	Populai	Volcanic high island	−10.6658	150.8892	198.33	121
Papua New Guinea	Louisiade Archipelago	Powell	Volcanic low island	−10.5811	151.2811	22.05	15
Papua New Guinea	Louisiade Archipelago	Rara (Redlick)	Reef island	−10.8233	152.4994	86.56	5

(continued)

(continued)

Region/country	Island group	Island name	Island type	Latitude	Longitude	Area (ha)	Maximum elevation
Papua New Guinea	Louisiade Archipelago	Sabara	Limestone high island	−11.1194	153.0922	514.89	40
Papua New Guinea	Louisiade Archipelago	Samarai (1)	Volcanic high island	−10.6122	150.6642	39.81	37
Papua New Guinea	Louisiade Archipelago	Sariba	Volcanic high island	−10.6167	150.7203	2550.35	283
Papua New Guinea	Louisiade Archipelago	Sideia	Volcanic high island	−10.5881	150.8636	10413.19	370
Papua New Guinea	Louisiade Archipelago	Skelton	Volcanic high island	−10.6103	151.2381	432.06	123
Papua New Guinea	Louisiade Archipelago	Sloss (1)	Reef island	−11.0397	152.3828	9.2	1
Papua New Guinea	Louisiade Archipelago	Sloss (2)	Reef island	−11.0419	152.4032	18.22	8
Papua New Guinea	Louisiade Archipelago	Tagula	Volcanic high island	−11.5069	153.4439	85323	806
Papua New Guinea	Louisiade Archipelago	Tinolan (Torlesse)	Reef island	−10.8067	152.2236	93.52	3
Papua New Guinea	Louisiade Archipelago	Tobaiam	Volcanic low island	−11.0339	152.5461	7.3	13
Papua New Guinea	Louisiade Archipelago	Toloi Awa Islet/Toloi Awa	Volcanic high island	−11.0417	152.5133	76.01	62
Papua New Guinea	Louisiade Archipelago	Uban (Brooker)	Volcanic high island	−11.0544	152.4475	180.53	92
Papua New Guinea	Louisiade Archipelago	Venama (Manudi)	Limestone low island	−11.6172	153.51	59.6	12
Papua New Guinea	Louisiade Archipelago	Wanim	Volcanic high island	−11.2619	153.1019	211.4	90

Papua New Guinea	Louisiade Archipelago	Wari	Volcanic high island	−10.9583	151.065	321.29	35
Papua New Guinea	Louisiade Archipelago	Yeina	Volcanic high island	−11.325	153.4339	1823.91	79
Papua New Guinea	Louisiade Archipelago	Yela (Rossel)	Volcanic high island	−11.3528	154.16	31216.57	838
Papua New Guinea	Main Group	Kadovar	Volcanic high island	−3.60889	144.5872	157.26	365
Papua New Guinea	North Solomon Islands	Bakawari (Pok Pok)	Composite high island	−6.20167	155.6647	730.85	344
Papua New Guinea	North Solomon Islands	Bougainville	Composite high island	−6.22333	155.305	882788.2	2715
Papua New Guinea	North Solomon Islands	Buka	Composite high island	−5.22417	154.6231	61576.46	458
Papua New Guinea	North Solomon Islands	Han (Carteret/Tulun)	Reef island	−4.78194	155.4667	60.14	2
Papua New Guinea	North Solomon Islands	Lolasa/Iofase	Reef island	−4.70944	155.4294	17.73	2
Papua New Guinea	North Solomon Islands	Madehas	Composite high island	−5.47	154.6325	367.7	75
Papua New Guinea	North Solomon Islands	Matsungan	Composite low island	−5.38833	154.5753	88.05	12
Papua New Guinea	North Solomon Islands	Namotu (2)	Reef island	−3.28306	154.7148	87.15	2
Papua New Guinea	North Solomon Islands	Nuguria	Reef island	−3.35917	154.6928	120.54	5
Papua New Guinea	North Solomon Islands	Nukumanu Atoll	Reef island	−4.57611	159.4856	487.13	2

(continued)

(continued)

Region/country	Island group	Island name	Island type	Latitude	Longitude	Area (ha)	Maximum elevation
Papua New Guinea	North Solomon Islands	Paona	Reef island	−3.11278	154.4381	60.19	2
Papua New Guinea	North Solomon Islands	Paopao (Nugarba)	Reef island	−3.44806	154.7453	624.33	7
Papua New Guinea	North Solomon Islands	Petats	Composite low island	−5.33667	154.5583	252.48	25
Papua New Guinea	North Solomon Islands	Piul (Pule)	Reef island	−4.81611	155.4656	21.97	2
Papua New Guinea	North Solomon Islands	Pororan	Composite low island	−5.20444	154.5225	203.17	21
Papua New Guinea	North Solomon Islands	Sale	Reef island	−5.42833	154.5731	11.61	2
Papua New Guinea	North Solomon Islands	Sohano	Composite low island	−5.44556	154.6667	53.29	16
Papua New Guinea	North Solomon Islands	Taiof	Composite high island	−5.51278	154.6506	2153.29	31
Papua New Guinea	North Solomon Islands	Tautsina	Composite high island	−6.19333	155.6511	102.68	128
Papua New Guinea	North Solomon Islands	Teop (Horan)	Composite low island	−5.57278	155.0789	66.03	21
Papua New Guinea	North Solomon Islands	Yame	Composite low island	−5.30361	154.5403	180.39	27
Papua New Guinea	Solomon Sea Islands	Bornapau/Bornapau	Limestone low island	−8.58556	151.0975	490.69	13
Papua New Guinea	Solomon Sea Islands	Budelun	Reef island	−9.29222	153.6928	239.62	9
Papua New Guinea	Solomon Sea Islands	Dugumenu	Limestone high island	−8.80111	151.9164	160.01	45

Papua New Guinea	Solomon Sea Islands	Gawa	Composite high island	−8.97111	151.9794	1538.65	175
Papua New Guinea	Solomon Sea Islands	Gilua/Unknown	Limestone low island	−8.56694	150.8353	72.72	12
Papua New Guinea	Solomon Sea Islands	Iwa	Limestone high island	−8.70417	151.675	286.52	112
Papua New Guinea	Solomon Sea Islands	Kadai	Limestone low island	−8.31861	150.8219	63.39	25
Papua New Guinea	Solomon Sea Islands	Kaduaga	Limestone high island	−8.52806	150.9536	4762.52	42
Papua New Guinea	Solomon Sea Islands	Kawa (2)	Reef island	−8.52417	150.2986	85.75	2
Papua New Guinea	Solomon Sea Islands	Kawa (3)	Reef island	−8.56306	150.2814	36.88	2
Papua New Guinea	Solomon Sea Islands	Kiriwina	Limestone high island	−8.58917	151.14	28829.24	36
Papua New Guinea	Solomon Sea Islands	Kitava	Limestone high island	−8.62806	151.3333	2412.1	95
Papua New Guinea	Solomon Sea Islands	Kwaiawatta	Composite high island	−8.91611	151.9136	606.34	154
Papua New Guinea	Solomon Sea Islands	Lulima	Reef island	−8.3925	150.6475	46.42	2
Papua New Guinea	Solomon Sea Islands	Madau	Composite high island	−8.99583	152.4069	3941.04	37
Papua New Guinea	Solomon Sea Islands	Mapas	Composite high island	−9.24278	152.7619	114.59	36
Papua New Guinea	Solomon Sea Islands	Munuwata	Limestone low island	−8.59639	150.8631	180.66	12

(continued)

(continued)

Region/country	Island group	Island name	Island type	Latitude	Longitude	Area (ha)	Maximum elevation
Papua New Guinea	Solomon Sea Islands	Muwo	Limestone low island	−8.74333	151.0056	326.63	6
Papua New Guinea	Solomon Sea Islands	Nuarguta/Unknown	Reef island	−8.3725	150.6639	11.53	2
Papua New Guinea	Solomon Sea Islands	Nubara	Reef island	−9.22389	153.1086	96.39	7
Papua New Guinea	Solomon Sea Islands	Simlindon	Limestone low island	−8.33056	150.5811	69.72	11
Papua New Guinea	Solomon Sea Islands	Simsim/Unknown	Reef island	−8.41611	150.4469	27.22	2
Papua New Guinea	Solomon Sea Islands	Tokona (Alcester)	Volcanic high island	−9.55556	152.4442	558.48	80
Papua New Guinea	Solomon Sea Islands	Tuma	Limestone low island	−8.35417	150.8636	498.96	28
Papua New Guinea	Solomon Sea Islands	Vakuta	Limestone low island	−8.86722	151.1869	2242.21	27
Papua New Guinea	Solomon Sea Islands	Waboma (Lisilus)/ Wabomat	Limestone low island	−9.27361	153.6839	113.63	8
Papua New Guinea	Solomon Sea Islands	Wiakau	Reef island	−9.27111	151.89	737.93	18
Papua New Guinea	Solomon Sea Islands	Woodlark (Muyua)	Volcanic high island	−9.1125	152.7819	89266.23	308
Pitcairn Islands		Ducie	Reef island	−24.6806	−124.788	142.9	4
Pitcairn Islands		Henderson Island	Limestone high island	−24.3742	−128.327	4534.58	33
Pitcairn Islands		Oeno	Reef island	−23.9228	−130.741	124.47	3
Pitcairn Islands		Pitcairn Island	Volcanic high island	−25.04	130.06	592.75	347

Samoa	Aleipata Islands	Nu'ulua	Volcanic high island	−14.0744	−171.409	54	47
Samoa	Aleipata Islands	Nu'utele	Volcanic high island	−14.0622	−171.424	148.86	200
Samoa	Apolima Strait	Apolima	Volcanic high island	−13.8217	−172.151	108.83	165
Samoa	Apolima Strait	Manono	Volcanic high island	−13.8481	−172.109	403.62	110
Samoa	Apolima Strait	Nu'ulopa	Volcanic high island	−13.8419	−172.129	9.23	50
Samoa	Outliers	Savai'i	Volcanic high island	−13.6414	−172.602	182347.9	1858
Samoa	Outliers	Upolu	Volcanic high island	−13.9787	−171.625	121546.4	1100
Solomon Islands	Main Group	Abuabua	Volcanic low island	−8.9325	160.7528	80.2	19
Solomon Islands	Main Group	Aelaunu	Reef island	−7.01139	155.6206	38.84	18
Solomon Islands	Main Group	Aghana	Volcanic low island	−8.46333	157.2814	55.24	24
Solomon Islands	Main Group	Ainuta Paina	Reef island	−9.1675	161.2631	270.75	2
Solomon Islands	Main Group	Aio	Reef island	−9.03	161.1986	187.66	2
Solomon Islands	Main Group	Aiura	Reef island	−9.20833	161.2689	69.06	2
Solomon Islands	Main Group	Akara	Limestone low island	−8.65083	157.8028	91.14	29
Solomon Islands	Main Group	Alasina (Aiaisina)	Volcanic high island	−6.85278	155.8408	62.09	125
Solomon Islands	Main Group	Ali'ite	Volcanic high island	−10.1389	161.9256	298.82	56

(continued)

(continued)

Region/country	Island group	Island name	Island type	Latitude	Longitude	Area (ha)	Maximum elevation
Solomon Islands	Main Group	Alite (Langa Langa)	Reef island	−8.90194	160.7328	192.45	1
Solomon Islands	Main Group	Alokan	Volcanic high island	−9.1475	159.1239	868.29	107
Solomon Islands	Main Group	Anuha (Anuah)	Composite high island	−8.99833	160.2233	77.31	44
Solomon Islands	Main Group	Aoa	Volcanic low island	−6.99139	156.0083	41.02	27
Solomon Islands	Main Group	Arnavon	Reef island	−7.48472	158.0508	31.05	2
Solomon Islands	Main Group	Asie	Volcanic high island	−6.87667	156.0233	339.78	134
Solomon Islands	Main Group	Aulaga	Volcanic low island	−8.93278	160.7422	60.84	5
Solomon Islands	Main Group	Avavasa	Limestone low island	−8.41083	157.9181	230.11	25
Solomon Islands	Main Group	Bakiava	Volcanic low island	−6.98139	156.0117	41.88	28
Solomon Islands	Main Group	Balalai (Ballalae)	Volcanic low island	−6.99111	155.8831	209.52	15
Solomon Islands	Main Group	Barora Fa I	Reef island	−7.48417	158.2911	8297.02	3
Solomon Islands	Main Group	Barora Ite I	Reef island	−7.61	158.4489	11607.7	2
Solomon Islands	Main Group	Bates	Reef island	−7.36583	158.2208	377.38	6
Solomon Islands	Main Group	Bembalama	Volcanic high island	−7.35028	157.5675	36.49	51
Solomon Islands	Main Group	Benana	Volcanic low island	−6.95639	156.0161	55.1	24
Solomon Islands	Main Group	Bethlehem	Reef island	−8.31944	157.3322	19.3	1
Solomon Islands	Main Group	Blanche	Volcanic high island	−8.60111	157.3953	50.64	50

Solomon Islands	Main Group	Bokala	Volcanic low island	−7.42611	157.5136	33.93	15
Solomon Islands	Main Group	Boloka	Limestone low island	−9.15167	159.0953	91.13	28
Solomon Islands	Main Group	Buena Vista (Vatilau)	Volcanic high island	−8.89333	160.0231	1606.15	279
Solomon Islands	Main Group	Chakakoaka (Mondomondo)	Reef island	−8.23639	157.7969	790.39	3
Solomon Islands	Main Group	Chalu	Reef island	−8.73944	158.1231	272.82	19
Solomon Islands	Main Group	Charapoana	Reef island	−8.43306	157.9606	51.91	2
Solomon Islands	Main Group	Chizunalapu	Reef island	−7.65111	158.4367	111.8	1
Solomon Islands	Main Group	Choiseul	Composite high island	−6.98556	156.895	315975.5	1066
Solomon Islands	Main Group	Chomborua	Limestone low island	−8.66722	158.1858	19.46	16
Solomon Islands	Main Group	Cross	Reef island	−8.13889	156.9178	22.72	2
Solomon Islands	Main Group	Dai	Volcanic high island	−7.90139	160.6242	1677.07	30
Solomon Islands	Main Group	Danisavo	Volcanic high island	−8.88833	159.9831	231.84	57
Solomon Islands	Main Group	Dololo	Volcanic high island	−7.41611	157.3794	134.8	52
Solomon Islands	Main Group	Elo Saghana	Limestone low island	−8.35639	157.2631	9.16	5
Solomon Islands	Main Group	Elonggava Hite	Reef island	−8.23111	157.1231	245.9	3
Solomon Islands	Main Group	Elonggava Nomana	Reef island	−8.24056	157.1244	40.76	12
Solomon Islands	Main Group	Endeve	Volcanic high island	−7.36806	157.2481	102.4	70

(continued)

(continued)

Region/country	Island group	Island name	Island type	Latitude	Longitude	Area (ha)	Maximum elevation
Solomon Islands	Main Group	Epangga (Nusatule)	Limestone low island	−8.09861	156.8642	32.52	27
Solomon Islands	Main Group	Erventa	Volcanic high island	−6.80167	155.8264	35.62	61
Solomon Islands	Main Group	Faila	Volcanic low island	−9.05722	159.2575	216.49	11
Solomon Islands	Main Group	Fauro	Volcanic high island	−6.88639	156.0794	7713.85	400
Solomon Islands	Main Group	Fera	Volcanic high island	−8.10861	159.5894	456.86	34
Solomon Islands	Main Group	Finuana	Volcanic high island	−8.05972	158.9872	169.74	32
Solomon Islands	Main Group	Gavutu	Volcanic high island	−9.11667	160.1883	20.41	45
Solomon Islands	Main Group	Ghaghe	Reef island	−7.44444	158.2292	4094.62	3
Solomon Islands	Main Group	Ghailava	Limestone low island	−8.36833	157.855	13.48	6
Solomon Islands	Main Group	Ghaomai	Volcanic low island	−7.06639	155.6442	214.32	30
Solomon Islands	Main Group	Ghateghe	Composite high island	−7.45417	158.3083	304.44	73
Solomon Islands	Main Group	Ghebira	Volcanic low island	−7.39472	158.2592	318.51	13
Solomon Islands	Main Group	Gheliana	Reef island	−8.29917	157.3331	14.37	4
Solomon Islands	Main Group	Ghinoa	Volcanic low island	−6.89278	156.625	11.72	18
Solomon Islands	Main Group	Ghizo (Gizo)	Volcanic high island	−8.08722	156.7975	3546.71	180

Solomon Islands	Main Group	Ghopuria/Ghopuna	Reef island	-7.41694	158.2967	39.88	8
Solomon Islands	Main Group	Ghurava	Volcanic high island	-8.03611	157.6131	180.63	42
Solomon Islands	Main Group	Guadalcanal	Composite high island	-9.61778	160.1397	554206.5	2449
Solomon Islands	Main Group	Hanawisi	Volcanic high island	-8.97306	159.1281	90.06	22
Solomon Islands	Main Group	Himbi	Limestone low island	-8.33444	157.3133	5.62	10
Solomon Islands	Main Group	Hombu	Reef island	-8.285	157.4958	17.92	1
Solomon Islands	Main Group	Honiavasa	Reef island	-8.30694	157.3842	653.08	3
Solomon Islands	Main Group	Honoa	Volcanic low island	-9.81111	160.8844	15.36	15
Solomon Islands	Main Group	Hopaseghe	Volcanic high island	-8.19028	157.7192	267.18	45
Solomon Islands	Main Group	Hotoanivena I	Limestone low island	-8.73889	158.1008	336.09	20
Solomon Islands	Main Group	Huleo	Reef island	-8.53361	158.0208	5.64	1
Solomon Islands	Main Group	Ighisi	Reef island	-8.3525	157.2728	26.42	3
Solomon Islands	Main Group	Ikiti	Reef island	-7.56639	158.7167	58.93	5
Solomon Islands	Main Group	Juakau	Volcanic low island	-8.14778	159.6153	10.2	8
Solomon Islands	Main Group	Kaghau	Composite high island	-7.32944	157.5844	44.38	34
Solomon Islands	Main Group	Kakarumu	Reef island	-8.30028	157.3417	20.02	1
Solomon Islands	Main Group	Kakia	Reef island	-8.26972	157.1539	12.33	5
Solomon Islands	Main Group	Kalala Nomana	Reef island	-8.27	157.1433	36.42	1

(continued)

(continued)

Region/country	Island group	Island name	Island type	Latitude	Longitude	Area (ha)	Maximum elevation
Solomon Islands	Main Group	Kambokinanggu	Limestone low island	−8.28722	157.5172	23.09	10
Solomon Islands	Main Group	Kambomola	Reef island	−8.27028	157.1142	4.73	1
Solomon Islands	Main Group	Kambotinoni	Reef island	−8.30778	157.1764	11.65	2
Solomon Islands	Main Group	Kapatene	Limestone low island	−8.66667	158.1153	14.15	10
Solomon Islands	Main Group	Karekarea	Limestone low island	−8.37694	157.8647	23.42	8
Solomon Islands	Main Group	Karikana I	Limestone low island	−8.44639	157.9978	160.57	29
Solomon Islands	Main Group	Karuhahe	Limestone low island	−8.35444	157.8556	25.24	8
Solomon Islands	Main Group	Karulingge Kiki	Limestone low island	−8.68361	158.1367	13.47	8
Solomon Islands	Main Group	Karumotusise (Islet)	Reef island	−8.39333	157.8831	3.05	8
Solomon Islands	Main Group	Karungarao	Limestone low island	−8.64444	158.1433	20.23	11
Solomon Islands	Main Group	Karunjou	Limestone high island	−8.62111	158.2014	53	50
Solomon Islands	Main Group	Karupenete	Reef island	−8.13194	156.9086	1.32	2
Solomon Islands	Main Group	Katuhi	Reef island	−7.01361	155.8297	14.99	3
Solomon Islands	Main Group	Kaukau	Limestone low island	−9.03722	159.2808	57.59	18
Solomon Islands	Main Group	Kauvi	Composite low island	−8.30833	157.1394	166.9	10
Solomon Islands	Main Group	Keala	Composite high island	−7.39361	157.2942	121.85	54

Solomon Islands	Main Group	Kerehikapa	Reef island	−7.4525	158.0225	152.09	3
Solomon Islands	Main Group	Kevana	Reef island	−8.32528	157.3036	4.75	9
Solomon Islands	Main Group	Kiaba	Reef island	−8.00611	159.4581	13.39	3
Solomon Islands	Main Group	Kingguru	Reef island	−8.77389	158.1269	9.93	16
Solomon Islands	Main Group	Kisan	Volcanic low island	−9.10111	159.3489	20.65	5
Solomon Islands	Main Group	Ko Tu Kuriana	Limestone low island	−8.31556	157.8292	558.53	18
Solomon Islands	Main Group	Kohinggo (Arundel)	Composite high island	−8.19972	157.145	9947.66	125
Solomon Islands	Main Group	Kohirio/Kohiro	Volcanic high island	−7.36389	158.1272	130.66	48
Solomon Islands	Main Group	Kokoana	Volcanic low island	−8.76111	158.0556	19.32	14
Solomon Islands	Main Group	Kokohale	Reef island	−8.33583	157.2661	4.84	6
Solomon Islands	Main Group	Kokopuna	Reef island	−7.49583	158.4403	74.79	7
Solomon Islands	Main Group	Kokoturo	Volcanic high island	−8.21194	157.7422	113.05	35
Solomon Islands	Main Group	Kolo Hite	Reef island	−8.29972	157.2044	36.21	4
Solomon Islands	Main Group	Kologhose	Reef island	−7.40861	158.2814	110.88	4
Solomon Islands	Main Group	Kologilo	Volcanic high island	−7.40306	158.1617	79.68	61
Solomon Islands	Main Group	Kolohengomo	Volcanic low island	−8.05472	157.6214	98.73	21
Solomon Islands	Main Group	Kolombangara	Volcanic high island	−7.99	157.0758	70009.84	1760
Solomon Islands	Main Group	Kombuana	Volcanic high island	−8.85194	160.0336	93.65	35

(continued)

(continued)

Region/country	Island group	Island name	Island type	Latitude	Longitude	Area (ha)	Maximum elevation
Solomon Islands	Main Group	Kondakanimboko	Volcanic low island	−6.67361	156.3906	56.39	15
Solomon Islands	Main Group	Koroia Ang	Volcanic low island	−6.94472	156.0331	9.87	23
Solomon Islands	Main Group	Kosa	Reef island	−9.85806	160.8756	13.34	3
Solomon Islands	Main Group	Kovuhika	Reef island	−8.98889	160.0336	8.32	3
Solomon Islands	Main Group	Kukuvulu	Volcanic high island	−6.87556	155.8761	7.13	33
Solomon Islands	Main Group	Kumarara Hira	Reef island	−7.50583	158.6725	80.79	4
Solomon Islands	Main Group	Kundu Hite	Reef island	−8.31556	157.1536	17.53	1
Solomon Islands	Main Group	Kurimarau	Reef island	−9.0175	159.2706	103.86	5
Solomon Islands	Main Group	Kwai	Reef island	−8.77222	160.9475	11.85	2
Solomon Islands	Main Group	Kwaro	Volcanic low island	−8.96167	160.7486	90.59	16
Solomon Islands	Main Group	Laena	Limestone high island	−7.31306	157.5989	223.74	195
Solomon Islands	Main Group	Lamu	Limestone low island	−9.09694	159.2964	32.08	18
Solomon Islands	Main Group	Langarana	Reef island	−8.3725	157.5119	30.86	1
Solomon Islands	Main Group	Laomana	Volcanic low island	−7.03278	155.6514	24.6	24
Solomon Islands	Main Group	Laumuan	Volcanic high island	−9.14083	159.3656	171.08	35
Solomon Islands	Main Group	Lavata	Reef island	−8.35333	157.2214	11.77	1
Solomon Islands	Main Group	Leli	Reef island	−8.74972	161.0322	452.91	2
Solomon Islands	Main Group	Leru	Reef island	−8.99722	159.0531	236.58	3

Solomon Islands	Main Group	Letuni	Volcanic low island	−7.35361	157.1472	8.95	22
Solomon Islands	Main Group	Liapari	Limestone high island	−7.95306	156.7106	82.43	34
Solomon Islands	Main Group	Lilihina	Reef island	−8.52139	158.0719	7.3	9
Solomon Islands	Main Group	Logha	Limestone low island	−8.0925	156.8458	66.62	28
Solomon Islands	Main Group	Lola	Reef island	−8.31278	157.1658	57.26	4
Solomon Islands	Main Group	Lologhan	Volcanic low island	−9.11944	159.3594	95.66	8
Solomon Islands	Main Group	Lolou	Limestone low island	−8.46167	158.0958	4.73	5
Solomon Islands	Main Group	Long	Composite low island	−7.47806	157.5714	125.2	17
Solomon Islands	Main Group	Loun	Volcanic high island	−9.11667	159.245	544.77	49
Solomon Islands	Main Group	Lughurua	Reef island	−8.32861	157.5681	22.95	1
Solomon Islands	Main Group	Lumalihe	Limestone low island	−8.45917	158.0572	76.74	26
Solomon Islands	Main Group	Lumbararo	Reef island	−8.22806	157.1003	3.74	8
Solomon Islands	Main Group	Lumbaria	Limestone low island	−8.405	157.3117	6.45	15
Solomon Islands	Main Group	Maana'oba	Reef island	−8.30639	160.7919	1440.79	2
Solomon Islands	Main Group	Mahighe	Volcanic high island	−8.515	159.9153	310.52	103
Solomon Islands	Main Group	Mahoro	Volcanic low island	−8.5125	158.0228	35.9	25

(continued)

(continued)

Region/country	Island group	Island name	Island type	Latitude	Longitude	Area (ha)	Maximum elevation
Solomon Islands	Main Group	Maifu	Volcanic high island	−6.89861	155.8372	31.78	38
Solomon Islands	Main Group	Malaita	Composite high island	−9.06417	161.005	397801	1435
Solomon Islands	Main Group	Malakobi	Composite high island	−7.38556	158.1369	725.67	91
Solomon Islands	Main Group	Malaulalo	Volcanic high island	−10.1856	161.9461	360	40
Solomon Islands	Main Group	Malaupaina	Volcanic high island	−10.2478	161.9703	603.34	43
Solomon Islands	Main Group	Malazeke	Limestone low island	−8.15444	157.075	60.99	15
Solomon Islands	Main Group	Malemale	Volcanic low island	−8.79028	158.2511	25.73	18
Solomon Islands	Main Group	Malumalu	Reef island	−8.57333	157.8161	30.84	4
Solomon Islands	Main Group	Mamalalunduranggo	Volcanic low island	−7.40528	157.3228	47.83	11
Solomon Islands	Main Group	Mana (3)	Volcanic high island	−8.96306	160.0375	21.81	47
Solomon Islands	Main Group	Mananguni	Reef island	−7.38972	157.6111	25.65	6
Solomon Islands	Main Group	Mandoliana	Volcanic low island	−9.20861	160.2861	69.65	23
Solomon Islands	Main Group	Mane	Reef island	−9.02833	159.0369	343.37	3
Solomon Islands	Main Group	Mangalonga (Maravagi)	Volcanic high island	−8.95306	160.0483	270.13	69
Solomon Islands	Main Group	Mania	Volcanic high island	−7.01639	155.9964	100.97	52
Solomon Islands	Main Group	Marakumbo	Reef island	−8.28417	157.1444	124.7	20

Solomon Islands	Main Group	Maramasike	Composite high island	-9.55833	161.4758	50625.32	200
Solomon Islands	Main Group	Marapa	Composite high island	-9.81611	160.8722	876.37	175
Solomon Islands	Main Group	Maraupia	Reef island	-9.85056	160.8378	5.51	3
Solomon Islands	Main Group	Marouo	Volcanic high island	-8.515	157.9961	890.12	153
Solomon Islands	Main Group	Maruiapa (Beagle)	Volcanic high island	-9.79639	160.8206	581.89	178
Solomon Islands	Main Group	Marulaon	Volcanic high island	-8.9875	159.0833	423.87	38
Solomon Islands	Main Group	Masamasa	Volcanic high island	-6.81472	156.1375	344.09	138
Solomon Islands	Main Group	Matenana	Limestone low island	-8.43694	157.9792	23.95	20
Solomon Islands	Main Group	Matiu	Limestone low island	-8.48194	158.1397	199.66	28
Solomon Islands	Main Group	Mauru	Volcanic low island	-8.57139	157.8381	17.61	20
Solomon Islands	Main Group	Mbagholae	Reef island	-8.28528	157.3369	99.96	19
Solomon Islands	Main Group	Mbaghumbaghu	Composite low island	-8.19389	157.6869	235.74	29
Solomon Islands	Main Group	Mbahoro	Reef island	-8.73167	157.8303	21.98	1
Solomon Islands	Main Group	Mbaleva	Limestone low island	-8.465	158.0719	26.39	5
Solomon Islands	Main Group	Mbalusu	Reef island	-8.28389	157.3614	61.64	4
Solomon Islands	Main Group	Mbambanga	Limestone low island	-8.12	156.8881	92.33	3

(continued)

(continued)

Region/country	Island group	Island name	Island type	Latitude	Longitude	Area (ha)	Maximum elevation
Solomon Islands	Main Group	Mbanga	Volcanic low island	−8.30056	157.2206	741.73	18
Solomon Islands	Main Group	Mbangai	Volcanic low island	−9.11417	160.1503	3.07	22
Solomon Islands	Main Group	Mbanika	Volcanic high island	−9.10361	159.1936	5058.23	227
Solomon Islands	Main Group	Mbanitambaika	Reef island	−8.70306	158.1592	8.89	1
Solomon Islands	Main Group	Mbarambuni	Limestone low island	−8.40167	157.3567	63.61	22
Solomon Islands	Main Group	Mbareho	Volcanic low island	−8.60056	158.1306	22.83	22
Solomon Islands	Main Group	Mbarikihi	Volcanic low island	−8.30667	157.2342	65.83	16
Solomon Islands	Main Group	Mbathakana	Reef island	−8.31139	160.5811	316.58	2
Solomon Islands	Main Group	Mbava (Baga)	Volcanic high island	−7.82278	156.5364	2378.96	235
Solomon Islands	Main Group	Mbelombelo	Reef island	−8.39417	157.5311	633.82	1
Solomon Islands	Main Group	Mbembea	Composite high island	−8.35556	157.3286	513.86	74
Solomon Islands	Main Group	Mbero	Volcanic high island	−7.69139	158.5236	2387.14	171
Solomon Islands	Main Group	Mbeta	Volcanic low island	−8.90417	160.0108	14.93	24
Solomon Islands	Main Group	Mbimbilusi	Reef island	−8.14583	156.8981	9.67	2
Solomon Islands	Main Group	Mbirimbiri	Reef island	−8.33583	157.1997	4.74	1
Solomon Islands	Main Group	Mbokonimbeti (Olevuga)	Volcanic high island	−8.96889	160.0914	3051.87	280

Solomon Islands	Main Group	Mboku	Volcanic low island	−8.39667	157.3247	81.83	28
Solomon Islands	Main Group	Mborokua	Volcanic high island	−9.02	158.7419	669.2	323
Solomon Islands	Main Group	Mbulo	Volcanic high island	−8.77111	158.2819	843.64	224
Solomon Islands	Main Group	Mbusana	Volcanic low island	−8.58306	157.3708	3.05	29
Solomon Islands	Main Group	Mbusihite	Reef island	−8.28389	157.5547	4.58	8
Solomon Islands	Main Group	Minjanga	Limestone high island	−8.68694	158.2164	576.71	51
Solomon Islands	Main Group	Moe	Volcanic high island	−9.1225	159.2897	166.26	33
Solomon Islands	Main Group	Mole (1)	Reef island	−8.28	157.1678	90.4	12
Solomon Islands	Main Group	Mole (2)	Volcanic low island	−6.82889	156.5178	68.72	28
Solomon Islands	Main Group	Naghotano	Volcanic high island	−8.91528	159.9647	28	34
Solomon Islands	Main Group	Nanango	Volcanic high island	−6.85639	156.9658	41.86	41
Solomon Islands	Main Group	Naruo (Rhodes Islands)	Volcanic low island	−7.7225	158.5742	105.1	8
Solomon Islands	Main Group	Ndokendoke	Reef island	−8.33917	157.4989	85.11	1
Solomon Islands	Main Group	Ndora	Volcanic high island	−8.30111	157.4872	1745.37	56
Solomon Islands	Main Group	Ndughiri	Volcanic low island	−8.44583	157.3983	9.87	10

(continued)

(continued)

Region/country	Island group	Island name	Island type	Latitude	Longitude	Area (ha)	Maximum elevation
Solomon Islands	Main Group	Nduhiri	Limestone low island	−8.34222	157.8517	165.26	15
Solomon Islands	Main Group	Ndurumena	Reef island	−8.24556	157.1125	69.64	4
Solomon Islands	Main Group	New	Limestone low island	−6.84833	156.1344	15.18	12
Solomon Islands	Main Group	New Georgia	Composite high island	−8.25528	157.5886	211751.5	851
Solomon Islands	Main Group	Ngarengare	Limestone low island	−8.63861	158.1156	19.29	25
Solomon Islands	Main Group	Nggatirana	Volcanic low island	−8.69583	157.8756	16.07	10
Solomon Islands	Main Group	Nggatokae (Gatukai)	Volcanic high island	−8.78778	158.1647	9956.12	887
Solomon Islands	Main Group	Nggela Pile	Volcanic high island	−9.13667	160.3197	13517.2	314
Solomon Islands	Main Group	Nggela Sule	Volcanic high island	−9.06194	160.2081	23325.57	274
Solomon Islands	Main Group	Nggulasa/Ngguiasa	Reef island	−8.29111	157.4183	39.24	1
Solomon Islands	Main Group	Ngongosila	Reef island	−8.77694	160.9411	6.65	2
Solomon Islands	Main Group	Nidero	Volcanic high island	−7.66917	158.4664	1099.76	55
Solomon Islands	Main Group	Nielai	Volcanic high island	−6.90778	155.93	70.69	127
Solomon Islands	Main Group	Njilatungu	Limestone low island	−8.41889	157.9122	10.67	8
Solomon Islands	Main Group	Njimiri (varu)/Njimin	Reef island	−8.01667	156.765	2.27	6
Solomon Islands	Main Group	Njingono	Reef island	−8.01444	156.7564	1.34	6

Solomon Islands	Main Group	Nuatambul	Volcanic high island	−7.125	157.1608	22.09	76
Solomon Islands	Main Group	Nudha	Reef island	−9.52917	160.7983	45.69	2
Solomon Islands	Main Group	Nughu	Reef island	−9.28556	160.3414	24.57	4
Solomon Islands	Main Group	Nusa Lokete	Reef island	−8.26972	157.1322	8.15	7
Solomon Islands	Main Group	Nusakova	Volcanic low island	−6.94778	155.9086	30.46	1
Solomon Islands	Main Group	Nusambekolo	Reef island	−8.28917	157.5306	5.79	9
Solomon Islands	Main Group	Nusarua	Reef island	−8.32639	157.34	265.38	4
Solomon Islands	Main Group	Nusave (Ilina)	Reef island	−6.92	155.8894	65.95	59
Solomon Islands	Main Group	Oema	Volcanic high island	−6.68194	156.0919	357.3	205
Solomon Islands	Main Group	Omona	Volcanic high island	−7.5225	158.6903	650.32	41
Solomon Islands	Main Group	O'orou	Reef island	−9.41444	161.3575	68.1	2
Solomon Islands	Main Group	Oroa (Philip)	Volcanic high island	−10.4825	161.5069	24.71	53
Solomon Islands	Main Group	Ovau	Volcanic high island	−6.79694	156.0147	1886.04	355
Solomon Islands	Main Group	Paleke	Reef island	−8.47778	158.0644	21.74	8
Solomon Islands	Main Group	Palutatamanaro	Volcanic low island	−7.78639	158.6264	145.08	22
Solomon Islands	Main Group	Papatura Fa	Reef island	−7.56222	158.7331	1488.25	2
Solomon Islands	Main Group	Papatura Ite	Reef island	−7.58472	158.7742	207.56	2
Solomon Islands	Main Group	Parama (2)	Volcanic low island	−6.68639	156.3969	35.8	25
Solomon Islands	Main Group	Paravoe	Reef island	−8.2125	157.1075	32.52	1

(continued)

(continued)

Region/country	Island group	Island name	Island type	Latitude	Longitude	Area (ha)	Maximum elevation
Solomon Islands	Main Group	Pareipoga	Reef island	−7.37722	158.1561	12.65	3
Solomon Islands	Main Group	Pavuvu	Volcanic high island	−9.08278	159.1064	13684.96	517
Solomon Islands	Main Group	Pelevo	Reef island	−8.48611	158.0728	5.76	1
Solomon Islands	Main Group	Pelosoe	Volcanic low island	−7.42639	157.4931	90.89	13
Solomon Islands	Main Group	Petani	Limestone high island	−8.33389	157.5575	320.37	39
Solomon Islands	Main Group	Pilo	Volcanic high island	−8.51389	159.8811	18.44	52
Solomon Islands	Main Group	Pio	Volcanic high island	−10.1747	161.6842	364.56	59
Solomon Islands	Main Group	Pioghi	Composite high island	−7.01222	155.8175	111.92	43
Solomon Islands	Main Group	Piraka	Reef island	−8.35833	157.3033	42.88	7
Solomon Islands	Main Group	Piru	Volcanic high island	−6.86111	156.1456	336.71	138
Solomon Islands	Main Group	Pirumeri	Volcanic high island	−7.11139	155.8833	385.86	41
Solomon Islands	Main Group	Popoghere	Limestone low island	−8.64	158.1367	20.22	15
Solomon Islands	Main Group	Popomunuana	Reef island	−8.06167	156.8292	1.35	6
Solomon Islands	Main Group	Poporang	Volcanic high island	−7.09472	155.8672	514.8	94
Solomon Islands	Main Group	Popotala/Popotaia	Volcanic high island	−6.80583	155.8328	24.52	50
Solomon Islands	Main Group	Popu	Volcanic high island	−7.40056	158.2111	1036.25	60

Solomon Islands	Main Group	Porepore	Limestone low island	−8.53778	158.1733	375.93	29
Solomon Islands	Main Group	Poro	Volcanic high island	−7.36278	157.1125	348.41	129
Solomon Islands	Main Group	Puruata	Volcanic low island	−6.24472	155.0297	31.63	17
Solomon Islands	Main Group	Putuo	Volcanic low island	−7.75583	158.5903	21.59	17
Solomon Islands	Main Group	Ramata	Limestone high island	−8.175	157.6536	194.84	33
Solomon Islands	Main Group	Ramos	Composite high island	−8.25861	160.1872	100.03	78
Solomon Islands	Main Group	Ranongga	Composite high island	−8.06333	156.5656	16146.5	869
Solomon Islands	Main Group	Rantan	Volcanic high island	−6.68306	155.9769	37.4	32
Solomon Islands	Main Group	Rauhi	Reef island	−9.87	160.8694	5.69	7
Solomon Islands	Main Group	Razu	Reef island	−8.28556	157.3972	17.01	1
Solomon Islands	Main Group	Renard	Volcanic high island	−8.58111	157.3831	46.48	60
Solomon Islands	Main Group	Rendova	Volcanic high island	−8.53861	157.2942	41550.93	974
Solomon Islands	Main Group	Rengge	Volcanic high island	−7.42667	157.4667	353.6	32
Solomon Islands	Main Group	Rereghana	Reef island	−8.30472	157.4367	686.82	7
Solomon Islands	Main Group	Rohae	Volcanic low island	−7.0025	156.0481	64.77	17
Solomon Islands	Main Group	Rokama Hite	Reef island	−8.28556	157.1786	21.99	8

(continued)

(continued)

Region/country	Island group	Island name	Island type	Latitude	Longitude	Area (ha)	Maximum elevation
Solomon Islands	Main Group	Rokoai	Reef island	−6.99778	155.8122	13.14	1
Solomon Islands	Main Group	Ropa	Volcanic high island	−7.36833	157.2044	217.11	93
Solomon Islands	Main Group	Roraimboko	Volcanic high island	−7.37083	157.2994	106.25	61
Solomon Islands	Main Group	Rovana	Volcanic high island	−8.13806	157.6406	244.95	36
Solomon Islands	Main Group	Rua Sura	Reef island	−9.50806	160.6161	190.33	3
Solomon Islands	Main Group	Ruingana	Volcanic high island	−6.97389	157.1117	40.24	38
Solomon Islands	Main Group	Ruria	Volcanic high island	−6.88194	156.0511	29.81	65
Solomon Islands	Main Group	Saerema	Volcanic high island	−7.34861	157.58	30.76	45
Solomon Islands	Main Group	Salakana	Volcanic low island	−7.44528	157.6544	84.2	27
Solomon Islands	Main Group	Samanagho	Volcanic low island	−6.97	156.0197	9.7	17
Solomon Islands	Main Group	Samarai (2)	Volcanic high island	−6.84333	156.1558	4.78	39
Solomon Islands	Main Group	Sambeke	Volcanic low island	−7.38306	157.2431	15.16	23
Solomon Islands	Main Group	San Cristobal (Makira)	Composite high island	−10.5544	161.7719	341340.6	1250
Solomon Islands	Main Group	San Jorge	Composite high island	−8.46556	159.5953	21037.06	454
Solomon Islands	Main Group	Sanihulumu	Limestone low island	−8.61944	158.1781	358.95	20

Solomon Islands	Main Group	Santa Ana	Limestone high island	−10.8344	162.4669	1762.64	153
Solomon Islands	Main Group	Santa Catalina	Limestone high island	−10.8925	162.4469	504.82	80
Solomon Islands	Main Group	Santa Isabel	Composite high island	−8.005	159.0869	388193.8	1219
Solomon Islands	Main Group	Sasaku	Volcanic high island	−7.36972	157.2331	42.67	45
Solomon Islands	Main Group	Sasare (Lieutenant)	Composite high island	−7.7075	158.5386	143.94	80
Solomon Islands	Main Group	Savanga	Reef island	−8.19417	157.0892	17.48	8
Solomon Islands	Main Group	Savo	Volcanic high island	−9.13389	159.8139	3237.01	485
Solomon Islands	Main Group	Sepo	Reef island	−8.04806	156.8178	3.72	5
Solomon Islands	Main Group	Sesehura	Volcanic low island	−7.8375	159.0192	96.25	2
Solomon Islands	Main Group	Shortland	Volcanic high island	−7.05583	155.7631	22422.41	185
Solomon Islands	Main Group	Sibau	Volcanic low island	−7.38472	158.1064	10.84	21
Solomon Islands	Main Group	Sigana	Volcanic low island	−8.51028	159.8642	5.46	23
Solomon Islands	Main Group	Sikaena	Volcanic high island	−7.35	157.1806	20.11	37
Solomon Islands	Main Group	Sikopo	Reef island	−7.445	157.9756	231.69	3
Solomon Islands	Main Group	Silapasope (Haycock)	Reef island	−7.475	157.8719	16.01	2
Solomon Islands	Main Group	Simbo	Volcanic high island	−8.29889	156.5339	99.38	70

(continued)

(continued)

Region/country	Island group	Island name	Island type	Latitude	Longitude	Area (ha)	Maximum elevation
Solomon Islands	Main Group	Simbo (Narovo)	Volcanic high island	−8.27167	156.5389	1245.1	335
Solomon Islands	Main Group	Sinevolo	Volcanic low island	−8.72472	158.1636	41.92	15
Solomon Islands	Main Group	Singgo	Volcanic high island	−8.10361	157.6325	180.62	38
Solomon Islands	Main Group	Sipozare	Volcanic low island	−6.70056	156.3961	99.54	27
Solomon Islands	Main Group	Sivata	Volcanic high island	−7.38083	157.5733	248.22	48
Solomon Islands	Main Group	Sivatae	Volcanic high island	−7.36278	157.1767	167.15	65
Solomon Islands	Main Group	Sivilua	Reef island	−7.01639	155.8375	1.32	1
Solomon Islands	Main Group	Siwairuka	Reef island	−9.81806	160.8386	35.07	9
Solomon Islands	Main Group	Soghonara	Volcanic high island	−8.97806	160.0217	112.84	86
Solomon Islands	Main Group	Sokara	Volcanic high island	−7.26333	157.3444	22.83	58
Solomon Islands	Main Group	Sokovoro	Volcanic low island	−6.645	156.6167	10.84	12
Solomon Islands	Main Group	Solokae	Limestone low island	−8.43833	157.9694	18.82	18
Solomon Islands	Main Group	Songonangona	Volcanic high island	−9.1236	160.155	21.03	53
Solomon Islands	Main Group	Stirling	Limestone high island	−7.41944	155.5628	617.82	52
Solomon Islands	Main Group	Sulei	Volcanic high island	−8.08056	159.535	289.85	73

Solomon Islands	Main Group	Susuku (Dillmore)	Volcanic high island	−7.44417	157.6339	481.15	56
Solomon Islands	Main Group	Taina	Volcanic high island	−9.13889	159.1397	132.09	60
Solomon Islands	Main Group	Tambiko	Volcanic low island	−8.56722	157.8494	44.28	21
Solomon Islands	Main Group	Tambumbirusoe	Volcanic high island	−7.41361	157.3628	24.66	31
Solomon Islands	Main Group	Tanambogo	Volcanic high island	−9.11111	160.185	6.54	37
Solomon Islands	Main Group	Tanambosu (Mondomondo)	Reef island	−8.26472	157.7928	166.75	9
Solomon Islands	Main Group	Tanasolo	Volcanic low island	−9.15528	160.3964	23.63	21
Solomon Islands	Main Group	Tanwoa	Volcanic high island	−5.57361	154.6608	322.36	169
Solomon Islands	Main Group	Taro	Reef island	−6.71083	156.3967	44.48	1
Solomon Islands	Main Group	Tatama	Limestone low island	−8.36833	157.8678	242.09	18
Solomon Islands	Main Group	Tatapuraka	Limestone low island	−8.33361	157.1861	37.62	10
Solomon Islands	Main Group	Taukuna	Volcanic low island	−7.04028	155.6144	41.98	15
Solomon Islands	Main Group	Tauna (2)	Volcanic high island	−6.86389	156.0947	96.11	100
Solomon Islands	Main Group	Taurato	Volcanic high island	−6.82694	155.8969	83	78
Solomon Islands	Main Group	Tava	Reef island	−8.32194	157.3261	13.6	1

(continued)

(continued)

Region/country	Island group	Island name	Island type	Latitude	Longitude	Area (ha)	Maximum elevation
Solomon Islands	Main Group	Tavanipupu	Reef island	−9.82667	160.8503	19.61	3
Solomon Islands	Main Group	Tekira	Volcanic low island	−8.57111	157.8603	5.51	5
Solomon Islands	Main Group	Telin	Volcanic high island	−9.10528	159.2636	222.18	31
Solomon Islands	Main Group	Tetepare	Limestone high island	−8.73861	157.5561	13101.44	400
Solomon Islands	Main Group	Toghovae	Volcanic high island	−8.07056	157.6283	60.98	40
Solomon Islands	Main Group	Towara'o (Tawa'ihi)	Volcanic high island	−9.835	160.84	201.36	46
Solomon Islands	Main Group	Tulaghi	Volcanic high island	−9.10167	160.1472	255.51	281
Solomon Islands	Main Group	Tupaerenge	Volcanic low island	−8.56278	157.8417	12.55	18
Solomon Islands	Main Group	Turovilu	Volcanic high island	−7.78611	156.5669	136.99	129
Solomon Islands	Main Group	Uepi	Reef island	−8.42222	157.9411	132.03	2
Solomon Islands	Main Group	Ufaon	Limestone low island	−9.07083	159.2467	141.86	19
Solomon Islands	Main Group	Uho	Volcanic high island	−6.83361	155.9733	63.46	73
Solomon Islands	Main Group	Uki Ni Masi	Volcanic high island	−10.2536	161.7386	4411.46	200
Solomon Islands	Main Group	Ulawa	Composite high island	−9.79528	161.9739	6761.6	181

Solomon Islands	Main Group	Uvinggete	Volcanic low island	-8.58611	157.8347	26.99	18
Solomon Islands	Main Group	Vaghena (Tasia)	Composite high island	-8.13889	159.6103	71.49	30
Solomon Islands	Main Group	Vaghena (Wagina)	Limestone high island	-7.43639	157.7644	8182.6	50
Solomon Islands	Main Group	Vaghidala Faa	Reef island	-7.60694	158.7556	60.74	6
Solomon Islands	Main Group	Vakao	Composite high island	-7.43583	158.2928	302.07	84
Solomon Islands	Main Group	Vangunu	Volcanic high island	-8.65056	157.995	54995.15	1040
Solomon Islands	Main Group	Varata	Limestone low island	-8.71944	158.1314	13.53	10
Solomon Islands	Main Group	Vatua	Reef island	-8.26333	157.1047	14.3	8
Solomon Islands	Main Group	Vavundamu	Limestone low island	-9.14417	160.4111	5.61	10
Solomon Islands	Main Group	Vealaviru (Rob Roy)	Composite high island	-7.41056	157.5942	7211.4	150
Solomon Islands	Main Group	Vella Lavella	Composite high island	-7.74194	156.6481	66156.51	780
Solomon Islands	Main Group	Vetenge	Reef island	-8.295	157.1492	37.66	9
Solomon Islands	Main Group	Vikenara/Vitora	Volcanic high island	-8.57389	159.8678	144.19	58
Solomon Islands	Main Group	Viketongana	Composite high island	-7.47	158.3414	186.47	55
Solomon Islands	Main Group	Vonavona (Parara)	Reef island	-8.21722	157.0697	7302.5	3
Solomon Islands	Main Group	Vulahana	Volcanic low island	-8.35	157.8336	19.04	12

(continued)

(continued)

Region/country	Island group	Island name	Island type	Latitude	Longitude	Area (ha)	Maximum elevation
Solomon Islands	Main Group	Vuraoto	Volcanic high island	−7.35111	157.1586	61.17	40
Solomon Islands	Main Group	Wahere	Volcanic high island	−9.82417	160.8139	387.6	139
Solomon Islands	Main Group	Wairokoi	Volcanic low island	−9.34389	161.1072	267.15	18
Solomon Islands	Main Group	Wickham	Volcanic low island	−8.7475	158.0647	34.82	24
Solomon Islands	Main Group	Yanuta	Volcanic high island	−10.3447	161.3611	102.35	84
Solomon Islands	Outliers	Avaha	Reef island	−5.16417	159.3872	29.87	6
Solomon Islands	Outliers	Bellona	Limestone high island	−11.2958	159.7906	2265.4	79
Solomon Islands	Outliers	Hale (Sikaiana)	Reef island	−8.41139	162.9411	160.57	3
Solomon Islands	Outliers	Ke Lla	Reef island	−5.45833	159.3781	380.98	3
Solomon Islands	Outliers	Kemalu	Reef island	−5.47444	159.4486	33.36	4
Solomon Islands	Outliers	Kukolu	Reef island	−5.48306	159.4794	138.27	9
Solomon Islands	Outliers	Luaniua	Reef island	−5.48111	159.7003	284.85	3
Solomon Islands	Outliers	Makalom	Reef island	−10.1686	166.2003	5.87	3
Solomon Islands	Outliers	Matuavi	Reef island	−8.42278	162.87	75.39	3
Solomon Islands	Outliers	Matuiloto	Reef island	−8.39694	162.8703	48.82	3
Solomon Islands	Outliers	Nalogo	Reef island	−10.0781	165.7442	17	3
Solomon Islands	Outliers	Pelau (Ontong Java)	Reef island	−5.09306	159.4047	179.63	3
Solomon Islands	Outliers	Rennell	Limestone high island	−11.6208	160.2039	83469.93	110
Solomon Islands	Outliers	Te Haole (Faore)	Reef island	−8.37917	162.8828	80.65	3

Solomon Islands	Outliers	Tuleki/Te Ako	Volcanic high island	−9.77583	167.0828	14.19	34
Solomon Islands	Santa Cruz Islands	Anuta	Volcanic high island	−11.61	169.8497	57.7	65
Solomon Islands	Santa Cruz Islands	Banie (Vanikoro)/Vanikolo	Volcanic high island	−11.6631	166.8844	19889.32	773
Solomon Islands	Santa Cruz Islands	Elingi (Obelisk)	Volcanic low island	−9.76694	167.0708	80.79	6
Solomon Islands	Santa Cruz Islands	Elo/Eio	Volcanic low island	−6.89389	155.8803	8.21	24
Solomon Islands	Santa Cruz Islands	Fenualoa	Limestone low island	−10.2303	166.31	765.17	10
Solomon Islands	Santa Cruz Islands	Kaa	Volcanic low island	−9.92139	167.2422	16.94	20
Solomon Islands	Santa Cruz Islands	Lakao	Volcanic high island	−9.80417	167.0986	333.36	200
Solomon Islands	Santa Cruz Islands	Lomlom	Limestone low island	−10.2769	166.3344	1657.36	8
Solomon Islands	Santa Cruz Islands	Loreva	Volcanic low island	−9.93111	167.2453	18.23	14
Solomon Islands	Santa Cruz Islands	Lua	Volcanic high island	−9.91444	167.2344	37.27	49
Solomon Islands	Santa Cruz Islands	Matema	Reef island	−10.2931	166.1842	32.14	8
Solomon Islands	Santa Cruz Islands	Matilavata/Mabiavata	Volcanic high island	−8.10111	157.6147	33.36	31
Solomon Islands	Santa Cruz Islands	Moluana Nggete	Volcanic low island	−8.58417	157.8539	49.45	23
Solomon Islands	Santa Cruz Islands	Mono	Composite high island	−7.3625	155.56	6888.72	31

(continued)

(continued)

Region/country	Island group	Island name	Island type	Latitude	Longitude	Area (ha)	Maximum elevation
Solomon Islands	Santa Cruz Islands	Nend./Nendo	Composite high island	−10.7281	165.9136	54214.92	549
Solomon Islands	Santa Cruz Islands	Ngawa	Limestone low island	−10.3056	166.3169	676.7	22
Solomon Islands	Santa Cruz Islands	Nibarga Temau	Limestone low island	−10.2664	166.3744	108.96	8
Solomon Islands	Santa Cruz Islands	Nifiloli (Niuve)/Nifiloli	Reef island	−10.1869	166.3	71.26	3
Solomon Islands	Santa Cruz Islands	Nukapu	Reef island	−10.0822	166.0503	38.74	3
Solomon Islands	Santa Cruz Islands	Nupani	Reef island	−10.0469	165.7206	55.73	3
Solomon Islands	Santa Cruz Islands	Pileni	Reef island	−10.1731	166.2475	36.81	3
Solomon Islands	Santa Cruz Islands	Taumako	Volcanic high island	−9.87778	167.1786	1089.77	366
Solomon Islands	Santa Cruz Islands	Teanu (Vanikolo)	Volcanic high island	−11.6333	166.9733	2374.85	851
Solomon Islands	Santa Cruz Islands	Tikopia	Volcanic high island	−12.2994	168.8231	644.08	374
Solomon Islands	Santa Cruz Islands	Tinakula	Volcanic high island	−10.3867	165.8056	1300.93	851
Solomon Islands	Santa Cruz Islands	Tomotu Neo (Te Motu)	Volcanic high island	−10.6786	165.8081	1325.61	82
Solomon Islands	Santa Cruz Islands	Tomotu Noi (Lord Howe)	Volcanic high island	−10.8103	166.0592	2020.61	117
Solomon Islands	Santa Cruz Islands	Tuleki (Anula)	Volcanic high island	−9.78028	167.0878	49.4	112
Solomon Islands	Santa Cruz Islands	Ulaka	Volcanic high island	−9.82111	167.1225	33.87	78

Solomon Islands	Santa Cruz Islands	Utupua	Volcanic high island	−11.2678	166.5233	7581.11	380
Tokelau	Outliers	Atafu	Reef island	−8.55829	−172.489	466.4	5
Tokelau	Outliers	Fakaofo	Reef island	−9.38	−171.219	589.05	5
Tokelau	Outliers	Nukunonu	Reef island	−9.2375	−171.86	550.97	5
Tonga	Eua Group	Eua	Limestone high island	−21.3842	−174.933	10480.67	312
Tonga	Eua Group	Kalau (1)	Limestone low island	−21.4706	−174.954	21.25	24
Tonga	Ha'apai Group	Foa	Limestone low island	−19.7489	−174.3	1571.78	24
Tonga	Ha'apai Group	Fotuha'a	Limestone high island	−19.8114	−174.722	134.53	45
Tonga	Ha'apai Group	Ha'ano	Limestone low island	−19.6731	−174.281	843.71	21
Tonga	Ha'apai Group	Hunga Ha'apai	Volcanic high island	−20.5503	−175.406	97.42	122
Tonga	Ha'apai Group	Hunga Tonga	Volcanic high island	−20.5369	−175.38	113.55	149
Tonga	Ha'apai Group	Kao	Volcanic high island	−19.6692	−175.018	1319.45	1033
Tonga	Ha'apai Group	Lifuka	Limestone low island	−19.8206	−174.348	1407.88	9
Tonga	Ha'apai Group	Limu (Uanukuhihifu)	Limestone low island	−19.9689	−174.498	12.85	11
Tonga	Ha'apai Group	Lofanga	Limestone low island	−19.8261	−174.551	127.79	3

(continued)

(continued)

Region/country	Island group	Island name	Island type	Latitude	Longitude	Area (ha)	Maximum elevation
Tonga	Ha'apai Group	Luahoko	Limestone low island	−19.6714	−174.394	9.8	2
Tonga	Ha'apai Group	Meama (1)	Limestone low island	−19.7558	−174.564	11.88	7
Tonga	Ha'apai Group	Mo'unga'one	Reef island	−19.6356	−174.483	169.67	3
Tonga	Ha'apai Group	Niniva	Limestone low island	−19.7675	−174.624	40.44	11
Tonga	Ha'apai Group	Nukunamo	Limestone low island	−19.7144	−174.276	12.89	12
Tonga	Ha'apai Group	Nukupule	Limestone low island	−19.7794	−174.537	5.01	2
Tonga	Ha'apai Group	Ofolanga	Reef island	−19.6003	−174.453	102.01	7
Tonga	Ha'apai Group	Tatafa	Limestone low island	−19.8747	−174.421	28.91	4
Tonga	Ha'apai Group	Tofanga	Limestone low island	−19.9586	−174.453	5.64	3
Tonga	Ha'apai Group	Tofua	Volcanic high island	−19.7481	−175.07	6223.32	515
Tonga	Ha'apai Group	Uanukuhahaki/ Uonukuhahake	Reef island	−19.9664	−174.485	46.3	4
Tonga	Ha'apai Group	Uiha	Limestone low island	−19.9047	−174.406	657.29	4
Tonga	Ha'apai Group	Uoleva	Limestone low island	−19.8489	−174.406	310.28	16
Tonga	Kotu Group	Fetoa	Limestone low island	−19.9653	−174.726	34.34	22
Tonga	Kotu Group	Fonuaika	Limestone low island	−20.1106	−174.702	3.39	9

Tonga	Kotu Group	Ha'afeva	Limestone low island	−19.9492	−174.712	135.45	14
Tonga	Kotu Group	Kito	Limestone low island	−19.9956	−174.787	7.67	2
Tonga	Kotu Group	Kotu	Limestone low island	−19.9506	−174.8	53.79	15
Tonga	Kotu Group	Lekeleka (2)	Limestone low island	−20.0675	−174.607	32.57	9
Tonga	Kotu Group	Luanamo	Limestone low island	−20.0211	−174.717	19.18	6
Tonga	Kotu Group	Matuku (2)	Limestone low island	−19.9597	−174.746	44.44	15
Tonga	Kotu Group	Nukulei	Limestone low island	−20.0508	−174.734	4.97	9
Tonga	Kotu Group	O'ua	Limestone high island	−20.0378	−174.681	78.76	30
Tonga	Kotu Group	Teaupa	Limestone low island	−19.975	−174.738	12.01	20
Tonga	Kotu Group	Tokulu	Limestone low island	−20.1019	−174.791	5.81	2
Tonga	Kotu Group	Tungua	Limestone low island	−20.0156	−174.766	163.98	22
Tonga	Nomuka Group	Fonoifua	Limestone low island	−20.2786	−174.629	36.67	18
Tonga	Nomuka Group	Kelefesia	Composite high island	−20.5036	−174.736	31.97	38
Tonga	Nomuka Group	Mango	Composite high island	−20.3292	−174.711	111.88	31

(continued)

(continued)

Region/country	Island group	Island name	Island type	Latitude	Longitude	Area (ha)	Maximum elevation
Tonga	Nomuka Group	Meama (2)	Limestone low island	−20.2686	−174.646	3.56	1
Tonga	Nomuka Group	Nomuka	Limestone high island	−20.2519	−174.796	1005.97	30
Tonga	Nomuka Group	Nomukeiki	Volcanic low island	−20.2836	−174.805	87.2	9
Tonga	Nomuka Group	Nuku (1)	Limestone low island	−20.4778	−174.76	1.81	2
Tonga	Nomuka Group	Tonumea	Limestone high island	−20.46	−174.761	43.73	42
Tonga	Otu Tolu Group	Fetokopunga	Limestone low island	−20.3036	−174.535	3.9	3
Tonga	Otu Tolu Group	Lalona	Reef island	−20.3481	−174.521	49.24	3
Tonga	Otu Tolu Group	Telekitonga	Reef island	−20.3967	−174.531	77.87	8
Tonga	Otu Tolu Group	Telekivava'u	Limestone low island	−20.3153	−174.521	36.57	3
Tonga	Outliers	Ata (Pylstaart)	Volcanic high island	−22.3325	−176.207	343.08	355
Tonga	Outliers	Fonualei	Volcanic high island	−18.0231	−174.317	414.58	183
Tonga	Outliers	Hakautu'utu'u	Limestone low island	−15.9347	−173.781	3.36	5
Tonga	Outliers	Hunganga	Limestone low island	−15.9531	−173.804	96.18	18
Tonga	Outliers	Niuafo'ou	Volcanic high island	−15.6022	−175.637	5655.14	260
Tonga	Outliers	Niuatoputapu	Composite high island	−15.9606	−173.778	1863.08	107

Tonga	Outliers	Tafahi	Volcanic high island	-15.8525	-173.749	420.15	610
Tonga	Outliers	Toku	Volcanic low island	-18.1619	-174.179	39.6	8
Tonga	Tongatapu Group	Ata/Ataa	Reef island	-21.0564	-175.002	14.22	5
Tonga	Tongatapu Group	Atata	Limestone low island	-21.0475	-175.254	79.31	2
Tonga	Tongatapu Group	Eua Iki/Euaiki	Limestone high island	-21.1178	-174.98	172.18	55
Tonga	Tongatapu Group	Fafaa	Reef island	-21.0875	-175.159	5.74	3
Tonga	Tongatapu Group	Fukave	Reef island	-21.0931	-175.031	10.49	9
Tonga	Tongatapu Group	Kanatea	Limestone low island	-21.1653	-175.21	30.75	2
Tonga	Tongatapu Group	Makaha'a	Reef island	-21.1144	-175.154	6.85	3
Tonga	Tongatapu Group	Manima	Reef island	-21.125	-175.149	2.28	2
Tonga	Tongatapu Group	Mata'aho	Reef island	-21.1613	-175.154	4.08	8
Tonga	Tongatapu Group	Monuafe	Reef island	-21.105	-175.14	6.95	4
Tonga	Tongatapu Group	Motutapu	Reef island	-21.0917	-175.057	8.81	2
Tonga	Tongatapu Group	Nuku (2)	Reef island	-21.0881	-175.023	4.94	2
Tonga	Tongatapu Group	Nukunukumotu	Reef island	-21.1403	-175.146	109.86	9
Tonga	Tongatapu Group	Oneata (2)	Reef island	-21.1294	-175.146	3.24	3
Tonga	Tongatapu Group	Onevai	Reef island	-21.0861	-175.113	28.91	3
Tonga	Tongatapu Group	Onevao	Reef island	-21.0917	-175.098	9.75	2
Tonga	Tongatapu Group	Pangaimotu	Limestone low island	-21.1242	-175.159	8.87	8
Tonga	Tongatapu Group	Polo'a	Limestone low island	-21.09	-175.244	11.71	8
Tonga	Tongatapu Group	Talakite	Reef island	-21.1594	-175.155	8.97	4

(continued)

(continued)

Region/country	Island group	Island name	Island type	Latitude	Longitude	Area (ha)	Maximum elevation
Tonga	Tongatapu Group	Tau	Reef island	−21.0214	−175.005	10.55	1
Tonga	Tongatapu Group	Toketoke	Limestone low island	−21.0625	−175.295	7.82	2
Tonga	Tongatapu Group	Tongatapu	Limestone high island	−21.1772	−175.18	30468.2	82
Tonga	Tongatapu Group	Tufaka	Reef island	−21.0661	−175.254	4.13	3
Tonga	Vava'u Group	A'a	Limestone high island	−18.705	−174.046	50.87	40
Tonga	Vava'u Group	Afo	Limestone low island	−18.7069	−173.997	10.07	20
Tonga	Vava'u Group	Euaeiki	Limestone high island	−18.7669	−174.019	22.33	38
Tonga	Vava'u Group	Euakafa	Limestone high island	−18.7508	−174.036	69.85	82
Tonga	Vava'u Group	Faioa (1)	Limestone low island	−18.6656	−173.923	54.54	29
Tonga	Vava'u Group	Fangasito	Limestone low island	−18.8192	−174.081	14.11	18
Tonga	Vava'u Group	Foeata (Fo'iata)	Limestone low island	−18.7167	−174.137	37.16	28
Tonga	Vava'u Group	Fofoa/Fofua	Limestone high island	−18.7031	−174.137	130.95	78
Tonga	Vava'u Group	Fonua'one'one	Limestone low island	−18.8181	−174.062	13.1	14
Tonga	Vava'u Group	Hunga	Limestone high island	−18.6806	−174.118	609.82	75
Tonga	Vava'u Group	Kalau (2)	Limestone low island	−18.6925	−174.136	12.65	4

Tonga	Vava'u Group	Kapa	Limestone high island	-18.7111	-174.027	665.16	96
Tonga	Vava'u Group	Katafanga	Limestone high island	-18.7386	-174.041	7.5	46
Tonga	Vava'u Group	Kenutu	Limestone low island	-18.6964	-173.926	63	8
Tonga	Vava'u Group	Koloa	Limestone high island	-18.6447	-173.928	266.59	41
Tonga	Vava'u Group	Kulo	Limestone low island	-18.7172	-174.098	3.3	14
Tonga	Vava'u Group	Lape	Limestone low island	-18.7194	-174.083	44.66	7
Tonga	Vava'u Group	Late	Volcanic high island	-18.8056	-174.647	2335.71	518
Tonga	Vava'u Group	Lekeleka (1)	Limestone low island	-18.7581	-173.998	2.29	1
Tonga	Vava'u Group	Luahiapo	Limestone low island	-18.7994	-174.04	3.3	1
Tonga	Vava'u Group	Luakapa	Limestone low island	-18.6925	-174.046	8.79	17
Tonga	Vava'u Group	Luamoko	Limestone low island	-18.6842	-174.104	14.43	3
Tonga	Vava'u Group	Mafana	Limestone low island	-18.6833	-173.953	57.68	4
Tonga	Vava'u Group	Mala	Limestone low island	-18.6911	-174.021	16.9	4
Tonga	Vava'u Group	Maninita	Limestone low island	-18.8575	-173.996	9.52	13

(continued)

(continued)

Region/country	Island group	Island name	Island type	Latitude	Longitude	Area (ha)	Maximum elevation
Tonga	Vava'u Group	Mounu (1)	Limestone low island	−18.7511	−174.068	8.23	4
Tonga	Vava'u Group	Mu'omu'a	Limestone low island	−18.7933	−174.111	62.99	13
Tonga	Vava'u Group	Nuapapu	Limestone high island	−18.6967	−174.074	464.95	64
Tonga	Vava'u Group	Nuku (3)	Limestone low island	−18.715	−174.043	5.82	12
Tonga	Vava'u Group	Ofu (2)	Limestone low island	−18.6992	−173.962	209.49	9
Tonga	Vava'u Group	Okoa	Limestone low island	−18.6542	−173.949	38.82	21
Tonga	Vava'u Group	Oto	Limestone high island	−18.7017	−174.056	64.79	47
Tonga	Vava'u Group	Ovaka	Limestone high island	−18.7442	−174.101	215.2	34
Tonga	Vava'u Group	Ovalau (Avalau)	Limestone high island	−18.7508	−174.078	28.1	626
Tonga	Vava'u Group	Sisia	Limestone low island	−18.7306	−174.052	27.31	21
Tonga	Vava'u Group	Tapana	Limestone low island	−18.7147	−173.992	79.91	9
Tonga	Vava'u Group	Taula	Limestone low island	−18.8478	−174.011	22.18	14
Tonga	Vava'u Group	Taunga	Limestone high island	−18.7444	−174.01	103.37	38
Tonga	Vava'u Group	Tauta	Limestone low island	−18.745	−174	5.88	18

Tonga	Vava'u Group	Totokafonua	Limestone low island	-18.7583	-174.124	18.47	29
Tonga	Vava'u Group	Umuna	Limestone high island	-18.6842	-173.923	90.52	38
Tonga	Vava'u Group	Utungake	Limestone high island	-18.6783	-174.023	131.58	88
Tonga	Vava'u Group	Vaka'eitu	Limestone high island	-18.7247	-174.104	166.32	60
Tonga	Vava'u Group	Vava'u	Limestone high island	-18.6833	-174.001	12719.52	88
Tuvalu	Outliers	Funafuti	Reef island	-8.52	179.1994	1026.78	3
Tuvalu	Outliers	Moriapepe (Lakina)	Reef island	-5.64722	176.0683	163.35	8
Tuvalu	Outliers	Nanumanga	Reef island	-6.28833	176.3206	329.45	3
Tuvalu	Outliers	Nanumea	Reef island	5.6825	176.1272	357.75	3
Tuvalu	Outliers	Niulakita	Reef island	-10.7281	179.4556	154.34	5
Tuvalu	Outliers	Niutao	Reef island	-6.10833	177.3425	286.7	3
Tuvalu	Outliers	Nui (Fenua Tapu)	Reef island	-7.24861	177.1542	421.99	3
Tuvalu	Outliers	Nukulaelae (Fangaua)	Reef island	-9.37222	179.8092	362.82	3
Tuvalu	Outliers	Savave/Nukufetau	Reef island	-8.02889	178.3142	564.63	3
Tuvalu	Outliers	Vaitupu	Reef island	-7.47861	178.6783	779.6	5
US Administered Islands in Central Pacific	Line Islands	Jarvis Island	Reef island	-0.37389	-159.998	438.17	7
US Administered Islands in Central Pacific	Line Islands	Kingman Reef	Reef island	6.433056	-162.387	226.77	1

(continued)

(continued)

Region/country	Island group	Island name	Island type	Latitude	Longitude	Area (ha)	Maximum elevation
US Administered Islands in Central Pacific	Line Islands	Palmyra Atoll	Reef island	5.878611	−162.073	797.49	2
US Administered Islands in Central Pacific	Outliers	Baker Island	Reef island	0.194722	−176.479	190.33	8
US Administered Islands in Central Pacific	Outliers	Howland Island	Reef island	0.808056	−176.613	232.43	3
US Administered Islands in Central Pacific	Outliers	Johnston Atoll	Reef island	16.73222	−169.531	322.12	5
US Administered Islands in Central Pacific	Outliers	Midway Atoll	Reef island	28.21	−177.376	645.8	4
US Administered Islands in Central Pacific	Outliers	Wake Island	Reef island	19.28528	166.6497	841.35	6
Vanuatu	Banks Islands	Kwakea	Limestone low island	−13.88	167.5981	104.51	27
Vanuatu	Banks Islands	Mere Lava	Volcanic high island	−14.4611	168.0431	1132.16	883
Vanuatu	Banks Islands	Merig	Volcanic high island	−14.31	167.7969	53.59	125
Vanuatu	Banks Islands	Mota	Volcanic high island	−13.8497	167.6947	1371.49	411
Vanuatu	Banks Islands	Mota Lava	Volcanic high island	−13.6742	167.6703	3038.57	411

Vanuatu	Banks Islands	Rah (Ra)	Reef island	−13.7172	167.63	51.71	3
Vanuatu	Banks Islands	Ravenga	Volcanic high island	−13.7881	167.5508	21.92	250
Vanuatu	Banks Islands	Santa Maria (Gaua)	Volcanic high island	−14.2586	167.5125	36505.91	797
Vanuatu	Banks Islands	Ureparapara	Volcanic high island	−13.5378	167.3272	4089.05	764
Vanuatu	Banks Islands	Vanua Lava	Volcanic high island	−13.8117	167.4692	34898.91	946
Vanuatu	Banks Islands	Vot Tande	Volcanic high island	−13.2592	167.6428	13.62	64
Vanuatu	Main Group	Aesi	Limestone high island	−15.4322	167.2542	834.95	36
Vanuatu	Main Group	Ambrym	Volcanic high island	−16.2728	168.1239	74805.42	1270
Vanuatu	Main Group	Anatom (Aneityum)	Volcanic high island	−20.19	169.8217	17928.56	852
Vanuatu	Main Group	Aniwa	Limestone high island	−19.2467	169.6028	2189.6	42
Vanuatu	Main Group	Aoba (Ambae)	Volcanic high island	−15.3861	167.8394	45737.23	1496
Vanuatu	Main Group	Aore	Limestone high island	−15.5822	167.1794	6568.39	99
Vanuatu	Main Group	Araki	Limestone high island	−15.6339	166.9547	302.22	227
Vanuatu	Main Group	Asuleka	Limestone low island	−15.6278	167.1881	24.32	14
Vanuatu	Main Group	Atchin	Volcanic low island	−15.9406	167.3431	63.28	18

(continued)

(continued)

Region/country	Island group	Island name	Island type	Latitude	Longitude	Area (ha)	Maximum elevation
Vanuatu	Main Group	Awei	Volcanic high island	−16.5361	167.7736	35.82	88
Vanuatu	Main Group	Bokissa	Limestone low island	−15.5914	167.2456	75.2	22
Vanuatu	Main Group	Efate	Composite high island	−17.6819	168.3911	99278.3	647
Vanuatu	Main Group	Emae	Volcanic high island	−17.0678	168.3756	3829.71	644
Vanuatu	Main Group	Emao	Volcanic high island	−17.4819	168.4883	871.49	416
Vanuatu	Main Group	Epi	Volcanic high island	−16.7311	168.2233	49601.73	833
Vanuatu	Main Group	Eretoka	Volcanic high island	−17.64	168.1533	90.54	96
Vanuatu	Main Group	Erromango	Volcanic high island	−18.8244	169.1358	99763.17	886
Vanuatu	Main Group	Espiritu Santo	Composite high island	−15.3942	166.9381	435490	1879
Vanuatu	Main Group	Ewose	Volcanic high island	−16.9592	168.5786	149.87	319
Vanuatu	Main Group	Falea	Volcanic high island	−16.9903	168.6014	52.09	100
Vanuatu	Main Group	Futuna (2)	Composite high island	−19.53	170.2181	1101.48	641
Vanuatu	Main Group	Ifira	Limestone low island	−17.7478	168.2953	44.85	25
Vanuatu	Main Group	Ile Buninga	Volcanic high island	−17.0239	168.585	139.33	216

Vanuatu	Main Group	Ile Ratua/Ratua	Composite low island	−15.6131	167.1822	80.05	20
Vanuatu	Main Group	Inyeung (Mystery)	Reef island	−20.2489	169.7711	12.49	7
Vanuatu	Main Group	Iririki	Volcanic high island	−17.7447	168.3094	7.85	43
Vanuatu	Main Group	Kakula	Composite low island	−17.5172	168.4058	19.4	16
Vanuatu	Main Group	Laika	Volcanic high island	−16.8381	168.5494	61.29	87
Vanuatu	Main Group	Lelepa	Volcanic high island	−17.5992	168.2108	876.54	202
Vanuatu	Main Group	Lembong (Ilot Bagatelle)	Volcanic high island	−16.5167	167.7833	65.14	67
Vanuatu	Main Group	Lopevi	Volcanic high island	−16.5086	168.3433	3157.26	1413
Vanuatu	Main Group	Maewo	Composite high island	−15.1736	168.1347	32699.84	811
Vanuatu	Main Group	Makira (Makura)	Volcanic high island	−17.1358	168.4375	232.75	297
Vanuatu	Main Group	Malakula	Composite high island	−16.2464	167.4831	224019.5	863
Vanuatu	Main Group	Malo	Composite high island	−15.6847	167.1714	20070.28	326
Vanuatu	Main Group	Mataso	Volcanic high island	−17.2531	168.4264	210.55	494
Vanuatu	Main Group	Mavea	Limestone high island	−15.3831	167.2292	516.58	48

(continued)

(continued)

Region/country	Island group	Island name	Island type	Latitude	Longitude	Area (ha)	Maximum elevation
Vanuatu	Main Group	Moso (Tranquility)	Volcanic high island	−17.5336	168.2564	2919.75	109
Vanuatu	Main Group	Nguna	Volcanic high island	−17.4542	168.3556	3146.15	472
Vanuatu	Main Group	Paama	Volcanic high island	−16.4728	168.2406	3648.78	544
Vanuatu	Main Group	Pele	Volcanic high island	−17.4939	168.4067	493.11	184
Vanuatu	Main Group	Pentecost	Composite high island	−15.7617	168.1933	53546.87	946
Vanuatu	Main Group	Pilat (Lataro)	Limestone high island	−15.2653	167.1875	334.42	98
Vanuatu	Main Group	Pilotin	Limestone high island	−15.2889	167.2152	166.77	65
Vanuatu	Main Group	Rano	Volcanic low island	−15.9875	167.3906	126.59	20
Vanuatu	Main Group	Sakao (Khoti)	Volcanic high island	−16.4975	167.8225	425.19	102
Vanuatu	Main Group	Sakao/Sakau	Limestone high island	−14.9644	167.1322	1391.35	108
Vanuatu	Main Group	Tanna	Volcanic high island	−19.4997	169.3369	63528.37	1084
Vanuatu	Main Group	Tefala	Volcanic high island	−16.8144	168.5344	11.13	93
Vanuatu	Main Group	Thion	Limestone high island	−15.0364	167.0933	312.74	202
Vanuatu	Main Group	Tomman	Volcanic high island	−16.5908	167.4744	244.36	84

Vanuatu	Main Group	Tongariki	Volcanic high island	-17.005	168.6272	638.26	521
Vanuatu	Main Group	Tongoa	Volcanic high island	-16.8159	168.5452	4569.9	487
Vanuatu	Main Group	Tutuba	Limestone high island	-15.5742	167.2808	1446.01	36
Vanuatu	Main Group	Ulilapo	Limestone high island	-15.5931	166.9908	53.93	36
Vanuatu	Main Group	Uliveo (Maskelyne)	Volcanic low island	-16.5242	167.83	44.07	25
Vanuatu	Main Group	Uri	Volcanic low island	-16.0989	167.4425	140.19	15
Vanuatu	Main Group	Uripiv	Volcanic high island	-16.0719	167.4517	105.8	38
Vanuatu	Main Group	Vao	Reef island	-15.8986	167.3094	134.11	4
Vanuatu	Main Group	Vula/Vulai	Volcanic high island	-16.5489	167.783	112.76	83
Vanuatu	Main Group	Wala	Volcanic low island	-15.9733	167.3786	60.81	26
Vanuatu	Reef Islands	Enwut (Ile Wosou)	Reef island	-13.6253	167.5222	42.13	2
Vanuatu	Reef Islands	Rowa	Reef island	-13.6111	167.525	115.98	3
Vanuatu	Reef Islands	Wosu	Reef island	-13.5972	167.5347	13.7	2
Vanuatu	Torres Islands	Hiu (Hiw)	Composite high island	-13.1325	166.5669	5246.9	366
Vanuatu	Torres Islands	Linua	Composite low island	-13.3258	166.6333	392.17	8
Vanuatu	Torres Islands	Loh	Composite high island	-13.3472	166.6442	1474.02	155

(continued)

(continued)

Region/country	Island group	Island name	Island type	Latitude	Longitude	Area (ha)	Maximum elevation
Vanuatu	Torres Islands	Metoma	Volcanic high island	−13.2067	166.6025	260.84	115
Vanuatu	Torres Islands	Tegua	Composite high island	−13.2458	166.6214	3195.12	254
Vanuatu	Torres Islands	Toga	Composite high island	−13.4169	166.6939	1986.9	240
Wallis and Futuna	Hoorn Islands (Futuna Islands)	Alofi	Volcanic high island	−14.3406	−178.041	3262.88	400
Wallis and Futuna	Hoorn Islands (Futuna Islands)	Futuna (1)	Volcanic high island	−14.2903	−178.119	6847.56	509
Wallis and Futuna	Wallis	Faioa (2)	Reef island	−13.3781	−176.171	91.88	8
Wallis and Futuna	Wallis	Fenua Fo'ou	Reef island	−13.3883	−176.214	5.72	5
Wallis and Futuna	Wallis	Fugalei	Volcanic high island	−13.2811	−176.146	26.97	60
Wallis and Futuna	Wallis	Luaniva	Volcanic high island	−13.2725	−176.148	27.03	50
Wallis and Futuna	Wallis	Nukuatea	Volcanic high island	−13.3744	−176.216	102.22	73
Wallis and Futuna	Wallis	Nukufotu	Limestone low island	−13.1844	−176.204	7.42	27
Wallis and Futuna	Wallis	Nukuhifala	Reef island	−13.2883	−176.126	9.07	8
Wallis and Futuna	Wallis	Nukuhione	Reef island	−13.2736	−176.131	5.75	7
Wallis and Futuna	Wallis	Nukuloa	Limestone low island	−13.1911	−176.193	63.26	12
Wallis and Futuna	Wallis	Nukutapu	Reef island	−13.2192	−176.168	9.98	7
Wallis and Futuna	Wallis	Nukuteatea	Reef island	−13.2042	−176.179	20.72	8
Wallis and Futuna	Wallis	Uvea (Wallis)	Volcanic high island	−13.295	−176.208	8533.63	145

References

Anthony SS (2004) Hydrogeology of selected islands of the Federated States of Micronesia. Dev Sedimentol 54:693–706

Ballmer MD, Ito G, Hunen J, Tackley PJ (2011) Spatial and temporal variability in Hawaiian hotspot volcanism induced by small-scale convection. Nat Geosci 4:457–460

Betzold C (2015) Adapting to climate change in small island developing states. Clim Chang 133:481–489

Bloom AL (1970) Holocene submergence in Micronesia as the standard for eustatic sea-level changes. Quaternaria 12:145–154

Bonatti E, Harrison C, Fisher D, Honnorez J, Schilling JG, Stipp J, Zentilli M (1977) Easter volcanic chain (southeast Pacific): a mantle hot line. J Geophys Res 82:2457–2478

Bonvallot J, Dupon J-F, Vigneron E, Gay J-C, Morhange C, Ollier C, Peugniez G, Reitel B, Yon-Cassat F (1993) Atlas de la Polynésie Française. ORSTOM, Paris

Brocher TM (ed) (1985) Geological investigations of the Northern Melanesian Borderland. Circum-Pacific Council for Energy and Mineral Resources, Houston

Calmant S, Cabioch G, Regnier M, Pillet R, Pelletier B (1999) Cosismic uplifts and interseismic subsidence recorded in corals at Malekula (Vanuatu, southwest Pacific). Comp R Acad Sciences 328:711–716

Cluzell D, Maurizot P, Collot J, Sevin B (2012) An outline of the geology of New Caledonia; from Permian-Mesozoic Southeast Gondwanaland active margin to Cenozoic obduction and supergene evolution. Episodes 35:72–86

Coleman P (1970) Geology of the Solomon and New Hebrides Islands, as part of the Melanesian re-entrant Southwest Pacific. Pacific Sci 24:289–314

Connell J (2013) Islands at risk? Environments, economies and contemporary change. Edward Elgar, Cheltenham

Dahl AL (1980) Regional ecosystems survey of the South Pacific area. Technical Paper 179. South Pacific Commission, Noumea

Dana JD (1875) Corals and coral islands. Sampson Low, London

Davis WM (1920) The islands and coral reefs of Fiji. Geogr J 55(34–45):200–220. 377–388

Derrick RA (1957) The Fiji Islands: a geographical handbook. Government Press, Suva

Dickinson WR (2004) Impacts of eustasy and hydro-isostasy on the evolution and landforms of Pacific atolls. Palaeogeogr Palaeoclimatol Palaeoecol 213:251–269

Dow DB (1977) A geological synthesis of Papua New Guinea. Australian Government Publishing Service, Australia

Duncan RA, McDougall I (1976) Linear volcanism in French Polynesia. J Volcanol Geotherm Res 1:198–227

Ferry J, Kumar PB, Bronders J, Lewis J (2004) Hydrogeology of carbonate islands of Fiji. In: Vacher HL, Quinn TM (eds) Geology and hydrogeology of carbonate Islands. Elsevier, Amsterdam

Furness L (2004) Hydrogeology of carbonate islands of the Kingdom of Tonga. In: Vacher HL, Quinn TM (eds) Geology and hydrogeology of carbonate Islands. Elsevier, Amsterdam

Gillespie RG, Clague DA (2009) Encyclopedia of Islands. University of California Press, Berkeley

Gillis JR (2014) Not continents in miniature: Islands as ecotones. Island Stud J 9:155–166

Greene HG, Wong FL (eds) (1988) Geology and offshore resources of Pacific Island Arcs – Vanuatu region. Circum-Pacific Council for Energy and Mineral Resources, Houston

Grossman EE, Fletcher CH III, Richmond BM (1998) The Holocene sea-level highstand in the equatorial Pacific: analysis of the insular paleosea-level database. Coral Reefs 17:309–327

Herzberg C (2011) Identification of Source Lithology in the Hawaiian and Canary Islands: implications for Origins. J Petrol 52:113–146

Hill PJ, Jacobson G (1989) Structure and evolution of Nauru Island, central Pacific Ocean. Aust J Earth Sci 36:365–381

Jost C (ed) (1998) The French-speaking Pacific. Boombana, Mount Nebo

Karolle BG (1993) Atlas of micronesia. Bess Press, Honolulu

Kayanne H, Yamano H, Randall RH (2002) Holocene sea-level changes and barrier reef formation on an oceanic island, Palau Islands, western Pacific. Sediment Geol 150:47–60

Keating BH (1992) The geology of the Samoan Islands. In: Keating BH, Bolton BR (eds) Geology and offshore resources of the Central Pacific Basin. Springer, New York

Keating BH, Bolton BR (eds) (1992) Geology and offshore mineral resources of the Central Pacific Basin. Springer, New York

Kench PS, Owen SD, Ford MR (2014) Evidence for coral island formation during rising sea level in the central Pacific Ocean. Geophys Res Lett 41:820–827

King SD, Adam C (2014). Hotspot swells revisited. *Physics of the Earth and Planetary Interiors*, 235:66–83.

Langdon R (1978) American whalers and traders in the Pacific: a guide to records on microfilm. Pacific Manuscripts Bureau, Canberra

Li Y-H (1988) Denudation rates of the Hawaiian Islands by rivers and groundwaters. Pac Sci 42:253–266

Lobban CS, Schefter M (1997) Tropical Pacific Island environments. University of Guam Press, Mangilao

Macdonald GA, Abbott AT, Peterson FL (1983) Volcanoes in the sea: the geology of Hawaii. University of Hawaii Press, Honolulu

Mcbirney AR, Williams H, Aoki K (1969) Geology and petrology of the Galápagos Islands. Geological Society of America, Boulder

McLean RF, Hosking PL (1991) Geomorphology of reef islands and atoll motu in Tuvalu. South Pacific J Nat Sci 11:167–189

McLean R, Kench P (2015) Destruction or persistence of coral atoll islands in the face of 20th and 21st century sea-level rise? Wiley Interdiscip Rev-Clim Change 6:445–463

McNamara KE (2013) Taking stock of community-based climate-change adaptation projects in the Pacific. Asia Pac Viewp 54:398–405

McNutt MK, Fischer KM (1987) The South Pacific Superswell. In: Keating BH, Fryer P, Batiza R, Boehlert GW (eds) Seamounts, islands, and atolls. American Geophysical Union, Washington

Menard HW (1983) Insular erosion, isostasy, and subsidence. Science 220:913–918

Menard HW (1986) Islands. Scientific American, New York

Mimura N, Nurse L, McLean RF, Agard J, Briguglio L, Lefale P, Payet R, Sem G (2007) Small islands. In: Parry ML, Canziani OF, Palutikof JP, Van der Linden PJ, Hanson CE (eds) Climatechange 2007: impacts, adaptation and vulnerability. Contribution of Working Group II to the Fourth Assessment Report of the Intergovernmental Panel on Climate Change. Cambridge University Press, Cambridge

Motteler LS (2006) Pacific Island names: a map and name guide to the New Pacific. B.P. Bishop Museum Press, Honolulu

Mueller-Dombois D, Fosberg FR (1998) Vegetation of the tropical Pacific Islands. Springer, New York

Neall VE, Trewick SA (2008) The age and origin of the Pacific islands: a geological overview. Philos Trans Royal Soc B: Biol Sci 363:3293–3308

Neef G, McCulloch M (2001) Pliocene–quaternary history of Futuna Island, south Vanuatu south-west Pacific. Aust J Earth Sci 48:805–814

Nunn PD (1994) Oceanic Islands. Blackwell, Oxford

Nunn PD (1995) Lithospheric flexure in Southeast Fiji consistent with the tectonic history of islands in the Yasayasa Moala. Aust J Earth Sci 42:377–389

Nunn PD (1998a) Late Cenozoic emergence of the islands of the northern Lau-Colville Ridge, southwest Pacific. Geological Society, London

Nunn PD (1998b) Pacific Island landscapes: landscape and geological development of Southwest Pacific Islands, especially Fiji, Samoa and Tonga. The University of the South Pacific, Suva

Nunn PD (1999) Environmental change in the Pacific Basin: chronologies, causes, consequences. Wiley, New York

Nunn PD (2009a) Responding to the challenges of climate change in the Pacific Islands: management and technological imperatives. Clim Res 40:211–231

Nunn PD (2009b) Vanished Islands and hidden continents of the Pacific. University of Hawai'i
Press, Honolulu

Nunn PD, Britton JMR (2004) The long-term evolution of Niue Island. In: TERRY J, MURRAY
W (eds) Geographical perspectives on the rock of polynesia. INSULA, Paris

Nunn PD, Kumar R (2017) Understanding climate-human interactions in Small Island Developing
States (SIDS): implications for future livelihood sustainability. Int J Clim Change Strategies
Manage 10(2):245–271

Nunn PD, Kumar L, Eliot I, McLean RF (2016) Classifying Pacific islands. Geosci Lett 3:1–19

Nurse L, McLean R, Agard J, Briguglio LP, Duvat V, Pelesikoti N, Tompkins E, Webb A (2014)
Small islands. In: Barros VR, Field CB, Dokken DJ, Mastrandrea MD, Mach KJ, Bilir TE,
Chatterjee M, Ebi KL, Estrada YO, Genova RC, Girma B, Kissel ES, Levy AN, Maccracken S,
Mastrandrea PR, White LL (eds) Climate change 2014: impacts, adaptation, and vulnerability.
Part B: Regional aspects. Contribution of Working Group II to the Fifth Assessment Report
of the Intergovernmental Panel on Climate Change. Cambridge University Press, Cambridge

O'Connor JM, Steinberger B, Regelous M, Koppers AAP, Wijbrans JR, Haase KM, Stoffers P,
Jokat W, Garbe-Schonberg D (2013) Constraints on past plate and mantle motion from new
ages for the Hawaiian-Emperor Seamount Chain. Geochem Geophy Geosyst 14:4564–4584

Pirazzoli PA, Montaggioni LF (1986) Late Holocene sea-level changes in the northwest Tuamotu
Islands, French Polynesia. Quat Res 25:350–368

Ramalho RS, Quartau R, Trenhaile AS, Mitchell NC, Woodroffe CD, Ávila SP (2013) Coastal
evolution on volcanic oceanic islands: a complex interplay between volcanism, erosion, sedimentation, sea-level change and biogenic production. Earth Sci Rev 127:140–170

Rapaport M (ed) (2013) The Pacific Islands: environment and society. University of Hawai'i Press,
Honolulu

Scholl DW, Vallier TL (eds) (1985) Geology and offshore resources of Pacific Island Arcs—Tonga
Region. Circum-Pacific Council for Energy and Mineral Resources, Houston

Scott GAJ, Rotondo GM (1983) A model to explain the differences between Pacific plate island-
atoll types. Coral Reefs 1:139–150

Shope JB, Storlazzi CD, Erikson LH, Hegermiller CA (2016) Changes to extreme wave climates
of islands within the Western Tropical Pacific throughout the 21st century under RCP 4.5 and
RCP 8.5, with implications for island vulnerability and sustainability. Glob Planet Chang
141:25–38

Smith DK, Schouten H, Montesi L, Zhu WL (2013) The recent history of the Galapagos triple
junction preserved on the Pacific plate. Earth Planet Sci Lett 371:6–15

Sorbadere F, Schiano P, Métrich N, Bertagnini A (2013) Small-scale coexistence of island-
arc- and enriched-MORB-type basalts in the central Vanuatu arc. Contrib Miner Petrol
166(5):1305–1321

Spencer T, Stoddart DR, Woodroffe CD (1987) Island uplift and lithospheric flexure: observations
and cautions from the South Pacific. Z Geomorphol Suppl 63:87–102

Stoddart DR, Spencer T (1987) Rurutu reconsidered: the development of makatea topography in
the Austral islands. Atoll Res Bull 297:1–19

Stratford JMC, Rodda P (2000) Late Miocene to Pliocene palaeogeography of Viti Levu, Fiji
Islands. Palaeogeogr Palaeoclimatol Palaeoecol 162:137–153

Taylor GR (1973) Preliminary observations on the structural history of Rennell Island, South
Solomon Sea. Geol Soc Am Bull 84:2795–2806

Taylor B (2006) The single largest oceanic plateau: Ontong Java–Manihiki–Hikurangi. Earth
Planet Sci Lett 241:372–380

Terry JP (1999) Kadavu island, Fiji: fluvial studies of a volcanic island in the humid tropical South
Pacific. Singap J Trop Geogr 20(1):86–98

Tian L, Castillo PR, Hilton DR, Hawkins JW, Hanan BB, Pietruszka AJ (2011) Major and trace element and Sr-Nd isotope signatures of the northern Lau basin lavas: implications for the composition and dynamics of the back-arc basin mantle. J Geophys Res 116:B11201

Tracey JI, Schlanger S, Stark J, Doan D, May H (1964) General geology of Guam. US Government Printing Office, Washington

Vacher LHL, Quinn TM (eds) (1997) Geology and hydrogeology of carbonate Islands. Elsevier, Amsterdam

Walsh RPD (1982) The influence of climate, lithology, and time drainage density and relief development in the volcanic terrain of the Windward Islands. In: Douglas I, Spencer T (eds) Environmental change and tropical geomorphology. Allen and Unwin, London

Whittaker RJ, Fernandez-Palacios JM (2007) Island biogeography: ecology, evolution, and conservation. Oxford University Press, Oxford

Wiens HJ (1962) Atoll environment and ecology. Yale University Press, New Haven

Wood BL (1967) Geology of the Cook Islands. N Z J Geol Geophys 10:1429–1445

Yamaguchi T, Kayanne H, Yamano H (2009) Archaeological investigation of the landscape history of an Oceanic atoll: Majuro, Marshall Islands. Pac Sci 63:537–565

Yasukochi T, Kayanne H, Yamaguchi T, Yamano H (2014) Sedimentary facies and Holocene depositional processes of Laura Island, Majuro Atoll. Geomorphology 222:59–67

Chapter 3
Climate Change Scenarios and Projections for the Pacific

Savin S. Chand

3.1 Introduction

Small island countries in the Pacific often experience changes and variability in their climate, for example, those associated with shifts in rainfall patterns, increasing frequency of extreme weather events such as increasingly intense tropical cyclones and rising sea levels (Nurse et al. 2014). However, distinguishing between natural variability and climate change due to human activity that alters composition of global atmosphere through greenhouse gas emissions can be extremely difficult in this region. This is in part due to lack of consistent long-term observed data records for climate change detection and attribution studies and in part due to limitations in climate models, such as insufficient model resolutions, to spatially resolve small islands (e.g. Australian Bureau of Meteorology and CSIRO 2011).

There is no doubt that the threats of climate change and sea-level rise are very real in the Pacific, even to an extent that the very existence of some atoll nations is threatened by rising sea levels attributed to global warming (Nurse et al. 2014; Church et al. 2013). Better understanding of the climate of the Pacific Island countries and how they reflect natural variability and change directly or indirectly due to human activity can have significant environmental and socio-economic implications. People living in the Pacific Island countries have a strong relationship with the land and ocean, so changes in climate can represent threat not only to the physical environment but also to their culture and customs.

In order to implement effective adaptation strategies to mitigate impacts of climate variability and change, the Australian Government implemented the *International Climate Change Adaptation Initiative* to meet high-priority adaptation needs of vulnerable Pacific Island countries. Through this initiative, the two

S. S. Chand (✉)
Center for Informatics and Applied Optimization, Federation University Australia,
Mt Helen, VIC, Australia
e-mail: s.chand@federation.edu.au

© Springer Nature Switzerland AG 2020
L. Kumar (ed.), *Climate Change and Impacts in the Pacific*, Springer Climate,
https://doi.org/10.1007/978-3-030-32878-8_3

successive major research programmes called the *Pacific Climate Change Science Program* (PCCSP, from 2009 to 2011) and the *Pacific-Australia Climate Change Science Adaptation Planning* (PACCSAP, from 2011 to 2014) programme were carried out to improve our understanding of the past, present and future climate of the Pacific Island countries. This chapter reviews some of the major findings of the research conducted as part of those two programmes, as well as other new research over recent years, to provide an up-to-date information on climate variability and change and associated scientific challenges for the Pacific Island countries. Particular emphasis is on the role of major climatic features and drivers (hereafter, collectively referred to as "features") of climate variability and change in the Pacific and how projected changes in these features are likely to affect ocean and atmospheric variables that are of significant concern for the people of the Pacific Island countries, such as extreme rainfall events and sea-level rise.

This chapter is structured into four parts. The first part looks at some of the major climatic features of climate variability in the Pacific, namely, the South Pacific Convergence Zone, El Niño-Southern Oscillation and Interdecadal Pacific Oscillation. The second part examines the observed climate variability and trends in the Pacific, with emphasis on rainfall and sea level. The third part focuses on results from climate model projections for the Pacific, including methods of climate projections and model evaluations. The last section gives the summary, including a discussion of uncertainties associated with climate projections over the Pacific.

3.2   Major Features of Climate Variability in the Pacific

There are several important features of the climate system that influence mean climate and variability in the Pacific. This section gives an overview of the main climatic features that are integral to the Pacific climate. Changes in these features as a result of human-induced global warming are discussed in latter sections.

3.2.1   South Pacific Convergence Zone

A prominent climatic feature in the Pacific is the South Pacific Convergence Zone (SPCZ) where convective activities such as thunderstorms and tropical cyclones are frequently spawned (e.g. Trenberth 1976; Vincent 1994). The SPCZ is characterized by a band of high cloudiness, strong convective precipitation and low-level convergence extending northwest-southeast diagonally from near the Solomon Islands (0°, 150°E) towards French Polynesia (30°S, 120°W) (Fig. 3.1a).

The SPCZ forms in the region of convergence between southeast trade winds and the easterly flow from the eastern South Pacific anticyclones. The western, tropical portion of the SPCZ lies over the region of relatively warmer sea surface temperature (SST) called the West Pacific Warm Pool, while the eastern portion undergoes

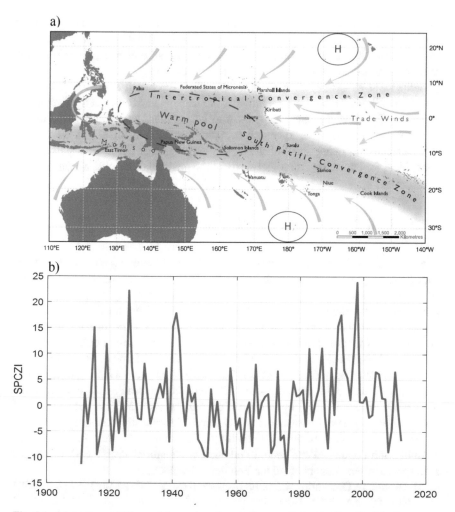

Fig. 3.1 (**a**) Average positions of the major climatic features in the Pacific. Blue shading represents convergence zones, yellow arrows show near-surface winds, and the red dashed oval indicates the West Pacific Warm Pool, and "H" represents the typical positions of moving high-pressure systems, and (**b**) November–April index of the South Pacific Convergence Zone (SPCZI, Salinger et al. 2014), calculated as the normalized November–April difference in mean sea-level pressure at Apia and Suva for the period 1932–1992

frequent mid-latitude interactions that contribute to its diagonal orientation (e.g. Vincent 1994; Widlansky et al. 2011). The SPCZ strongly contributes to the seasonal cycle of the rainfall in the South Pacific. It is more clearly defined during the months of December–February accounting for higher than average annual rainfall during these months and weaker and less well defined in June–August.

The interannual variability of the SPCZ is dominated by the impact of the El Niño-Southern Oscillation phenomenon (ENSO, defined in the following section) with the SPCZ moving north and east during El Niño events and south and west during La Niña events (e.g. Trenberth 1976; Vincent 1994; Folland et al. 2002; Salinger et al. 2014). As a result, convective activities such as tropical cyclones and heavy rainfall move accordingly (e.g. Vincent et al. 2011).

Salinger et al. (2014) derived an index called the South Pacific Convergence Zone index (SPCZI) to monitor interannual to decadal variability in the Pacific climate. The SPCZI is computed using the normalized mean sea level pressure difference between the two stations based in Apia, Samoa, and Suva, Fiji, as they lie symmetrically on either sides of the SPCZ making them (Apia-Suva pair of stations) ideal for capturing latitudinal shifts in the position of the SPCZ (Fig. 3.1b).

3.2.2 El Niño-Southern Oscillation

The El Niño-Southern Oscillation (ENSO) phenomenon is a major mode of interannual (year-to-year) climate variability in the Pacific (e.g. Troup 1965; Trenberth 1997). The ENSO cycle is irregular, and most of its variability has periods of 2–7 years. The term "El Niño", which is Spanish for "the boy" or "the Christ child", was traditionally used to refer to the annual occurrence of a warm ocean current that flowed southward along the west coast of Peru and Ecuador around Christmas time. By the mid-twentieth century, scientists realized that El Niño is far more than a coastal phenomenon and that it is associated with basin-scale warming of the tropical Pacific Ocean. Nowadays the term "El Niño" is commonly used to refer to the occurrence of anomalously high sea surface temperature (SST) in the central and eastern equatorial Pacific Ocean every few years. The opposite "La Niña" ("the girl" in Spanish) consists of basin-wide cooling of the tropical Pacific. This anomalous warming and cooling of the central and eastern equatorial Pacific SST drives the atmospheric phenomenon called the Southern Oscillation.

The Southern Oscillation, initially discovered by Sir Gilbert Walker in the 1920s and 1930s (Walker 1923, 1924), is characterized by a seesaw in tropical sea-level pressure (SLP) between the Western and Eastern Hemispheres (e.g. Trenberth 1976; Trenberth and Shea 1987). During El Niño, the SLP falls in the central and eastern Pacific and rises in the western Pacific; the reverse occurs during La Niña. The El Niño and the Southern Oscillation are two coupled aspects of the same phenomenon. The zonal atmospheric circulation that arises as a result of this coupling is called the "Walker circulation".

Normally, the rising air associated with the Walker circulation is located in the equatorial western Pacific near the warm Indonesian region and sinking air near the cold equatorial eastern Pacific. These rising and sinking branches of the Walker cell are connected by easterlies in the lower troposphere and westerlies in the upper troposphere. During El Niño events when the central and eastern Pacific SST becomes anomalously warmer than the western Pacific SST, the rising branch of the Walker cell shifts accordingly to the central or eastern Pacific, and the sinking branch is located over the western Pacific. This rising branch of the cell is often associated with convective activity such as rainfall and tropical cyclones (e.g. Trenberth and Caron 2000; Chand and Walsh 2009; Vincent et al. 2011).

Effects of ENSO are not only confined to the equatorial Pacific alone but are also observed in many parts of the world through "teleconnections" (e.g. García-Serrano et al. 2017). Numerous studies have documented the influence of ENSO on various weather and climate variables around the world, but our focus in this chapter is on how ENSO affects climate variability in the Pacific Ocean basin. Note that the terms "El Niño" and "La Niña" are sometimes used interchangeably with "warm phase" and "cold phase", respectively. The term "neutral phase" describes conditions when the equatorial SSTs are near climatological averages.

Numerous indices have been developed and used to monitor the status of ENSO (e.g. Trenberth and Stepaniak 2001). The two commonly used indices are called the Southern Oscillation index (SOI) and the Niño3.4 index. The SOI is calculated using the barometric pressure difference between Tahiti and Darwin. A strong, persistently negative SOI is typical of El Niño conditions, while a strong and persistently positive SOI is indicative of La Niña. Similarly, the Niño3.4 index measures the SST anomaly in the central and eastern Pacific (5°N-5°S; 170°W-120°W). A strong, persistently positive Niño3.4 index indicates an El Niño event. Note that SOI and Niño3.4 index change simultaneously, indicative of strong ocean-atmospheric coupling during ENSO events.

Over the past years, another type of El Niño [referred to as the "El Niño Modoki", as in Ashok et al. 2007] is observed. Unlike traditional El Niño events, El Niño Modoki events have above-normal SSTs that are confined more to the central Pacific region flanked by below-normal SSTs on the eastern and western sides (Fig. 3.2). Some scientists hypothesize that this might be related to anthropogenic global warming (e.g. Yeh et al. 2009), and if so, then this type of El Niño may become more frequent in the future (see latter sections).

3.2.3 Pacific Decadal Oscillation and Interdecadal Pacific Oscillation

Climate in and around the Pacific Ocean also shows substantial "ENSO-like" patterns of variability on decadal and interdecadal time scales (e.g. Power et al. 1999a, 1999b; Callaghan and Power 2011). Much of this variability has been linked to the Pacific Decadal Oscillation (PDO, Mantua et al. 1997) and Interdecadal Pacific Oscillation (IPO, Power et al. 1999a). The PDO is a characteristic of the North Pacific Ocean, whereas IPO is the Pacific-wide manifestation that includes the Southern Hemisphere, and so the interdecadal variability in PDO and IPO indices are very similar (Power et al. 1999b). When the IPO is in a positive phase, SST anomalies over the North Pacific are negative, as are anomalies near New Zealand, while SST anomalies over the tropical Pacific are positive. An index, termed the IPO tripole index (TPI), developed by Henley et al. (2015), can be used as a measure of

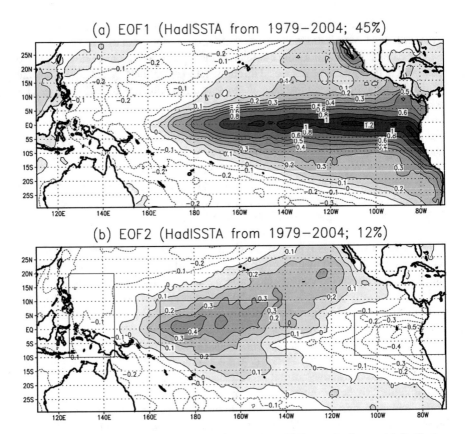

Fig. 3.2 First two modes of empirical orthogonal function (EOF) representing spatial distribution of monthly sea surface temperature anomalies in the Pacific (multiplied by respective standard deviations of respective principal components to give unit in °C): (**a**) the first EOF mode represents traditional El Niño phenomenon, and (**b**) the second EOF mode represents El Niño Modoki (Source: Ashok et al. 2007). Boxes show areas over which the (**a**) Niño 3.4 and (**b**) El Niño Modoki indices are calculated as a measure of traditional El Niño and Modoki-type El Niño events, respectively

interdecadal variability in the Pacific. This index is based on the difference between the sea surface temperature anomalies averaged over the central equatorial Pacific and in the Northwest and Southwest Pacific (Fig. 3.3).

The IPO has a strong influence on the Pacific climate by modulating teleconnections with ENSO (e.g. Salinger et al. 2001). For example, the rapid shift from negative to positive IPO during mid-1970s was associated with a shift to an El Niño-dominated period, whereas the shift to a negative IPO after around the year 2000 was associated with La Niña-dominated period. Some studies have indicated that the synergetic match of positive IPO and El Niño would strengthen the effects of either mode's impact on climate, whereas the opposite phases (i.e. negative IPO and La Niña) would weaken the impact (e.g. Gershunov and Barnett 1998; Grant and Walsh 2001).

The occurrence of slow, natural oceanic processes can make some of the decadal variability linked to the IPO and PDO more predictable than ENSO (e.g. Power and Colman 2006; Mochizuki et al. 2010). However, the extent to which this translates into predictability of atmospheric variables, such as rainfall and tropical cyclones in the Pacific, is subject to ongoing research.

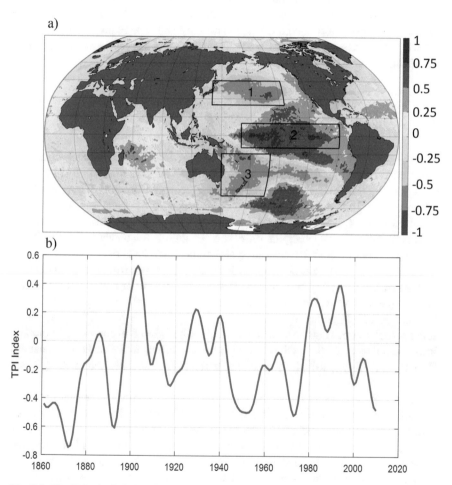

Fig. 3.3 The IPO tripole index (TPI) is based on the difference between the SSTA averaged over the central equatorial Pacific and the average of the SSTA in the Northwest and Southwest Pacific. It is a measure of interdecadal variability in the Pacific. The map (**a**) shows the correlations of the low-pass index TPI (**b**) with filtered HadISST2.1. Data to construct these figures are obtained from Henley et al. (2015), freely available on the website https://www.esrl.noaa.gov/psd/data/time-series/IPOTPI/

3.3   Observed Climate Variability and Change in the Pacific

Climate variability and change in the Pacific region can occur at different time scales and involve different contributing factors. Prior to the Industrial Revolution (around 1750), the climate of the Pacific underwent large variations mainly associated with changes in intensity and frequency of ENSO (e.g. Nunn 2007; Gergis and Fowler 2009). However, it is now highly likely that the climate is also influenced directly or indirectly by human activities (e.g. Cubasch et al. 2013). This section reviews the impact of observed climate variability and change on ocean and atmospheric variables, such as rainfall, tropical cyclones and sea-level rise, which are of significant concern to the people of the Pacific Island countries.

3.3.1   General Perspective

There is no doubt that global climate is changing and evidence of such change is broad and compelling (e.g. Cubasch et al. 2013). Key indicators of global climate change include increasing concentrations of greenhouse gases in the atmosphere, which drive significant changes in physical responses of ocean and atmospheric variables such as rising global average near-surface air temperature and humidity, increasing intensity of precipitation events, changing frequency and intensity of severe weather events and accelerating global mean sea-level rise.

Further evidence of changes in the global climate comes from natural indicators such as earlier flowering and ripening dates, coral bleaching and poleward migration of plants and animals (e.g. Rosenzweig et al. 2007; Chand et al. 2014). Reconstructed paleoclimate temperature records over the past 2000 years from sources such as tree rings, ice cores and corals, when placed in context with modern instrumental records, also indicate a rapid rate of warming in the backdrop of natural climate variability, particularly since the early twentieth century (e.g. Gergis and Fowler 2009; Mann et al. 2009).

For the Pacific Island countries, there is a general agreement among the communities that changes in weather and climate have occurred in their region more significantly over the past decade than ever before. Such perceptions arise mainly from local observations such as shifts in seasonal patterns of rainfall and tropical cyclones, more frequent and extreme rainfall causing flooding and mudslides, increasing frequency of droughts, fires and number of hot days, and more storm surges, coastal erosion and salt water contaminations of freshwater springs. In order to determine whether these perceived claims are scientifically valid and, if so, how to quantify relative contributions from human-induced and natural variability, a major concerted research effort was implemented through the PCCSP and PACCSAP projects by the Australian Bureau of Meteorology and the Commonwealth Scientific and Industrial Research Organisation (CSIRO) in partnership with several research institutes in the Pacific Island countries over the period 2009–2014 (Australian

Bureau of Meteorology and CSIRO 2011). This section summarizes some of the main results of those findings.

Note that lack of sufficient high-quality data, as well as the presence of large natural climate variability, makes it difficult to scientifically confirm the extent of human impacts on some oceanic and atmospheric variables such as rainfall and tropical cyclones. This highlights the need for more research on detection and attribution of climate change in the Pacific as new and updated data become available in the future.

3.3.2   Temperature

Station data from meteorological services show that mean surface air temperatures have generally increased throughout the Pacific during the twentieth century, with most stations recording trends around +0.08–0.20 °C per decade (Fig. 3.4a). Trends in maximum and minimum temperatures are generally similar to those of mean temperature, and the amount of warming in wet (November–April) and dry (May–October) is similar for most stations. Overall, the magnitude of background warming in the Pacific since the mid-twentieth century is consistent with human-induced global warming (Fig. 3.4b).

3.3.3   Rainfall

Rainfall variability in the Pacific Island countries is strongly linked to ENSO and the IPO phenomena and directly attributable to resulting shifts in the SPCZ (e.g. Folland et al. 2002; Salinger et al. 2001, 2014). On average, the mean position of the SPCZ gets displaced northeastward during El Niño events, thus causing enhanced rainfall activity around most of the Pacific Island countries that lie northeast of the SPCZ, extending to French Polynesia (e.g. Salinger et al. 2014). On the contrary, the mean position of the SPCZ gets displaced southwestward during La Niña, causing suppressed rainfall activity in the Pacific region. The IPO also modulates rainfall in the South Pacific by shifting the SPCZ northeastward during the positive phase (and southwestward during the negative phase), causing enhanced rainfall activity northeast of the SPCZ during the positive phase (e.g. Salinger et al. 2001).

Unlike changes in temperature, long-term rainfall trends in the Pacific are not very clear mainly due to strong background natural variability. Some previous studies (e.g. Griffiths et al. 2003) have shown a general increase in rainfall totals for countries that lie northeast of the SPCZ (and decrease for countries in the southwest) over the period 1960–2010 (Fig. 3.5a). This pattern of change is reflected in both wet and dry seasons. However, the pattern of trends has changed markedly in the southwest Pacific since 1990, consistent with a shift of the SPCZ back to its climatological position since 1990 (Fig. 3.5b).

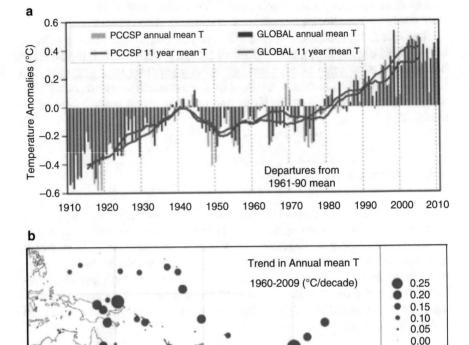

Fig. 3.4 (**a**) Annual mean surface temperature anomalies for the globe and for the Pacific (PCCSP, 120°E-150°E; 25°S-25°N). (**b**) Sign and magnitude of trends in annual mean temperatures at Pacific Island meteorological stations for 1960–2009. Australian stations are included for comparisons. (Source: Australian Bureau of Meteorology and CSIRO 2011)

3.3.4 *Tropical Cyclones*

Small island countries in the Pacific are among some of the worst affected by tropical cyclone events due to their high vulnerability and low adaptive capacity. For this reason, a separate chapter (Chap. 6) is dedicated entirely to the impact of climate variability and change to tropical cyclones in the Pacific.

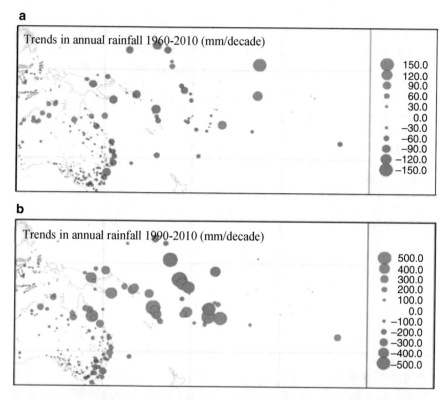

Fig. 3.5 Trends in annual total rainfall at Pacific meteorological stations (including over Australia) for (**a**) 1960–2010 and (**b**) 1990–2010. (Source: Australian Bureau of Meteorology and CSIRO 2011)

3.3.5   Sea Level

Sea-level rise poses one of the major threats to small island countries in the Pacific (e.g. Nicholls and Cazenave 2010; Church et al. 2006; Zhang and Church 2012; Nunn 2013), particularly for the low-lying coastal areas where most of the communities and infrastructure are located. There are various factors that contribute to changes in sea level such as tides and changes in weather and climate variables. A small increase in overall, long-term sea-level rise due to climate change can compound the effects of natural variability and cause extreme sea levels to occur more frequently. However, it should be noted that sea-level changes are usually not spatially uniform as many regions can experience a higher or lower rate of sea-level change than the global average (e.g. Church et al. 2010; Becker et al. 2012).

ENSO, as a dominant source of interannual variability, has a major influence on year-to-year variability of sea level across the Pacific (Zhang and Church 2012). For example, strengthening trade winds during La Niña events shove more water towards the west resulting in higher than normal sea surface in the western tropical

Pacific. On the contrary, weakening trade winds during El Niño events are unable to maintain the normal gradient of sea level, leading to a drop in sea level in the west and rise in the east (Fig. 3.6a). The IPO phenomenon also has an ENSO-like (but distinct) impact on sea-level variability in the Pacific at decadal time scale (Zhang and Church 2012) with positive sea-level variations in the central and eastern tropical Pacific and negative sea-level variation in a narrow "horseshoe-like" pattern in the western tropical Pacific (Fig. 3.6b).

In addition to the influence of ENSO and the IPO, sea level is also rising globally and in the Pacific. Satellite altimeter records and in situ measurements indicate that global averaged sea-level rate was 1.7 ± 0.2 mm per year between 1901 and 2010 and that it has significantly increased to 3.2 ± 0.4 mm per year between the period 1993 and 2010 (Church et al. 2013). This rise has occurred everywhere in the Pacific (Fig. 3.7c) and even at a faster rate in the western and central tropical Pacific, northeast Pacific and south Pacific (Zhang and Church 2012; Becker et al. 2012).

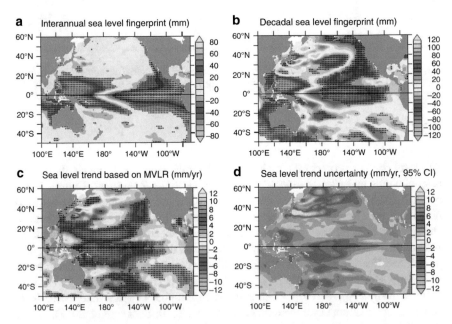

Fig. 3.6 (**a**) Interannual sea-level fingerprint associated with ENSO, (**b**) interdecadal sea-level fingerprint associated with Decadal Pacific Oscillation, (**c**) linear trend in sea-level over the period 1993–2011 after taking into account interannual and interdecadal variability through regression analysis and (**d**) uncertainty in sea-level linear trend at 95% confidence interval. (Source: Zhang and Church 2012)

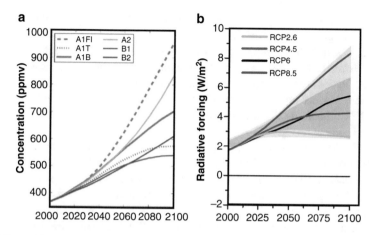

Fig. 3.7 (**a**) Global average carbon dioxide emission in gigatonnes for six future emission scenarios used as part of CMIP3 experiments (Source: IPCC 2007) and (**b**) trends in radiative forcing relative to pre-industrial values used as part of CMIP5 experiments. (Source: van Vuuren et al. 2011)

3.4 Climate Projections

3.4.1 Overview

Earth's climate is a complex system that undergoes significant variability and change as a result of multiple linear and non-linear processes operating at various spatial and temporal scales. This means that past climate trends cannot be simply extrapolated to understand future climate variability and change. Several non-linear processes must be taken into account, along with a range of plausible future greenhouse gas and aerosol concentration scenarios and pathways. Climate models are the primary tools available for investigating how climate system responds to these scenarios and pathways and for making projections of future climate over the coming century and beyond in order to help us better understand how the climate system evolves.

The models used in climate research can be as simple as an energy balance model or as complex as global climate models (GCMs) and regional climate models (RCMs) requiring state-of-the-art high-performance computing (Flato et al. 2013). GCMs can be either "standard" atmosphere-ocean general circulation models (AOGCMs) or Earth system models (ESMs) that expand on AOGCMs to include representation of various biogeochemical cycles such as those involved in the carbon cycle or ozone as well (Flato 2011). AOGCMs are extensively used to understand the dynamics of the physical components of the climate system and for making projections based on future greenhouse gas and aerosol forcing. RCMs, on the other hand, are limited-area models with representations of climate processes comparable to those in the atmospheric and land surface components of AOGCMs, often used to dynamically "downscale" global model simulations for some particular geographical region to provide more detailed information.

Many research institutions around the world develop and maintain their own GCMs. While these models are similar in many ways, subtle differences exist with respect to factors such as spatial resolution, parametrization and model components (e.g. some models represent atmospheric chemistry, while others may not). This means that climate model simulations arising from these GCMs may be different from each other.

In order to facilitate a community-based infrastructure in support of intercomparison of results from GCMs and data access, the Coupled Model Intercomparison Project (CMIP) framework was established in 1995 under the auspices of the Working Group on Coupled Modelling. The CMIP intercomparison project provides up-to-date information on climate science and adaptation policies to the Intergovernmental Panel on Climate Change (IPCC). Results from the two last CMIP phases, CMIP3 (Meehl et al. 2007) and CMIP5 (Taylor et al. 2012), are reported in the IPCC's Fourth Assessment Report (completed in 2007) and Fifth Assessment Report (completed in 2014), respectively, and form the basis of climate projections presented in this chapter for the Pacific.

3.4.2 Emission Scenarios and Pathways

CMIP3 models use emission scenarios to estimate the plausible future concentration of greenhouse gases in the atmosphere based on assumptions about demographic changes, economic development and technological change, as well as taking into consideration the natural source and sink of these gasses (details of emission scenarios are available from the IPCC Special Report on Emissions Scenarios; Nakicenovic et al. 2000). These emission scenarios are grouped into four "storylines": A1, A2, B1 and B2. The A1 storyline describes a future world of rapid population growth, a global population that peaks in mid-century and declines thereafter and a rapid introduction of new and more efficient technologies. The technological change may be fossil intensive (A1FI), non-fossil intensive (A1T) or a balance across all sources (A1B). The A2 storyline is based on "business-as-usual" case where population increases continuously with fragmented economic and technological growth. The B1 storyline describes a world with population growth the same as A1 but with rapid change in economic structure and introduction of clean and efficient technologies. Finally, the B2 storyline describes a world with increasing global population at a rate lower than A2, intermediate levels of economic development and less rapid and more diverse technological change than in the B1 and A1 storylines. These storylines lead to different levels of global average carbon dioxide emissions in the atmosphere, as well as different levels of carbon dioxide concentrations after taking into consideration natural sources and sinks, by the end of the twenty-first century (Fig. 3.7a).

On the other hand, CMIP5 models are based on a set of four plausible future greenhouse gas concentrations (not emissions), called the Representative Concentration Pathways (RCPs), that were developed for the climate modelling community as a basis for long-term and near-term climate modelling experiments (see van Vuuren et al. 2011 for details). These four RCPs, namely, RCP2.6, RCP4.5, RCP6.0 and RCP8.5, are defined according to the radiative forcing (i.e. cumulative measure of human emissions of greenhouse gasses from all sources expressed in Watts per square metre) target levels for 2100 relative to their pre-industrial levels. RCP2.6 is a low forcing level that assumes global average forcing levels peak between 2010 and 2020, then declining substantially thereafter, reaching 2.6 W m^{-2} by 2100 relative to pre-industrial levels. In the two stabilization trajectories RCP4.5 and RCP6.0, forcing levels peak around 2040 and 2080 before declining to 4.5 W m^{-2} and 6 W m^{-2}, respectively, by 2100 relative to their pre-industrial levels. The rising radiative forcing RCP8.5, which depicts a relatively conservative business as usual case, is a very high baseline scenario where radiative forcings continue to rise throughout the twenty-first century to around 8.5 W m^{-2} by 2100 (Fig. 3.7b).

CMIP3 and CMIP5 datasets each contain outputs from a large number of GCMs. These data are freely available from the Program for Climate Model Diagnosis and Intercomparison at Lawrence Livermore National Laboratory (www-pcmdi.llnl. gov). Note a direct comparison of CMIP3 and CMIP5 results is not possible as these models use different ways of describing the amount of greenhouse gases in the atmosphere in the future and that CMIP5 models are more advanced in terms of increasing model complexity. Regardless, results from both the projects indirectly simulate low, medium and high emission futures, and so some comparisons of projection results are possible.

Several different experiments were conducted as part of CMIP3 (Meehl et al. 2007) and CMIP5 (Taylor et al. 2012) phases. Overall, the long-term climate simulations for both the phases were essentially similar in that both included the climate of the twentieth-century simulations (also referred to as the "historical" simulations) and the climate of the twenty-first-century simulations (or the "future-climate" simulations) in their experiment design. Results from the twentieth-century simulations were extensively used in model evaluation and validation, while the results from the twenty-first-century simulations were used in determining climate projections.

3.4.3   Climate Model Evaluations

In order to use GCMs for scientifically robust and confident projections, it first has to be demonstrated that these models are sufficiently realistic in simulating the present climate. The skill of a model depends on its ability to represent the long-term average and seasonally varying cycles of various atmospheric and ocean variables such as temperature, rainfall and sea level, as well on its ability to represent important large-scale features such as ENSO, SPCZ and IPO that modulate natural climate variability. In addition, climate models should be stable and free from substantial drift that might lead to spurious departures in simulations in the absence of factors that would otherwise be responsible to induce the change.

The level of agreement between model simulations and observations (or gridded reanalysis products as estimates of observations) is an indicator of model reliability. Note that no one model is the best in representing all aspects of the climate system, and so in climate studies, collective results from a group of models (also known as model ensembles) are often used for validating climate model results and for making projections of the future climate. Over the past years, several studies have comprehensively evaluated the performance of CMIP3 and CMIP5 model ensembles over the tropical western Pacific (e.g. Grose et al. 2014; Wang et al. 2015; Moise et al. 2015). They found that the ability of GCMs to realistically reproduce several key climatic variables and features of the late twentieth century has improved significantly over the tropical Pacific.

A recent study by Grose et al. (2014) assessed and compared the performance of CMIP3 and CMIP5 models for the western tropical Pacific. Their study reported that while models from both these phases are able to capture important large-scale climatic features with a certain degree of fidelity, the CMIP5 models have shown some improvements in performance over CMIP3 models. For example, they showed that the observed mean SST and precipitation, respectively, compare well with those in CMIP3 and CMIP5 model ensembles for the tropical western Pacific (see Figs. 1 and 4 of Grose et al. 2014). However, despite the similarities in the zonal orientation of the mean SST between observations and model ensembles, the cold tongue (defined by the 28.5 °C isotherm) extends too far westward (often referred to as the cold-tongue bias). This cold-tongue bias reduces the extent of the Indo-Pacific Warm Pool, making it generally too cold, thus having implications on wind distributions, atmospheric convergence and rainfall. Their study showed the impact of this cold-tongue bias on tropical precipitations. Climate models typically simulate too little precipitation along the equator and too much precipitation to the north and south of the cold tongue in the ITCZ and SPCZ regions. GCMs from both CMIP3 and CMIP5 experiments can have an overly zonal SPCZ that can be too far north in the austral winter (June–August) and too far east in the austral summer (December–February). Consequently, this can create potential biases in rainfall patterns for the small island countries that lie along or on either sides of the SPCZ in the South Pacific (Widlansky et al. 2013).

As discussed earlier, ENSO is a major component of natural climate variability in the Pacific, and to have confidence in climate projections for the Pacific Island countries, it is essential that ENSO is well simulated in climate models. In order to simulate ENSO, models should be able to not only simulate the mean climate conditions but also ocean-atmosphere interactions (such as associated changes in sea surface temperature and atmospheric pressure) at various time and spatial scales. Several studies have shown that a number of climate models from the CMIP experiments can simulate ENSO-like variability reasonably well, with models from CMIP5 having slightly better performance than those from CMIP3 (e.g. Guilyardi et al. 2012; Grose et al. 2014; Bellenger et al. 2014). These improvements include better simulation of the magnitude and frequency of ENSO events, as well as improvements in simulating ENSO seasonal phase locking and the location of the strongest SST anomalies during the onset and peak phases of El Niño and La Niña. However, significant development is still needed in climate models to accurately simulate the basic characteristics of ENSO (such as amplitude, frequency, seasonal phase lock, etc.), as well as the underlying physical processes (e.g. atmospheric Bjerknes feedback) that control ENSO evolution. Moreover, several models still also have challenges in simulating the evolving nature of ENSO (such as the Modoki-type events) identified in the past investigations (e.g. Kim and Yu 2012).

In general, the global climate models from the CMIP experiments can represent essential aspects of the most important large-scale climate features of the Pacific region. These include representing the geographic and temporal patterns of sea surface temperature, location and seasonality of the major convergence of the SPCZ (which is the dominant climatic feature of the South Pacific) and the associated rainfall. This provides confidence in the use of models for regional climate projections. However, a number of common model biases and errors are apparent which lead to important limits in this confidence. Perhaps the most significant of these arise from the cold-tongue bias that impacts realistic simulations of several key ocean-atmospheric variables in the Pacific. It is critical that such biases and shortcomings are borne in mind when interpreting results from climate model projections for practical applications within the region.

3.4.4 Climate Model Projections for the Pacific

As highlighted earlier, climate models are the primary tools available for investigating how the climate system responds to different climate scenarios and pathways and for making projections of future climate over the coming century and beyond in order to help us better understand how climate system evolves. In this section, we review projected changes in the two major climate features and variability, ENSO and the SPCZ, as well as some key atmospheric and oceanic variables, using results from the CMIP experiments.

El Niño-Southern Oscillation

As ENSO is the dominant mode of interannual natural climate variability in the Pacific, any substantial change in the character of ENSO in response to anthropogenic global warming will have major implications on regional climate of the small island countries in the Pacific. Recent studies provide some indications of projected future changes in certain aspects of ENSO using current-generation climate models (Kim and Yu 2012; Power et al. 2013; Cai et al. 2014). This includes an increased frequency of extreme El Niño events (such as the events of 1982/1983 and 1997/1998) due to more occurrences of atmospheric convection in the eastern Pacific (Cai et al. 2014), as well as a potential increase in the frequency of the

"Modoki-type" central Pacific El Niño events (Kim and Yu 2012). Moreover, some studies also give an indication of the weakening of the Walker circulation and the associated decrease in the pressure gradient across the Pacific (e.g. Vecchi et al. 2006).

However, it should be emphasized that there is a large degree of inconsistency among climate models on future projections of these changes (e.g. Collins et al. 2010). Therefore, care must be exercised when interpreting climate projection results. Regardless, there is a strong consensus that ENSO variability will continue to dominate regional-scale climate in the future (Power et al. 2013; Chand et al. 2017) and strongly influence weather-related variables in the changing climate (Stevenson et al. 2012).

South Pacific Convergence Zone

The SPCZ is the largest rainband in the Southern Hemisphere and provides most of the summer rainfall to the southwest Pacific Island countries. Therefore, any changes in the characteristic of the SPCZ in response to greenhouse warming will have major implications on communities of the small island countries in the Pacific.

A study by Widlansky et al. (2013) describes the likely projected changes in the SPCZ using hierarchy of CMIP3 and CMIP5 climate models and idealized experiments. They propose two competing mechanisms "wet gets wetter" and "warmest get wetter" in response to greenhouse warming (Fig. 3.8). Mean specific humidity is projected to increase over the entire tropical Pacific in response to greenhouse warming, supporting an enhanced future hydrological cycle (Seager et al. 2010), sometimes referred to as the "wet gets wetter" thermodynamic response to greenhouse warming. Even though the simulated moisture increase in the SPCZ region (Fig. 3.8a) is weaker than along the equator, it is substantially greater than that in the southeast Pacific, a region that warms least and where drying is projected by nearly all climate models (Brown et al. 2012). On the other hand, this effect is partially offset, in regions such as the SPCZ that experience relatively minor warming, by the anomalous divergence of mean moisture (Fig. 3.8b). The corresponding anomalous circulation accounts for anomalous moisture convergence towards the warmest waters, resulting in the increased rainfall within the ITCZ region (i.e. a "warmest gets wetter" dynamic response to greenhouse warming).

As such, some islands in the SPCZ region could see a rainfall increase if temperatures rise high enough, while those that lie along the southeastern margin of the SPCZ (15°S-30°S; 135°W-105°W) would experience more robust drying (e.g. Widlansky et al. 2013). A potentially weaker austral summer SPCZ would result in a diminished rainy season for most southwest Pacific Island nations. According to the hierarchy of bias-corrected atmospheric model experiments presented by Widlansky et al. (2013), projected summer rainfall may decrease in Samoa and other neighbouring islands on average by 10–20% during the twenty-first century. Less future rainfall, combined with increasing surface temperatures and enhanced potential evaporation, increases the potential for longer-term droughts in the region.

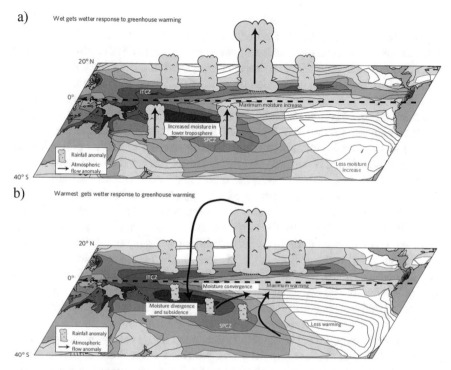

Fig. 3.8 Illustration of two opposing mechanisms responsible for SPCZ rainfall response to projected twenty-first-century greenhouse warming. (**a**) "Wet gets wetter" and (**b**) "warmest get wetter" response to greenhouse warming. (Source: Widlansky et al. 2013)

However, as with projections for ENSO, it should be emphasized here that that the bias-corrected models considered in Widlansky et al. (2013) may still be prone to large uncertainties in the representation of convective processes and hence in the representation of the dynamic response to greenhouse warming. Thus, care must be exercised when interpreting results from climate model experiments.

Rainfall

Projections of rainfall for Pacific Island countries are not only dependent on the ability of the climate models to realistically simulate major climatic features and variability such as ENSO and the SPCZ but also on the spatial resolution of the models. Generally, climate models from the CMIP3 and CMIP5 experiments are too coarse to resolve island-scale rainfall, and so the actual amount of rainfall might be grossly underestimated. Regardless, some insights into projected changes in rainfall can be obtained from these models after accounting for model biases in climate features and variability and utilizing techniques such as statistical or dynamical downscaling to resolve island-scale rainfall patterns.

Findings from the PCCSP and PACSSAP (Australian Bureau of Meteorology and CSIRO 2011) and the updated climate projection results for different Pacific Island countries (Australian Bureau of Meteorology and CSIRO 2014) reveal that on average, wetter conditions are projected over most of the small island countries, particularly those that lie in the vicinity of the SPCZ such as the Solomon Islands and Papua New Guinea due to increased moisture convergence in warmer climate. Small island countries that lie farther southeast or south of the SPCZ mean position, between Vanuatu and the Cook Islands, are likely to experience decreases in rainfall in the future, warming climate (see Fig. 3.9 for examples of rainfall projections in selected Pacific Island countries). On seasonal basis, projected rainfall increases are widespread during November–April associated with intensification of the SPCZ. Rainfall increases are also projected during May–October in the deep tropics.

Moreover, a study by Power et al. (2013) evaluated El Niño-related rainfall variability in the Pacific. Typically, the rainfall activity in the central and eastern equatorial Pacific is enhanced during El Niño conditions and suppressed during La Niña conditions, whereas the activity in the western Pacific is enhanced during La Niña and suppressed during El Niño conditions. Power et al. (2013) found that this pattern of El Niño-driven drying in the western Pacific Ocean and rainfall increases in the central and eastern equatorial Pacific is likely to further intensify by the mid- to late-twenty first century in response to greenhouse warming.

In another study, Power et al. (2017) examined the year-to-year disruptions in ENSO rainfall over the Pacific to determine whether the likelihood of the frequency of this disruption has already increased and whether the projected twenty-first-century increase can be avoided or moderated through sustained reduction in greenhouse gas emissions. They found using latest generation of climate models that humans may have already contributed to the major disruption that occurred in the real world during the late twentieth century. They also demonstrated that, although marked and sustained reductions in twenty-first-century anthropogenic greenhouse gas emissions can greatly moderate the likelihood of major disruption, elevated risk of occurrence seems locked in now and for at least the remainder of the twenty-first century.

Note that while these projected changes in rainfall patterns are physically plausible and relatively consistent between climate models, the small-scale details of these projections should be interpreted with caution given the known biases in model simulations. In particular, the overly zonal orientation of the SPCZ in most simulations, as well as the presence of the cold-tongue bias, can have implications on regional-scale rainfall projections.

Extreme Rainfall

Changes in short-term extreme rainfall events in response to climate change will have major implications, particularly for the vulnerable small Pacific Island countries that are subject to flash flood, erosions and landslides. While CMIP3 and CMIP5 climate models have limitations in resolving extreme daily and sub-daily

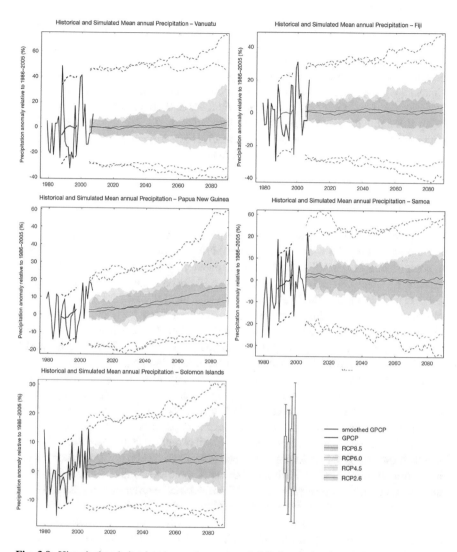

Fig. 3.9 Historical and simulated annual average rainfall time series for selected Pacific Island countries. The graph shows the anomaly (from the base period 1986–2005) in rainfall from observations (the Global Precipitation Climatology Project dataset, in purple) and for the CMIP5 models under the very high (RCP8.5, in red) and very low (RCP2.6, in blue) emission scenarios. The solid red and blue lines show the smoothed (20-year running average) multi-model mean anomaly in rainfall, while shading represents the spread of model values (5–95th percentile). The dashed lines show the 5–95th percentile of the observed interannual variability for the observed period (in black) and added to the projections as a visual guide (in red and blue). This indicates that future rainfall could be above or below the projected long-term averages due to interannual variability. The ranges of projections for a 20-year period centred on 2090 are shown by the bars for RCP8.5, 6.0, 4.5 and 2.6. (Source: Australian Bureau of Meteorology and CSIRO 2014)

rainfall events, there is an indication from other fine-resolution modelling studies that extreme rainfall events are likely to intensify for the tropics in the future, warming climate (e.g. O'Gorman 2015). It is generally accepted that extreme daily events will increase at a rate of about 6–7% per degree warming, consistent with the Clausius-Clapeyron relation. Furthermore, a recent regional climate modelling study by Bao et al. (2017) found a robust increase in daily precipitation (~5.7–15% °C^{-1}) throughout the major Australian cities, including those in the tropics. This study can provide some indication of what can also be expected for the small Pacific Island countries in the future, warming climate.

Sea-Level Rise

Reliable projections of sea-level change depend critically on improved understanding and modelling of a wide range of contributing factors. The primary contributors to contemporary sea-level change are the expansion of the ocean as it warms and the transfer of water currently stored on land to the ocean, particularly from melting glaciers and ice sheets. The IPCC's Fourth Assessment Report (AR4) comprising of models from CMIP3 included the estimates of ocean thermal expansion, melting of glaciers and ice caps as well as increased melting of the Greenland ice sheet. Results from AR4 gave a wide range in global averaged projections of about 20 to 80 cm by 2100 under several illustrative scenarios (Fig. 3.10) (Church et al. 2011).

However, current rate of sea-level rise is already near the upper end of these projections. Since AR4, climate models have been improved substantially. These improvements include bias corrections in historical ocean temperature observations resulting in improved estimates of ocean thermal expansion, a better ability of models to estimate contributions from melting glaciers and ice caps as well as improvements in modelling the underpinning processes of sea-level rise. These changes are incorporated in the IPCC's Fifth Assessment Report (AR5) that uses models from the CMIP5 experiments, giving improved projections in order of about 28 to 98 cm by 2100 under different RCPs (Fig. 3.11) (Church et al. 2013).

Regional sea-level changes may differ substantially from a global average, indicative of complex spatial patterns that result from ocean dynamical processes, movements of the sea floor and changes in gravity due to water mass redistribution in the climate system (Church et al. 2013). The regional distribution is associated with natural or anthropogenic climate modes rather than factors causing changes in the global average value and includes such processes as a dynamical redistribution of water masses and a change of water mass properties caused by changes in winds and air pressure, air-sea heat and freshwater fluxes and ocean currents. Thus, estimating mean sea-level rise at regional scale can be very challenging. Figure 3.12 shows ensemble mean relative sea-level change between 1986–2005 and 2081–2100 for RCPs 2.6, 4.5, 6.0 and 8.5 (Church et al. 2013). Based on this analysis, it is very likely that regional sea level in the Pacific will be higher in the future-climate conditions relative to the current-climate conditions (in order of about 20 cm for RCP2.6 to well over 60 cm for RCP8.5).

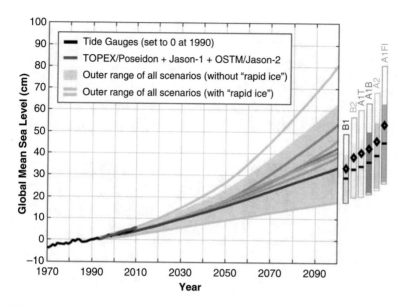

Fig. 3.10 Global averaged projections of sea-level rise during the twenty-first century for different IPCC (2007) emission scenarios. (Source: Church et al. 2011)

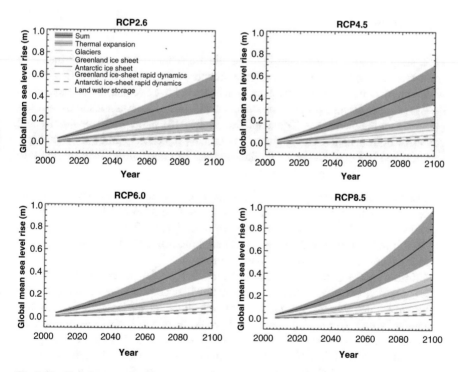

Fig. 3.11 Global averaged projections of sea-level rise during the twenty-fist century for different RCPs. (Source: Church et al. 2013)

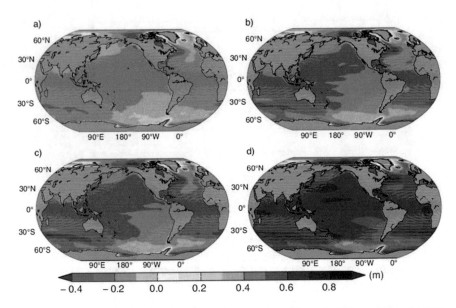

Fig. 3.12 Ensemble mean regional sea-level change evaluated from CMIP5 models for (**a**) RCP 2.6, (**b**) RCP 4.5, (**c**) RCP 6.0 and (**d**) RCP 8.5 scenarios. (Source: Church et al. 2013)

3.5   Summary

Estimating impacts from human-induced climate change often rely on projections from climate models. Coordinated experiments such as CMIP3 and CMIP5, in which many climate models run a set of scenarios, have become the de facto standard to produce climate projections (Meehl et al. 2007). Uncertainties in these models are a limiting factor (e.g. Knutti and Sedláček 2013), particularly for small island countries in the Pacific. Uncertainties in climate projections can be from multiple sources, and below are some of the key sources of uncertainties that are likely to affect climate projections of the Pacific.

- **Emission scenarios and RCPs:** It is uncertain how society will evolve over this century, and therefore it is not possible to know exactly how anthropogenic emissions of greenhouse gases and aerosols will change. Emission scenarios and RCPs produced by the IPCC are considered plausible, with the range of uncertainty increasing over the twenty-first century.
- **Climate model deficiencies:** Climate models have deficiencies in representing key physical processes. Many important small-scale processes cannot be represented explicitly in models and so must be included in approximate form as they interact with larger-scale features. This could be due to limitations in scientific understanding of those processes and lack of detailed observations of some physical processes. Subtle differences between models associated with this deficiency result in a range of climate projections for a given scenario. Climate models are

based on physical laws, and so they are not perfect representations of the real world. While most models are able to capture the broad-scale climate features of the Pacific, a number of deficiencies still remain at local scales. When a country is located in a region with model deficiencies, less confidence can be placed on associated climate projections.

- **Natural climate variability:** Some of the most difficult aspects of understanding and projecting changes in regional climate relate to possible changes in the circulation of the atmosphere and oceans and their patterns of variability. When interpreting projected changes in the mean climate, it is important to remember that natural climate variability (such as ENSO) will be superimposed and can cause conditions to vary substantially from the long-term mean from 1 year to the next and sometimes from one decade to the next.

The efforts for current-generation CMIP5 models are enormous, with a larger number of more complex models run at higher resolution and with more complete representations of external forcings over CMIP3 models. Therefore, it is widely expected to provide more detailed and more certain projections. However, this is not necessarily the case as demonstrated by Knutti and Sedláček (2013). Improving model complexity and resolution does not essentially reduce model uncertainty. Some uncertainties will always remain, and these should be carefully considered when making climate projections. For example, the presence of cold-tongue bias can have impacts on realistic simulations of several key ocean-atmospheric variables, such as rainfall, over the small island countries in the Pacific. Adjustments of these biases, for example, through statistical or dynamical downscaling approaches, are vital to have more confidence in the projections.

While there has been progress on several fronts to monitor, understand and project climate change relevant to the Pacific Island countries (e.g. through PCCSP and PACCSAP projects), there are still many challenges. Further work to strengthen scientific understanding of Pacific climate change is required to inform adaptation and mitigation strategies. Ongoing research and collaboration is necessary to advance climate science in the Pacific, particularly through improving and expanding the network of ocean and atmospheric observations to advance understanding of current climate, climate variability and trends, as well as improving climate models and climate model projections.

Acknowledgement This work is supported through funding from the Earth Systems and Climate Change Hub of the Australian Government's National Environmental Science Programme (NESP).

References

Ashok K, Behera S, Rao AS, Weng HY, Yamagata T (2007) El Niño Modoki and its teleconnection. J Geophys Res 112:C11007. https://doi.org/10.1029/2006JC003798

Australian Bureau of Meteorology and CSIRO (2011) Climate change in the Pacific: scientific assessment and new research. Volume 1: regional overview. Volume 2: country reports. Hennessy K, Power S and Cambers G (Scientific Editors), CSIRO, Canberra

Australian Bureau of Meteorology and CSIRO (2014) Climate variability, extremes and change in the Western Tropical Pacific: new science and updated country reports. Pacific-Australia Climate Change Science and Adaptation Planning Program Technical Report, Australian Bureau of Meteorology and Commonwealth Scientific and Industrial Research Organisation, Melbourne, Australia

Bao J, Sherwood SC, Alexander LV, Evans JP (2017) Future increases in extreme precipitation exceed observed scaling rates. Nat Clim Chang 7:128–132

Becker M, Meyssignac B, Letetrel C, Llovel W, Cazenave A, Delcroix T (2012) Sea level variations at tropical Pacific islands since 1950. Glob Planet Chang 80–81(1):85–98

Bellenger H, Guilyardi E, Leloup J, Lengaigne M, Vialard J (2014) ENSO representation in climate models: from CMIP3 to CMIP5. Clim Dyn 42:1999–2018

Brown JR, Moise AF, Delange FP (2012) Changes in the South Pacific convergence zone in IPCC AR4 future climate projections. Clim Dyn 39:1–19

Cai W et al (2014) Increasing frequency of extreme El Niño events due to greenhouse warming. Nat Clim Chang 4:11–116

Callaghan J, Power SB (2011) Variability and decline in severe landfalling tropical cyclones over eastern Australia since the late 19th century. Clim Dyn 37:647–662

Chand SS, Walsh KJE (2009) Tropical cyclone activity in the Fiji region: spatial patterns and relationship to large-scale circulation. J Clim 22:3877–3893

Chand SS, Chambers LE, Waiwai M, Malsale P, Thompson E (2014) Indigenous knowledge for environmental prediction in the Pacific Island countries. Wea Clim Soc 6:445–450

Chand SS, Tory KJ, Ye H, Walsh KJE (2017) Projected increase in El Niño-driven tropical cyclone frequency in the Pacific. Nat Clim Chang 7:123–127

Church JA et al (2010) Sea-level rise and variability: synthesis and outlook for the future. In: Church JA (ed) Understanding sea-level rise and variability. Wiley-Blackwell, Chichester, pp 402–419

Church JA, Gregory JM, White NJ, Platten SM, Mitrovica JX (2011) Understanding and projecting sea level change. Oceanography 24:130–143

Church JA, Clark PU, Cazenave A, Gregory JM, Jevrejeva S, Levermann A, Merrifield MA, Milne GA, Nerem RS, Nunn PD, Payne AJ, Pfeffer WT, Stammer D, Unnikrishnan AS (2013) Sea level change. In: Stocker TF, Qin D, Plattner G-K, Tignor M, Allen SK, Boschung J, Nauels A, Xia Y, Bex V, Midgley PM (eds) Climate change: the physical science basis. Contribution of Working Group I to the fifth assessment report of the intergovernmental panel on climate change. Cambridge University Press, Cambridge, pp 1137–1216

Church JA, White NJ, Hunter JR (2006) Sea level rise at tropical Pacific and Indian Ocean islands. Global Planet Change 53:155–168. https://doi.org//10.1029/2005GL024826

Collins M et al (2010) The impact of global warming on the tropical Pacific Ocean and El Niño. Nat Geosci 3:391–397

Cubasch U, Wuebbles D, Chen D, Facchini MC, Frame D, Mahowald N, Winther J-G (2013) Introduction. In: Stocker TF, Qin D, Plattner G-K, Tignor M, Allen SK, Boschung J, Nauels A, Xia Y, Bex V, Midgley PM (eds) Climate change. The physical science basis. Contribution of Working Group I to the Fifth Assessment Report of the Intergovernmental Panel on Climate Change. Cambridge University Press, Cambridge/New York, NY

Flato G (2011) Earth system models: an overview. WIREs Clim Change 2:783–800

Flato G, Marotzke J, Abiodun B, Braconnot P, Chou SC, Collins W, Cox P, Driouech F, Emori S, Eyring V, Forest C, Gleckler P, Guilyardi E, Jakob C, Kattsov V, Reason C, Rummukainen M (2013) Evaluation of climate models. In: Stocker TF, Qin D, Plattner G-K, Tignor M, Allen SK, Boschung J, Nauels A, Xia Y, Bex V, Midgley PM (eds) Climate change. The physical science basis. Contribution of Working Group I to the Fifth Assessment Report of the

Intergovernmental Panel on Climate Change. Cambridge University Press, Cambridge, United Kingdom and New York, NY, USA

Folland CK, Renwick JA, Salinger MJ, Mullan AB (2002) Relative influences of the interdecadal Pacific oscillation on the South Pacific convergence zone. Geophys Res Lett 29(13):21-1–21-4

García-Serrano J, Cassou C, Douville H, Giannini A, Doblas-Reyes FJ (2017) Revisiting the ENSO teleconnection to the tropical North Atlantic. J Clim 30:6945–6957

Gergis JL, Fowler AM (2009) A history of ENSO events since a.D. 1525: implications for future climate change. Clim Chang 92:343–387

Gershunov A, Barnett TP (1998) Interdecadal modulation of ENSO teleconnections. Bull Am Meteorol Soc 79:2715–2725

Grant AP, Walsh KJE (2001) Interdecadal variability in north-east Australian tropical cyclone formation. Atmos Sci Lett 2:9–17

Griffiths GM, Salinger MJ, Leleu I (2003) Trends in extreme daily rainfall across the South Pacific and relationship to the South Pacific convergence zone. Int J Climatol 23:847–869

Grose MR et al (2014) Assessment of the CMIP5 global climate model simulations of the western tropical Pacific climate system and comparison to CMIP3. Int J Climatol 34:3382–3399

Guilyardi E, Bellenger H, Collins M, Ferrett S, Cai W, Wittenberg W (2012) A first look at ENSO in CMIP5. Clivar Exchanges 58:29–32

Henley BJ, Gergis J, Karoly DJ, Power SB, Kennedy J, Folland CK (2015) A tripole index for the interdecadal Pacific oscillation. Clim Dyn 45:3077–3090

IPCC, (2007): Climate Change 2007: The Physical Science Basis. Contribution of Working Group 1 to the Fourth Assessment Report of the Intergovernmental Panel in Climate Change [Solomon, S, D. Qin, M. Manning, Z. Chen, M. Marquis, K.B. Ayert, M. Tignor and H.L. Miller (eds.)]. Cambridge University Press, Cambridge, United Kingdom and New York, NY, USA, 996 pp.

Kim ST, Yu J-Y (2012) The two types of ENSO in CMIP5 models. Geophys Res Lett 39:L11704. https://doi.org/10.1029/2012GL052006

Knutti R, Sedláček J (2013) Robustness and uncertainties in the new CMIP5 climate model projections. Nat Clim Chang 3:369–373

Mann ME, Zhang Z, Rutherford S, Bradley RS, Hughes MK, Shindell D, Ammann C, Faluvegi G, Ni F (2009) Global signatures and dynamical origins of the little ice age and medieval climate anomaly. Science 326:1256–1260

Mantua NJ, Hare SR, Zhang Y, Wallace JM, Francis RC (1997) A Pacific interdecadal climate oscillation with impacts on salmon production. Bull Am Meteorol Soc 78:1069–1079

Meehl GA et al (2007) The WCRP CMIP3 multimodel dataset: a new era in climate change research. Bull Am Meteorol Soc 88:1383–1394

Mochizuki T, Ishii M, Kimoto M, Chikamoto Y, Watanabe M, Nozawa T, Sakamoto TT, Shiogama H, Awaji T, Sugiura N, Toyoda T, Yasunaka S, Tatebe H, Mori M (2010) Pacific decadal oscillation hindcasts relevant to near-term climate prediction. PNAS 107:1833–1837

Moise A et al (2015) Evaluation of CMIP3 and CMIP5 models over the Australian region to inform confidence in projections. Aust Meteorol Oceano J 65:19–53

Nakicenovic N et al (2000) Special report on emissions scenarios: a special report of working Group III of the Intergovernmental Panel on Climate Change. Cambridge University Press, Cambridge, p 599

Nicholls RJ, Cazenave A (2010) Sea-level rise and its impact on coastal zones. Science 328:1517–1520

Nunn P (2007) Climate, environment, and society in the Pacific during the last millennium. Elsevier, Amsterdam, p 316

Nunn PD (2013) The end of the Pacific? Effects of sea level rise on Pacific Island livelihoods. Singap J Trop Geogr 34(2):143–171

Nurse LA, McLean RF, Agard J, Briguglio LP, Duvat-Magnan V, Pelesikoti N, Tompkins E, Webb A (2014) Small islands. In: Barros VR, Field CB, Dokken DJ, Mastrandrea MD, Mach KJ, Bilir TE, Chatterjee M, Ebi KL, Estrada YO, Genova RC, Girma B, Kissel ES, Levy AN, MacCracken S, Mastrandrea PR, White LL (eds) Climate change: impacts, adaptation, and vulnerability. Part B: Regional aspects. Contribution of Working Group II to the fifth assessment report of the intergovernmental panel on climate change. Cambridge University Press, Cambridge, pp 1613–1654

O'Gorman PA (2015) Precipitation extremes under climate change. Curr Clim Change Rep 1:49–59

Power S, Colman R (2006) Multi-year predictability in a coupled general circulation model. Clim Dyn 26:247–272

Power S, Casey T, Folland C, Colman A, Mehta V (1999a) Inter-decadal modulation of the impact of ENSO on Australia. Clim Dyn 15:319–324

Power S, Tsetikin F, Mehta V, Lavery B, Torok S, Holbrook N (1999b) Decadal climate variability in Australia during the twentieth century. Int J Climatol 19:169–184

Power S, Delage F, Chung C, Kociuba G, Keay K (2013) Robust twenty-first-century projections of El Niño and related precipitation variability. Nature 502:541–545

Power SB, Delage FPD, Chung CTY, Ye H, Murphy BF (2017) Humans have already increased the risk of major disruptions to Pacific rainfall. Nat Commun 8:14368. https://doi.org/10.1038/ncomms14368

Rosenzweig C, Casassa G, Karoly DJ, Imeson A, Liu C, Menzel A, Rawlins S, Root TL, Seguin B, Tryjanowski P (2007) Assessment of observed changes and responses in natural and managed systems. In: Parry ML, Canziani OF, Palutikof JP, van der Linden PJ, Hanson CE (eds) Climate change: impacts, adaptation and vulnerability. Contribution of Working Group II to the Fourth Assessment Report of the Intergovernmental Panel on Climate Change. Cambridge University Press, Cambridge, pp 79–131

Salinger MJ, Renwick JA, Mullan AB (2001) Interdecadal Pacific oscillation and South Pacific climate. Int J Climatol 21:1705–1721

Salinger MJ, McGree S, Beucher F, Power SB, Delage F (2014) A new index for variations in the position of the South Pacific convergence zone 1910/11–2011/2012. Clim Dyn 43(3–4):881–892

Seager R, Naik N, Vecchi GA (2010) Thermodynamic and dynamic mechanisms for large-scale changes in the hydrological cycle in response to global warming. J. Clim 23: 4651–4668.

Stevenson S, Fox-Kemper B, Jochum M, Neale R, Deser C, Meehl G (2012) Will there be a significant change to El Niño in the twenty-first century? J Clim 25:2129–2145

Taylor KE, Stouffer RJ, Meehl GA (2012) An overview of CMIP5 and the experiment design. Bull Am Meteorol Soc 93:485–498

Trenberth KE (1976) Spatial and temporal variations in the southern oscillation. Quart J R Meteor Soc 102:639–653

Trenberth KE (1997) The definition of El Niño. Bull Am Meteorol Soc 78:2771–2777

Trenberth KE, Caron JM (2000) The southern oscillation revisited: sea level pressures, surface temperatures, and precipitation. J Clim 13:4358–4365

Trenberth KE, Shea DJ (1987) On the evolution of the southern oscillation. Mon Weather Rev 115:3078–3096

Trenberth KE, Stepaniak DP (2001) Indices of El Niño evolution. J Clim 14:1697–1701

Troup AJ (1965) The southern oscillation. Q J R Meteorol Soc 91:490–506

Vecchi GA, Soden BJ, Wittenberg AT, Held IA, Leetma A, Harrison MJ (2006) Weakening of the topical atmospheric circulation due to anthropogenic forcing. Nature 419:73–76

Vincent DG (1994) The South Pacific convergence zone (SPCZ): a review. Mon Weather Rev 122:1949–1970

Vincent EM, Lengaigne M, Menkes CE, Jourdain NC, Marchesiello P, Madec G (2011) Interannual variability of the South Pacific convergence zone and implications for tropical cyclone genesis. Clim Dyn 36:1881–1896

van Vuuren DP et al (2011) The representative concentration pathways: an overview. Clim Chang 109:5–31

Walker GT (1923) Correlations in seasonal variations of weather VIII. Mem India Meteorol Dept 24:75–131

Walker GT (1924) Correlations in seasonal variations of weather IX. Mem India Meteorol Dept 24:333–345

Wang G, Dommenget D, Frauen C (2015) An evaluation of the CMIP3 and CMIP5 simulations in their skill of simulating the spatial structure of SST variability. Clim Dyn 44:95–114

Widlansky M, Webster P, Hoyos C (2011) On the location and orientation of the South Pacific convergence zone. Clim Dyn 36:561–578

Widlansky MJ, Timmermann A, Stein K, McGregor S, Schneider N, England MH, Lengaigne M, Cai W (2013) Changes in South Pacific rainfall bands in a warming climate. Nat Clim Chang 3:417–423

Yeh S-W, Kug J-S, Dewitte B, Kwon M-H, Kirtman BP, Jin F-F (2009) El Niño in a changing climate. Nature 461:511–674

Zhang X, Church JA (2012) Sea level trends, interannual and decadal variability in the Pacific Ocean. J Geophys Res 39:L21701. https://doi.org/10.1029/2012GL053240

Chapter 4
Comparison of the Physical Susceptibility of Pacific Islands to Risks Potentially Associated with Variability in Weather and Climate

Lalit Kumar, Ian Eliot, Patrick D. Nunn, Tanya Stul, and Roger McLean

4.1 Introduction

Events such as marine inundation and coastal erosion are associated with natural variation in weather and climate, the former at short and the latter over longer time scales. The events unequivocally demonstrate some places are more susceptible to particular types of events than others. Arguably some are more vulnerable than others, making them more or less susceptible to changes in climate. However, vulnerability is defined and interpreted in many ways (Hinkel 2011). The Intergovernmental Panel on Climate Change (IPCC) defines vulnerability as "the extent to which a system is sensitive to climate change, including climate variability and extremes, or unable to cope with it" (IPCC 2007). While this definition assumes there are many forms and physical causes of vulnerability and its impact, the definition commonly

L. Kumar (✉)
School of Environmental and Rural Science, University of New England,
Armidale, NSW, Australia
e-mail: lkumar@une.edu.au

I. Eliot
University of Western Australia and Damara WA Pty Ltd., Innaloo, WA, Australia
e-mail: ian.eliot@bigpond.com

P. D. Nunn
University of the Sunshine Coast, Maroochydore, QLD, Australia
e-mail: pnunn@usc.edu.au

T. Stul
Damara WA Pty Ltd., Innaloo, WA, Australia
e-mail: tanya@seaeng.com.au

R. McLean
School of Physical, Environmental and Mathematical Sciences, University of New South
Wales at the Australian Defence Force Academy, Canberra, ACT, Australia
e-mail: r.mclean@adfa.edu.au

© Springer Nature Switzerland AG 2020
L. Kumar (ed.), *Climate Change and Impacts in the Pacific*, Springer Climate,
https://doi.org/10.1007/978-3-030-32878-8_4

encompasses a range of concepts, such as harmful effects on biota and lack of adaptability, particularly by humans (IPCC 2014).

Issues related to vulnerability are commonly discussed in a similar context to risk assessment, in which risk may result from short- to long-term variability in weather and climate. The risks include floods, droughts, cyclones, heat waves and other extreme climate-associated events to which ecosystems, people and the economy are exposed (Mambo 2017). In this context, vulnerability is a useful analytical tool for defining the exposure to damage, impotence and marginality of physical and social systems. The multidimensional nature of risk, vulnerability and impact of physical, social, economic and political frameworks make vulnerability assessment dynamic, complex and constantly changing (Adger 2006).

Recognition of interactions between physical and biologic factors as components of natural systems is a significant part of any assessment of vulnerability in its broad sense. However, it is not the approach adopted here. Instead, the information reported here is a first step to a more detailed comparison of island vulnerabilities made possible through the application of GIS techniques. Hence the term *susceptibility* has been used to identify the narrower, physical assessment of the relative extent to which small Pacific islands may be affected by climate and oceanographic processes. The objective of our project was to determine geographic diversity in susceptibility at a regional scale suited to strategic planning and relate that to marine and climate processes known to present a risk to island inhabitants.

The approach adopted is partly consistent with the IPCC (2007) definition of vulnerability in as much as it examines the physical attributes and oceanographic setting of 1532 islands across 15 Pacific island countries from a database developed for the purposes of comparing the relative susceptibility of islands to geographic variation in ocean water level, the annual average significant wave height and the frequency of tropical cyclones. The spread of island countries in the database is illustrated in Fig. 4.1. The database included information such as island name, island type, country, area, perimeter, maximum elevation and lithology. Most of the information was extracted from various country reports. The reference for island names was Motteler and Bryan (1986); however Google Earth and country maps were also used. Time series information describing change in water levels, wave heights and tropical cyclone frequencies was obtained from the CSIRO (CSIRO 2015).

4.2 Aim

The aim of the project was to develop of an index for broad-scale comparison of the susceptibility of small islands in the Pacific to climate change. At a whole-island scale, some island types are inherently more susceptible or resistant to change resulting from external processes. Islands or groups of islands have unique physical characteristics and geographic attributes. Depending on where they are located, they are exposed to differing climatic and oceanographic conditions. Thus, each island will either be relatively more at risk or less at risk from climate-related changes.

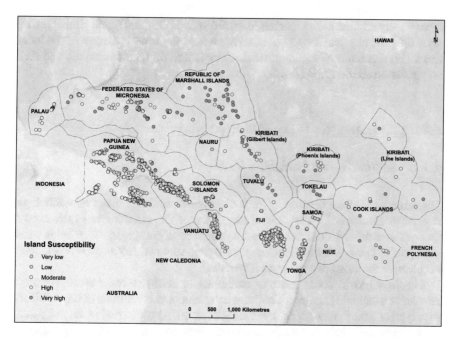

Fig. 4.1 Indicative susceptibility of the 1532 islands based on criteria shown in Table 4.1

Table 4.1 Ranking and cut-off values for the variables used to determine the indicative susceptibility of islands

1. Lithology		2. Circularity		3. Height		4. Area	
Material	Rank	Roundness index	Rank	Maximum elevation (m)	Rank	Area (km²)	Rank
Continental or volcanic high or volcanic low	1	Round 0.75–1	1	>100	1	>100	1
Composite high or composite low	2	Sub-rounded 0.5 to <0.75	2	30–100	2	10–100	2
Limestone high or limestone low	3	Sub-angular 0.25 to <0.5	3	10 to <30	3	1 to <10	3
Reef island	4	Angular 0 to <0.25	4	<10	4	<1	4

Here three indices were used to organise and objectively interpret the available information. The first, *indicative susceptibility*, refers to island type and combines several physical attributes for each island. This aspect of the analysis has been described in more detail by Kumar et al. (2018). The indicative susceptibility index was extended by combining it with a second index, an *exposure index*, describing broad-scale climate and oceanic processes. The combination constituted a third index which arguably describes the *geomorphic susceptibility* of Pacific islands to climate and oceanographic change. The term susceptibility is used instead of the more widely applied term vulnerability defined by IPCC (2007) and Hinkel (2011) since only the physical aspects of islands are used in the development of the index and no human dimensions are utilised. Compilation of the three indices and their interpretation are described below.

4.3 The Indicative Susceptibility Index

4.3.1 Methods and Variables Used

The variables used for developing the indicative susceptibility index were lithology, maximum elevation, area and circularity. Each of the variables was ranked on a four-point scale, with one being least susceptible to change and four being most susceptible. These individual rankings were then summed across all four variables to calculate the indicative susceptibility index.

Lithology provides a measure of erodibility or ability to resist change through erosion or weathering. It describes the relative hardness or softness of the dominant rock type of a particular island. For example, an island comprised mainly of volcanic rocks is less likely to readily change its form compared to an island made up primarily of unconsolidated sediments, such as sandy or reef islands. The island categories used here are described in detail in Chap. 2. For lithology, continental and volcanic high and low islands were ranked as one (least susceptible to change), composite low and high islands were ranked as two, limestone high and low islands were ranked as three, and reef islands ranked as four (most susceptible to change (see Table 4.1).

Maximum elevation was used as a variable since it provides a surrogate measure of an island to marine inundation. While it would have been better to use a median or mean value for elevation, accurate elevation data for whole islands for all islands in the Pacific is not available. In the absence of such data, maximum elevation values were used.

The islands in the database had elevations ranging from 0 m to 2715 m. The islands having the lowest elevations would be most susceptible to marine inundation and change and thus were given the lowest rank. The divisions were subjective and are explained in more detail in Chap. 2. The rankings used for maximum elevation were greater than 100 ranked one, 100 m to 30 m ranked two, 30 m to 10 m ranked three and less than 10 m ranked four (Table 4.1).

Island area was used as one of the variables since area can be related to susceptibility of an island to change. All other variables being equal, a larger island would be more stable than a smaller island. Island areas in the database ranged from 0.013 sq. km to 35,780 sq. km. These were divided into four categories, again subjectively, as greater than 100 sq. km ranked one (least susceptible), 100 sq. km to 10 sq. km ranked two, 10 sq. km to 1 sq. km ranked three, and less than 1 sq. km ranked four (most susceptible) (Table 4.1).

Circularity essentially describes the plan shape of an island. The variable was used as a measure in development of the susceptibility index because the shape of an island can also determine how vulnerable it is, although this proposition may require closer investigation. The shape of an island arguably affects factors such as wave focussing due to refraction, amplification of storm surge in embayments and patterns of nearshore water movement. Circularity was calculated as the ratio of the shape of a circle to the shape of an island polygon. A circle has a shape factor of 3.54 ($P_{circle}/\sqrt{A_{circle}} = 2\pi r/\sqrt{(\pi r^2)} = 3.54$), with the circularity of an island calculated as $3.54/(P_{island}/\sqrt{A_{island}})$. If an island was perfectly circular, the ratio would be one, and for all other islands, it would be less than one, approaching zero for the least circular islands. This index was then divided into four classes in 0.25 intervals and ranked from lowest to highest susceptibility according to 1, ≥0.75; 2, 0.5 to <0.75; 3, 0.25 to <0.5; and 4, <0.25 (Table 4.1).

The four variables were summed without any weightings being applied. There were suggestions that lithology is a more important factor and so could be given a higher weight than the other factors; however, for this exercise, it was decided to stay with equal weightings. The sums of the four variables gave scores from 4 to 16; these were then categorised into five susceptibility classes as 4–6 (very low susceptibility), 7–8 (low susceptibility), 9–11 (moderate susceptibility), 12–13 (high susceptibility) and 14–16 (very high susceptibility).

4.3.2 Results: Susceptibility Index

The indicative susceptibility of the 1532 islands in the database is shown in Fig. 4.1. All indicative susceptibility classes are represented in the distribution. Figure 4.1 also shows that the distribution of susceptibility classes across the Pacific is not uniform; there are clusters of susceptibility classes. A high percentage of high and very high susceptibility classes occur in an arc from Palau in the northwest to the Tuamotus in the southeast. Many of the islands in Micronesia and Polynesia fall into high and very high susceptibility classes. On the other hand, most of the Melanesian islands and islands in the eastern Pacific fall into low and very low susceptibility classes. Overall, 12% of islands are in the very low susceptibility class, 23% in the low, 25% in the medium, 31% in the high and 9% in the very high class (Fig. 4.2). The overall distribution closely follows a normal distribution curve.

The distribution of the indicative susceptibilities for the different island-type groups is illustrated in Fig. 4.3. There is a marked difference in the distributions for different island types. The majority of volcanic islands have either very low or low indicative susceptibility, with no islands falling in the high or very high susceptibility classes. Conversely, the majority of reef islands fall into the very high or high indicative susceptibility classes, with no reef island falling in the low or very low susceptibility classes. Limestone low islands generally fall into the medium and high susceptibility classes, while most of the limestone high islands fall in the

Fig. 4.2 Overall
distribution of indicative
susceptibility

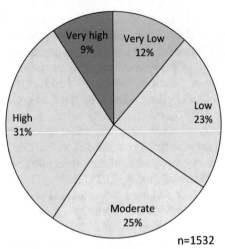

Indicative susceptibility - Overall distribution

n=1532

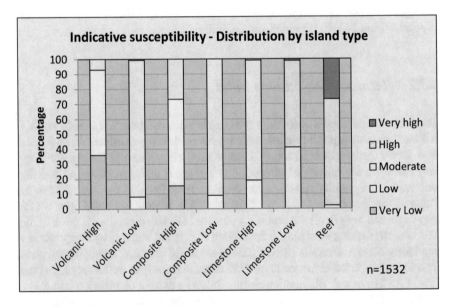

Fig. 4.3 Distribution of indicative susceptibility by island type

medium susceptibility class, similar to the volcanic low islands. Composite high islands are distributed across the very low, low and medium susceptibility classes.

Island susceptibility breakdown by country is given in Table 4.2 and Fig. 4.4. The table and figure reveal there are marked differences in how different countries fare in this distribution. Some of the countries (such as Nauru, Niue and Samoa) have all their islands in the very low and low indicative susceptibility classes, whereas others (such as Marshall Islands, Tokelau and Tuvalu) have all their islands

Table 4.2 Counts of indicative susceptibility by country with modal indicative susceptibility shown in bold and colour coded

Country (n)	Very Low		Low		Moderate		High		Very High		Total	
	Count	% of Country	Count	% of Country	Count	% of Country	Count	% of Country	Count	% of Country	No. Islands	% of Islands
Cook Islands (15)	2	13%	3	20%	1	7%	4	27%	**5**	**33%**	15	1%
F.S. Micronesia (127)	8	6%	19	15%	10	8%	**65**	**51%**	25	20%	127	8%
Fiji (211)	39	18%	**80**	**38%**	47	22%	44	21%	1	0.50%	211	14%
Kiribati (33)	0	0%	0	0%	6	18%	**15**	**45%**	12	36%	33	2%
Marshall Islands (34)	0	0%	0	0%	0	0%	5	15%	**29**	**85%**	34	2%
Nauru (1)	0	0%	**1**	**100%**	0	0%	0	0%	0	0%	1	0%
Niue (1)	0	0%	**1**	**100%**	0	0%	0	0%	0	0%	1	0%
Palau (33)	1	3%	3	9%	**16**	**48%**	12	36%	1	3%	33	2%
Papua New Guinea (437)	61	14%	93	21%	126	29%	**130**	**30%**	29	7%	439	29%
Samoa (7)	**5**	**71%**	2	29%	0	0%	0	0%	0	0%	7	0%
Solomon Islands (415)	29	7%	114	28%	107	26%	**137**	**33%**	26	6%	413	27%
Tokelau (3)	0	0%	0	0%	0	0%	0	0%	**3**	**100%**	3	0%
Tonga (124)	6	5%	5	4%	**56**	**45%**	51	41%	6	5%	124	8%
Tuvalu (10)	0	0%	0	0%	0	0%	5	50%	**5**	**50%**	10	1%
Vanuatu (81)	27	33%	**28**	**35%**	18	22%	8	10%	0	0%	81	5%
TOTAL	**178**	**12%**	**349**	**25%**	**387**	**23%**	**476**	**31%**	**142**	**9%**	**1532**	

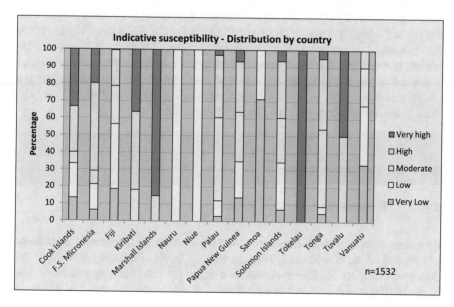

Fig. 4.4 Breakdown of indicative susceptibility by country

in the very high or high susceptibility classes. The Cook Islands, Federated States of Micronesia, Fiji, Palau, Papua New Guinea, Solomon Islands and Tonga have islands distributed across all the indicative susceptibility classes. Of the 15 countries in this study, 1 country has a modal indicative susceptibility class as very low, 4 as low, 2 as moderate, 4 as high and 4 as very high (Table 4.2).

4.3.3   Discussion: Indicative Susceptibility

This is a first-pass broad-scale assessment of island susceptibility at a whole-island scale, and many other attributes will affect the stability of islands. The variables used to compile the indicative susceptibility index were available for all 1532 islands. Despite the apparent limitations of scale and availability of information, the index provides a comprehensive measure of the indicative susceptibility of the 1532 islands in the Pacific to projected climate change. While the measure is coarse and is based on a limited set of physical variables of islands, it nonetheless allows us to rank the islands in terms of their physical susceptibility. This may not be a perfect measure, but it is an important step towards developing an index of direct relevance to the Pacific region.

The physical variables used have facilitated compilation of regional-scale maps of indicative susceptibility of islands independent of changes in climate or oceanic forces. The maps reveal the diversity and distribution of relative susceptibility of islands. They indicate which islands or island groups are more susceptible than others. The results also provide the first comprehensive breakdown of which countries are more at risk compared to others based on how many islands from each country fall in the high and very high susceptibility classes. Tokelau has all its islands in the very high susceptibility class, while Marshall Islands and Tuvalu have all their islands in the high and very high susceptibility classes. Kiribati has all its islands in the moderate, high and very high susceptibility classes. This in itself is useful information that can be used in planning and support of such countries. It provides an indication of where more support is needed. For example, the above results can be readily combined with population data for each of the islands to identify those sections of the communities who may be more vulnerable.

4.4   An Exposure Index

4.4.1   Methods and Variables Used

Processes associated with climate and oceanic factors all have major impacts on islands and island components, albeit to varying levels of intensity, frequency and duration. The most significant climate and ocean processes driving coastal change in the Pacific region are associated with prevailing (most common) and dominant (tropical storm) winds, wave action by sea and extra-tropical swell, tide type and range, sea-level variability associated with ENSO phase and longer-term sea-level change. All are occurring naturally and, perhaps with the exception of tides, may change in response to projected change in climate.

Three suites of process variables were considered for selection of parameters to indicate the vertical range of water level and wave activity, together with the influence of tropical cyclones on coastal dynamics. A range of parameters was available for each suite. However, due to questions of whether the parameter was meaningful for coastal response at an island scale, it had variability across the Pacific, and whether a dataset fully covering the Pacific region was available, only one parameter was selected from each suite to allow a ranking to be developed. For example, frequency, intensity, duration and approach direction are all relevant for wave activity and tropical cyclones, but only average annual significant wave height and tropical cyclone frequency were selected.

The three parameters selected for ranking exposure were a composite water-level range (tide and ENSO), average annual significant wave height and tropical cyclone frequency. Each parameter was separated into five categories to ensure sufficient spatial variation across the Pacific. Here, exposure was determined as the location of an island with respect to comparative variation of each parameter and expressed through a combination of their rankings.

A *composite water-level* parameter was developed to incorporate the vertical range of frequent (tide) and inter-annual (ENSO) variations in water level. Emphasis is placed on the vertical range of water-level fluctuations to demonstrate the significance of any projected long-term rise in sea level which would have its largest effects in areas of low water-level ranging. The parameter selected for total tidal range was a numerical model output of lowest astronomical tide to highest astronomical tide (LAT to HAT) as this is a measure that is not dependent on tide type, which is variable through the region, but indicates the maximum likely tidal excursion.

However, tidal range is not the sole consideration in the Pacific due to variations in water level attributable to ENSO, particularly in areas with a low tidal range. The absolute magnitude of water-level range due to ENSO is included when considering future effects of potential sea-level rise as it provides an indication of inter-annual water-level ranging and some likely resilience to landforms to small longer-term variations in mean sea level. To ensure the ENSO signal is incorporated appropriately in the ranking, it was added at twice its value (double weighted). This is particularly important as ENSO phases sustain higher or lower mean sea level for months or years as opposed to more frequent tidal movements. Thus the composite water-level parameter is a sum of the tidal range and twice the ENSO range. The LAT, HAT and ENSO raster layers were obtained from CSIRO numerical modelling.

The *composite water-level* parameter was split into five categories from very low (<1 m) to very high (>2.5 m) in 0.5 m intervals ensuring there was gradation across the Pacific (Table 4.3). The geographical variation in composite water-level categories is shown in Fig. 4.5.

Annual average significant wave height (H_s) was selected as the parameter for representing wave energy. This is an annual average of the wave height that is greater than two thirds of all modelled wave heights. The parameter is a proxy for the average wave energy available to move sediments within the average water-level range (e.g. composite water level; Fig. 4.6). The wave parameter is considered in conjunction with tropical cyclones to compare ambient and extreme conditions, as well as for consideration of the capacity of a system to be resilient to potential sea-level rise.

Table 4.3 Three variables used for the exposure index

		Variable		
		1. Composite water-level ranging	2. Annual average H_s	3. Tropical cyclone frequency
	Description	Composite WL = (HAT-LAT) + 2 × (ENSO ranging)	Annual average significant wave height	Based on the number of tropical cyclone tracks in longest dataset available
Value	Very low	<1.0 m	<1.0 m	None in available dataset
	Low	1.0 to <1.5 m	1.0 to <1.5 m	1 (<1 in 20 years)
	Moderate	1.5 to <2 m	1.5 to <1.75 m	2–8
	High	2.0–2.5 m	1.75–2.0 m	9–15
	Very high	>= 2.5 m	>2.0 m	>15 (>1 in 3 years)

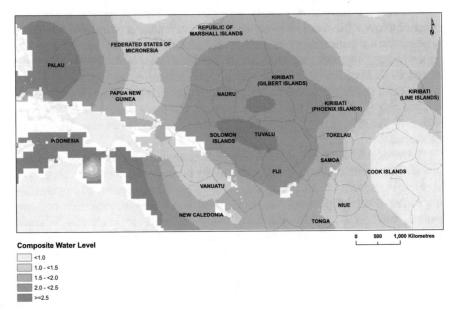

Composite Water Level

- <1.0
- 1.0 - <1.5
- 1.5 - <2.0
- 2.0 - <2.5
- >=2.5

Fig. 4.5 Composite water-level categories in the Pacific Islands region

The wave height parameter (H_s) is a raster layer prepared by CSIRO (2015) operating a numerical model to generate 30 years (1979–2009) of wave information at 30 km spatial resolution. The model was run hourly, with a monthly average applied, with a final average applied to each monthly average. The parameter was split into five categories from very low (<1 m) to very high (>2 m) according to the values in Table 4.4. Values for categories were selected to incorporate physical meaning for sheltering provided by island chains and ridges, as well as to correlate with wave energy (H_s^2). The geographical variation in categories in Fig. 4.6 clearly shows an east to west decline in wave height as well as a zonal decline towards the equator in both hemispheres.

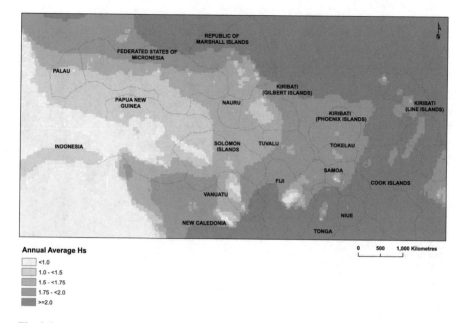

Fig. 4.6 Annual average significant wave height (H_s) categories in the Pacific Islands region

Table 4.4 Combining annual average significant wave height and tropical cyclone frequency variables

		TC				
		Very low	Low	Moderate	High	Very high
H_s	Very low	1	3.5	5	7	8
	Low	1.75	5.5	6.75	6.75	6.75
	Moderate	4	7.25	8	6.5	5
	High	2	5	5.5	3.5	2.5
	Very high	1	1.25	1.25	1	1

Tropical cyclone frequency was selected to represent extreme weather events. Frequency indicates whether an island experiences tropical cyclones. Here it is used to provide an indication of potential inundation related to storm surge and landform response to extreme events based on estimates of resilience and likelihood of disturbance. A map of tropical cyclone tracks from 1985 to 2005 (Wikipedia 2019) was annotated to separate the Pacific into five categories of tropical cyclone frequency from very low to very high (Table 4.3). This annotated figure was converted to a raster layer at 30 km spatial resolution. The very low category represents no tropical cyclones in the available dataset. This was required as the interaction of waves and water level would only be considered in these areas to ensure registration of a score on the index. A very high category indicates areas where tropical cyclones occur more frequently than one in 3 years. Geographical variation in the rankings is shown in Fig. 4.7.

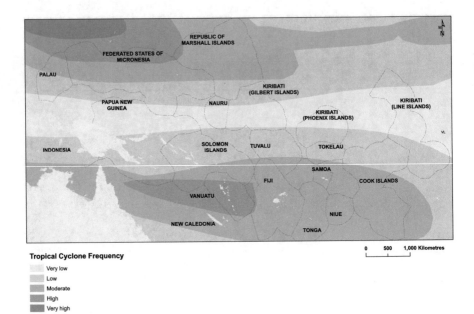

Tropical Cyclone Frequency
- Very low
- Low
- Moderate
- High
- Very high

Fig. 4.7 Tropical cyclone frequency categories in the Pacific Islands region

4.4.2 Compilation of the Exposure Index

The three parameters were combined in three steps to yield a single value for ranking exposure. First, the rankings of tropical cyclone frequency and annual average significant wave heights (H_s) were combined. Tropical cyclone influence is integrated with wave response. These parameters are linked in terms of the island landform resilience and likely disturbance as a result of potential sea-level rise, particularly storm surge and marine inundation of lowlands. Areas with exposure to frequent tropical cyclones are highly likely to have a high resilience to changing environmental parameters, and an area with exposure to no record of tropical cyclones will respond to waves only. Areas with low wave heights and high tropical cyclone frequency are more sensitive to changes in mean sea level as the coastal landforms have a limited capacity to rebuild when they are eroded or deflated during extreme events. Areas with high wave heights and low tropical cyclone frequency are likely to have a large hydraulic zone with capacity for rebuilding. However, such areas are most susceptible to human modification on coastal landforms and reef structures.

Based on this, the rankings of tropical cyclone frequency and annual average significant wave heights (H_s) were combined into a matrix to obtain a score (Table 4.4) for combination with water level in the second step. The results of applying this matrix to the Pacific are demonstrated in Fig. 4.8.

The second step determined the relationship of water level and sensitivity to changing environmental variables. The result is monotonic and inverse to the

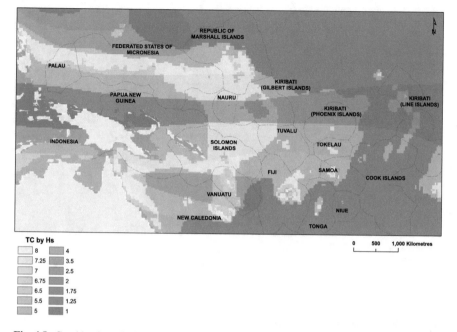

Fig. 4.8 Combined tropical cyclone frequency and annual average significant wave height (H_s)

Table 4.5 Composite water-level multiplier

		Multiplier
Composite water level	Very low	2
	Low	1.5
	Moderate	1
	High	0.5
	Very high	0.05

Table 4.6 Cut-off values for the exposure index

		Range
Process-based index	Very low	0–0.625
	Low	0.63–2
	Moderate	2.1–4
	High	4.1–8
	Very high	8.1–16

water-level range. A very low water-level range is most sensitive to changing environmental variables. A composite water-level multiplier (Table 4.5) was applied to the results in Fig. 4.5.

The third step involved multiplication of the values in Step 1 (Table 4.4; Fig. 4.8) and Step 2 (Table 4.5, Fig. 4.5) to generate values from 0.05 to 16 for the 125 unique combinations of the 3 parameters. Cut-off values were applied to this score to obtain the five-point process ranking (Table 4.6). The cut-off values were selected to gen-

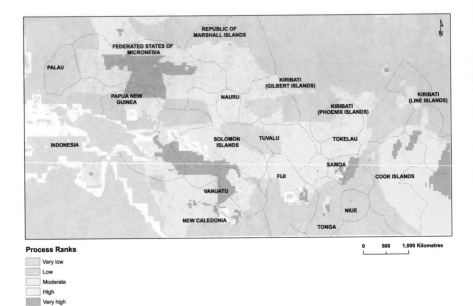

Fig. 4.9 Process sensitivity spatial distribution

erate a physically meaningful process ranking according to the authors' understanding of dynamics in the Pacific Ocean. The process rank is presented in Fig. 4.9 as a five-class raster. For each of the islands, the process ranking value that was the closest to that island in terms of Euclidean distance was selected and attributed with a process-based index.

The process-based index for the Pacific islands is presented in Fig. 4.10. The index was mainly moderate (26%), high (27%) and very high (30%) with less islands falling into the lower ranks of very low (5%) and low (12%). This skewness towards the moderate to higher rankings is attributed to the location of islands in areas of a combination of lower composite water-level ranging, moderate to high wave heights and some tropical cyclone activity zones.

4.5 Compilation of the Geomorphic Sensitivity Index

The term geomorphic sensitivity is applied to the combination of island susceptibility and process-based indices. Although the processes under consideration would have their greatest effect close to the shorelines of the islands, this is a means of considering the sensitivity and exposure of whole islands to potential environmental change by integrating both the processes and susceptibility. The five-point rankings of indicative susceptibility were combined with the five-point process-based index as shown in Table 4.7.

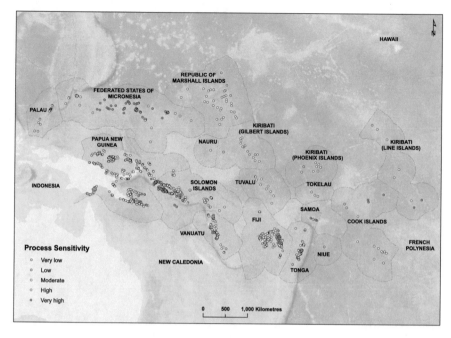

Fig. 4.10 Exposure index for the islands in the Pacific

Table 4.7 Combining indicative susceptibility and process-based indices to obtain a geomorphic sensitivity index

		Process-based index				
		Very low	**Low**	**Moderate**	**High**	**Very high**
Indicative susceptibility index	**Very low**	Very low	Very low	Very low	Low	Low
	Low	Very low	Low	Low	Moderate	Moderate
	Moderate	Low	Moderate	Moderate	High	High
	High	Moderate	High	High	Very high	Very high
	Very high	Moderate	High	Very high	Very high	Very high

4.6 Results

The geomorphic sensitivity for all islands is presented in Fig. 4.11. The geomorphic sensitivities for whole islands were mainly moderate (23%), high (28%) and very high (25%) with less falling into the lower ranks of very low (5%) and low (19%) (Fig. 4.12a). Even though many of the islands fall in the lower indicative suscepti- bility classes (Fig. 4.12b), they are located in the areas with higher process-based

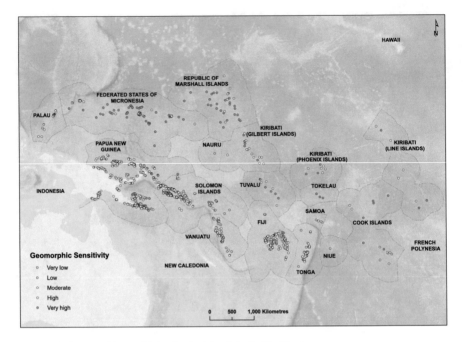

Fig. 4.11 Geomorphic sensitivity for the islands in the Pacific

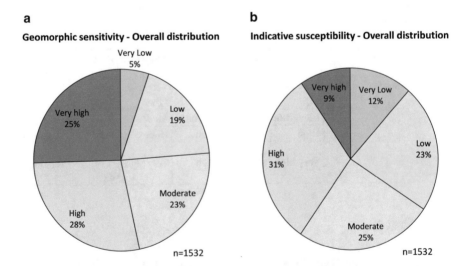

Fig. 4.12 Distributions of (**a**) geomorphic sensitivity and (**b**) indicative susceptibility (previous section)

index; hence they get bumped up in the rankings when considering geomorphic sensitivities. As an example, only 9% of the islands were in the very high indicative susceptibility class, but in geomorphic sensitivity ranking this increased to 25%.

The distribution of geomorphic sensitivity with island type is included in Fig. 4.13a. Volcanic high and composite high islands tend to have lowest geomorphic sensitivity. The islands with the highest geomorphic sensitivity are reef islands and the limestone low islands.

If an island is a volcanic high or composite high, it is most likely to be in the low or moderate geomorphic sensitivity categories. If an island is volcanic low, it is most likely to be in the moderate or high category. For composite low or limestone high islands, it is most likely to be in the high category. If an island is limestone low or a reef island, it is most likely to be in the very high category. This distribution is mainly related to their location in the Pacific.

The distribution of geomorphic sensitivity for each country is included in Fig. 4.14a and Table 4.8. The figure and table demonstrate a range of geomorphic sensitivities for all but single island countries, along with Tokelau. The five countries with the modal category of very high geomorphic sensitivity are the Federated States of Micronesia, Marshall Islands, Solomon Islands, Tokelau (all three islands) and Tuvalu. None of the countries have a very low modal geomorphic sensitivity. Three countries (Palau, Samoa and Vanuatu) have modal category as low geomorphic sensitivity. Three countries (Fiji, Papua New Guinea and Vanuatu) have a relatively even distribution of geomorphic sensitivity ranks, with islands in all five categories.

Table 4.8 Counts of geomorphic sensitivity by country with modal geomorphic sensitivity shown in bold and colour coded

| | Geomorphic sensitivity | | | | | | | | | | Total | |
| | Very Low | | Low | | Moderate | | High | | Very High | | | |
Country (n)	Count	% of Country	Count	% of Country	Count	% of Country	Count	% of Country	Count	% of Country	No. Islands	% of Islands
Cook Islands (15)	2	13%	3	20%	1	7%	3	20%	6	40%	15	1%
F.S. Micronesia (127)	1	1%	11	9%	16	13%	28	22%	71	56%	127	8%
Fiji (211)	11	5%	57	27%	62	29%	53	25%	28	13%	211	14%
Kiribati (33)	0	0%	0	0%	10	30%	12	36%	11	33%	33	2%
Marshall Islands (34)	0	0%	0	0%	0	0%	7	21%	27	79%	34	2%
Nauru (1)	0	0%	1	100%	0	0%	0	0%	0	0%	1	0.1%
Niue (1)	0	0%	1	100%	0	0%	0	0%	0	0%	1	0.1%
Palau (33)	4	12%	16	48%	13	39%	0	0%	0	0%	33	2%
Papua New Guinea (437)	52	12%	86	20%	111	25%	139	32%	49	11%	437	29%
Samoa (7)	0	0%	5	71%	2	29%	0	0%	0	0%	7	0.5%
Solomon Islands (415)	5	1%	70	17%	106	26%	107	26%	127	31%	415	27%
Tokelau (3)	0	0%	0	0%	0	0%	0	0%	3	100%	3	0.2%
Tonga (124)	1	1%	8	6%	3	2%	57	46%	55	44%	124	8%
Tuvalu (10)	0	0%	0	0%	0	0%	5	50%	5	50%	10	1%
Vanuatu (81)	3	4%	28	35%	27	33%	16	20%	7	9%	81	5%
TOTAL	79	5%	286	19%	351	23%	427	28%	389	25%	1532	

a

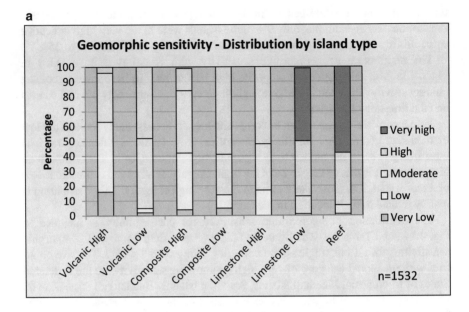

b

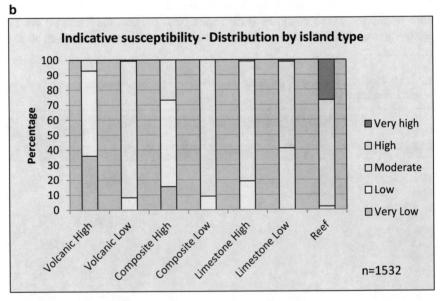

Fig. 4.13 Rankings separated by island type for (**a**) geomorphic sensitivity and (**b**) indicative susceptibility (previous section)

a

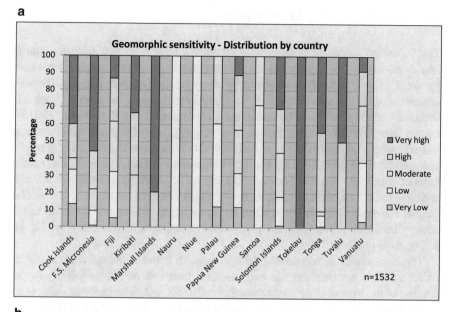

b

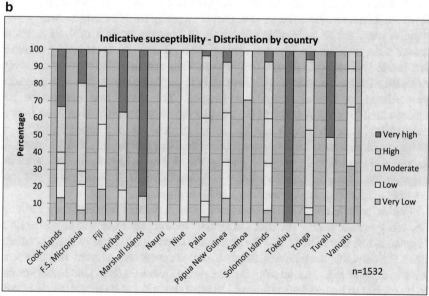

Fig. 4.14 Rankings separated by country for (**a**) geomorphic sensitivity and (**b**) indicative susceptibility (previous section)

4.7   Discussion and Conclusions

The index of geomorphic sensitivity considers how relatively sensitive whole islands are to changes in potential changes in climatic and oceanographic processes. Its use is illustrated in this assessment by comparison of geomorphic sensitivity of islands which combines indicative susceptibility (physical structure of islands including lithology, circularity, maximum elevation and area; Kumar et al. (2018)) and an exposure-based index (water-level ranging, wave height, tropical cyclone frequency) at a whole-island scale. Although this is a coarse assessment of geomorphic sensitivity, it provides a comparative perspective of islands in the Pacific on the basis of consistent information across the region. In particular, the study provides a regional-scale analysis and maps of areas where coastal fringes of, if not whole, islands are more likely to be sensitive to potential changes in sea level and storm activity based on physical island characteristics and natural variability of climate and oceanic processes.

The exposure index was developed to consider whole-island sensitivity to potential changes in weather, climate and sea level based on natural variability in the ranging of the hydraulic zone and potential resilience and disturbance of the system by extreme events, with less emphasis on present-day risk. This index incorporated three parameters and was intended to provide a proxy for other drivers, with investigation restricted to three parameters only to limit the number of outcomes and ensure a range of sensitivities are achieved. The composite water-level parameter represented the vertical range of frequent and inter-annual variations in water level, as a proxy for the hydraulic zone. Sensitivity to projected sea-level rise is inversely proportional to natural water-level ranging. The other two parameters are incorporated to compare ambient (annual average wave height) and extreme (tropical cyclone frequency) conditions. This approach does not incorporate the influence of human modification on the system resilience. For example, areas with high wave heights and low tropical cyclone frequency already have a large hydraulic zone with capacity for rebuilding and have a low exposure index, with these areas most susceptible to modified coastal landforms and reef structures.

Further parameters should be considered when investigating sensitivity and vulnerability at finer scales. The focus of which landforms are likely to be susceptible to change, together with the driving processes, significantly alters as scale becomes finer. For example, the whole-island approach becomes less relevant, and it is possible to exclude high ground from the analyses as focus shifts to land levels known to be subject to marine inundation and likely to be affected by projected change in sea level. Similarly, the roles of sea and swell could be expected to vary around an island or in different parts of an archipelago. Additionally, the exposure index was developed using modelled information at a regional scale due to sparsity of actual datasets. A 30 km spatial resolution was available for the modelled information, and therefore the exposure index was prepared at this resolution. An example of a limitation of this resolution is that it is unlikely to capture the variability in processes influencing small islands within an archipelago, with all islands attributed the same exposure rank due to scale effects.

Categorisation cut-offs for the three process parameters, the combined tropical cyclone and wave height rank and the geomorphic sensitivity rank, were subjective, albeit with some consideration of spatial controls. It is not anticipated that slightly altering the cut-offs would alter the broader spatial trends observed, with potential changing of one level in rank for islands located near boundaries in the process ranking. Indicative susceptibility was weighted with a non-linear skewness in the geomorphic sensitivity ranking as it is the notional "receptor" or response variable. This results in islands with very low indicative susceptibility only able to result in very low or low geomorphic sensitivity rankings. Islands with very high indicative susceptibility rankings are skewed to have a higher likelihood of a very high geomorphic sensitivity ranking.

Despite the limitations of scale and availability of information, the exposure ranking in Fig. 4.11 provides an indication of regional variability in relative sensitivity to projected changes in climate and oceanographic processes. Focal high and very high exposure ranks occur in the area of low composite water level through Federated States of Micronesia, through Papua New Guinea and Tuvalu (Fig. 4.9). This is the area with low tidal ranges and is the node for ENSO fluctuations. The majority of the area considered has a low exposure ranking because of the coverage of either high wave heights or very low (no) tropical cyclones; the latter is a central band through the area. Very low exposure ranks are in the west and in patches near Kiribati and Tuvalu due to the very low water-level rank associated with high tidal ranges. Areas of moderate exposure are either attributed to high water-level ranging or very low water-level ranging in the south-east with very high wave heights and low to moderate tropical cyclones. Many divisions of the exposure ranking are associated with wave sheltering by large islands and island ridges, such as leeward of Kiribati, and sheltering within island groups, such as Fiji. When the exposure ranking is attributed to islands as an exposure index (Fig. 4.6), there is a skewness towards moderate to higher rankings due to the location of many islands in areas with lower composite water-level ranging, moderate wave heights and some tropical cyclone activity.

When the exposure and indicative sensitivity indices are combined, most (93%) of the reef islands are ranked as high and very high geomorphic sensitivity (Fig. 4.13a) since they have moderate indicative susceptibility with high to very high exposure, high indicative susceptibility with low to moderate exposure or very high indicative susceptibility with low to very high exposure. Reef and limestone low islands are the only island types with very high geomorphic sensitivity largely attributed to the physical susceptibility of the islands. Composite low and limestone high islands have a modally high geomorphic sensitivity (48–59%) with most (83–86%) categorised as moderate to high. The high geomorphic sensitivity for these island types is attributed to moderate susceptibility and high to very high exposure. Volcanic low islands have mostly (95%) moderate and high geomorphic sensitivities, with equal distributions of the two ranks due to the moderately ranked susceptibility of the islands located in areas ranging from low to very high exposure.

Composite high islands are mainly (80%) low to moderate ranks of geomorphic sensitivity with low susceptibility with higher exposure or moderate susceptibility with low to moderate exposure. Volcanic high islands are mostly very low to moderate (96%), with the modal rank of low (47%). The low ranking is attributed to a very low susceptibility with high to very high exposure or low susceptibility with low to moderate exposure. Very low geomorphic sensitivity is restricted to volcanic and composite island types because of the low to very low indicative susceptibility and very low to moderate exposure.

Geomorphic sensitivity varied regionally, between and within countries. On a regional scale, the northern and eastern areas of the Pacific region are geomorphically sensitive with 80% high and very high rankings. Most (92%) of these are reef islands of small size and contribute to less significant wave sheltering of adjacent islands. Elsewhere, many islands with very low and low geomorphic sensitivity are attributed to wave sheltering due to their location in the lee of larger islands or island ridges (e.g. Papua New Guinea) or within island groups (e.g. Fiji).

On a country scale, Palau is an exception to the higher sensitivity islands in the north and east, with lower geomorphic sensitivities due to very low exposure (high natural variability of all processes). The Federated States of Micronesia have mostly very high geomorphic sensitivity (61%) due to the location of susceptible small, reef islands in areas of very high process-based index. The very low to low water-level ranging in areas of moderate wave conditions and subject to tropical cyclones suggests a small projected sea-level rise would disturb the coastal landforms. Kiribati has modal moderate geomorphic sensitivity, with some highly susceptible reef and limestone low islands located in an area of low exposure. A combination of relatively large water-level ranging, no tropical cyclone activity and low wave heights contributes to the low exposure index. Solomon Islands and Fiji demonstrate the influence of sheltering on geomorphic sensitivity for islands located in the lee of ridge of islands or large islands (Solomon Islands) and internal sheltering in clustered islands (Fiji). Both Solomon Islands and Fiji are located on boundaries of water-level ranging, with sheltering demonstrated by areas of lowest wave energy rank. The local wave sheltering contributes to exposure indices in all five categories.

The modal sensitivity to projected changes in environmental conditions for volcanic islands is one rank higher than the susceptibility for all island types other than volcanic high islands. Volcanic low, composite low and limestone high islands had minimal high susceptibilities, with 50–60% high in geomorphic sensitivity. The proportion of very high ranks for limestone low and reef islands changed from 1–27% for indicative susceptibility to 49–58% for geomorphic sensitivity. A greater proportion of islands in the very low category occurred for geomorphic sensitivities for volcanic low and composite low islands compared to indicative susceptibility, with decrease in proportion for volcanic high and composite high islands.

Countries where the distribution of rankings within it changed for geomorphic sensitivity compared to indicative susceptibility were the Federated States of Micronesia, Kiribati, Palau, Samoa, Tonga and Vanuatu (Fig. 4.14a, b). The Federated States of Micronesia had a modal (51%) high rank for indicative susceptibility and is located in an area with modal (61%) very high exposure ranking, generating an upward shift to a modal (56%) very high geomorphic sensitivity. Kiribati had a decrease in the rankings for geomorphic sensitivity compared to indicative susceptibility due to some areas of low process ranking as explained above. Palau had 40% of islands with high or very high indicative vulnerability with no geomorphic sensitivities in these categories due to a downward shift from a very low exposure ranking in an area of high exposure. The islands of Samoa all increased rank by one when applying the high and very high exposure ranks with very low and low indicative susceptibilities and low and moderate geomorphic sensitivities. Tonga has a modal (79%) very high exposure ranking (generally lower water levels and high wave energy), generating geomorphic sensitivities of 90% high or very high, for islands with 45% high or very high indicative susceptibility. Vanuatu had 33% of islands with very low indicative susceptibility with no very high ranks. Application of modally high (69%) exposure ranking reduced the ranks of very low to 4% and resulted in 9% very high geomorphic sensitivities compared to no very high indicative susceptibilities.

Compilation of the index of geomorphic sensitivity establishes how relatively sensitive whole islands are to changes in potential weather, climate and oceanic conditions through consideration of their physical characteristics (Kumar et al. 2018) and exposure to existing regional climate and oceanic processes. The value of assessing geomorphic sensitivity is that it establishes a method to consider which islands are more sensitive to projected environmental change, relative to other islands in the Pacific region. Although further detailed assessment is required for each island, the geomorphic sensitivities potentially provide a basis for regional strategic planning by countries within the region and external donor organisations supporting mitigation of the impacts of climate and ocean processes driving coastal change.

References

Adger W (2006) Vulnerability. Glob Environ Chang 16(3):268–281. https://doi.org/10.1016/j.gloenvcha.2006.02.006

CSIRO (2015) Tidal dataset— CAMRIS—lowest astronomical tide. v1. CSIRO. Data collection. Accessed 25 May 2019, CSIRO https://doi.org/10.4225/08/55148535DD183

Hinkel J (2011) "Indicators of vulnerability and adaptive capacity": towards a clarification of the science–policy interface. Glob Environ Chang 21(1):198–208. https://doi.org/10.1016/j.gloenvcha.2010.08.002

IPCC (2007) Climate change 2007: impacts, adaptation and vulnerability. Contribution of working group II to the fourth assessment report of the intergovernmental panel on climate change. Accessed https://www.ipcc.ch/site/assets/uploads/2018/03/ar4_wg2_full_report.pdf

IPCC (2014) IPCC 2014: Summary for policymakers in climate change 2014: impacts, adaptation, and vulnerability. Part A: Global and sectoral aspects. Contribution of Working Group II to the Fifth Assessment Report of the Intergovernmental Panel on Climate Change. Accessed https://www.ipcc.ch/site/assets/uploads/2018/03/ar5_wgII_spm_en-1.pdf

Kumar L, Eliot I, Nunn PD, Stul T, McLean R (2018) An indicative index of physical susceptibility of small islands to coastal erosion induced by climate change: an application to the Pacific islands. Geomat Nat Haz Risk 9(1):691–702. https://doi.org/10.1080/19475705.2018.1455749

Mambo J (2017) Risk and vulnerability to global and climate change in South Africa. In: Mambo J, Faccer K (eds) South African risk and vulnerability atlas, 2nd edn. Media, Africa, p 160. https://www.csir.co.za/sites/default/files/Documents/CSIR%20Global%20Change%20eBOOK.pdf:AfricanSun

Motteler L, Bryan E (1986) Pacific island names: a map and name guide to the new Pacific, vol 34, 2nd edn. Bishop Museum Press, Honolulu

Wikipedia (2019) WikiProject tropical cyclones. Wikipedia. Accessed https://en.wikipedia.org/wiki/Wikipedia:WikiProject_Tropical_cyclones

Chapter 5
Downscaling from Whole-Island to an Island-Coast Assessment of Coastal Landform Susceptibility to Metocean Change in the Pacific Ocean

Ian Eliot, Lalit Kumar, Matt Eliot, Tanya Stul, Roger McLean, and Patrick D. Nunn

5.1 Introduction

This chapter describes coastal vulnerability assessment downscaling for Pacific Islands, from a primary-scale, regional assessment (Kumar et al. 2018, Chap. 4) to an intermediate secondary-scale assessment, intended to inform coastal management at an archipelagic or country level. It forms part of a hierarchical suite of assessments with potential to downscale in a consistent manner from regional to local scale. The step described in this chapter involves relating vulnerability of whole islands to changing climate and ocean conditions, appropriate for regional strategic planning, to holistic assessment of the marine and terrestrial landforms that form island shores.

I. Eliot (✉)
University of Western Australia and Damara WA Pty Ltd., Innaloo, WA, Australia
e-mail: ian.eliot@bigpond.com

L. Kumar
School of Environmental and Rural Science, University of New England,
Armidale, NSW, Australia
e-mail: lkumar@une.edu.au

M. Eliot · T. Stul
Damara WA Pty Ltd., Innaloo, WA, Australia
e-mail: matt.eliot@damarawa.com; tanya@seaeng.com.au

R. McLean
School of Physical, Environmental and Mathematical Sciences, University of New South Wales at the Australian Defence Force Academy, Canberra, ACT, Australia
e-mail: r.mclean@adfa.edu.au

P. D. Nunn
University of the Sunshine Coast, Maroochydore, QLD, Australia
e-mail: pnunn@usc.edu.au

© Springer Nature Switzerland AG 2020
L. Kumar (ed.), *Climate Change and Impacts in the Pacific*, Springer Climate,
https://doi.org/10.1007/978-3-030-32878-8_5

At a broad-level, regional-scale analysis of the physical character of 1779 islands outlined in Chap. 4, Nunn et al. (2015) and Kumar et al. (2018) enabled an indicative assessment of island susceptibility based on their lithologic and geometric characteristics. Such assessment is relevant to high-level strategic planning in that the measures of indicative susceptibility provide insight allowing regional and country-wide evaluation; yet it does not identify problems of direct relevance at a local level, either within a particular island subgroup or for the coastal fringe around a specific island. However, such foci are necessary adjuncts to the regional approach because environmental problems most acutely occur on coastal landforms skirting the main structural body of an island. For example, retreat of the shoreline along a narrow, low-lying coastal plain adjoining a steeply rising hinterland is likely to affect built infrastructure. It is therefore desirable to downscale from the high-level primary assessment to more detailed levels for country, district (subnational) and local community planning and management purposes.

The overall approach comprises a series of steps, relating different scales in a management hierarchy ranging from regional, through country and island scales to local governance. The first downscaling step was developed through consideration of the separate but related concepts of *susceptibility* and *instability*, which determine the overall vulnerability of an island to changing conditions. *Susceptibility* is used here for physical characteristics that describe if a coast will respond to changing conditions, whereas *instability* is used to characterise the relative response to those changes. A simple illustration of this distinction is that a sandy coast is more *susceptible* to change than a rock coast, and a cliffed coast is more *unstable* than a gently graded coast. Examined at a broad, indicative scale, the physical structure of whole islands together with landforms of the coastal fringe around them is therefore characterised in terms of lithology, gross structural features and landform features.

When considered over a hierarchy of spatial scales, assessment of coastal vulnerability to changing conditions focuses on island geology at a coarse, regional scale, grading to interpretation of geomorphic features as the focus becomes finer. The envisaged transition encompasses increased detail of interaction between coastal processes and morphology but also reflects a transition from characteristics mainly describing susceptibility, increasingly towards those describing instability. The corresponding change to coastal vulnerability assessment with downscaling may alter the perception of marine and climate processes driving coastal dynamics as more subtle interactions between landform and process become apparent.

The following objectives were set in order to downscale: (1) establish a pathway for downscaling landform assessments at a conceptual level suited to the use of sparse or coarse information; (2) extend the range of variables used by Kumar et al. (2018) (and Chap. 4) to estimate the indicative *susceptibility* of island structure to changing climate and ocean; (3) develop criteria to comparatively estimate the *relative instability* of coastal landforms; and (4) apply the analysis to a variety of island types sufficient to demonstrate utility of the framework at a whole-island scale and, separately, for sections of an island coast sharing a common landform assemblage.

5.2   Downscaling and Upscaling

Downscaling and upscaling are techniques to deal with information and application occurring at multiple scales. Downscaling (upscaling) refers to the process of relating information derived from characterisation over larger (finer) scales to application at a finer (coarser) scale. The two techniques are important in coastal vulnerability assessment, as information is clustered, either as broadly available coarse information such as satellite-derived shoreline change (Luijendijk et al. 2018) or more sparsely available detailed measurement and evaluation (e.g. Davis 2013). In some situations, this results in separate methods of assessment based on information levels (Duong et al. 2017, 2018); however, this provides potential for inconsistency. For coastal systems, meaningful downscaling or upscaling typically is not simple interpolation or aggregation of information, with spatial relationships typically non-linear or even fractal in character and the process of aggregation being complicated by coherence or exchange. Downscaling and upscaling therefore require an introduction or loss of information with the change of scale. A hierarchical, internally consistent scaling framework can therefore be developed through progressive addition of information with each step of downscaling.

Although the dynamics of island coasts are broadly relatable to the dynamics of continental shelf coasts, differences in behaviour need to be considered at all scales of coastal management (PRIF 2017; Govan 2011; Giardino et al. 2018). Consequently, the process of downscaling and upscaling requires a framework that is specifically relevant to islands, with relevance to applied levels of decision-making.

A hierarchical scaling framework is ultimately proposed for management of Pacific islands, with the upper three scales indicated in Fig. 5.1, and the relationship between the two largest scales described in this paper. At all scales, a consistent procedure should be used to estimate the susceptibility of the most common geologic and morphologic features apparent: regional, whole-island, whole-island coast, coastal segment and individual landform scale (Fig. 5.1). Selection of these features is intended to provide consistency in methodology bridging the gap between regional-scale assessments of island susceptibility and fine-scale analyses applicable at a community level. At each scale, lithologic and morphologic features are used as criteria to determine the susceptibility of an island, its coastal fringe or part of the coast. The criteria vary from scale to scale as the degree of detail required in estimates of susceptibility or instability increases with downscaling in the hierarchy.

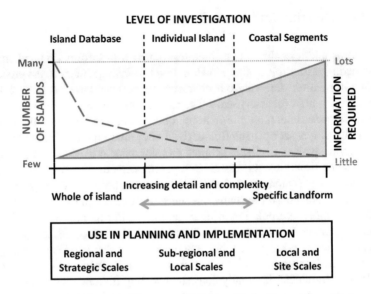

Fig. 5.1 The three upper scales of a hierarchical framework for downscaling of information at multiple scales

5.3   Methodology

5.3.1   Overview

A sample of 36 islands from the Pacific region was selected for estimation of the indicative vulnerability of each island as a whole entity. The islands selected include all island types in the Pacific listed by Nunn et al. (2015) and in Chap. 2.

In analysis for the first step in the downscaling sequence, the term *indicative vulnerability* specifically refers to possible change in whole-island structural response to changing environmental conditions. It is determined by combining the estimate of susceptibility based on the lithology and geometry of islands described by Kumar et al. (2018) with the relative instability of the most common landforms around its coastal fringe. The combination is made through compilation of a matrix and derivation of a value indicating whether the indicative vulnerability of the island should be considered as being very high, high, moderate, low or very low.

Explanation of Concepts

Coastal fringe is comprised of submarine and subaerial (terrestrial) landforms adjoining the intertidal shore.

Coastal vulnerability is the term commonly applied to the results of hazard and risk assessments involving people and land use in coastal areas.

Long-term changes to coastal landforms and the processes driving them are those occurring interdecadally, extending over a planning horizon of at least 100 years and longer.

Metocean is a combination of the words meteorologic and oceanographic. The term is used to infer both types of processes occurring separately or in combination.

Midterm changes are those occurring at an interannual to intra-decadal scales, including those in response to ENSO events.

Physical character describes the structural features of an island, its physical shape and the material of which it is mainly comprised. The description refers to each island as a single entity and not to the various landforms the structure supports.

Regional scale refers to interpretation of information at a very broad geographic level. It is commonly a scale used for strategic planning purposes at an international or national administrative level.

Short-term changes occur at seasonal and higher frequencies, such as changes in response to the passage of tropical cyclones.

Relative instability refers to the degree to which an island's coastal landforms, the marine and terrestrial landforms adjoining the shore, respond to change in climate-ocean conditions. A comparative qualitative scale of relative instability is used; for example, a low sandy beach is more sensitive to change from external processes than a high limestone cliff. The comparison is *indicative* rather than absolute or predictive.

5.3.2 Calculation of Susceptibility

The procedure used to determine whole-island susceptibility follows that described by Kumar et al. (2018), to which four new parameters were added: circularity, insularity, isolation (proximity) and gradient to deep ocean seabed. For each of the islands, we obtained information pertaining to island lithology, maximum elevation and area from literature and database compiled by Kumar et al. (2018). These and the additional four parameters were calculated for all 36 islands.

Circularity describes the plan shape of an island. The variable was used as a measure in development of the susceptibility index because the shape of an island may locally affect vulnerable parts of an island coast through determination of factors such as wave focusing due to refraction, amplification of storm surge in embayments and patterns of nearshore water movement. Following Kumar et al. (2018), circularity was calculated as the ratio of the shape of a circle to the shape of an island polygon. A circle has a shape factor of 3.54 ($P_{circle}/\sqrt{A_{circle}} = 2\pi r/\sqrt{(\pi r^2)} = 3.54$), with the circularity of an island calculated as $3.54/(P_{island}/\sqrt{A_{island}})$. If an island was perfectly circular, the ratio would be 1, and for all other islands, it would be less than 1, approaching 0 for the least circular islands. This index was then divided into four classes at 0.25 intervals and ranked from the lowest to the highest susceptibility according to (1) ≥ 0.75, (2) 0.5 to <0.75, (3) 0.25 to <0.5 and (4) <0.25 (see Chap. 4).

Insularity was calculated as the ratio of the island perimeter to the square root of its area. This combines a measure of shoreline length, including its irregularity, with the subaerial area of the island. If insularity is small, the island is considered to be less susceptible to less focusing of storm surge in embayments. It is noted that this parameter is almost the same as circularity but differs in that it is a direct measure of island area rather than comparison of island perimeter with a circle through application of the shape factor 3.54.

Proximity to nearby islands was determined by nearest neighbour check. This had two stages: first, a count of the number of neighbouring islands within a 20-km radius around the island was completed and, second, the island setting in relation to its neighbours was established using GIS software and proximity analytic tools. It is noted that this parameter could be strengthened by a more robust nearest neighbour analysis where the degree of shelter provided by neighbouring islands related to their size and elevation is considered.

The next parameter used was the gradient to deep ocean seabed or shoreface gradient. This refers to the slope from an island shore to the edge of any submarine platform or shelf. It was determined either centrally on the ocean side of an island within an archipelago surrounded by barrier reef or for the apparently steepest shore of an isolated island. Wave height and surge height are both affected by seabed gradient. Characterisation of gradient for Pacific islands is considered limited, due to sparse availability of bathymetric data, with estimates alternatively derived through interpretation of Google Earth imagery. This limitation could be addressed through collation of more comprehensive bathymetry data and refined consideration of reef characteristics.

Criteria describing each parameter were ranked on a five-point scale (Table 5.1). The final ranking for susceptibility is a ranking of 1–3 (low susceptibility, moderate susceptibility and high susceptibility). Weightings were also considered for specific variables: as a generic weighting and as variable weights based on lithology. The weighting of variables is inevitably subjective, but it may be used as an exploratory tool to establish the relative importance of the variables being used.

5.3.3   Determination of Island Instability

Parameters used to estimate instability of the coastal fringe around each island describe landform characteristics of the backshore, intertidal zone, inshore and reef. These characteristics are based on conceptual models of landform change over

Table 5.1 Criteria for the secondary assessment of instability

1. Backshore

(IA) Backshore elevation within 25 m of HWL	Rank	(1B) Backshore sediment	Rank	(IC) Backshore landform component on >50% of coast	Rank
Elevation >20 m	1	Soil over rock or stepped platforms and terraces	1	High (>10 m elevation) coastal plain or terrace with a steep gradient to landward	1
Elevation 15–20 m	2	Soil over coastal plains or partly infilled embayments, including alluvial fans and deltas	2	Coastal platform or rocky terrace (<10 m elevation)	2
Elevation 10–14.9 m	3	Soil on an atoll island (core is unknown)	3	Coastal plains (including beach ridge plains, outwash plains, deltas and alluvial fans)	3
Elevation 5–9.9 m	4	Washover features on >25% coast or mangrove forests	4	Coastal flats (including partly infilled embayments and mangrove forests) or cuspate forelands or spits	4
Elevation <5 m	5	Unconsolidated sediment	5	Tectonically unstable island with active volcanoes	5

2. Intertidal

(2A) Intertidal sediment type on >50% of coast	Rank	(2B) Intertidal landforms on >50% of coast	Rank
High cliffs or steep rock ramps	1	Cliffs or bluffs, possibly adjoining rock platforms or ramps	1
Coral rubble (gravel and cobble) beach	2	Rocky headlands and small bay beaches	2
Mixed sand and cobble beaches	3	Mainly rocky shores, including beachrock ramps and rock platforms	3
Sandy beaches	4	Island or islet with sandy beaches and washover features on an atoll coast *or* Long sandy beaches separated by rocky topography such as headlands, cliffs and bluffs on a non-atoll coast	4
Mudflats (mangrove forests)	5	Bare reef platform with passages and washover features (e.g. depositional fan) on an atoll coast *or* Irregular shoreline with mangroves, tidal creeks and partly infilled inlets or long sandy beaches abutting coastal flats, coastal plains or cuspate forelands on non-atoll coast	5

(continued)

Table 5.1 (continued)

3. Inshore zone	
(3A) Inshore morphology for >50% of coast	Rank
Stepped subtidal sand terrace grading to beach	1
Inshore lagoon with patch reef and sand sheets	2
Inshore lagoon with reef pavement and bare sand sheets	3
Discontinuous subtidal platform and reef pavement	4
Continuous subtidal platform and boulder ramp	5

4. Reef					
(AA) Seaward reef type	Rank	(4B) Seaward reef width	Rank	(4C) Reef coverage (reef width > 50 m)	Rank
Plunging cliffs or boulder ramps	1	Width > 200 m	1	Continuous reef with >90% sheltering of shore	1
Fringing reef attached to a rocky island	2	Width 150–200 m	2	Nearly continuous reef with 70–90% sheltering of shore	2
Barrier reef or attached reef	3	Width 100–149.9 m	3	Discontinuous reef with 30–69.9% of sheltering of shore	3
Mixed fringing and barrier reef	4	Width 50–99.9 m	4	Discontinuous reef with 10–29.9% sheltering of shore	4
Fringing reef attached to an atoll or reef island	5	No reef or width < 50 m	5	No reef, <10% sheltering the shore by reef or boulder ramp	5

interdecadal periods or observations at more frequent intervals. The parameters include an estimate of modal elevation of the backshore within 25 m of the shoreline; description of the type of sediment apparent on the backshore within 25 m of the shoreline; identification of the modal (most common) landform comprising more than 50% of the land within 25 m of the shoreline; description of the apparent/likely type of sediment on more than 50% of the intertidal0 shore; determination of subtidal landforms present on more than 50% of the inshore seabed more than 25 m seaward of the shoreline; identification of reef type most commonly sheltering the shore—where two or more reef types are present in the offshore, the most seaward reef type was used; and estimation of the proportion of island shore sheltered by reef.

5.3.4 The Indicative Vulnerability of Whole Islands

Criterion of low, moderate and high for the island and coastal fringe susceptibility rankings (Table 5.2), respectively, were combined in a matrix to estimate the *indicative vulnerability* at an island scale. The matrix is shown in Table 5.3.

Table 5.2 Classification of susceptibility and instability into categories and implications for management

Susceptibility scores (6–30)	Indicative susceptibility	Implications for management	Instability scores (9–45)	Indicative instability	Implications for management
6–14.9	Low	A mainly structurally sound geologic or geomorphic feature likely to require limited investigation of minor sites	9–20.9	Low	Resilient natural system occasionally requiring minimal maintenance
15–22.9	Moderate	Some natural structural features are unsound. Detailed assessment of coastal hazards and risks is advised	21–32.9	Moderate	Management responses to metocean events are currently required and may involve stabilisation work in the future
23–30	High	Natural structural features are extensively unsound. Major engineering works are likely to be required	33–45	High	Management responses require repeated installation or repair of major, established stabilisation works

Table 5.3 Combining susceptibility and instability into a measure of indicative vulnerability

		Instability		
		Low (score: 9–20.9)	Moderate (score: 21–32.9)	High (score: 33–45)
Susceptibility	Low (score: 6–14.9)	Very low	Low	Moderate
	Moderate (score: 15–22.9)	Low	Moderate	High
	High (score: 23–30)	Moderate	High	Very high

Weightings for island susceptibility and coastal fringe instability were applied after assignment of the rankings for each criterion and prior to compilation of the matrices (Table 5.4). This was done because not every variable is equally important in defining the three rankings. It is recognised that different weightings might be applied to different island lithologies, although this is not done in the present analysis. However, it is a point that warrants further investigation should more robust information become available for the classifications.

Table 5.4 Weightings for secondary assessment criteria

Susceptibility criteria	Weighting (%)	Instability criteria	Weighting (%)
Geology	35	Backshore—slope	20
Roundness	15	Backshore—sediment	5
Perimeter to $\sqrt{(area)}$	25	Backshore—landform component	15
Maximum elevation	5	Intertidal—sediment type	15
Proximity	15	Intertidal—landforms	15
Gradient to deep ocean seabed	5	Inshore—morphology	15
Total	100	Reef—type	5
		Reef—width	5
		Reef—coverage	5
		Total	100

5.4 Results

5.4.1 Weighted and Non-weighted Estimates of Island Susceptibility

Susceptibility levels estimated for the 36 islands are shown for the 6 criteria together with their susceptibility ranking in Table 5.5, for both non-weighted and weighted (as in Table 5.4) criteria. Eight islands had a high susceptibility, four had low susceptibility, and the majority (24) show moderate susceptibility. For the weighted criteria, this shifted to 14 with high susceptibility, 19 with moderate susceptibility and 3 with low susceptibility.

5.4.2 Weighted and Non-weighted Estimates of the Instability of the Coastal Fringe of Islands

Instability scores for the 36 islands are shown for the 9 criteria that contribute to susceptibility together with the final instability ranking in Table 5.6, for both non-weighted and weighted (as in Table 5.4) criteria. The rationale behind each of the rankings is included for each of the nine criteria in Table 5.1. Two islands had a high instability, and 12 had low instability, while the majority (22) showed moderate instability. For the weighted criteria, this shifted to 12 of high instability, 17 with moderate instability and 7 with low instability.

Table 5.5 Susceptibility rankings for the 36 illustrative islands

Island name	Rank						Susceptibility score	Susceptibility	Weighted susceptibility (see Table 5.4 for weightings)
	Geology	Roundness	Perimeter to square root area ratio	Maximum elevation	Proximity	Gradient to deep ocean seabed			
Aitutaki Island, Southern Group, Cook Islands	3	3	3	1	5	5	20	Moderate susceptibility	Moderate susceptibility
Aniwa Island, Vanuatu	4	2	4	2	5	4	21	Moderate susceptibility	High susceptibility
Aore Island, Vanuatu	4	2	4	2	2	2	16	Moderate susceptibility	Moderate susceptibility
Atafu Island, Tokelau	5	4	1	5	5	5	25	High susceptibility	High susceptibility
Atiu, Cook Islands	3	1	5	2	5	5	21	Moderate susceptibility	High susceptibility
Banaba (Ocean), Kiribati	4	1	4	2	5	4	20	Moderate susceptibility	High susceptibility
Bellona Island, Solomon Islands	4	2	4	2	5	4	21	Moderate susceptibility	High susceptibility
Eluvuka, Fiji	5	1	5	5	3	2	21	Moderate susceptibility	High susceptibility
Emananus Island, Saint Matthias (Mussau) Group, Papua New Guinea	3	3	3	2	4	4	19	Moderate susceptibility	Moderate susceptibility
Kiritimati (Christmas Island), Line Islands Group, Kiribati	5	4	1	5	5	4	24	High susceptibility	Moderate susceptibility

Table 5.5 (continued)

| Island name | Rank | | | | | | Susceptibility score | Susceptibility | Weighted susceptibility (see Table 5.4 for weightings) |
	Geology	Roundness	Perimeter to square root area ratio	Maximum elevation	Proximity	Gradient to deep ocean seabed			
Lifuka Island, Ha'apai Group, Tonga	4	3	4	4	3	4	22	Moderate susceptibility	Moderate susceptibility
Loun Island, Russell Group, Solomon Islands	2	1	4	2	2	4	15	Moderate susceptibility	Moderate susceptibility
Mangaia, Cook Islands	3	1	4	1	5	5	19	Moderate susceptibility	Moderate susceptibility
Manihiki, Cook Islands	5	5	1	5	5	5	26	High susceptibility	High susceptibility
Manono Island, Samoa	2	2	4	1	3	2	14	Low susceptibility	Moderate susceptibility
Manuae (Scilly) Island, French Polynesia	5	4	2	4	5	5	25	High susceptibility	High susceptibility
Mauke, Cook Islands	3	1	4	3	5	5	21	Moderate susceptibility	Moderate susceptibility
Nauru	4	1	5	2	5	5	22	Moderate susceptibility	High susceptibility
Niuafo'ou Island, Tonga	2	1	5	1	5	5	19	Moderate susceptibility	Moderate susceptibility
Niue	4	1	4	2	5	5	21	Moderate susceptibility	High susceptibility
Onotoa Island, Gilbert Group, Kiribati	5	4	1	5	5	2	22	Moderate susceptibility	Moderate susceptibility

Oreor (Koror), Palau	3	4	3	1	1	1	13	Low susceptibility	Low susceptibility
Ovalau, Fiji	2	2	4	1	4	1	14	Low susceptibility	Moderate susceptibility
Penrhyn, Cook Islands	5	5	1	5	5	4	25	High susceptibility	High susceptibility
Pohnpei, FSM	2	1	5	1	4	2	15	Moderate susceptibility	Moderate susceptibility
Pukapuka, Cook Islands	5	2	4	5	5	4	25	High susceptibility	High susceptibility
Rakahanga, Cook Islands	5	2	4	5	5	4	25	High susceptibility	High susceptibility
Rarotonga, Cook Islands	3	1	4	1	5	4	18	Moderate susceptibility	Moderate susceptibility
Savai'i, Samoa	2	2	4	1	2	4	15	Moderate susceptibility	Moderate susceptibility
Tarawa, Kiribati	5	5	1	5	3	4	23	High susceptibility	Moderate susceptibility
Tongatapu, Tonga	4	4	2	2	2	4	18	Moderate susceptibility	Moderate susceptibility
Tonowas (Tonoas) Island, Chuuk Group, Federated States of Micronesia	2	3	3	1	1	1	11	Low susceptibility	Low susceptibility

(continued)

Table 5.5 (continued)

Island name	Rank						Susceptibility score	Susceptibility	Weighted susceptibility (see Table 5.4 for weightings)
	Geology	Roundness	Perimeter to square root area ratio	Maximum elevation	Proximity	Gradient to deep ocean seabed			
Upolu, Samoa	2	3	3	1	3	4	16	Moderate susceptibility	Moderate susceptibility
Utupua Island, Eastern outer Solomon Islands	2	4	2	1	3	4	16	Moderate susceptibility	Moderate susceptibility
Vaitupu Island, Tuvalu	5	2	4	5	4	2	22	Moderate susceptibility	High susceptibility
Vogali (Mbuke; Wogali) Island, Manus Group, Papua New Guinea	2	2	4	3	2	1	14	Low susceptibility	Low susceptibility

Table 5.6 Instability rankings for the 36 illustrative islands

| Island name | Rank | | | | | | | | | Instability score | Instability | Weighted instability (see Table 5.4 for weightings) |
	Backshore slope	Backshore sediment	Backshore landforms	Intertidal shore—sediment type	Intertidal shore—landforms	Inshore—morphology	Reef—type	Reef—width	Reef—coverage			
Aitutaki Island, Southern Group, Cook Islands	3	2	3	4	5	2	4	1	1	25	Moderate instability	Moderate instability
Aniwa Island, Vanuatu	3	2	3	3	5	2	2	4	1	25	Moderate instability	Moderate instability
Aore Island, Vanuatu	3	1	1	4	4	2	2	4	3	24	Moderate instability	Moderate instability
Atafu Island, Tokelau	4	3	4	4	4	1	5	3	1	29	Moderate instability	Moderate instability
Atiu, Cook Islands	4	1	2	3	3	2	2	4	1	22	Moderate instability	Moderate instability
Banaba (Ocean), Kiribati	3	1	1	3	2	1	2	4	1	18	Low instability	Low instability
Bellona Island, Solomon Islands	3	1	2	2	1	2	2	4	2	19	Low instability	Low instability
Eluvuka, Fiji	5	5	4	4	4	1	5	1	1	30	Moderate instability	High instability

(continued)

Table 5.6 (continued)

| Island name | Rank | | | | | | | | | Instability score | Instability | Weighted instability (see Table 5.4 for weightings) |
	Backshore slope	Backshore sediment	Backshore landforms	Intertidal shore—sediment type	Intertidal shore—landforms	Inshore—morphology	Reef—type	Reef—width	Reef—coverage			
Emananus Island, Saint Matthias (Mussau) Group, Papua New Guinea	4	2	4	5	1	2	2	1	2	23	Moderate instability	Moderate instability
Kiritimati (Christmas Island), Line Islands Group, Kiribati	5	5	4	4	4	2	5	4	3	36	High instability	High instability
Lifuka Island, Ha'apai Group, Tonga	4	1	3	4	4	1	2	1	2	22	Moderate instability	Moderate instability
Loun Island, Russell Group, Solomon Islands	3	2	3	4	3	2	2	4	2	25	Moderate instability	Moderate instability
Mangaia, Cook Islands	3	1	1	1	3	1	2	3	1	16	Low instability	Low instability

Manihiki, Cook Islands	4	3	4	4	4	1	4	5	4	1	30	Moderate instability	High instability
Manono Island, Samoa	4	1	2	4	3	1	3	3	1	2	21	Low instability	Moderate instability
Manuae (Scilly) Island, French Polynesia	5	4	4	4	4	3	5	5	4	1	34	High instability	High instability
Mauke, Cook Islands	3	1	1	3	2	1	2	2	4	1	18	Low instability	Low instability
Nauru	3	1	2	4	4	1	2	2	2	1	20	Low instability	Moderate instability
Niuafo'ou Island, Tonga	2	1	1	1	3	2	1	1	5	5	21	Low instability	Low instability
Niue	2	1	1	1	3	2	2	2	4	4	20	Low instability	Low instability
Onotoa Island, Gilbert Group, Kiribati	4	3	4	4	5	1	5	5	1	3	29	Moderate instability	Moderate instability
Oreor (Koror), Palau	3	1	1	1	5	3	2	2	1	2	19	Low instability	Moderate instability
Ovalau, Fiji	4	2	4	5	2	2	4	4	4	3	30	Moderate instability	High instability

(continued)

Table 5.6 (continued)

Island name	Rank									Instability score	Instability	Weighted instability (see Table 5.4 for weightings)
	Backshore slope	Backshore sediment	Backshore landforms	Intertidal shore—sediment type	Intertidal shore—landforms	Inshore—morphology	Reef—type	Reef—width	Reef—coverage			
Penrhyn, Cook Islands	4	3	4	4	4	1	5	4	1	30	Moderate instability	High instability
Pohnpei, FSM	5	2	4	5	5	2	4	1	3	31	Moderate instability	High instability
Pukapuka, Cook Islands	5	3	4	4	4	3	5	1	1	30	Moderate instability	High instability
Rakahanga, Cook Islands	4	3	4	4	4	3	5	3	2	32	Moderate instability	High instability
Rarotonga, Cook Islands	4	2	3	3	4	3	2	1	2	24	Moderate instability	Moderate instability
Savai'i, Samoa	3	1	1	1	2	1	2	5	4	20	Low instability	Low instability
Tarawa, Kiribati	4	4	4	4	4	1	5	3	2	31	Moderate instability	High instability
Tongatapu, Tonga	3	1	1	1	3	3	2	4	3	21	Low instability	Moderate instability

Island												
Tonowas (Tonoas) Island, Chuuk Group, Fed. States of Micronesia	4	2	4	5	2	2	3	1	1	24	Moderate instability	Moderate instability
Upolu, Samoa	3	1	3	3	2	3	2	1	2	20	Low instability	Moderate instability
Utupua Island, Eastern outer Solomon Islands	5	1	4	5	2	2	4	1	2	26	Moderate instability	Moderate instability
Vaitupu Island, Tuvalu	4	3	4	4	4	5	5	1	2	32	Moderate instability	High instability
Vogali (Mbuke; Wogali) Island, Manus Group, Papua New Guinea	5	2	4	5	4	2	2	1	2	27	Moderate instability	High instability

5.4.3   Indicative Vulnerability at an Island Scale

Following the rankings for susceptibility and instability above, rankings for the indicative vulnerability of the 36 islands were calculated, and the results are shown in Table 5.7. The rationale for how susceptibility (Table 5.5) and instability (Table 5.6) results were combined was shown in Table 5.3. Additionally in Table 5.7, the primary assessment ranking for susceptibility is included for comparison.

The results warrant comparison with those from the primary assessment. Differences may indicate inclusion of more criteria may result in a higher incidence of moderate values. Hence weightings were used to test the outcome. For the 36 islands considered in the secondary assessment, the number of islands with very low (2), low (13), moderate (13), high (6) and very high (2) rankings was different to those in the primary assessment (8, 9, 8, 5 and 6, respectively) for the same islands. This may be due to the increased number of criteria.

Weightings (as in Table 5.4) were applied to ensure that the most important criteria have greater importance. The weighted results are also included in Table 5.7 with counts of the very low (0), low (6), moderate (15), high (8) and very high (7) categories demonstrating a skewness to higher rankings compared to the non-weighted criteria for the 36 islands selected.

5.5   Discussion and Conclusions

Assessments of susceptibility and landform instability for Pacific islands were undertaken at two scales to demonstrate a technique for downscaling and establish the availability of coastal landform information suitable for wider application. The assessment demonstrates the transition from relative susceptibility to instability and was performed because determinations of coastal landform vulnerability at different scales are relevant for disparate management purposes. The broadest level examines whole-island structure. It is relevant for development of strategic, high-level policy and regional management. Additionally, production of regional-scale maps of island susceptibility and landform instability can be used by regional organisations and external donor agencies managing aid expenditure. On the other hand, analysis of landforms comprising the coastal fringe of islands, using finer-scale information, is considered more directly applicable to policy, planning and management at country and island scales.

Although the application described here was restricted to 36 islands, the broadest level of analysis used information from a dataset describing island structure for 1532 islands (Kumar et al. 2018). The 36 were selected because detailed information describing the coastal fringe (landform characteristics adjoining and marine features affecting the shore) is not readily available for all islands in the dataset. The only geomorphic information consistently spanning the Pacific region of interest is Google Earth imagery, which requires a degree of informed interpretation and ide-

Table 5.7 Final secondary assessment of indicative vulnerability

Island	Non-weighted			Weighted			Primary assessment
	Susceptibility	Instability	Landform vulnerability	Susceptibility	Instability	Landform vulnerability	
Aitutaki Island, Cook Islands	Moderate susceptibility	Moderate instability	Moderate	Moderate susceptibility	Moderate instability	Moderate	Low
Aniwa Island, Vanuatu	Moderate susceptibility	Moderate instability	Moderate	High susceptibility	Moderate instability	High	Moderate
Aore Island, Vanuatu	Moderate susceptibility	Moderate instability	Moderate	Moderate susceptibility	Moderate instability	Moderate	Moderate
Atafu Island, Tokelau	High susceptibility	Moderate instability	High	High susceptibility	Moderate instability	High	Very high
Atiu, Cook Islands	Moderate susceptibility	Moderate instability	Moderate	High susceptibility	Moderate instability	High	Low
Banaba (Ocean), Kiribati	Moderate susceptibility	Low instability	Low	High susceptibility	Low instability	Moderate	Moderate
Bellona Island, Solomon Islands	Moderate susceptibility	Low instability	Low	High susceptibility	Low instability	Moderate	Moderate
Eluvuka, Fiji	Moderate susceptibility	Moderate instability	Moderate	High susceptibility	High instability	Very high	High
Emananus I., Papua N. Guinea	Moderate susceptibility	Moderate instability	Moderate	Moderate susceptibility	Moderate instability	Moderate	Moderate
Kiritimati, Kiribati	High susceptibility	High instability	Very high	Moderate susceptibility	High instability	High	High
Lifuka Island, Tonga	Moderate susceptibility	Moderate instability	Moderate	Moderate susceptibility	Moderate instability	Moderate	Moderate
Loun Island, Solomon Islands	Moderate susceptibility	Moderate instability	Moderate	Moderate susceptibility	Moderate instability	Moderate	Low

(continued)

Table 5.7 (continued)

Island	Non-weighted			Weighted			Primary assessment
	Susceptibility	Instability	Landform vulnerability	Susceptibility	Instability	Landform vulnerability	
Mangaia, Cook Islands	Moderate susceptibility	Low instability	Low	Moderate susceptibility	Low instability	Low	Very low
Manihiki, Cook Islands	High susceptibility	Moderate instability	High	High susceptibility	High instability	Very high	Very high
Manono Island, Samoa	Low susceptibility	Low instability	Very low	Moderate susceptibility	Moderate instability	Moderate	Very low
Manuae Island, French Polynesia	High susceptibility	High instability	Very high	High susceptibility	High instability	Very high	Very high
Mauke, Cook Islands	Moderate susceptibility	Low instability	Low	Moderate susceptibility	Low instability	Low	Low
Nauru	Moderate susceptibility	Low instability	Low	High susceptibility	Moderate instability	High	Low
Niuafo'ou Island, Tonga	Moderate susceptibility	Low instability	Low	Moderate susceptibility	Low instability	Low	Very low
Niue	Moderate susceptibility	Low instability	Low	High susceptibility	Low instability	Moderate	Low
Onotoa Island, Kiribati	Moderate susceptibility	Moderate instability	Moderate	Moderate susceptibility	Moderate instability	Moderate	Very high
Oreor (Koror), Palau	Low susceptibility	Low instability	Very low	Low susceptibility	Moderate instability	Low	Low
Ovalau, Fiji	Low susceptibility	Moderate instability	Low	Moderate susceptibility	High instability	High	Very low
Penrhyn, Cook Islands	High susceptibility	Moderate instability	High	High susceptibility	High instability	Very high	Very high

Pohnpei, FSM	Moderate susceptibility	Moderate instability	Moderate	Moderate susceptibility	High instability	High	Very low
Pukapuka, Cook Islands	High susceptibility	Moderate instability	High	High susceptibility	High instability	Very high	High
Rakahanga, Cook Islands	High susceptibility	Moderate instability	High	High susceptibility	High instability	Very high	High
Rarotonga, Cook Islands	Moderate susceptibility	Moderate instability	Moderate	Moderate susceptibility	Moderate instability	Moderate	Very low
Savai'i, Samoa	Moderate susceptibility	Low instability	Low	Moderate susceptibility	Low instability	Low	Very low
Tarawa, Kiribati	High susceptibility	Moderate instability	High	Moderate susceptibility	High instability	High	Very high
Tongatapu, Tonga	Moderate susceptibility	Low instability	Low	Moderate susceptibility	Moderate instability	Moderate	Moderate
Tonowas I., Fed. St. of Micronesia	Low susceptibility	Moderate instability	Low	Low susceptibility	Moderate instability	Low	Low
Upolu, Samoa	Moderate susceptibility	Low instability	Low	Moderate susceptibility	Moderate instability	Moderate	Very low
Utupua I., Solomon Islands	Moderate susceptibility	Moderate instability	Moderate	Moderate susceptibility	Moderate instability	Moderate	Low
Vaitupu Island, Tuvalu	Moderate susceptibility	Moderate instability	Moderate	High susceptibility	High instability	Very high	High
Vogali I., Papua New Guinea	Low susceptibility	Moderate instability	Low	Low susceptibility	High instability	Moderate	Moderate

ally some familiarity with actual field situations (Nunn et al. 2015). Whereas island elevations derived from this source are not always accurate, other measurements can be made from the imagery. For example, length of a shoreline segment, width of a landform and distance from the beach to an offshore reef can be determined. Additionally, many of the vertical images have associated landscape photographs that are especially useful for landform interpretation.

Use of Google Earth imagery in the project presents an opportunity to Pacific island countries for full development of the downscaling framework from a regional to a local site scale.[1] The advantages of Google Earth imagery are it is freely and readily available, includes historical imagery and could be used in-country and in-region by staff with expertise in GIS, assuming availability of resources and personnel to undertake the work. However, there would need to be expert agreement on the definition of landform features to be used as criteria for assessing susceptibility and instability and how they might be interpreted. Although all islands have different suites of landforms, geological controls, reef structures and connectivity with adjacent islands around them, some features are common to each of the island types described in Nunn et al. (2015). Additionally, more accurate measures of variables, such as elevation of a landform feature, are required in finer analyses, where it is appropriate to factor in the margin of error in the estimates.

Methods for fine-scale coastal vulnerability assessments are commonly applied to hazard and risk assessments involving people and land use in coastal areas. They are not reviewed here but have been elsewhere (Hay and Mimura 2013) due to our focus on whole of island exposure to metocean processes (PIANC 2014). While of value at a local area and site scale, fine-scale assessments are based on regionally sparse albeit locally detailed information, including field surveys, interpretation of imagery and numerical modelling of landform-process interactions. Owing to the scarcity of such information at regional and country scales, and the time required to capture it on an island-by-island basis, there is limited capacity for incorporating such information in assessment hierarchies where broader scales may be used for regional or country management purposes, which establish a context for detailed analyses.

In the present analysis, selection of parameters from which criteria could be developed for susceptibility and instability was determined by the availability of data suitable for comparison between islands, countries and landforms. In many instances, the only data readily available without a time-consuming search of archival material, is Google Earth imagery. This meant many of the criteria, particularly the images informing rankings on the five-point scale used for each criterion, had to be discernible from aerial photography. Although this carried potential problems of

---

[1] Google is leveraging off a Climate Data Initiative by the US Government to make climate and other data (http://www.dailymail.co.uk/sciencetech/article-2552948/Google-Earth-shows-CLIMATE-CHANGE-Regional-temperatures-1850-added-mapping-service.html) available for free to the public via Google Earth. The company has committed to provide one petabyte (1000 terabytes) of cloud storage to house satellite observations, digital elevation data and climate and weather model datasets drawn from government open data for public access.

misinterpretation, it also offered an opportunity for tighter definition of the criteria at a country or island scale where authorities could subsequently verify the interpretations through direct field observation.

Comparison of results for the coastal fringe with those for the island structure indicates a higher incidence of moderate values with downscaling, with reduction in the number of very low and very high classes. The differences may indicate unforeseen bias in selection of the 36 islands examined, although all island types were represented; the need for a review of the criteria selected, particularly similarity of the inclusion of circularity and insularity in the one level of analysis; or the likelihood of more criteria resulting in a higher incidence of moderate values. These limitations were especially apparent when weightings were applied to criteria considered to be most important. The weighted results demonstrated a skewness to higher rankings compared to the non-weighted coastal fringe criteria for the 36 islands selected.

Despite apparent limitations of these results, the analysis demonstrates downscaling in a consistent manner relating assessment of susceptibility of whole-island structure to holistic assessment of relative instability of marine and terrestrial landforms adjoining their shores. The downscaling technique was developed to address problems of sparse data availability and coverage of islands within a very large area. It involves using geologic and morphologic information derived from readily available satellite data, which can be interpreted in the context of known or modelled metocean conditions. In this respect, the technique offers scope to develop planning and management approaches which are consistent across scales ranging from broad strategic to local area and ultimately site plans.

References

Davis DM. (2013) Distinguishing processes that induce temporal beach profile changes using principal component analysis: a case study at Long Key, West-Central Florida. Graduate Theses and Dissertations. University of South Florida

Duong TM, Ranasinghe R, Luijendijk A, Walstra D, Roelvink D (2017) Assessing climate change impacts on the stability of small tidal inlets: part 1-data poor environments. Mar Geol 390:331–346

Duong TM, Ranasinghe R, Thatcher M, Mahanama S, Wang ZB, Dissanayake PK, Hemer M, Luijendijk A, Bamunawala J, Roelvink D, Walstra D (2018) Assessing climate change impacts on the stability of small tidal inlets: part 2-data rich environments. Mar Geol 395:65–81

Giardino A, Nederhoff K, Vousdoukas M (2018) Coastal hazard risk assessment for small islands: assessing the impact of climate change and disaster reduction measures on Ebeye (Marshall Islands). Reg Environ Chang 18(8):2237–2248

Govan H (2011) Good coastal management practices in the Pacific: experiences from the field. SPREP, Samoa

Hay JE, Mimura N (2013) Vulnerability, risk and adaptation assessment methods in the Pacific Islands Region: Past approaches, and considerations for the future. Sustainability Science, 8(3):391–405. https://doi.org/10.1007/s11625-013-0211-y

Kumar L, Eliot I, Nunn PD, Stul T, McLean R (2018) An indicative index of physical susceptibility of small islands to coastal erosion induced by climate change: an application to the Pacific islands. Geomat Nat Haz Risk 9(1):691–702

Luijendijk A, Hagenaars G, Ranasinghe R, Baart F, Donchyts G, Aarninkhof S (2018) The state of the world's beaches. Sci Rep 8(1):6641

Nunn P, Kumar L, Eliot I, McLean RF (2015) Regional coastal susceptibility assessment for the Pacific Islands: technical report. Australian Government and Australian Aid, Canberra, p 123

PRIF (2017) Guidance for coastal protection works in Pacific island countries. Design Guidance Report. Pacific Region Infrastructure Facility

PIANC (2014) PIANC yearbook 2014. The World Association for Waterborne Transport Infrastructure, PIANC, Belgium

Chapter 6
A Review of South Pacific Tropical Cyclones: Impacts of Natural Climate Variability and Climate Change

Savin S. Chand, Andrew Dowdy, Samuel Bell, and Kevin Tory

6.1 Introduction

Tropical cyclones are one of the costliest natural disasters impacting communities in the Pacific Island countries due to their high vulnerability and low adaptive capacity to tropical cyclone events. Strong winds coupled with heavy rainfall and coastal hazards (such as large waves and high seas) often have devastating consequences for life and property. The damage and mitigation costs associated with these events have increased in recent decades and will continue to increase due to growing coastal settlement and infrastructure development as well as increasing construction and replacement costs (e.g. Kumar and Taylor 2015). For example, severe tropical cyclone Pam in 2014 caused a total economic loss of over US$449.4 million in Vanuatu (Esler 2015). This is equivalent to 64.1% of Vanuatu's annual gross domestic product. Similarly, severe tropical cyclone Winston in February 2016 crippled Fiji's economy, causing devastating damages to infrastructure and social security (Esler 2016).

Physical theory and numerical simulations suggest that human-induced global warming should increase the severity of tropical cyclones around the globe, and signals of an increasing trend of severe tropical cyclones may already be evident in the recent historical observations (e.g. Knutson et al. 2010). However, detecting anthropogenic influence on tropical cyclone trends from historical observational records, particularly for the South Pacific, is often complicated by several confounding factors. These include a lack of long-term consistent data records for trend analyses (e.g. Landsea et al. 2006; Landsea and Franklin 2013; Klotzbach and Landsea

S. S. Chand (✉) · S. Bell
Center for Informatics and Applied Optimization, Federation University Australia,
Mt Helen, VIC, Australia
e-mail: s.chand@federation.edu.au

A. Dowdy · K. Tory
Bureau of Meteorology, Melbourne, VIC, Australia

© Springer Nature Switzerland AG 2020
L. Kumar (ed.), *Climate Change and Impacts in the Pacific*, Springer Climate,
https://doi.org/10.1007/978-3-030-32878-8_6

2015), as well as the competing influence of anthropogenic aerosol cooling (which opposes the effect of greenhouse warming, e.g. Ting et al. 2009; DelSole et al. 2011) and the presence of large natural climate variability (which masks any potential trend, e.g. Dowdy 2014).

Several efforts have been made over the recent years to improve tropical cyclone data quality for the South Pacific (e.g. Kuleshov et al. 2008, 2009; Diamond et al. 2012) and to better understand the impact of natural climate variability on tropical cyclones (e.g. Chand and Walsh 2009, 2010; Dowdy et al. 2012; Chand et al. 2013; Diamond et al. 2012, 2013, 2015). In addition, we now have 10 more years of data since the last comprehensive study on tropical cyclone trends for the South Pacific (Kuleshov et al. 2010), providing an opportunity to re-examine tropical cyclone trends using updated data records for the South Pacific basin. Advances in climate modelling—such as those from the Climate Model Intercomparison Project Phase 3 (CMIP3, Meehl et al. 2007) and Phase 5 (CMIP5, Taylor et al. 2012)—have provided another platform to examine future changes in tropical cyclone characteristics under different global warming scenarios for the Pacific Island countries (e.g. Chand et al. 2017).

A recent study by Chand (2018) has provided a review of past studies on tropical cyclones over the South Pacific basin. In this chapter, we not only review past studies on tropical cyclones but also consolidate new information derived from updated tropical cyclone data records and state-of-the-art climate model results. The first part of this chapter examines historical South Pacific tropical cyclone data and reviews the improved data records, enabling more robust research on climate variability and change. The second part looks at the impact of natural climate variability on South Pacific tropical cyclone activity. The third part emphasises the impact of climate change on tropical cyclone activity as evidenced from observational and climate modelling studies. The final part provides a summary of the review and gives recommendations for future work.

6.2 Historical Data Records and Homogeneity

The potential risks from tropical cyclone events are huge and significant. Therefore the accuracy of historical tropical cyclone records for quantitative risk assessments cannot be overemphasised, particularly for the vulnerable small island countries in the South Pacific. A number of past high-impact tropical cyclone events have been documented for the South Pacific Island countries, some extending back many hundreds of years. An archival database of historical tropical cyclone records over the period 1558–1970 contains tropical cyclone records in the form of historical notes (d'Aubert and Nunn 2012). An example is a likely cyclone near *Ontong Java* in the Solomon Islands in the year 1558:

> "Cyclone, 1558 February 1st week, Ontong Java. On the first February 1558, two ships. 'Los Reyes' (250 tons), and the 'Todos Santos' (107 tons), were sailing under the captaincy of Alvaro de Mendana. The ships narrowly avoided being shipwrecked on a reef, almost

certainly the one near Ontong Java. Immediately after this, the vessels were swept away by a cyclone and driven south for six days. On the seventh day the weather cleared".

There are several other accounts of historical tropical cyclones in this database and elsewhere (e.g. Visher 1925; Kerr 1976; Ramage and Hori 1981; Revell and Goulter 1986). However, it should be emphasised that these accounts are scattered and incomplete as they date back to the pre-satellite era (i.e. before the 1970s) when the comprehensive monitoring of tropical cyclones, particularly over the open oceans, was not possible. Although efforts have been made in the past to create enhanced records of pre-satellite historical tropical cyclones for the South Pacific (e.g. Diamond et al. 2012), homogeneity issues still remain. Therefore, it has been recommended that South Pacific tropical cyclone datasets prior to the satellite era should be used with due diligence for climate variability and change analyses (e.g. Buckley et al. 2003).

Comprehensive compilation of observational tropical cyclone records for the South Pacific began after the 1970s when state-of-the-art satellite technologies became operational on a routine basis in this region. Estimates of tropical cyclone intensities improved after the 1980s when objective tools and methods (such as the Dvorak scheme) were established (e.g. Harper et al. 2008). These technological and methodological improvements have paved the way for several scientific investigations that advanced various areas of tropical cyclone research. Areas that are of particular significance, and therefore form basis of this review, include studies on climatological characteristics of tropical cyclones and the impact of natural climate variability and climate change on tropical cyclone activity.

In the next section, we examine climatological characteristics of tropical cyclones in the South Pacific basin (defined as the region between 0–25°S and 145°E-120°W), with particular emphasis on genesis locations, frequency, tracks and intensity. Tropical cyclone data used in this work is from the Southwest Pacific Enhanced Archive for Tropical Cyclones (SPEArTC, Diamond et al. 2012) database. Systems that reached at least the gale strength classification are considered in the analysis. This classification scheme (Table 6.1), which uses the maximum 10-min sustained wind speed, is the same as one proposed by Revell (1981) and adopted by Holland (1984), Thompson et al. (1992), Sinclair (2002) and Chand and Walsh (2010) for tropical cyclone studies in the South Pacific basin. It varies slightly from the Saffir-Simpson scale or the one used by the Australian Bureau of Meteorology. As a result, some descriptive statistics of tropical cyclone climatology may differ from those of other studies that may have included weaker storms.

Table 6.1 Tropical cyclone classification in the South Pacific basin

Intensity class	Description	Speed range (m s^{-1})
1	Tropical depression	<17
2	Gale	17–24
3	Storm	25–32
4	Hurricane	>32

Speed range is defined using 10-min sustained wind speed

6.3 Climatological Characteristics

Every year roughly 80 tropical cyclones form globally, with about one-third of them occurring in the Southern Hemisphere. In the South Pacific basin (east of 145°E), tropical cyclones can form as far east as French Polynesia (Fig. 6.1a) with an annual average of about nine tropical cyclones forming between the seasons 1981/1982 and 2016/2017. Of these, about four cyclones per year reach the severe intensity category (i.e. those that attained hurricane strength as per Table 6.1). However, the annual numbers for individual seasons can vary substantially, for example, from as low as four in 1994/1995 to as high as 16 in 1997/1998 season (Fig. 6.1b).

Tropical cyclones in the South Pacific Ocean basin mainly occur between the months of November and April, which defines a typical cyclone season with the peak activity (~ 70% of annual numbers) occurring during January–March (Fig. 6.1c). However, there are cases when tropical cyclones develop on either side of this period, including as early as October and as late as June. These cases are often tied to early onset and late decay of the El Niño-Southern Oscillation phenomenon (e.g. Chand and Walsh 2010), which is a major driver of tropical cyclone variability in the South Pacific at interannual timescale.

Tropical cyclone motion in the South Pacific basin can have different characteristics to those in other basins of the world. Here most tropical cyclones have an eastward component of motion during their lifetime or quickly recurve to the east

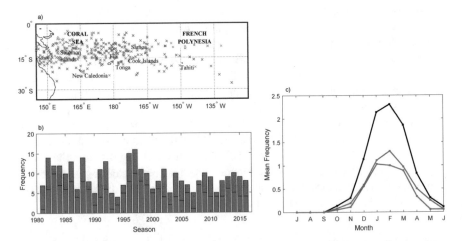

Fig. 6.1 (a) Tropical cyclone genesis locations in the South Pacific Ocean basin, defined as east of 145°E, (b) tropical cyclone counts over the period 1981–2016 and (c) tropical cyclone seasonality. Blue and orange indicate non-severe and severe tropical cyclones, respectively, and black line in (c) represents their total. Note non-severe tropical cyclones correspond to categories 1 and 2, whereas severe cyclones correspond to categories 3–5 according to the Australian intensity classification scheme. Because the Southern Hemisphere tropical cyclone season spans two calendar years (i.e. November and December of the first year and January to April of the second year), the first of the years is used in to refer to a particular season

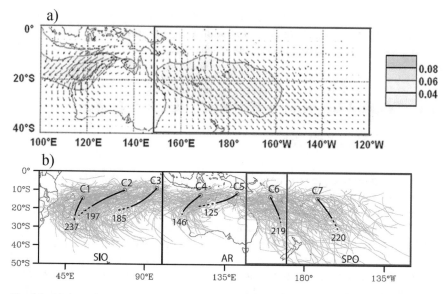

Fig. 6.2 (**a**) Annual average tropical cyclone transport fields measured in degrees per day (Source: Dowdy et al. 2012, used with permission from the American Meteorological Society) and (**b**) mean tropical cyclone track clusters (Source: Ramsay et al. 2012). The South Pacific Ocean basin is indicated by red boxes

after initially moving west (Fig. 6.2a, Dowdy et al. 2012), while the mean trajectories of tropical cyclones in other basins are generally westward. Overall, there are two main eastward moving track regimes identified in the South Pacific basin, each with their characteristic geographical domain: one is located west of the dateline and includes the Coral Sea region and the other east of the dateline (Fig. 6.2b, Ramsay et al. 2012). In a more localised study over the Fiji islands, Chand and Walsh (2009) found another track regime where westward motion exists initially in a small region west of the dateline and equatorward of 10°S before recurving eastward.

Moreover, a prominent climatic feature in the South Pacific is the South Pacific Convergence Zone (SPCZ; see Chap. 3 for details) where tropical cyclones are frequently spawned. The SPCZ is characterised by a band of high cloudiness, strong convective precipitation and low-level convergence extending from the West Pacific warm pool southeastward towards French Polynesia. It has been shown that variability of the SPCZ at different timescales, and thus the variability of tropical cyclone activity in the South Pacific, is primarily modulated by natural drivers such as the Madden-Julian Oscillation (MJO, e.g. Chand and Walsh 2010; Diamond and Renwick 2015) and the El Niño-Southern Oscillation (ENSO, Chand and Walsh 2009; Dowdy et al. 2012; Vincent et al. 2011; Jourdain et al. 2011).

In the next section, we discuss how the MJO and ENSO modulate tropical cyclone activity in the South Pacific. We limit our discussion to these modes as they are well resolved in the existing tropical cyclone data records. Other modes of variability (e.g. Interdecadal Pacific Oscillation) require longer time series of data to deduce any meaningful conclusion regarding their impacts on the South Pacific tropical cyclones, and so are not discussed hereafter.

6.4   Tropical Cyclones and Natural Climate Variability

6.4.1   Impact of ENSO

The ENSO phenomenon is a dominant mode of natural climate variability that operates at interannual timescales (see Chap. 3 for details). The term "El Niño" is commonly used to refer to the occurrence of anomalously high sea surface temperature (SST) in the central and eastern equatorial Pacific Ocean every few years (Trenberth 1997). The opposite "La Niña" phase consists of basin-wide cooling of the tropical Pacific. This anomalous warming and cooling of the central and eastern equatorial Pacific SST is coupled with the atmospheric phenomenon called the Southern Oscillation, which is characterised by a seesaw in tropical sea-level pressure (SLP) between the Western and Eastern Hemispheres. During El Niño events, the SLP falls (rises) in the central and eastern Pacific (western Pacific); the reverse occurs during La Niña events. The term "neutral phase" describes conditions when SST and SLP are near climatological averages. The zonal atmospheric circulation that arises as a result of SST and SLP coupling is called the "Walker circulation" (Walker 1923, 1924).

Initially some contradictory views existed on the relationship between ENSO and tropical cyclone activity in the South Pacific basin, primarily due to data homogeneity issues in those earlier studies. For example, Ramage and Hori (1981) and Hastings (1990) could not detect any significant relationship between tropical cyclone numbers and ENSO in the entire South Pacific basin. However, when Basher and Zheng (1995) explored the link between the Southern Oscillation Index (SOI; Troup 1965) and tropical cyclone incidence in various subregions of the South Pacific basin, they found that tropical cyclone incidence in the Coral Sea region and the region east of about 170°E significantly correlates with ENSO, although these two regions have opposite phases. Similarly, Kuleshov et al. (2009) found a statistically significant relationship between tropical cyclone numbers and various ENSO indices in the South Pacific basin.

A well-documented influence of ENSO on tropical cyclone activity in the South Pacific basin is the mean location of tropical cyclone genesis positions and tracks (see also a review by Chand 2018). In El Niño years, tropical cyclone activity systematically shifts northeastward to the Cook Islands and French Polynesia with the greatest incidence around the dateline, extending east-southeast of the Fiji islands. Simultaneously, low activity dominates the Coral Sea and Australian regions. In contrast, the reverse occurs during La Niña years when tropical activity is displaced southwestward into the New Caledonia, Coral Sea and Australian regions with relatively low activity east of about 170°E. Ramsay et al. (2012), in their study encompassing the entire Southern Hemisphere, found that tropical cyclone tracks are significantly modulated by ENSO in the South Pacific basin. Consistent with other studies, they also documented an equatorward shift of the mean genesis locations of cyclones during El Niño and a poleward shift during La Niña, in addition to large changes in mean numbers. The regions of increase or decrease in tropical cyclone numbers for a given phase of ENSO are influenced by the shift in the SPCZ due to ENSO, with a southwestward shift in the SPCZ for La Niña conditions corresponding to more TCs forming around the far-west South Pacific, and an northeastward shift in the SPCZ for El Niño conditions corresponding to more TCs forming around the central South Pacific region (Fig. 6.3, Dowdy et al. 2012).

In a more localised study, Chand and Walsh (2009) found three track clusters that showed substantial modulation of tropical cyclone genesis locations and tracks over Fiji, Samoa and Tonga regions as a result of ENSO. During the El Niño phase, for example, they found tropical cyclones that formed poleward of 10°S and west of the dateline were frequently steered southeastward into the northern part of the Fiji islands and the Tonga regions by a predominant northwesterly mean flow regime. However, those that formed east of the dateline were usually steered north of the Samoa region. Cyclones, that on average formed in the mean northeasterly flow regime between 5–10°S and 170°E–180°, recurved west-southwest of the Fiji islands. For La Niña phase, they found that cyclones were often steered over the Fiji islands and the Tonga region with relatively little or no threat to the Samoa region. Furthermore, Sinclair (2002) and Chand and Walsh (2011) also examined the influence of ENSO on mean cyclone intensity in the South Pacific basin and concluded that cyclone intensity decreased rapidly around 20°–25°S latitudes during El Niño years but was often maintained as far as 40°S into the Tasman Sea in La Niña years.

A number of studies (e.g. Trenberth and Stepaniak 2001; Ashok et al. 2007; Kug et al. 2009; Kao and Yu 2009) have also identified a "non-traditional" type of El Niño event (hereafter referred to as the "El Niño Modoki" as in Ashok et al. 2007) with the above-normal SSTs confined more to the central Pacific region flanked by below-normal SSTs on the eastern and western sides. Over recent years, investigations relating to the impact of El Niño Modoki on tropical cyclones have also garnered attention in several tropical cyclone basins around the globe (e.g. Kim et al. 2009, 2011; Chen and Tam 2010; Chen 2011; Hong et al. 2011). For the South Pacific basin, Chand et al. (2013) identified four separate ENSO regimes with distinct impact on tropical cyclone genesis location and frequency. Two of the regimes were associated with traditional El Niño and La Niña events, while the other two regimes, which they termed "positive-neutral" and "negative-neutral", showed Modoki-type patterns. All of these ENSO regimes have a large impact on tropical cyclone genesis over the central southwest Pacific, with enhanced tropical cyclone activity during El Niño and positive-neutral years and reduced tropical cyclone activity during La Nina and negative-neutral years (Fig. 6.4).

Furthermore, a separate study by Diamond and Renwick (2015) looked at the impacts of the Southern Annular Mode (SAM) on climatological characteristics of the South Pacific tropical cyclones during different phases of ENSO. They found

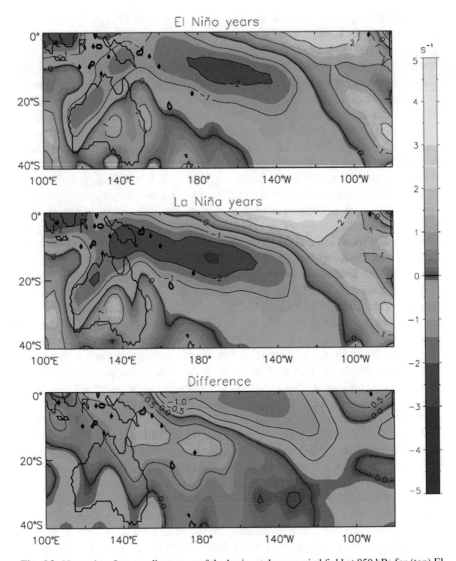

Fig. 6.3 November–January divergence of the horizontal mean wind field at 850 hPa for (top) El Niño years, (middle) La Niña years and (bottom) the difference between El Niño and La Niña years. (Source: Dowdy et al. 2012, used with permission from the American Meteorological Society)

that a synergetic relationship between the positive phase of SAM (i.e. when the belt of westerly winds contracts towards Antarctica and anomalously higher pressure dominates southern Australia) and La Niña events results in more cyclones reaching farther south, thus increasing the likelihood of tropical cyclones undergoing extra-tropical transition near New Zealand. While this relationship is statistically

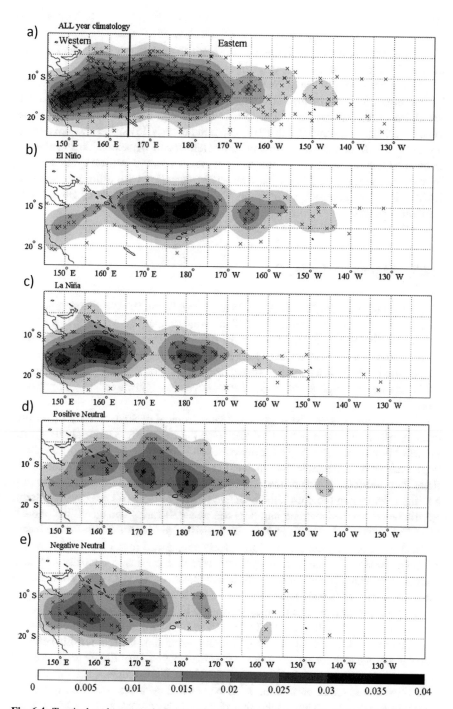

Fig. 6.4 Tropical cyclone genesis density for overall climatology and for different ENSO regimes over the period 1969/1970–2011/2012. The number of tropical cyclone genesis per year and per 2.5° × 2.5° boxes are represented as anisotropic Gaussian density distribution (shaded) and as actual genesis positions (crosses). (Source: Chand et al. 2013, used with permission from the American Meteorological Society)

significant, they found no clear mechanism that can explain this link between mid-latitude SAM and South Pacific tropical cyclones.

The relationship between ENSO and tropical cyclones in the South Pacific Ocean basin is also modulated by the MJO, which in itself is a major driver of intraseasonal variability of tropical cyclones in the South Pacific (e.g. Hall et al. 2001; Leroy and Wheeler 2008; Chand and Walsh 2010; Diamond et al. 2015) and elsewhere around the globe (e.g. Maloney and Hartmann 2000; Bessafi and Wheeler 2006 and others). In the next section, we examine how MJO modulates overall tropical cyclone activity over the South Pacific basin, as well as in interaction with different phases of ENSO given that characteristics of the MJO are linked to phases of ENSO (e.g. Hendon et al. 1999).

6.4.2   Impact of the MJO

The Madden-Julian Oscillation (MJO; Madden and Julian 1971) is characterised by an eastward propagating disturbance with a period of about 30–90 days. It is a leading mode of intraseasonal variability in the tropical atmosphere, particularly during austral summer months (see Chap. 3 for details). The propagating disturbance associated with the MJO is a centre of strong deep convection ("active phase"), flanked on both sides by regions of weak deep convection ("inactive" or "suppressed phases").

The MJO phenomenon strongly modulates tropical cyclone activity in various cyclone basins at intraseasonal timescale (see the review by Klotzbach 2014). As the MJO progresses eastward over the equatorial South Pacific basin, it modulates large-scale environmental factors such as vertical wind shear, low-level relative vorticity and mid-level moisture that are known to affect tropical cyclone formation and intensity. When the convectively active phases of the MJO are over the South Pacific basin, tropical cyclone numbers can be significantly enhanced. In contrast, when the convectively inactive phases of the MJO are over the South Pacific basin, tropical cyclone numbers can be significantly suppressed (e.g. Fig. 6.5, Leroy and Wheeler 2008). The number of cyclones reaching hurricane intensity can also undergo significant enhancement in the active phases of the MJO when compared to the inactive phases (Chand and Walsh 2010; Klotzbach 2014).

A number of past studies have established a link between the MJO and ENSO, though the exact causative mechanism between the two is still not well understood (e.g. Hendon et al. 1999). Consequently, studies examining the relationship between tropical cyclone activity and the MJO in the different phases of ENSO have emerged for several tropical cyclone basins around the globe. For example, Chand and Walsh (2010) examined the impact of the MJO on tropical cyclones over the Fiji region, spanning the dateline. They found that if the enhanced phases of the MJO occur during El Niño events, more tropical cyclones are likely to form in the region when compared to the enhanced phases of the MJO occurring during La Niña events. Similar results were found in an earlier study by Hall et al. (2001) for the Australian

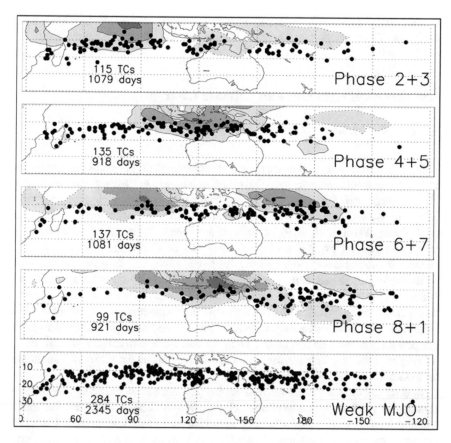

Fig. 6.5 Tropical cyclone genesis locations (dots) according to the phase of the MJO, as defined by Wheeler and Hendon (2004) over the period 1969–2004. Also shown are contours of outgoing longwave radiation (OLR) anomalies for each averaged MJO phase with negative contours solid and positive contours dashed; positive contours indicate enhanced convective activities, and negative contours indicate suppressed convective activities. Also listed are the number of tropical cyclones (TCs) counted within each phase and the number of days for which that MJO phase category occurred. (Source: Leroy and Wheeler (2008), used with permission from the American Meteorological Society)

region that showed that the relationship between the MJO and tropical cyclones is strengthened during El Niño periods.

Overall, the impacts of ENSO and the MJO are well documented for South Pacific tropical cyclones. Impacts of lower-frequency mode variability, for example, the Interdecadal Pacific Oscillation, require longer temporal records of tropical cyclones and so are not well understood for the basin. This poses a serious challenge for climate change and attribution studies as such understanding is important to deduce any meaningful information on the impact of climate change on tropical cyclone activity. Over the past decades, new tools and methods have been developed to circumvent some of these challenges in order to provide information on the impact of climate change on tropical cyclones (see the following sections).

6.5 Tropical Cyclones and Climate Change

6.5.1 Overview

There is a strong scientific consensus that anthropogenic activities are unequivocally significant contributors to global climate change (e.g. Christensen et al. 2013). There is also substantial evidence that environmental conditions that support tropical cyclones are changing as a result of anthropogenic climate change (e.g. Knutson et al. 2010; Walsh et al. 2016). However, estimating the effect of anthropogenic climate change on tropical cyclone activity can be challenging as the observed changes in the tropical cyclone activity due to anthropogenic influences are masked by the variability expected through natural causes. Additionally, systematic assessments of observed changes are further limited by insufficient long-term homogeneous data records. Whether one can develop reliable simulations of change in these environmental factors affecting tropical cyclones—and hence, changes in tropical cyclone metrics such as frequency, intensity and track distribution—is also challenging as many climate models still have biases and deficiencies in simulating these factors at local and basin scales.

Several efforts have been made over the past decade to strengthen the understanding of the links between climate change and tropical cyclones. For example, recent advances in the production of improved and more homogeneous tropical cyclone datasets (e.g. Kossin et al. 2013), as well as long-term reanalysis products (e.g. Saha et al. 2010; Compo et al. 2011), have provided some confidence in the detection and attribution of observed changes in tropical cyclone characteristics due to climate change. Increasing use of geological proxies to determine historic and prehistoric tropical cyclone variability patterns has also provided a longer climate baseline for exploring the dependence of tropical cyclone activity on climate change. Considerable effort has been made over the past few years to improve climate model performance in simulations of the climate system. This has provided opportunities to examine future projections of tropical cyclone activity, even at regional scales, in more detail than before (e.g. Chand et al. 2017).

6.5.2 Paleotempestology

In recent decades, several global studies have been undertaken to determine past tropical cyclone activity by examining evidence from historical documents and geological proxies such as oxygen isotopes in sediment cores extracted from lake beds and cave stalagmites, as well as tree-ring chronologies. Such studies, often referred to as "paleotempestology", have extended the tropical cyclone records as far back as the early Holocene (about 10,000 years ago) in many basins globally.

For example, a North Atlantic study reconstructed tropical cyclone activity over the last millennium using oxygen isotopes in deep seawater sediments (Woodruff et al. 2012). More recently, Haig et al. (2014) derived a tropical cyclone index for Australia using stalagmite records obtained from Queensland and Western Australia. They showed, on the basis of this index, that the present low levels of storm activity on the Australian coasts are unprecedented over the past 550–1500 years. Similarly, Callaghan and Power (2011) used historical documentations to show a decline in the number of severe tropical cyclone making landfall over eastern Australia since the late nineteenth century. These studies have provided scientists with a better understanding of tropical cyclone activity and natural climate variability and there-fore help to determine the anthropogenic influence of climate change on tropical cyclones from natural variability.

However, research on historic and prehistoric reconstruction of tropical cyclones in the South Pacific is lacking. The focus of limited past studies in the South Pacific has primarily been on the reconstruction of general climate variability patterns and features such as El Niño Southern Oscillation, Pacific Decadal Oscillation and South Pacific Convergence Zone (e.g. Bagnato et al. 2004; LeBec et al. 2000). While ENSO is the major driver of interannual variability of tropical cyclones in the Pacific in the present climate, the extent to which it impacted tropical cyclones in the historic and prehistoric climate is not known, leaving unanswered questions on the sensitivity of tropical cyclones to ENSO before the modern era. The link between natural climate variability and tropical cyclones in the historic and prehistoric con-text can provide climate scientists with crucial information on tropical cyclones and climate change in the South Pacific. Particularly, site-specific geological proxies can form important indicators of extreme landfalling tropical cyclone impacts on individual island countries as they often preserve high-resolution time series infor-mation of extreme events.

6.5.3   Trends from Observations

A previous review (Knutson et al. 2010) concluded that there was no significant change in the total number of cyclones over the period 1970–2004 both globally and for individual basins with the exception of the North Atlantic. However, a substan-tial increase in the global number of the most intense tropical cyclones was reported from 1975 to 2004 (Webster et al. 2005). This finding was later contested by Klotzbach and Landsea (2015) who argued that the trend was attributed to improve-ments in tropical cyclone observational practices at various tropical cyclone warn-ing centres, primarily in the first two decades of that study.

For the South Pacific basin, Kuleshov et al. (2010) found no apparent trend in the total number of tropical cyclones over the period 1981–2007, nor any significant

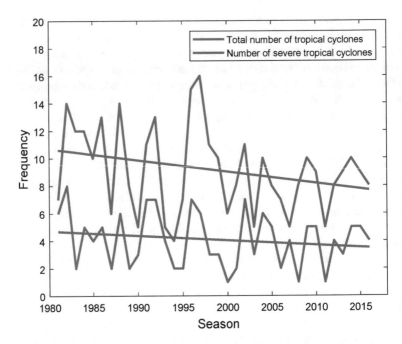

Fig. 6.6 Frequency and associated trends for the total number of tropical cyclones (blue) and severe tropical cyclones (orange) in the South Pacific over the period 1981/1982 to 2016/2017

trend in the number of severe tropical cyclones (i.e. cyclones with central pressure below 970 hPa). We have repeated the trend analysis for the South Pacific basin with 10 years of additional data from the SPEArTC dataset (Diamond et al. 2012), taking into consideration influences of various modes of natural variability (Fig. 6.6). Consistent with the earlier work of Kuleshov et al. (2010), we also found a weak and statistically insignificant trend both in the total number of cyclones and in the frequency of severe tropical cyclones. It is important to emphasise here that lack of trends in the observed record over the period 1981–2016 does not negate the presence of trend. As more data becomes available in the future to better quantify low-frequency variability, such as Interdecadal Pacific Oscillation (IPO), more confidence can be expected in the analyses of tropical cyclone trends and climate change.

Regardless, while temporal homogeneity of tropical cyclone data is of concern, studies are now emerging on metrics that are not very sensitive to past data records. For example, Kossin et al. (2014) examined the average latitude when tropical cyclones reach their lifetime maximum intensity (LMI) over the period 1982–2009. They found a pronounced trend in poleward migration of tropical cyclone LMI in both the Northern and the Southern Hemispheres. However, this was contradicted by Moon et al. (2015) who argued that poleward migration of the LMI is basin-dependent and can be greatly influenced by multi-decadal variability, particularly for the Northern Hemisphere basins. For the Southern Hemisphere basins, both studies agree on a significant poleward migration of the LMI, which could be a result of a poleward expansion of the tropics due to anthropogenic climate change as suggested by theoretical (Held and Hou 1980) and modelling (Lu et al. 2007) studies. This may imply that tropical cyclone exposure is likely to increase in the future climate for the small island countries that are located farther south of the equator in the South Pacific basin. However, more research is needed to determine and quantify the extent of the exposure.

6.5.4   Results from Climate Modelling Experiments

In the absence of homogeneous observed records for robust conclusions on tropical cyclones and climate change, analytical tools such as Global Climate Models (GCMs) are used to understand how climate variability and change may impact tropical cyclones in various regions around the globe. Over the past several years, considerable effort has been made to improve climate model performance in the simulation of various aspects of the climate system including tropical cyclones. Methods of detecting tropical cyclones in climate model simulations have also improved substantially (e.g. Tory et al. 2013a). However, some caveats still remain in resolving certain aspects of tropical cyclones, particularly intensity due to coarse horizontal resolutions that can vary from about ~100–300 km for CMIP3 and CMIP5 models to ~10–50 km for new generation of high-resolution models. Several downscaling strategies (e.g. Emanuel et al. 2008) and theoretical approaches (e.g. Emanuel 1987; Holland 1997) have been applied to mitigate coarse resolution issues and, therefore, better understand the impact of climate change on tropical cyclone intensities.

A number of climate modelling studies have a consensus projection of a likely decrease in the globally averaged tropical cyclone frequency (~5–30%) by the late twenty-first century. There is also a clear tendency among the high-resolution models to project a global increase in the frequency of stronger tropical cyclones (~5–30% for Category 4 and 5 cyclones on the Saffir-Simpson scale), as well as an increase in tropical cyclone LMI (~0–5%) and tropical cyclone rainfall rate (~5–20%), by the late twenty-first century (see a review by Christensen et al. 2013). However, large variations in projected changes of tropical cyclone characteristics can occur between different climate models at a regional scale, particularly for the Northern Hemisphere basins (Fig. 6.6). This could be potentially attributed to climate model deficiencies and biases in simulating regional changes in conditions known to affect tropical cyclone variability and change.

For the South Pacific basin, a multi-model study reported a consistent decrease (~3–27%) in tropical cyclone frequency by the late twenty-first century across all models (Fig. 6.7, Tory et al. 2013b) giving more confidence in the projection results. However, there is no substantial projected change in the spatial distribution of genesis locations and tracks over the South Pacific (Fig. 6.8). Note that projected changes in TC track density between the current- (1970–2000) and the future-climate (2070–2100) simulations are constructed using the high emissions

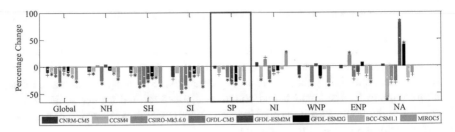

Fig. 6.7 Percentage change in the mean tropical cyclone frequency between the late twentieth (1970–2000) and late twenty-first (2070–2100) centuries for selected CMIP5 models. Changes that are significant at 95% and 90% confidence levels are indicated by asterisk and plus symbols, respectively. The South Pacific basin is highlighted in the red box. (Source: Tory et al. 2013b, used with permission from the American Meteorological Society)

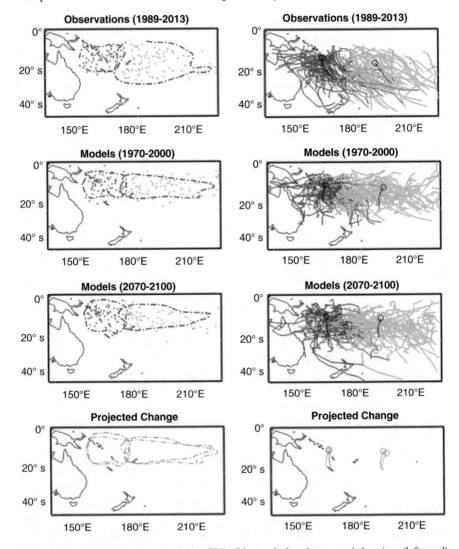

Fig. 6.8 Kernel density contours enclosing 75% of the tropical cyclone genesis locations (left panel) and mean track trajectories (right panel) for the South Pacific basin. A random sample, consistent with observational climatology, for model tropical cyclone genesis locations and tracks are selected and plotted on the figures. Projected changes of genesis locations and tracks between the late twenty-first-century (orange) and late twentieth-century (grey) simulations are also shown (bottom panel)

Representatives Concentration 8.5 Pathways (RCP8.5; e.g. Riahi et al. 2011). RCP8.5 represents an approximately 8.5 W m^{-2} increase in radiative forcing values in the year 2100 as compared to the pre-industrial emission levels and was chosen to best elucidate any changing TC behaviour in a warmer climate.

In addition, a recent study by Chand et al. (2017) used state-of-the-art climate models from the CMIP5 experiments to show robust changes in regional-scale ENSO-driven tropical cyclone variability by the end of the twenty-first century (Fig. 6.9). In particular, they showed that tropical cyclones become more frequent (~20–40%) during future-climate El Niño events compared to present-climate El Niño events—and less frequent during future-climate La Niña events—around a group of small island nations (e.g. Fiji, Vanuatu, Marshall Islands and Hawaii) in the Pacific. Their results have important implications for climate change and adaptation pathways for the vulnerable small island nations in the South Pacific. The study

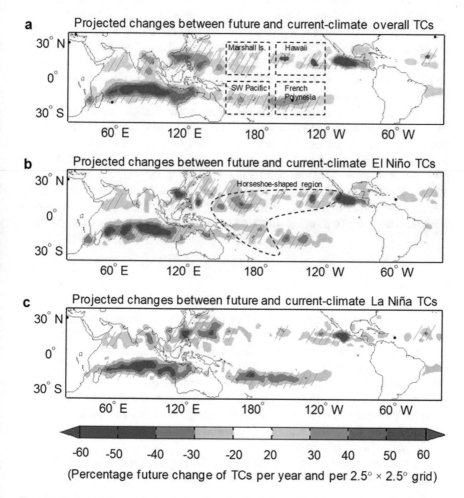

Fig. 6.9 Projected changes in tropical cyclone density between the late twentieth and late twenty-first centuries for (**a**) overall climatology, (**b**) El Niño years and (**c**), La Niña years. Red shading indicates projected increases in tropical cyclone frequency. Stippling denotes changes that are statistically significant at the 95% level. (Source: Chand et al. 2017)

shows that even though the overall frequency of tropical cyclones is likely to decrease in the future climate, large-scale drivers of climate variability (such as ENSO) are crucial in determining regional-scale changes in tropical cyclone characteristics.

6.6 Socio-economic Impacts of Tropical Cyclones

Social and economic impacts of tropical cyclones—often arising from strong winds, heavy rainfall and storm surges—can be severe in the South Pacific Island countries due their low adaptive capacity and high vulnerability. When tropical cyclones pass over an island nation, the extent of damage extends largely well beyond the coasts, affecting significant proportion of the population and infrastructure at national scale. This makes small island countries in the South Pacific one of the most exposed groups to tropical cyclones (Nurse et al. 2014). Severe tropical cyclones *Pam* over Vanuatu and *Winston* over Fiji give an insights into the extent of destructions that can be caused by tropical cyclone events over the small island countries in the South Pacific. This not only includes damages to critical infrastructure such as roads, dwellings, hospitals and environment but also loss of life and livelihood even for several weeks in the aftermath of a tropical cyclone event.

With growing threats from human-induced climate change, impacts of TCs are likely to exacerbate further in the future climate by, for example, increasing coastal flood risks through enhanced TC-induced rainfall rates (e.g. Knutson et al. 2010; Patricola and Wehner 2018), higher sea-level rise exacerbating the coastal hazards (e.g. Woodruff et al. 2013) and enhancing TC wind speed intensity (Patricola and Wehner 2018). These changes are likely to put additional pressures on the Pacific Island communities in their effort towards building adaptive capacity and climate resilience, as well as enhancing disaster risk reduction (e.g. McGray et al. 2007).

6.7 Summary

We presented a review of the impact of natural climate variability and climate change on tropical cyclone activity in the South Pacific basin. Where possible, we have also consolidated past literature with new results using updated data and climate model results. We first discussed how the impact of different modes of natural climate variability modulates South Pacific tropical cyclones, with emphasis on the Madden-Julian Oscillation (MJO) and El Niño-Southern Oscillation (ENSO) phenomena. We then presented results from the historical analyses and climate modelling studies on the extent climate change has impacted tropical cyclones in the South Pacific Island countries.

It is well-established that in the South Pacific, tropical cyclone activity systematically shifts northeastward during El Niño events, reaching as far as the Cook

Islands and French Polynesia with the greatest incidence around the dateline and the Fiji islands (e.g. Basher and Zheng 1995; Chand and Walsh 2009; Vincent et al. 2011). Simultaneously, low activity dominates the Coral Sea and Australian regions (e.g. Basher and Zheng 1995; Ramsay et al. 2012). In contrast, the reverse occurs during La Niña events when tropical cyclone activity is displaced southwestward into the Coral Sea and Australian region with relatively low activity east of the dateline. In addition to conventional El Niño and La Niña events, two other regimes are found to modulate tropical cyclones in the South Pacific: "positive-neutral" and "negative-neutral" regimes that showed ENSO Modoki-type SST patterns (Chand et al. 2013).

Similarly, the MJO phenomenon has a significant impact on intraseasonal variability of tropical cyclones in the South Pacific. For example, when the convectively active phases of the MJO are over the South Pacific basin, tropical cyclone numbers can be significantly enhanced. In contrast, the reverse occurs during convectively inactive phases of the MJO when tropical cyclone numbers are substantially suppressed (Leroy and Wheeler 2008; Chand and Walsh 2010; Klotzbach 2014).

Moreover, tropical cyclone activity in the South Pacific can be substantially influenced by global warming. While no significant attributable trend has yet appeared in the observed number of tropical cyclones over the last few decades due to insufficient data records, climate modelling studies show significant changes in the overall frequency of tropical cyclones in the South Pacific basin in future (e.g. Tory et al. 2013b; Chand et al. 2017). Overall, tropical cyclones are projected to become less frequent but potentially more intense, by the late twenty-firs-century compared to late twentieth-century climatology (e.g. Knutson et al. 2010). A recent study by Chand et al. (2017) showed that tropical cyclones become more frequent (~20–40%) during future-climate El Niño events compared to present-climate El Niño events—and less frequent during future-climate La Niña events—around a group of small island countries (e.g. Fiji, Vanuatu, Marshall Islands and Hawaii) in the Pacific.

While contributions of past studies are substantial, more work is still needed to better quantify the impacts of natural climate variability and climate change on South Pacific tropical cyclones. Rising sea levels due to global warming will also increase many of the coastal impacts of tropical cyclones in small islands (e.g. Nurse et al. 2014). However, much further work is required on this theme in small island situations, especially comparative research. Important information, data gaps and many uncertainties still exist on impacts of tropical cyclones in small islands. This research will be feasible when longer temporal records of updated tropical cyclone data, as well as new generations of climate models with fewer biases and deficiencies, become available in the future. Such information would raise the level of confidence in the adaptation planning and implementation process in small islands in the South Pacific.

Acknowledgements This project is supported through funding from the Earth Systems and Climate Change Hub of the Australian Government's National Environmental Science Program.

2

References

Ashok K, Behera S, Rao AS, Weng HY, Yamagata T (2007) El Niño Modoki and its teleconnection. J Geophys Res 112:C11007. https://doi.org/10.1029/2006JC003798

Bagnato S, Linsley B, Howe S, Wellington G, Salinger J (2004) Evaluating the use of the massive coral *Diploastrea heliopora* for paleoclimate reconstruction. Paleoceanography 19:1–12

Basher RE, Zheng X (1995) Tropical cyclones in the Southwest Pacific: spatial patterns and relationships to southern oscillation and sea surface temperature. J Clim 8:1249–1260

Bessafi M, Wheeler MC (2006) Modulation of South Indian Ocean tropical cyclones by the Madden–Julian oscillation and convectively coupled equatorial waves. Mon Weather Rev 134:638–656

Buckley BW, Leslie LM, Speer MS (2003) The impact of observational technology on climate database quality: tropical cyclones in the Tasman Sea. J Clim 16:2640–2645

Callaghan J, Power SB, (2011) Variability and decline in the number of severe tropical cyclones making land-fall over eastern Australia since the late nineteenth century. Clim Dyn, 37:647–662

Chand SS (2018) Impact of climate variability and change on tropical cyclones in the South Pacific. In: DellaSala DA, Goldstein MI (eds) The Encyclopedia of the anthropocene, vol 2. Elsevier, Oxford, pp 217–225

Chand SS, Walsh KJE (2009) Tropical cyclone activity in the Fiji region: spatial patterns and relationship to large-scale circulation. J Clim 22:3877–3893

Chand SS, Walsh KJE (2010) The influence of the Madden-Julian oscillation on tropical cyclone activity in the Fiji region. J Clim 23:868–886

Chand SS, Walsh KJE (2011) Influence of ENSO on tropical cyclone intensity in the Fiji region. J Clim 24:4096–4108

Chand SS, McBride JL, Tory KJ, Wheeler MC, Walsh KJE (2013) Impact of different ENSO regimes on Southwest Pacific tropical cyclones. J Clim 26:600–608

Chand SS, Tory KJ, Ye H, Walsh KJE (2017) Projected increase in El Niño-driven tropical cyclone frequency in the Pacific. Nat Clim Chang 7:123–127

Chen G (2011) How does shifting Pacific Ocean warming modulate on tropical cyclone frequency over the South China Sea? J Clim 24:4695–4700

Chen G, Tam CY (2010) Different impacts of two kinds of Pacific Ocean warming on tropical cyclone frequency over the western North Pacific. Geophys Res Lett 37:L01803. https://doi.org/10.1029/2009GL041708

Christensen JH, Krishna K, Aldrian E, An S-I, Cavalcanti IFA, de Castro M, Dong W, Goswami P, Hall A, Kanyanga JK et al (2013) Climate phenomena and their relevance for future regional climate change. In: Stocker TF, Qin D, Plattner G-K, Tignor M, Allen SK, Boschung J, Nauels A, Xia Y, Bex V, Midgley PM (eds) Climate change 2013: The physical science basis. Contribution of Working Group I to the Fifth Assessment Report of the Intergovernmental Panel on Climate Change (IPCC AR5). Cambridge University Press, Cambridge, UK and New York, NY

Compo GP et al (2011) The twentieth century reanalysis project. Quart J R Meteor Soc 137:1–28. https://doi.org/10.1002/qj.776

d'Aubert A, Nunn PD (2012) Furious winds and parched islands: tropical cyclones (hurricanes) 1558–1970 and droughts 1722–1987 in the Pacific. XLibris, Bloomington, p 358

DelSole T, Tippett MK, Shukla J (2011) A significant component of unforced multidecadal variability in the recent acceleration of global warming. J Clim 24:909–926

Diamond HJ, Renwick JA (2015) The climatological relationship between tropical cyclones in the Southwest Pacific and the southern annular mode. Int J Climatol 35:613–623

Diamond HJ, Lorrey AM, Knapp KR, Levinson DH (2013) A Southwest Pacific tropical cyclone climatology and linkages to the El Niño–southern oscillation. J Clim 32:3–25

Diamond HJ, Lorrey AM, Renwick JA (2012) Development of an enhanced tropical cyclone tracks database for the Southwest Pacific from 1840 to 2010. Int J Climatol 32:2240–2250

Diamond HJ, Lorrey AM, Renwick JA (2015) The climatological relationship between tropical cyclones in the southwest pacific and the Madden–Julian oscillation. Int J Climatol 35:676–686

Dowdy AJ (2014) Long-term changes in Australian tropical cyclone numbers. Atmos Sci Lett 15:292–298. https://doi.org/10.1002/asl2.502.

Dowdy AJ, Qi L, Jones D, Ramsay H, Fawcett R, Kuleshov Y (2012) Tropical cyclone climatology of the South Pacific Ocean and its relationship to El Niño–southern oscillation. J Clim 25:6108–6122

Emanuel KA (1987) The dependence of hurricane intensity on climate. Nature 326:483–485

Emanuel K, Sundararajan R, Williams J (2008) Hurricanes and global warming: results from downscaling IPCC AR4 simulations. Bull Am Meteorol Soc 89:347–367

Esler S (2015) Vanuatu post disaster needs assessment: Tropical Cyclone Pam, March 2015. Government of Vanuatu Report, Vanuatu, p 172

Esler S (2016) Fiji post disaster needs assessment: Tropical Cyclone Winston, February 2016. Government of Fiji, Fiji Islands, p 160

Haig J, Nott J, Reichart G (2014) Australian tropical cyclone activity lower than at any time over the past 550–1,500 years. Nature 505:667–671

Hall JD, Matthews AJ, Karoly DJ (2001) The modulation of tropical cyclone activity in the Australian region by the Madden-Julian oscillation. Mon Weather Rev 129:2970–2982

Harper BA, Stroud SA, McCormack M, West S (2008) A review of historical tropical cyclone intensity in northwestern Australia and implications for climate change trend analysis. Aust Meteorol Oceano J 57:121–141

Hastings PA (1990) Southern oscillation influence on tropical cyclone activity in the Australian/south-West Pacific region. Int J Climatol 10:291–298

Held IM, Hou AY (1980) Nonlinear axially symmetric circulations in a nearly inviscid atmosphere. J Atmos Sci 37:515–533

Hendon HH, Zhang C, Glick JD (1999) Interannual variation of the Madden–Julian oscillation during austral summer. J Clim 12:2538–2550

Holland GJ (1984) On the climatology and structure of tropical cyclones in the Australian/Southwest Pacific region: I. data and tropical storms. Aust Meteorol Mag 32:1–15

Holland GJ (1997) The maximum potential intensity of tropical cyclones. J Atmos Sci 54:2519–2541

Hong C-C, Li Y-H, Li T, Lee M-Y (2011) Impacts of Central Pacific and eastern Pacific El Niños on tropical cyclone tracks over the western North Pacific. Geophys Res Lett 38:L16712. https://doi.org/10.1029/2011GL048821

Jourdain NC et al (2011) Mesoscale simulation of tropical cyclones in the South Pacific: climatology and interannual variability. J Clim 24:3–25

Kao H-Y, Yu J–Y (2009) Contrasting eastern-Pacific and Central-Pacific types of ENSO. J Clim 22:615–632

Kerr IS (1976) Tropical storms and hurricanes in the Southwest Pacific, November 1939 to April 1969. New Zealand Ministry of Transport, Wellington, p 114

Kim H-M, Webster PJ, Curry JA (2009) Impact of shifting patterns of Pacific Ocean warming on North Atlantic tropical cyclones. Science 325:77–80

Kim H-M, Webster PJ, Curry JA (2011) Modulation of North Pacific tropical cyclone activity by three phases of ENSO. J Clim 24:1839–1849

Klotzbach PJ (2014) The Madden–Julian Oscillation's impacts on worldwide tropical cyclone activity. J Clim 27:2317–2330

Klotzbach PJ, Landsea CW (2015) Extremely intense hurricanes: revisiting Webster et al. (2005) after 10 years. J Clim 28:7621–7629

Knutson TR et al (2010) Tropical cyclones and climate change. Nat Geosci 3:157–163

Kossin JP, Olander TL, Knapp KR (2013) Trend analysis with a new global record of tropical cyclone intensity. J Clim 26:9960–9976

Kossin JP, Emanuel KA, Vecchi GA (2014) The poleward migration of the location of tropical cyclone maximum intensity. Nature 509:349–352

Kug J–S, Jin F–F, An S–I (2009) Two types of El Niño events: cold tongue El Niño and warm Pool El Niño. J Clim 22:1499–1515

Kuleshov Y, Qi L, Fawcett R, Jones D (2008) On tropical cyclone activity in the southern hemisphere: trends and the ENSO connection. Geophys Res Lett 35:L14S08. https://doi.org/10.1029/2007GL032983

Kuleshov Y, Chane Ming F, Qi L, Chouaibou I, Hoareau C, Roux F, (2009) Tropical cyclone genesis in the Southern Hemisphere and its relationship with the ENSO. Ann Geophys 27(6):2523–2538

Kuleshov YR, Fawcett R, Qi L, Trewin B, Jones D, McBride J, Ramsay H (2010) Trends in tropical cyclones in the South Indian Ocean and the South Pacific Ocean. J Geophys Res Atm 115:D01101. https://doi.org/10.1029/2009JD012372

Kumar L, Taylor S (2015) Exposure of coastal built assets in the South Pacific to climate risks. Nat Clim Chang 5:992–996

Landsea CW, Franklin JL (2013) Atlantic hurricane database uncertainty and presentation of a new database format. Mon Weather Rev 141:3576–3592

Landsea CW, Harper BA, Hoarau K, Knaff JA (2006) Can we detect trends in extreme tropical cyclones? Science 313:452–454

LeBec N, Juillet-Leclerc A, Corrège T, Blamart D, Delcroix T (2000) A coral $\delta^{18}O$ record of ENSO driven sea surface salinity variability in Fiji (southwestern tropical Pacific). Geophys Res Lett 27:3897–3900

Leroy A, Wheeler MC (2008) Statistical prediction of weekly tropical cyclone activity in the southern hemisphere. Mon Weather Rev 136:3637–3654

Lu J, Vecchi GA, Reichler T (2007) Expansion of the Hadley cell under global warming. Geophys Res Lett 34:L06805. https://doi.org/10.1029/2006GL028443

Madden RA, Julian PR (1971) Detection of a 40–50 day oscillation in the zonal wind in the tropical Pacific. J Atmos Sci 28:702–708

Maloney ED, Hartmann DL (2000) Modulation of hurricane activity in the Gulf of Mexico by the Madden–Julian oscillation. Science 287:2002–2004

McGray H, Hammil A, Bradley R (2007) Weathering the storm: options for framing adaptation and development. World Resources Institute, Washington, p 57

Meehl GA et al (2007) The WCRP CMIP3 multimodel dataset: a new era in climate change research. Bull Am Meteorol Soc 88:1383–1394

Moon I-J, Kim S-H, Klotzbach P, Chan JCL (2015) Roles of interbasin frequency changes in the poleward shifts of the maximum intensity location of tropical cyclones. Env Res Lett 10. https://doi.org/10.1088/1748-9326/10/10/104004

Nurse LA, McLean RF, Agard J, Briguglio LP, Duvat-Magnan V, Pelesikoti N, Tompkins E, Webb A (2014) Small islands. In: Barros VR, Field CB, Dokken DJ, Mastrandrea MD, Mach KJ, Bilir TE, Chatterjee M, Ebi KL, Estrada YO, Genova RC, Girma B, Kissel ES, Levy AN, MacCracken S, Mastrandrea PR, White LL (eds) Climate change: impacts, adaptation, and vulnerability. Part B: Regional aspects. Contribution of working group II to the fifth assessment report of the intergovernmental panel on climate change. Cambridge University Press, Cambridge, pp 1613–1654

Patricola CM, Wehner M (2018) Anthropogenic influences on major tropical cyclone events. Nature 563:339–346. https://doi.org/10.1038/s41586-018-0673-2

Ramage CS, Hori AM (1981) Meteorological aspects of El Niño. Mon Weather Rev 109:1827–1835

Ramsay HA, Camargo SJ, Kim D (2012) Cluster analysis of tropical cyclone tracks in the southern hemisphere. Clim Dyn 39:897–917

Revell CG (1981) Tropical cyclones in the southwest Pacific, Nov. 1969 to April 1979. New Zealand Meteorological Service Misc. Pub. 170, p 53

Revell CG, Goulter SW (1986) South Pacific tropical cyclones and the southern oscillation. Mon Weather Rev 114:1138–1145

Riahi K, Rao S, Krey V, Cho C, Chirkov V, Fischer G, Kindermann G, Nakicenovic N, Rafaj P (2011) RCP 8.5—a scenario of comparatively high greenhouse gas emissions. Clim Chang 109:33–57. https://doi.org/10.1007/s10584-011-0149-y

Saha S et al (2010) The NCEP climate forecast system reanalysis. Bull Am Meteorol Soc 91:1015–1057

Sinclair MR (2002) Extratropical transition of Southwest Pacific tropical cyclones. Part I: Climatology and mean structure changes. Mon Weather Rev 130:590–609

Taylor KE, Stouffer RJ, Meehl GA (2012) An overview of CMIP5 and the experiment design. Bull Am Meteorol Soc 93:485–498

Thompson CS, Ready S, Zheng X (1992) Tropical cyclones in the Southwest Pacific: November 1979 to May 1989. New Zealand Meteorological Service, Wellington, p 35

Ting M, Kushnir Y, Seager R, Li C (2009) Forced and natural 20th century SST trends in the North Atlantic. J Clim 22:1469–1481

Tory KJ, Dare RA, Davidson NE, McBride JL, Chand SS (2013a) The importance of low-deformation vorticity in tropical cyclone formation. Atmos Chem Phys 13:2115–2132

Tory KJ, Chand SS, McBride JL, Dare R, Ye H (2013b) Projected changes in late 21st century tropical cyclone frequency in 13 coupled climate models from the coupled climate model Intercomparison project. J Clim 26:9946–9959

Trenberth KE (1997) The definition of El Niño. Bull Am Meteorol Soc 78:2771–2777

Trenberth KE, Stepaniak DP (2001) Indices of El Niño evolution. J Clim 14:1697–1701

Troup AJ (1965) The southern oscillation. Q J R Meteorol Soc 91:490–506

Vincent EM, Lengaigne M, Menkes CE, Jourdain NC, Marchesiello P, Madec G (2011) Interannual variability of the South Pacific convergence zone and implications for tropical cyclone genesis. Clim Dyn 36:1881–1896

Visher S (1925) Tropical Cyclones of the Pacific. Bernice P. Bishop Museum, Bulletin 20, p 163

Walker GT (1923) Correlations in seasonal variations of weather VIII. Mem India Meteorol Dept 24:75–131

Walker GT (1924) Correlations in seasonal variations of weather IX. Mem India Meteorol Dept 24:333–345

Walsh KJE et al (2016) Tropical cyclones and climate change. WIRES Clim Change 7:65–89. https://doi.org/10.1002/wcc.371

Webster PJ, Holland GJ, Curry JA, Chang HR (2005) Changes in Tropical Cyclone Number, Duration, and Intensity in a Warming Environment. Science 309(5742):1844–1846

Wheeler MC, Hendon HH, (2004) An All-Season Real-Time Multivariate MJO Index: Development of an Index for Monitoring and Prediction. Mon Weather Rev 132(8):1917–1932

Woodruff J, Sriver RL, Lunf DC (2012) Tropical cyclone activity and western North Atlantic stratification over the last millennium: a comparative review with viable connections. J Quaternary Sci 27:337–343

Woodruff JD, Irish JL, Camargo SJ (2013) Coastal flooding by tropical cyclones and sea-level rise. Nature 504:44–52

Chapter 7
Impacts of Climate Change on Coastal Infrastructure in the Pacific

Lalit Kumar, Tharani Gopalakrishnan, and Sadeeka Jayasinghe

7.1 Introduction

Adverse impacts of climate change on coastal societies and their affiliate infrastructure will have a significantly greater impact than inland regions owing to the restricted land area, dispersed population along the coast, remoteness and inherent obstacles of topography factors in the Pacific region. Such factors lead to challenges that are specific and considered normal for the region, even without considering climate change impacts. Most of the Pacific island countries have relatively low economic stability, and weak monetary and fiscal policies, reflecting on the poor infrastructure performance. The major challenge is the Pacific island countries' vulnerability to changing climate scenarios and natural hazards.

Small population sizes dispersed in Pacific island countries make it difficult to obtain economies of scale in infrastructure service provisions. Geographical isolations of Pacific islands with natural resources that are limited, economies that are dependent on only a few commodities, major markets that are quite distant and vulnerability to external shocks impact on growth and infrastructure instability (Granger 1999). Small Pacific islands are a long way from major trading centres which makes it difficult to transport fuel, machinery, equipment and materials, leading to the high cost of service provision. For example, Samoa, Tonga, Kiribati and the Marshall Islands are all over 3500 km from the nearest major port. Pacific island countries (PICs), such as Fiji, Samoa, Vanuatu and Timor-Leste, are relatively

L. Kumar (✉) · T. Gopalakrishnan
School of Environmental and Rural Science, University of New England,
Armidale, NSW, Australia
e-mail: lkumar@une.edu.au; tgopalak@myune.edu.au

S. Jayasinghe
Department of Export Agriculture, Faculty of Animal Science and Export Agriculture,
Uva Wellassa University, Badulla, Sri Lanka
e-mail: ljayasi2@myune.edu.au

© Springer Nature Switzerland AG 2020
L. Kumar (ed.), *Climate Change and Impacts in the Pacific*, Springer Climate,
https://doi.org/10.1007/978-3-030-32878-8_7

mountainous, making it difficult and costly to link infrastructure networks with hinterland communities. Volcanic islands have high slopes inland, and so risks of landslides from heavy rainfall are greater, resulting in higher risk of damage to infrastructure. On the other hand, countries that are mainly reefal in origin, such as Tuvalu and Kiribati, and consist of many small islands or atolls have to bear an additional cost of linking services between islands. Atolls are low-lying islands which are especially exposed to sea-level rise. For instance, in Kiribati, there is a lack of elevated lands where important infrastructure can be constructed to avoid impacts from rising sea levels or king tides (Alison et al. 2011); on the other hand, infrastructure on raised atolls can face coastal erosion. In the Pacific, small islands experience more frequent natural disasters and vulnerability to extreme events such as tropical cyclones, earthquakes, tsunami and drought, and these significantly increase infrastructure-related costs (Granger 1999).

The Pacific area is considered to be one of the most susceptible to natural hazards and climate change because it is situated in an extremely vibrant ocean-atmosphere interface (Kumar and Taylor 2015). Climate change introduces a variety of serious impacts on infrastructure in the Pacific island countries and has worsened effects as these infrastructures are connected to extremely populate urban centres on the coastal margins. Further, infrastructure in PICs is particularly susceptible to climate change effects, with rapid rural-urban migration over the past century and significant population centres now located in low-lying and vulnerable coastal regions. The stress posed by coastal inundation due to sea-level rises, changes in air temperature, frequency and intensity of tropical cyclones, wind and wave patterns and changes in precipitation on infrastructure facilities like water supplement, ICT, energy, building and transportation is additional stress to that already present. Increased risks of coastal flooding and erosion result in damage to built infrastructure, transport network and cultural sites, along with coastal retreat and beach loss (Hu et al. 2015). As we pass through the twenty-first century, it is anticipated that climate change impacts will increase and be more severe, with the likelihood of increased physical changes around coastal areas and thus higher levels of threat to populations and infrastructure in the Pacific island countries (Alex et al. 2019).

PICs are subjected to a broad range of natural hazards that include cyclones, earthquakes, tropical storms, floods, droughts, storm surges and tsunamis. Climate change will exacerbate some of these risks. In particular, PICs are usually made up of coral/sandy formations, have a relatively small area and low elevations, are distributed over a large ocean area and fall in an active tropical cyclone zone, resulting in them being more susceptible to climate-related risks. According to the risk assessment of 2030 (Alex et al. 2019), the climate change impacts that affect infrastructure in the Pacific region include a mix of slow-onset events such as rising temperatures and sea-level rise (e.g. the Western Pacific Ocean sea level rises at approximately three times the worldwide pace of around 3 mm per year), more variable patterns of rainfall and extended droughts, and higher intensity and frequency of extreme weather events such as storm surges and tropical cyclones. In particular, it also shows the potential hazards applicable to Pacific island countries and territories' infrastructure in the future (Alex et al. 2019). Specifically, infrastructure is

Table 7.1 Overview of major impacts of climate change and extreme events on key coastal infrastructure assets in PICs

Key infrastructure	Impacts of climate change and extreme events								
	Storm surge	Sea-level rise	King tide	Wave action	Rainfall	Drought	Prolonged rain	Flood	Cyclone wind
Energy	M	M	M	M	W	M	M	M	S
Water					W				
• Supply	S	S	S	S		S	S	S	S
• Wastewater	S	S	S	S		S	S	S	S
• Drainage	S	S	S	S		S	S	S	S
Solid waste	M	M	M	M	W	W	S	S	S
Transport					W				
• Road	S	S	S	S		M	M	S	S
• Ports	S	S	S	S		M	M	S	S
• Airports	S	S	S	S		M	M	S	S
ICT	M	M	W	W	W	W	W	M	S
Buildings					W				
• Settlement	S	S	S	S		S	S	S	S
• Health	S	S	S	S		S	S	S	S

Adapted from Alison et al. (2011)
Key: *S* strong, *M* moderate, *W* weak/none

more critical in the context of climate change in Pacific island nations, and this phenomenon is more acutely affected when natural disasters strike as more infrastructure is spread along the coastline. Table 7.1 shows the effect of significant climate impacts and extreme events in the PICs on important components of infrastructure.

There are still many remote islands with no or limited economic infrastructure, minimal or intermittent government social services and surviving with a mixture of traditional and imported infrastructure for primary wellbeing. These populations need to be considered in climate change scenarios, as well as the urban centres which will tend to have the more centralised infrastructure to provide sanitary facilities, water, drainage and wastewater treatment space and accessibility to materials for housing and other constructions. Another important aspect of infrastructure is to maintain health services to support communities, particularly during disaster events (World Bank 2006). As most of the infrastructure in small islands is located on the coast, this is of particular concern. Infrastructure to be impacted by the increase in sea level involves wharfs, jetties and ports and harbors and associated support infrastructure. These infrastructures will be impacted in a multitude of ways, including by the effects of inundation, seawater intrusion as well as flooding occurrences (i.e. alkali-silica reaction, concrete cancer, erosion and corrosion) and by the more instant and disastrous harm caused by wave action (Alex et al. 2019). Transport is a critical infrastructure in the Pacific as there is a highly dispersed population and there is a need to move goods and services across long distances (Alison et al. 2011;

Hanson 2008). This becomes even more critical in times of disasters when emergency services are essential. ICT costs tend to be high per capita, for the same matter. Energy is another crucial infrastructure, and Pacific island countries are more vulnerable as fossil fuel prices are tied to world markets. Furthermore, socioeconomic activities such as agriculture, tourism, fisheries and forestry associated with infrastructure are another important piece of infrastructure for ongoing development and sustainable livelihood in the Pacific regions.

In addition, there are predominant weaknesses in infrastructure in the Pacific which is widely considered to be of relatively low quality (World Bank 2006). Some of the reasons for this are the accessibility to remote islands, access to raw material, poor construction quality, limited resources, lack of infrastructure professionals compared to requirements, a high dependency on foreign aid, and sectorial fragmentation of responsibilities to maintenance and policy implementations, land tenure conditions and poor infrastructure management. On low-lying islands in the Pacific, important public facilities and infrastructure, together with commercial and residential property, are especially vulnerable. Even with these challenges, some Pacific countries demonstrate better infrastructure performances with greatest inherent challenges. For example, Vanuatu exhibits a better infrastructure performance than other smaller islands which have a relatively dispersed population, high ethnic diversity and a high degree of resilience on foreign aid (Alex et al. 2019). This suggests that some poor performances occur not merely due to the above factors, but due to poor policies and institutional capacity. In the given background, three major constraints that have been identified as infrastructure barriers in the Pacific are less accessibility to infrastructure in rural areas, inefficient services provisions and pricing for infrastructure which is inappropriate (World Bank 2006). Lack of coordination among government institutions and poor policies restrict appropriate performance in infrastructure maintenance. Many governments in Pacific island countries have focused more on building new infrastructure than on the maintenance side. Lack of coordination between policies and fiscal regulations are not aligned well, and it tends to lead to failures in the desired outcomes in infrastructure. Further, in many Pacific island countries, infrastructure services are provided and authorised by the same entity, the government. Pacific island countries generally have lower accessibility to electricity, telecommunications, water and sanitation than in comparator countries (such as the Caribbean islands and the Philippines) which have similar income levels. However, mobile phone usage has significantly increased access to telecommunications in recent years in rural areas (Alex et al. 2019).

Strong policies and better institutional management can help to improve the overall progress of infrastructure facilities in Pacific island countries. Since major infrastructures are generally located along the coastal fringes of most PICs, they are particularly vulnerable to the impacts of climate change. With the pace of climate change set to increase over the next 50 to 100 years and projections of more frequent extreme events, infrastructure needs to be made as resilient as possible to projected climate change, especially in low-lying PIC regions.

7.2   Case Study: Exposure of Coastal Built Infrastructure Assets in the South Pacific to Climate Risks

7.2.1   Introduction

Infrastructure plays a vital role in development, so understanding the relationship between climate change and infrastructure is critical and crucial. In many cases throughout the Pacific, the stress on built infrastructure posed by climate change is an increase in stresses already present. The South Pacific region covers the Pacific Ocean that is located south of the equator and comprises 23 countries and territories, with thousands of islands and islets. The South Pacific region will continue to be one of the areas that will be most affected by the multiple challenges raised by climate change. Projections suggest that the low-lying atolls and islands that are highly dependent on coastal areas for socio-economic activity are highly vulnerable to climate and other extreme events (Nurse et al. 2014). Indeed, the region is characterised by diversity between its different islands having total land area of 88,000 km^2 and a population of three million, with the average area of these islands being 90 km^2 and the median being only 1.3 km^2 (Kumar et al. 2018). This shows that island size is highly skewed towards smaller islands. Due to the high number of low-lying atolls and islets with large growing population centres, sea-level rise is one of the most pressing concerns in the South Pacific region, which has a detrimental effect on infrastructure. In addition, the impacts on infrastructure in the South Pacific region are likely to be exacerbated by an increase in the probability of high-intensity cyclones and associated storm surges (Handmer and Nalau 2019; Preston et al. 2006).

A comprehensive analysis is critical given the high exposure of built assets in the PICs to climate change risk. Infrastructure in the Pacific has characteristics that affect its vulnerability to climate change. Therefore, this study sought to assess the potential impacts of current and future climate change risk on built infrastructure for 12 Pacific island countries (PICs). A comprehensive database of all infrastructure, such as industrial, commercial and residential buildings, airports, wharfs, hospitals, school and any other built infrastructure, was assembled using data from multiple sources. The proximity of the infrastructure to coastal areas was analysed using Geographic Information System (GIS) to determine the percentage of infrastructure in close proximity to the coast. The infrastructure was also valued to see potential replacement costs of those within close proximity.

7.2.2   Methods

The Pacific Risk Information System (PacRIS), one of the largest collections of PICs geospatial risk data (ACP-EU 2017), was used in this study to acquire most of the raw data. PacRIS was established through the Pacific Catastrophe Risk

Assessment and Financing Initiative (PCRAFI), a joint initiative of the Secretariat of the Pacific Community, the World Bank and the Asian Development Bank. As part of this initiative, data was gathered on 15 PICs (the Cook Islands, Federated States of Micronesia (FSM), Fiji, Kiribati, Marshall Islands, Nauru, Niue, Palau, Papua New Guinea, Samoa, Solomon Islands, Timor-Leste, Tonga, Tuvalu and Vanuatu) (ACP-EU 2017). Some countries had complete coverage, while others only had a partial coverage. For this study we only used those countries that had complete or near-complete coverage in terms of infrastructure. These were the Cook Islands, Federated States of Micronesia, Kiribati, Marshall Islands, Nauru, Niue, Palau, Samoa, Solomon Islands, Tonga, Tuvalu and Vanuatu (Fig. 7.1). The 12 selected countries represent the small island states in the South Pacific in terms of lithology, maximum elevation range, remoteness, demography and economic status (Table 7.2). The island's lithology is mainly volcanic, limestone, reef or a composite of these three; however, about 67% of the islands are of reef or sandy in origin (Kumar et al. 2018). The reef islands are usually less than 3 m in maximum elevation, while some of the volcanic islands have maximum elevations greater than 2000 m. The area of the islands varies from 0.01 km^2 to 5500 km^2, with a median of just 0.9 km^2. The total population of the 12 countries is 1.4 million, with population for individual countries ranging from 1480 to 547,540 residents. The 12 PICs have a total land area of 50,212 km^2, whereas the Exclusive Economic Zone covers 13 million square kilometres, showing the islands' large spread. The 12 PICs comprise

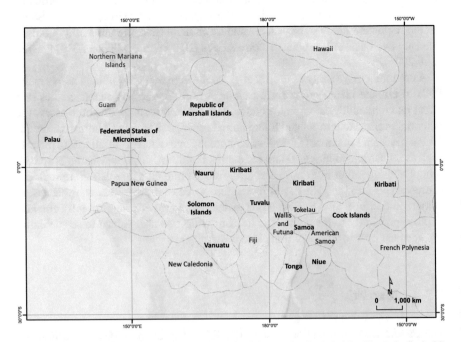

Fig. 7.1 Study region showing the locations of the 12 countries included in this study (in bold) and the other countries of the region for context

Table 7.2 Characteristics of the 12 countries from the South Pacific used in this study

Country	No. of islands	Coastline (km)	Total land area (km²)	Population (2014)	GDP 2011 (US$M)	Max elevation (m)	Av. land area/ island (km²)	Dominant lithology
Cook Is.	15	454	296	19,800	183	658	19.7	Reef
FSM	607	1036	702	111,560	331	791	1.2	Reef
Kiribati	33	1845	995	104,488	180	81	30.2	Reef
Marshall Is.	34	2172	286	54,820	178	6	8.4	Reef
Nauru	1	19	23	10,800	60	71	23.0	Limestone
Niue	1	75	298	1480	10	60	298.0	Limestone
Palau	250	514	495	20,500	272	207	2.0	Limestone
Samoa	7	482	3046	182,900	995	1858	435.1	Volcanic
Solomon Is.	413	8848	29,672	547,540	1046	2449	71.8	Volcanic
Tonga	176	929	847	103,350	523	1033	4.8	Limestone
Tuvalu	9	233	26	9561	35	8	2.9	Reef
Vanuatu	82	3234	13,526	245,860	687	1879	165.0	Volcanic

a total of 1628 islands (counting only islands greater than one hectare), with a coastline of 19,841 km.

The assessment undertaken as part of this study was based on accurate geographic coordinates of individual built infrastructure assets; such data is rarely available on a regional scale for various countries in the South Pacific region. The PCRAFI database offers a detailed collection of properties such as residential, commercial, public and industrial and other built infrastructure such as airports, communications infrastructure, energy generation, docks and ports, bridges, storage facilities and water infrastructure such as storage tanks. The database comprises location, occupancy, type of construction and asset replacement value information. While analysis was undertaken at the detailed infrastructure type, in this chapter they were grouped together as built infrastructure for reporting.

In addition to the PCRAFI database, other information was gathered from a variety of sources including field visits, manual inspection of high-resolution satellite imagery, GIS databases, Australian government data, reports and publications, government databases and reports on catastrophe recognition. For quality control, all point data were overlaid in a GIS package against a georeferenced background, and points that were not aligned or fell outside the coastline of the island were removed due to recording errors. The number of points removed through this exercise ranged from 0 to 2% for individual countries; hence overall a low number of points were lost. After error corrections, there were 451,726 built infrastructure points left in the database.

To determine distances of infrastructure assets from the coast, accurate coastline data is also necessary. Coastline data was acquired from countries where accurate data were accessible; however for many countries and smaller islands, the coastline boundary data was fairly generalised and thus could not be used for the detailed analysis required as part of this study. Therefore many island boundaries were re-digitised at a scale of 1:20,000 or better from high-resolution imagery and topographic maps. As this assessment was designed to determine the exposure of built infrastructure to coastal hazards, four intervals from the coast were used in the analysis: 0–50 m, 50–100 m, 100–200 m and 200–500 m. A maximum distance of 500 m from the coastline was selected as the majority of the islands in the South Pacific are small in area, and for some countries (i.e. Marshall Islands and Tuvalu) about 99% of the entire country's land area falls in this zone. For the Eastern Caribbean region, Lewsey et al. (2004) used a range of 2 km; however, this distance would be unsuitable for the small South Pacific island countries.

The number and replacement value of built infrastructure was obtained for each of these intervals. In addition, the built infrastructure point data was overlaid for each country on a soil layer, and the soil type was determined for each asset's location. The soil type was reclassed into two categories: (1) soft to hard rock and (2) soft to firm soil, including sandy soil.

Road infrastructure, including bridges, was not included in this analysis since detailed information about road surface and number of lanes of road was not possible to determine for all roads.

7.2.3   Results and Discussion

The results indicate that a large proportion of the infrastructure for the 12 PICs is situated in close proximity to the coast. Overall, for the 12 PICs, 57% of the built infrastructure is within 500 m of the coastline. For the individual bands assessed here, 9%, 11%, 16% and 21% fall in intervals of 0–50 m, 50–100 m, 100–200 m and 200–500 m intervals, respectively. Therefore only 43% of built infrastructure is beyond 500 m of the coastline when all 12 PICs are considered together.

In terms of individual countries, we see large variations in the distributions of the infrastructure in terms of the distance from the coastline. Figure 7.2 shows this quite clearly, with Fig. 7.2a showing the distribution by individual bands, while Fig. 7.2b shows the same information but in cumulative terms. For some countries, such as Niue, Samoa, Solomon Islands, Tonga and Vanuatu, the majority of infrastructure is beyond 500 m of coastline. For countries such as Kiribati, Marshall Islands and Tuvalu, only a very small fraction of built infrastructure is beyond 500 m of the coastline (Fig. 7.2a). For Kiribati this is only 3%, for Marshall Islands it is 2% and Tuvalu this is only 1%, placing almost all infrastructure in very close proximity to the coast and to the immediate impacts of climate change. Kiribati, Marshall Islands and Tuvalu are also noteworthy in the sense that the largest proportion of built infrastructure is within 50 m of the coastline (Fig. 7.2a), the percentages being 36, 36 and 35, respectively. For the above three countries, the percentage of built infrastructure within 100 m of the coastline is 67, 72 and 66, respectively, placing a very high

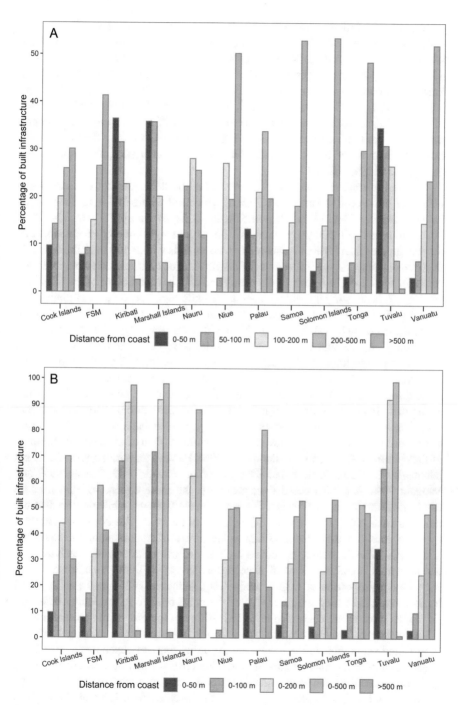

Fig. 7.2 Counts of built infrastructure (percentage of country total) within 0–50 m, 50–100 m, 100–200 m and 200–500 m intervals from the coastline. Figure (**a**) shows percentages in individual bands, while (**b**) shows this as cumulative values (0–50 m, 0–100 m, 0–200 m and 0–500 m)

number of infrastructure assets in very close proximity to the coast. This is to be expected, as these countries have islands that are relatively small; the two largest islands by area in Kiribati, Marshall Islands and Tuvalu are 478 km² and 55 km², 29 km² and 24 km² and 10 km² and 8 km² respectively. On the other hand, for Niue, Samoa, Solomon Islands and Vanuatu, the two largest island sizes are 298 km² (one island country), 1823 km² and 1215 km², 5542 km² and 3978 km² and 4355 km² and 2240 km², respectively. Compared to other countries, Niue has a much smaller proportion of built infrastructure within 100 m of the coastline. Tonga, Vanuatu, Solomon Islands and Samoa also exhibit similar characteristics, although the percentages are higher (Fig. 7.2b).

In terms of replacement value, the total of all built infrastructure assessed was US$ 27.7 billion (2011 valuation), 79% of which fell within 500 m of the coast by value. Moreover, 11%, 14%, 34% and 20% fell in the intervals of 0–50 m, 50–100 m, 100–200 m and 200–500 m, respectively. For every single country except Niue and Samoa, greater than 50% of infrastructure assets falls within 500 m of the coastline in terms of value. Figure 7.3 shows the distribution of built infrastructure assets by value for the 12 PICs in each of the bands and also beyond 500 m of the coastline, with Fig. 7.3a showing individual band data, while Fig. 7.3b shows cumulative values up to 500 m of the coastline. For Kiribati and Palau, the highest proportion of built infrastructure by value is within 50 m of the coastline (Fig. 7.3a). Niue, Samoa, Solomon Islands, Tonga and Vanuatu are on the opposite end of the spectrum, with the lowest proportion of built infrastructure by value within 50 m of the coastline.

Kiribati, Marshall Islands and Tuvalu have replacement values of 95%, 98% and 99%, respectively, of built infrastructure within 500 m of the coast (Fig. 7.3b), indicating that almost the entire proportion of built infrastructure by value is located in very close proximity to the coast. From Fig. 7.3b we see that a very high proportion of the built infrastructure by value is situated within 500 m of the coastline for most of the countries. For the Cook Islands it is 90%, FSM 71%, Kiribati 95%, Marshall Islands 99%, Nauru 93%, Palau 89%, Solomon Islands 74%, Tuvalu 99% and Vanuatu 90%. It is also noted that, generally, the more expensive infrastructure assets are along the coastal fringe. For example, for Vanuatu, while only 48% of the built infrastructure by count was within 500 m of the coast, this accounted for 90% by value, implying that Vanuatu's largest built infrastructure is along the coastline. Most PICs follow the same pattern as high-value built infrastructure, such as ports and refineries, are always located close to the coast to facilitate transportation.

A higher proportion of commercial (71%), industrial (62%) and public (63%) infrastructure is located within 500 m of the coast relative to residential buildings (52%), as most urban centres in nearly all PICs are located on the coastal border. While transportation has been the main factor in setting up commercial centres, most island economies in this region are primarily tourism-dependent as they have appealing beaches, coral reef habitats and other coastal amenities. As a result, much of the infrastructure connected with this sector is therefore concentrated on the coast.

Of the built infrastructure within 500 m of the coast, 69% is situated on soft to firm soil (including sandy soil), while the remaining 31% is on soft to hard rock.

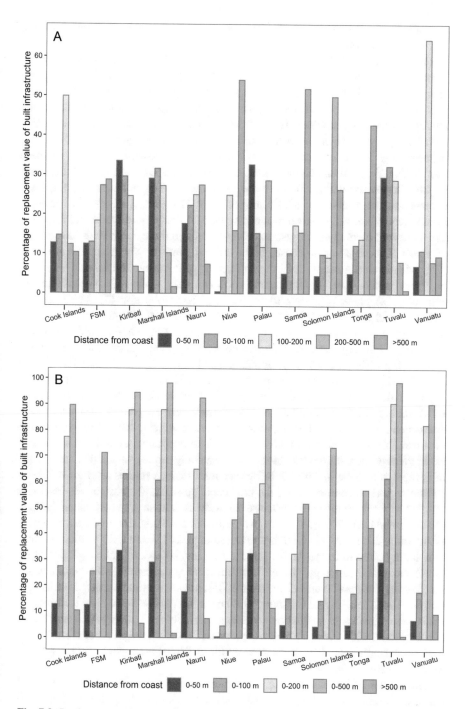

Fig. 7.3 Replacement value of built infrastructure (percentage of country total) within 0–50 m, 50–100 m, 100–200 m and 200–500 m intervals from the coastline. Figure (**a**) shows percentages in individual bands, while (**b**) shows this as cumulative values (0–50 m, 0–100 m, 0–200 m and 0–500 m)

Kiribati, Marshall Islands and Tuvalu have over 90% of the built infrastructure in this zone located on soft sandy soil.

Figures 7.4, 7.5, 7.6, 7.7 and 7.8 show the distribution of all built infrastructure assets on the islands of Koror (Palau), Rarotonga (Cook Islands), Tongatapu (Tonga), Funafuti (Tuvalu) and Upolu (Samoa), respectively. Each of these examples shows how the built infrastructure assets are clustered around the coast and mostly at very low elevations. Each of the figures includes satellite images to provide some perspective on the distribution. Table 7.3 gives the proportion of the built infrastructure assets within 100 m and 500 m of the coastline for each of these islands. For Funafuti and Koror, 98.8% and 96.3%, respectively, of the built infrastructure assets are within 500 m of the coastline. For Tongatapu and Upolu, these percentages are 41.8% and 39.4%, respectively. However, for both these islands, infrastructure is mostly spread around the coastal fringe, and the inland region is almost devoid of any built infrastructure, even though these two islands have plenty of space to accommodate these in the central parts of the islands. Rarotonga (Fig. 7.5) also has most of the hinterland devoid of infrastructure assets.

These results have consequences for disruption under (1) present climate extremes, (2) present geo-hazard events (e.g. tsunamis) and (3) future climate change, whichever exacerbates these extremes, posing significant threats to economies in the Pacific's extremely focused coastal area. The rising sea levels, tropical cyclones, storm surges and extreme weather events present a significant climate risk to low-lying PICs. As most of the population and urban centres are located along the coastline, the infrastructure risks are considerable. Thus, assessments of vulnerability, as conducted in this research, along with other variables such as elevation and GDP, should play an important part in formulating adaptation measures. Such assessments in the South Pacific, however, have previously been hampered by the non-availability of such comprehensive data. A comprehensive coastal vulnerability assessment of the nature presented in this case study requires a detailed knowledge of the spatial distribution of built infrastructure. Previous studies on climate risks to coastal regions have investigated the impacts of specific risks such as sea-level rise on coastal ecosystems, infrastructure as well as human populations in Europe (Nicholls and Klein 2005), America (Arkema et al. 2013), Australia (Commonwealth of Australia 2011) and the Caribbean (Lorde et al. 2013) and risks to food production and agriculture in the Pacific Islands (Barnett 2001) and India (O'Brien et al. 2004). Adaptation strategies in terms of hurricane preparedness have been investigated in the Cayman Islands (Tompkins 2005); however, climate risks to infrastructure have been less studied (Mimura 1999) even though they have been identified as major impediments to adaptation planning, particularly in low-lying island states (Lewsey et al. 2004). Many previous research focused on the effect of climate hazards on a single island infrastructure within a PIC (Forbes and Solomon 1998; Storey and Hunter 2010).

The results of this study have highlighted the potential for severe impacts of disastrous events caused by climate change on the coastal infrastructure of the low-lying islands in the Pacific. A combination of small land area with low elevations

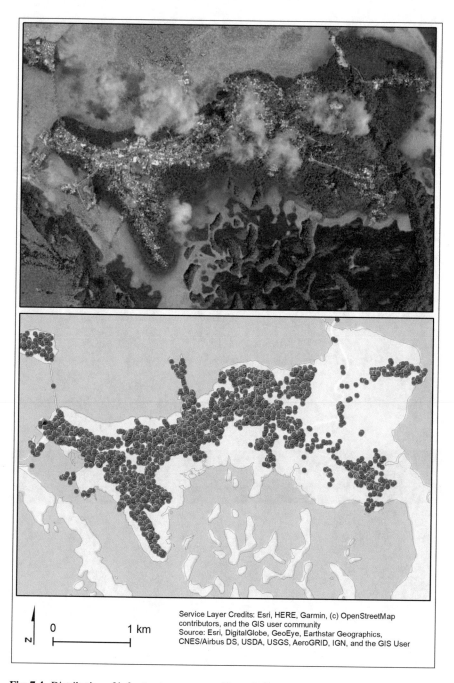

Fig. 7.4 Distribution of infrastructure assets on Koror, Palau

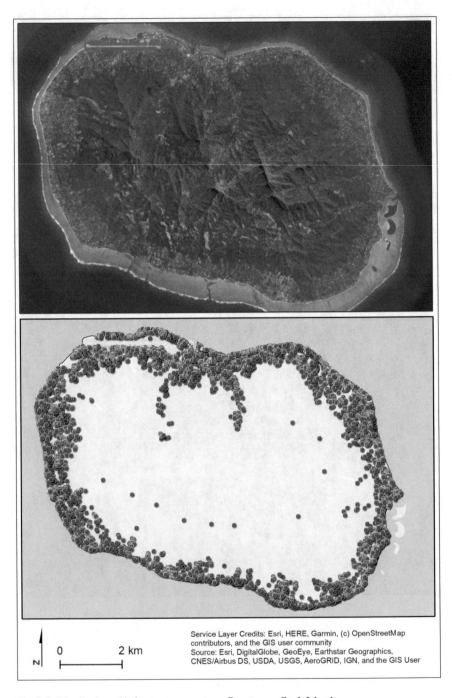

Fig. 7.5 Distribution of infrastructure assets on Rarotonga, Cook Islands

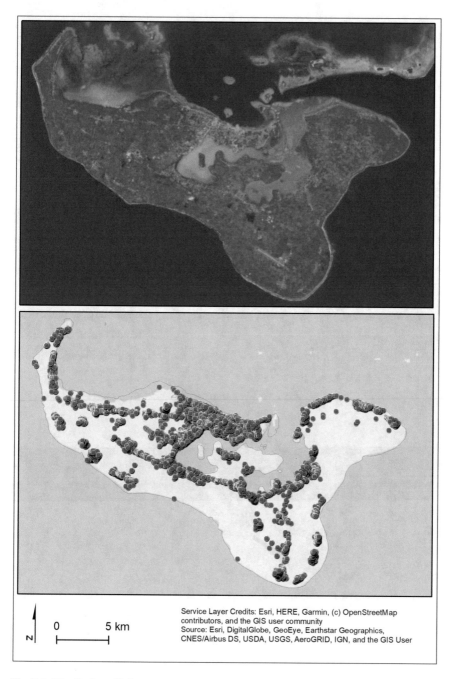

Fig. 7.6 Distribution of infrastructure assets on Tongatapu, Tonga

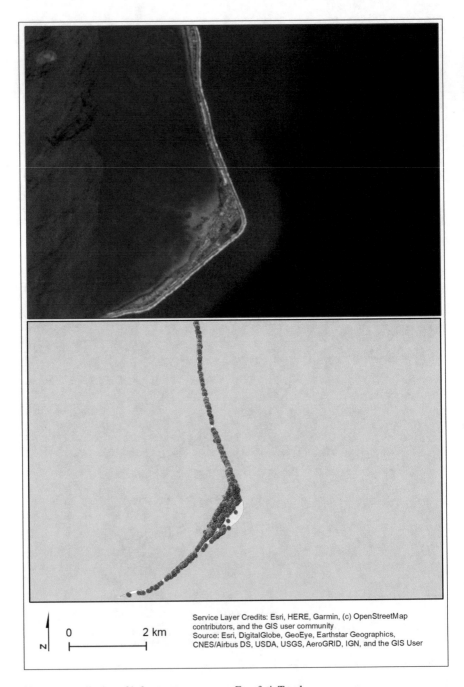

Fig. 7.7 Distribution of infrastructure assets on Funafuti, Tuvalu

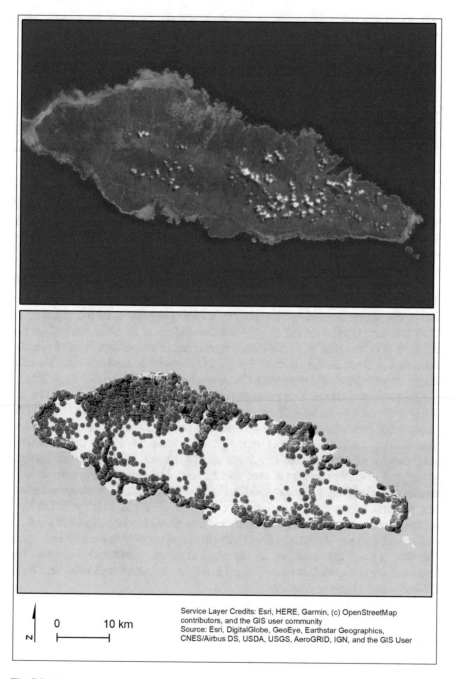

Fig. 7.8 Distribution of infrastructure assets on Upolu, Samoa

Table 7.3 Distribution of infrastructure on the islands of Koror, Rarotonga, Tongatapu, Funafuti and Upolu, as shown in Figs. 7.4 7.5, 7.6, 7.7 and 7.8

Island	% of infrastructure <100 m from the coastline	% of infrastructure <500 m from the coastline
Koror	9.7	96.3
Rarotonga	11.5	76.6
Tongatapu	1.9	41.8
Funafuti	35.3	98.8
Upolu	5.2	39.4

makes these countries' built infrastructure assets intensely susceptible to climate hazards. Based on the predicted increase in sea level, a larger proportion of built infrastructure resources that are closer to the coast will increase their vulnerability to coastal hazards. Thus, in the context of sea-level rise, these findings could be used to assess how much built infrastructure would be affected by numerous catastrophic events, present or future, related to climate.

Climate risk exposure will differ across the region for the distinct PICs and will therefore have varying effects on the islands and the built infrastructure. For example, in this region tropical cyclone activity is highly dependent on the cycles of El Niño Southern Oscillation (ENSO) (Henderson-Sellers et al. 1998), and across the Pacific it is quite variable. In addition, the planning of adaptations is quite complicated and difficult in the context of projected changes in sea-level rise and tidal ranges varying across the different PICs (Australian Bureau of Meteorology 2011).

Since sea-level rise in the region could range from 26 to 55 cm or 45 to 82 cm by 2081–2100, relative to the RCP2.6 and RCP8.5 emission scenarios for 1986–2005, respectively, this research would have benefited from the incorporation of elevation information; however, vertical elevation precision requires to be at the centimetre scale to have any significance in climate change debates. For most of the Pacific, elevation data at this accuracy are not available and, due to remoteness and the costs involved, are unlikely to be collected in the foreseeable future. A concerted effort may be needed to target such information collection for regions recognised as having the greatest exposure. More effort should also be made to systematically collect data on extreme events and associated harm to the region's infrastructure since such data can improve the results of similar research as conducted here. However, the remoteness and distribution of the islands in the Pacific makes such tasks extremely difficult and prohibitive cost-wise.

References

ACP-EU (2017) Pacific Catastrophe Risk Assessment and Financing Initiative (PCRAFI) Component 2: Exposure Data Collection and Management. ACP-EU Natural Disaster Risk Reduction Program. Retrieved from https://www.gfdrr.org/en/pacific-catastrophe-risk-assessment-and-financing-initiative-phase-3

Alex B, Baillat A, Gemenne F (2019) Implications of climate change on defence and security in the South Pacific by 2030. Institute of International and Strategic Affairs, p 72

Alison J, Baker G, Week D, Consult A (2011) Infrastructure and climate change in the Pacific. Pacific Australia Climate Change Science and Adaptation Planning (PACCSAP), 74. https://www.environment.gov.au/system/files/resources/1f81b90c-1581-4460-865a-7b7306f7f528/files/infrastructure-report.pdf

Arkema KK, Guannel G, Verutes G, Wood SA, Guerry A, Ruckelshaus M, Silver JM (2013) Coastal habitats shield people and property from sea-level rise and storms. Nat Clim Chang 3(10):913. https://doi.org/10.1038/NCLIMATE1944

Australian Bureau of Meteorology (2011) Scientific assessment and new research. In: Climate change reports, volume 1: regional view. Australian Bureau of Meteorology and CSIRO, Melbourne. Retrieved 1 July 2019, from http://www.bom.gov.au/climate/data/

Barnett J (2001) Adapting to climate change in Pacific Island countries: the problem of uncertainty. World Dev 29(6):977–993. https://minerva-access.unimelb.edu.au/handle/11343/34507

Commonwealth of Australia (2011) Climate change risks to coastal buildings and infrastructure: A supplement to the first pass national assessment. Retrieved from Department of Climate Change and Energy Efficiency, Commonwealth of Australia. https://www.environment.gov.au/system/files/resources/0f56e5e6-e25e-4183-bbef-ca61e56777ef/files/risks-coastal-buildings.pdf

Forbes DL, Solomon S (1998) Approaches to vulnerability assessment on pacific island coasts: examples from Southeast Viti Levu (Fiji) and South Tarawa (Kiribati). SOPAC Sectariat, Kiribati

Granger K (1999) An information infrastructure for disaster management in Pacific island countries. Austrailian Geological Survey Organisation, Commonwealth of Australia, Canberra

Handmer J, Nalau J (2019) Understanding loss and damage in Pacific Small Island developing states. In: Mechler R, Bouwer LM, Schinko T, Surminski S (eds) Loss and damage from climate change. Springer, Cham, pp 365–381

Hanson, F. (2008). The dragon in the Pacific: more opportunity than threat. Retrieved from Lowy Institute for International Policy https://archive.lowyinstitute.org/sites/default/files/pubfiles/Hanson%2C_the_dragon_in_the_pacific_web_1.pdf

Henderson-Sellers A, Zhang H, Berz G, Emanuel K, Gray W, Landsea C, Holland G, Lighthill J, Shieh S-L, Webster P, and McGuffie K (1998) Tropical cyclones and global climate change: A post–IPCC assessment. Bull. Am. Meteorol. Soc 79:19–38

Hu D, Wu L, Cai W, Gupta AS, Ganachaud A, Qiu B, Gordon AL, Lin X, Chen Z, Hu S (2015) Pacific western boundary currents and their roles in climate. Nature 522(7556):299. https://doi.org/10.1038/nature14504

Kumar L, Taylor S (2015) Exposure of coastal built assets in the South Pacific to climate risks. Nat Clim Chang 5(11):992. https://doi.org/10.1038/nclimate2702

Kumar L, Eliot I, Nunn PD, Stul T, McLean R (2018) An indicative index of physical susceptibility of small islands to coastal erosion induced by climate change: an application to the Pacific islands. Geomat Nat Haz Risk 9(1):691–702. https://doi.org/10.1080/19475705.2018.1455749

Lewsey C, Cid G, Kruse E (2004) Assessing climate change impacts on coastal infrastructure in the eastern Caribbean. Mar Policy 28(5):393–409

Lorde T, Gomes C, Alleyne D, Phillips W (2013) An assessment of the economic and social impacts of climate change on the coastal and marine sector in the Caribbean. Retrieved from https://repositorio.cepal.org/bitstream/handle/11362/38519/LCCARL395.pdf?sequence=1&isAllowed=y

Mimura N (1999) Vulnerability of island countries in the South Pacific to sea level rise and climate change. Clim Res 12(2–3):137–143. https://www.int-res.com/articles/cr1999/12/c012p137.pdf

Nicholls RJ, Klein RJ (2005) Climate change and coastal management on Europe's coast. In: Vermaat JE, Bouwer LM, Turner RK, Salomons W (eds) Managing European coasts. Springer, Berlin Heidelberg, pp 199–226

Nurse LA, McLean RF, Agard J, Briguglio LP, Duvat-Magnan V, Pelesikoti N, Tompkins E, Webb A (2014) Small islands. In: Climate change 2014: Impacts, adaptation, and vulnerability. Part

B: Regional Aspects. Contribution of Working Group II to the Fifth Assessment Report of the Intergovernmental Panel on Climate Change. Cambridge University Press, Cambridge, p 42

O'Brien K, Leichenko R, Kelkar U, Venema H, Aandahl G, Tompkins H, Javed A, Bhadwal S, Barg S, Nygaard L, West J (2004) Mapping vulnerability to multiple stressors: climate change and globalization in India. Glob Environ Chang 14(4):303–313. https://doi.org/10.1016/j.gloenvcha.2004.01.001

Preston BL, Suppiah R, Macadam I, Bathols JM (2006) Climate change in the Asia/Pacific region. A consultancy report prepared for the climate change and development roundtable. CSIRO Marine & Atmospheric Research, Aspendale, Vic

Storey D, Hunter S (2010) Kiribati: an environmental 'perfect storm'. Aust Geogr 41(2):167–181. https://doi.org/10.1080/00049181003742294

Tompkins EL (2005) Planning for climate change in small islands: insights from national hurricane preparedness in the Cayman Islands. Glob Environ Chang 15(2):139–149. https://doi.org/10.1016/j.gloenvcha.2004.11.002

World Bank (2006) The Pacific infrastructure challenge: a review of obstacles and opportunities for improving performance in Pacific Islands. World Bank, Washington

Chapter 8
Population Distribution in the Pacific Islands, Proximity to Coastal Areas, and Risks

Lalit Kumar, Tharani Gopalakrishnan, and Sadeeka Jayasinghe

8.1 Introduction

Higher sea level is one of the most obvious consequences of climate change that is already affecting coastal populations, and the expectation is that larger populations will be affected with higher rates of increase projected for the future. The IPCC's 'Business-As-Usual' scenario estimates that by 2100, the global sea-level rise will reach 65 cm or 41 cm should 'atmospheric stabilisation' occur (Matsuoka et al. 1995; Warrick and Oerlemans 1990). A sea-level rise of 50–100 cm is likely to have a significant impact on the majority of Pacific island nations. These impacts will mainly be through a higher frequency of extreme events, storm damage, flooding, inundation, damage to coral reefs and atolls, increased salinity of rivers and coastal soils, loss of agricultural land and loss of coastal structures (houses and infrastructure). While sea-level rise will result in direct inundation of low-lying coastal areas, it will also have indirect impacts further inland, including saltwater intrusion and loss of freshwater resources, leading to destruction of habitats. A significant population displacement is anticipated owing to sea-level rise and coastal erosion, mainly in developing countries (Leatherman and Beller-Simms 1997). The Solomon Islands has been a hotspot for sea-level rise-related news for the past 20 years. It has been reported that five islands in the Solomon Islands have been lost due to sea-level rise and coastal erosion, and several more have been severely eroded (Nunn 2013). The islands, ranging in size from 1 ha to 5 ha, supported 300-year-old tropical vegetation.

L. Kumar (✉) · T. Gopalakrishnan
School of Environmental and Rural Science, University of New England,
Armidale, NSW, Australia
e-mail: lkumar@une.edu.au; tgopalak@myune.edu.au

S. Jayasinghe
Faculty of Animal Science and Export Agriculture, Department of Export Agriculture,
Uva Wellassa University, Badulla, Sri Lanka
e-mail: ljayasi2@myune.edu.au

© Springer Nature Switzerland AG 2020
L. Kumar (ed.), *Climate Change and Impacts in the Pacific*, Springer Climate,
https://doi.org/10.1007/978-3-030-32878-8_8

Nuatambu Island, home to 25 families, has seen 11 houses washed into the sea since 2011, and its habitable area has shrunk by more than 50%. These are just a few anecdotal evidences of the impacts of climate change on the people across the Pacific.

Many islands in the Pacific are adversely affected by sea-level rise and extreme climate events, including storm surge, tropical cyclones, floods, and droughts. Such events, particularly cyclones, unusually high waves, and storm surges, have the greatest impact on the livelihood of the people. For settlements and infrastructure, this is likely to lead to a loss of habitat and salinisation of soils, thereby causing changes in the distribution of plants and animals (Watson et al. 1998). A rise in sea level would also adversely affect the health, well-being, and economy of the area's inhabitants. Within 1 km of the sea, approximately 90% of inhabitants rely on coastal zone resources for their livelihoods. Owing to the rapid changes in temperature and rainfall, especially heavy rainfall, flooding has resulted in outbreaks of diarrhoea and other water- and vector-borne diseases. People in coastal areas frequently face natural disasters and the detrimental effects of climate change; they generally have low-quality houses, which are often unable to withstand the impacts of cyclones, flooding, high winds, and storm surges (Nicholls et al. 2007). Therefore, part of this vulnerability stems from the low resilience of their living standards.

The responses of the atmosphere and ocean to greenhouse gas emissions are already apparent, as evidenced by increasing sea temperature, extreme events, ocean acidification, and sea-level rise. These changes occurring over the tropical Pacific have been having major impacts on marine ecosystems and inhabitants; in particular, the impacts of climate change on the livelihoods of residents, such as agriculture, fisheries, and food security, are well reported (Barnett 2011). As temperatures increase in the future, climate models project that the Pacific will experience more frequent and stronger El Niño events (Mimura 1999). El Niño years cause significant changes for all Pacific islands, such as changes in rainfall, winds, drought, waves, erosion processes, and water temperature (Campbell and Barnett 2010). As an example, in 1997, Kiritimati (Christmas Island) was severely impacted by an El Niño event that brought heavy rainfall, flooding, and storm surges, which resulted in a 0.5 m rise in sea level. As a result of this, around 40% of the island's coral died, and its 14 million birds, reputed to be among the world's richest bird population, fled the island (Duvat et al. 2013).

As a result of climate change, water resources and food supplies are being seriously affected by prolonged dry spells and drought (McCubbin et al. 2015). This can result in food shortages on some islands and requires catering for the food demand of dense populations in coastal belts. In addition, contamination of water resources has occurred in the past owing to a rise in sea level and abnormally high waves due to cyclones and the El Niño effect (Ramsay 2011). Food and water security is a key determinant of the sustainability of coastal systems. With the increasing population in coastal areas, lack of sufficient food and water or the lack of a reliable supply of such goods has the potential to compel communities to move to areas having more reliable sources (McCubbin et al. 2015).

The population distribution is becoming increasingly skewed and concentrated along or near coasts. Ocean, coastal, and marine resources are very important for people living in coastal communities. More than 600 million people reside in coastal areas that are below 10 m elevation, while about 40% of the world's population live within 100 km of the coast (Factsheet 2017). Attributes such as economic, social, recreational, and cultural benefits make coastal living more attractive, resulting in the average population density in coastal areas being 80 people per square kilometre, which is two times greater than the world's average population density (Martínez et al. 2007).

The rates of urban area expansion in the coastal zones have been found to be higher than those in non-coastal regions in similar time periods (Seto et al. 2011) and are anticipated to continue or even increase in the future (Nicholls et al. 2007). The skewed coastal benefits lead a majority of people to choose to live in coastal areas. Coastal regions also host critical infrastructure, such as ports, and provide a wide range of sporting and recreational opportunities, such as fishing and diving (Post and Lundin 1996). The other reasons why people tend to be more concentrated along coastal areas are that cities near the coast provide more facilities in terms of infrastructure, transport, education, health, and employment opportunities. Half of the large cities worldwide (with over 100,000 inhabitants) are within 100 km of the coast (Barragán and de Andrés 2015). The majority of the world's megacities (cities with over ten million inhabitants) are located in coastal areas (DESA 2015). The population of coastal megacities is expected to increase from 220.7 million in 2009 to 301.7 million by 2025 (Von Glasow et al. 2013).

The three major island groups in the Pacific Ocean are Polynesia, Micronesia, and Melanesia. According to the 2016 census (PRISM 2019), the total population in Polynesia was 655,893, while there were 506,680 and 8,915,584 people in Micronesia and Melanesia, respectively (Table 8.1). Overall, the population density and the average annual growth rate of the Pacific islands are increasing. The population in the Pacific is projected to increase by around 4.4 million people within the next 30 years, reaching around 14.5 million by 2030 and increasing to 19 million by 2050 (Table 8.1). Islands, including the Pacific islands, have much larger coastlines proportional to their total land areas than continental countries, and all have populations that have settled near the shore (Nunn and Mimura 1997). For example, Kiribati is a small island country in the central Pacific Ocean that stretches over 3.5 million km^2. Kiribati consists of 33 islands, but most of the population is concentrated on South Tarawa. South Tarawa is a narrow strip of land between the lagoon and the ocean; it is home to 50,182 people, thereby making it an overcrowded city with a population density similar to that of Tokyo or Hong Kong (Duvat et al. 2013).

Climate change has resulted in observable effects on regions worldwide, but the Pacific islands have been more vulnerable than other regions, as the entire population is facing existential challenges (Mertz et al. 2009). The impacts of climate change (i.e. mainly global warming and accelerated sea-level rise) on coastal ecosystems and human populations are uncertain; however, the countries of the Pacific islands are notably vulnerable to such impacts (Mimura 1999). This vulnerability

Table 8.1 Estimates and projections of demographic indicators for selected years in the Pacific islands

Region/ country	Most recent census	Population count at last census	Land area (km²)	Projected population (mid-year)								Population density (people/ km²)		Average annual population growth rate (%)
				2020	2025	2030	2035	2040	2045	2050		2016	2030	Last intercensal period
Melanesia			540,030	11,124,500	12,229,500	13,340,700	14,479,000	15,692,500	16,977,000	18,282,900		19	25	
Fiji	2007	837,271	18,333	895,400	909,300	918,700	924,900	927,900	927,700	924,700		48	50	0.8
New Caledonia	2014	268,767	18,576	293,700	313,200	331,600	348,200	362,900	375,800	387,100		15	18	1.8
Papua New Guinea	2011	7,059,653	462,840	8,901,200	9,846,300	10,790,800	11,757,100	12,798,600	13,915,200	15,057,600		18	23	2.8
Solomon Islands	2009	515,870	28,230	714,800	802,800	902,300	1,010,900	1,123,600	1,237,200	1,351,600		23	32	2.8
Vanuatu	2009	234,023	12,281	319,500	357,900	397,300	437,900	479,500	521,100	561,900		24	32	2.5
Micronesia		506,680	3,156	422,800	433,000	441,200	447,800	452,700	455,900	458,500				
Federated States of Micronesia	2010	102,843	701	106,000	107,700	108,900	109,400	109,500	109,400	109,300		149	155	0.3
Guam	2010	159,358	541	175,200	181,300	186,100	190,000	193,100	195,700	198,300		313	344	0.3
Kiribati	2015 (p)	109,693	811	r										1.2
Marshall Islands	2011	53,158	181	55,900	56,800	57,900	59,300	60,700	61,700	62,400		304	320	0.4
Nauru	2011	10,084	21	11,200	11,700	12,100	12,600	13,200	13,800	14,200		514	576	1.8

Island group	Year												
Northern Mariana Islands	2010	53,883	457	56,600	57,500	58,200	58,600	58,600	58,100	57,500	122	127	−2.5
Palau	2015 (p)	17,661	444	17,900	18,000	18,000	17,900	17,600	17,200	16,800	40	41	−1.2
Polynesia		**655,893**	**8,126**	**677,100**	**692,700**	**708,400**	**722,700**	**734,900**	**744,000**	**749,800**			
American Samoa	2010	55,519	199	57,100	58,100	59,300	60,300	61,300	62,200	63,200	283	298	−0.3
Cook Islands	2011	14,974	237	15,300	15,400	15,500	15,500	15,400	15,300	15,000	64	65	−0.5
French Polynesia	2012	268,270	3,521	280,600	289,700	297,700	304,100	309,100	312,700	315,000	78	85	1.8
Niue	2011	1,611	259	1,500	1,400	1,500	1,500	1,500	1,500	1,500	6	6	−0.2
Pitcairn Islands	2012	57	47	n.a.									
Samoa	2011	187,820	2,934	199,300	205,700	212,700	220,300	227,800	234,200	239,100	66	72	0.8
Tokelau	2011	1,411	12	1,400	1,400	1,400	1,400	1,400	1,400	1,400	117	117	0.9
Tonga	2011	103,252	749	100,000	99,200	98,400	97,700	96,700	95,300	93,600	134	131	0.2
Tuvalu	2012	10,782	26	10,300	10,500	10,600	10,700	10,600	10,600	10,600	388	408	1.2
Wallis and Futuna	2013	12,197	142	11,600	11,300	11,300	11,200	11,100	10,800	10,400	83	80	−1.9
Total			551,312										

Adapted from PRISM (2019)

The values in bold are for island groups of Melanesia, Micronesia and Polynesia

Note: (p) preliminary census results, n.a. not available or not projected, r under review, pending clearance of migration data and 2015 census growth

stems from the large amounts of valuable and productive land in low-lying areas with populations located near the water in coastal areas, including populations in major coastal urban centres and natural ecosystems that are already subjected to other environmental stresses (Mertz et al. 2009; Mote and Salathe 2010). For developing nations like those in the Pacific islands, the competency to deal with these problems is limited by population dynamics, fragile island ecosystems, traditional land management systems, limited natural resource bases in the coast, heavy reliance on foreign aid, and geographical isolation (Haberkorn 2008).

The consequences of climate change have made islanders climate change refugees. Kiribati has a population of 110,000 people on 33 small islands that are at risk of being extensively submerged and becoming uninhabitable. Thus, the President of Kiribati purchased 20 km^2 of land on Vanua Levu in the Fiji islands, which is about 2000 km away, as a backup plan for the relocation of its citizens (Caramel 2014). Several communities in the Solomon Islands have been relocated due to rapidly eroding coastlines, communities that occupied these lands for generations (Birk 2012). In addition, Taro, the capital of Choiseul Province, is already drawing plans to move its residents and services to higher ground in response to sea-level rise (Albert et al. 2016). A study conducted in 12 Pacific island countries to investigate the level of exposure of built infrastructure to climate risk found that 57% of the total built infrastructure is located within 500 m of the coastline and has a total replacement value of USD21.9 billion (Kumar and Taylor 2015). In Kiribati, the Marshall Islands, and Tuvalu, over 95% of the built infrastructure is located within 500 m of the coastline (Kumar and Taylor 2015); these countries have been identified as countries most at risk to sea-level rise (Barnett and Adger 2003).

Natural disasters inevitably lead to increased problems for coastal populations, and this is no different in the Pacific islands. Natural disasters disrupt agricultural systems, as mentioned previously, and this scenario is even more accentuated by sea-level rise (Small and Nicholls 2003). The prospect of even moderate sea-level rise has the potential to displace a significant number of people. Natural disasters (i.e. tsunamis or tidal waves) have had serious consequences for a large number of people in the Pacific region; for example, Vanuatu experienced severe Tropical Cyclone Pam in 2015 (Mohan and Strobl 2017), and Fiji experienced severe Tropical Cyclone Winston in 2016 (Terry and Lau 2018). The livelihood of the coastal population in the Pacific is based around agricultural land, tourism, forests, water resources, fisheries, mangroves, and coral reefs; however, these are all climate-sensitive sources of income. The majority of commercial and recreational fisheries of Pacific nations are dependent on coastal marshes, which are also some of the most vulnerable sectors to biotic and abiotic stresses (Parker 2018). The occurrence of climate change and natural disasters has a significant negative impact on agriculture and aquacultural productivity. During cyclones, heavy rainfall, high winds, and floods cause damage to agricultural crops and influence the soil in low-lying areas (Olsthoorn et al. 2002; Van Aalst 2006), which have a direct detrimental effect on crop productivity and thus human welfare.

Pacific island states are small with limited natural resources. The limited and fragile resource base allows less room for error in their utilisation and management (Leatherman and Beller-Simms 1997). They are highly susceptible to natural environmental events because of their size, location, and isolation. Populations concentrated along the coastlines and coastal plains are highly dependent on marine resources or secondary activities that are directly related to the marine environment, such as tourism-related activities. With the impacts of climate change highly skewed against coastal regions, people residing along the coasts will be affected the most. Unfortunately, in most Pacific island countries, the majority of people reside close to the coast, some due to reasons already discussed while others have no choice due to the physical limitation of some islands. The following case study is meant to highlight how fragile the Pacific community is to climate change and rising sea levels due to the very large proportion of population living in close proximity to the coast.

8.2 Case Study: Exposure of Coastal Populations to Climate Risks in 12 Pacific Island Countries

8.2.1 Introduction

The location and construction of residential buildings in coastal areas can contribute to the vulnerability of communities to coastal hazards and extreme events. Climate change, particularly sea-level rise, is likely to increase the exposure of these assets and the vulnerability of communities. A detailed analysis of the exposure of coastal populations for 12 Pacific island countries using locational data of residential buildings was conducted to assess the level of exposure of the inhabitants to coastal climate-related impacts. The project assembled a comprehensive database of residential buildings for the 12 countries, namely, the Cook Islands, Kiribati, the Marshall Islands, Nauru, Niue, Palau, Samoa, the Solomon Islands, Tonga, Tuvalu, and Vanuatu. The aim was to determine the percentage of each country's population that resides in the coastal fringes and to understand the vulnerabilities of the population to the effects of climate change, such as sea-level rise, coastal inundation, storm surges, and extreme events such as tropical cyclones.

8.2.2 Methodology

Population distribution data are not available on an individual household basis for Pacific island countries. Some data are available on a district or provincial basis for a number of countries, but these data are not suitable for proximity analysis. For this study, we required population distribution data on a household basis (locational data). In the absence of such population data, we used residential buildings as a proxy for the actual population distribution. Building data can be sourced from several sources, including satellite imagery. We then assumed that on average, each residential building housed the same number of occupants. While the population distribution from the residential buildings data and the above assumption would not be 100% accurate, it provided the best population distribution data on a point-by-point basis. Therefore, in this study, the location and distribution of residential buildings were taken as a proxy for population distribution. This assumption was based on the premise that, on average, the number of people residing in a building in a coastal area was similar to that of an inland region. There was no reason to believe that this was not true for the Pacific island countries.

Data required for the analysis of population distribution were sourced from several sources and then supplemented through digitisation. The Pacific Catastrophe Risk Assessment and Financing Initiative (PCRAFI) (Air Worldwide 2011) database was the main source of data used in this analysis. This was supplemented with high-resolution satellite imagery as well as ESRI base map, where needed. The PCRAFI database contains all infrastructure locations for most of the countries in the Pacific. The residential buildings were extracted from the building exposure database. For the purposes of this analysis, a residential building was defined as a structure comprised of a roof and walls permanently situated in one location and used for family dwellings. Individual buildings were manually digitised from high-resolution satellite imagery and then field-verified through inspections. Clusters of buildings extracted from moderate- to high-resolution imagery were manually outlined by polygons and counted.

A thorough check of the coastlines for each country was undertaken using ESRI's base map layer in ArcGIS software. Any missing coastlines and misalignments were corrected before the analysis. Missing coastline data were either digitised or supplemented using data from the Global Self-Consistent, Hierarchical, and High-Resolution Geography Database (GSHHG). The National Geophysical Data Center, which is part of the National Oceanic and Atmospheric Administration, is the official distribution point for the GSHHG dataset. The GSHHG is a high-resolution geography dataset that includes coastlines, political borders, and rivers. The files are provided in two formats, namely, ESRI shapefile format or native binary format. All datasets were in the WGS84 geographic (simple latitudes and longitudes; decimal degrees) horizontal datum.

The appropriate projection for each country was determined. For this exercise, we demarcated four intervals from the coast, namely, 0–50 m, 50–100 m, 100–200 m, and 200–500 m. Buffers were created at the required distances from the coastline at 50 m, 100 m, 200 m, and 500 m. Residential buildings that fell within each interval were identified, and the required details, as outlined above, were extracted. For the building layer, the point data rather than footprint data were used, as this obtained a more conservative estimate. In addition to the calculations in individual bands, we also calculated the total residential buildings for each country and produced maps to show their overall distribution. We also looked at the overall distribution of the population for the 12 countries in this case study. A few islands per country were also selected to show the distribution of the population. All the analyses were performed using ArcGIS software.

Table 8.2 Breakdown of the population percentage in specific buffer bands for all 12 countries

Country	Population in 0–50 m band (%)	Population in 50–100 m band (%)	Population in 100–200 m band (%)	Population in 200–500 m band (%)	Population beyond 500 m (%)
Cook Islands	5.8	11.4	19.6	29.6	33.6
FSM	6.6	8.3	13.9	26.2	45.1
Kiribati	35.0	32.1	23.9	7.3	1.8
Marshall Islands	36.1	36.2	19.9	5.7	2.0
Nauru	12.2	22.6	30.3	26.2	8.8
Niue	–	1.2	27.0	21.1	55.5
Palau	7.0	9.0	16.9	25.2	40.3
Samoa	5.2	9.5	15.3	18.5	51.5
Solomon Islands	3.2	5.2	11.6	19.5	60.6
Tonga	3.0	5.8	11.5	29.7	50.1
Tuvalu	33.6	30.7	28.0	7.1	0.6
Vanuatu	2.6	5.8	12.3	23.2	56.2

8.2.3   Results

The population distribution within each of the distance bands used for all 12 countries is shown in Table 8.2.

The Cook Islands

The Cook Islands is made up of 15 small island groups with an estimated population of 14,974. It is located in the South Pacific with a land area of 296 km² and 454 km of coastline. The Cook Islands' main population centre is on the island of Rarotonga, which has an international airport. In the Cook Islands, nearly 67% of the population resides within 500 m of the coastline (Table 8.2). In addition, 17.2% of the population resides within 100 m of the coastline, while 36.8% resides within 200 m of the coastline. This is a relatively high percentage of the total population living in close proximity to the coast and exposed to sea-level rise and other climate change impacts. Figure 8.1 shows the population distribution on two islands in the Cook Islands, namely, Rarotonga and Aitutaki. Almost the entire population in Rarotonga is around the coast, with the inland area being almost devoid of settlements. On Aitutaki, the population is generally concentrated around the coast and mostly on the west coast.

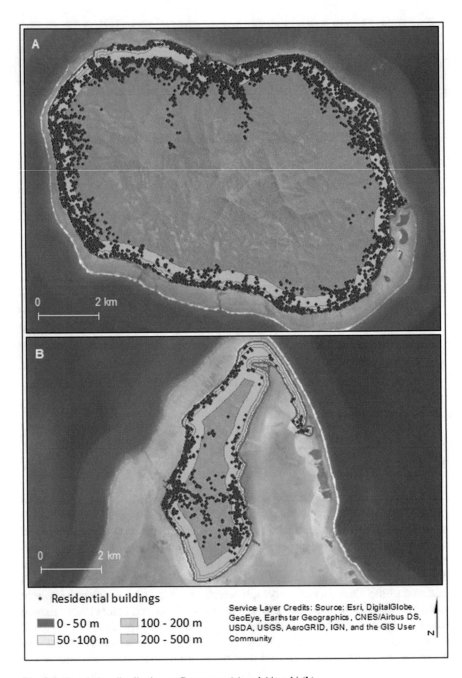

Fig. 8.1 Population distribution on Rarotonga (**a**) and Aitutaki (**b**)

Federated States of Micronesia

The Federated States of Micronesia (FSM) is made up of 607 islands with an estimated population of 111,560. It has a small land area of 702 km² with 1036 km of coastline. The capital city of the FSM is Palikir, which is located on Pohnpei Island, while the largest city is Weno, which is located in the Chuuk Atoll. The FSM consists of four states, namely, Yap, Chuuk, Pohnpei, and Kosrae. Approximately 55% of the population of the FSM lives within 500 m of the coastline and 14.9% is located within 100 m of the coastline (Table 8.2). Figure 8.2 shows the distribution of the population on Pohnpei. The settlements are mainly scattered around the coast all around the island. The population is not in very close proximity to the coast on most of the island, except for the north and northeast of the island where a majority of the population is closer to the coast.

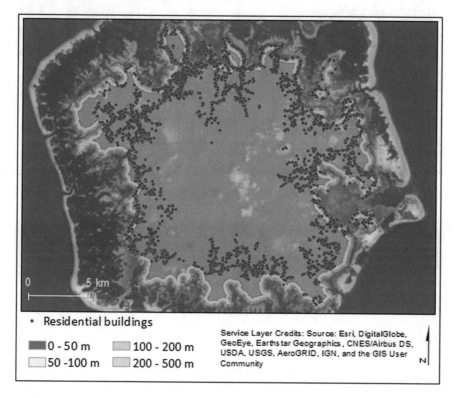

Fig. 8.2 Population distribution on Pohnpei in the Federated States of Micronesia

Kiribati

Kiribati is made up of 33 atolls and islands with an estimated population of 109,693. It has a land area of 995 km² with 1845 km of coastline. Kiribati's main population centre is on the island of Tarawa, which is where the capital South Tarawa is located.

Fig. 8.3 Population distribution on Tamana in Kiribati

Approximately 98.2% of the population of Kiribati resides within 500 m of its coastline, and nearly 67.1% is within 100 m (Table 8.2). These values show the high vulnerability of the population of Kiribati to sea-level rise and climate change. Figure 8.3 shows the island of Tamana in Kiribati where the majority of the population resides primarily within 200 m of the coast.

The Marshall Islands

The Marshall Islands is an island country located in the northern Pacific Ocean. It is spread out over 24 low-lying coral atolls and is comprised of more than 1150 small islands and islets. It has a much smaller land area of 286 km² and a relatively large coastline of 2172 km. The population of the Marshall Islands is approximately 53,158. Majuro is the capital of the Marshall Islands and is also the most populous atoll. Approximately 98% of the population of the Marshall Islands lives within 500 m of the coastline and 72.3% is located within 100 m (Table 8.2). This makes a very large proportion of the country's population vulnerable to climate change

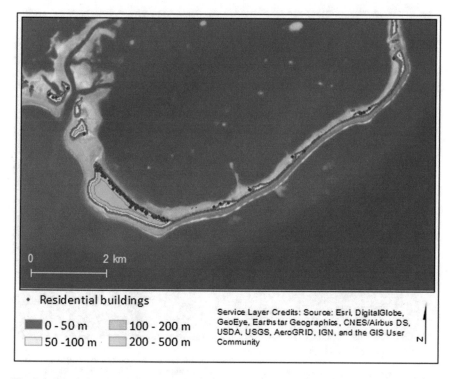

Fig. 8.4 Population distribution on Majuro Island in the Marshall Islands

impacts. Majuro Island is a typical example of the setting in the Marshall Islands (Fig. 8.4). The population here is in very close proximity to the coast and is mostly on the lagoon side for protection from the open sea.

Nauru

Nauru is made up of one island with an estimated population of 10,084. It has a much smaller land area of 22.6 km^2 with 18.7 km of coastline. Approximately 91.2% of the population of Nauru is located within approximately 500 m of the coastline (Table 8.2). In addition, 34.8% of the population lives within 100 m of the coastline, making these people vulnerable to coastal climate change impacts. Figure 8.5 shows that the vast majority of the people of Nauru live very close to the coast, with only one large settlement inland at Arenibek.

Fig. 8.5 Population distribution of Arenibek in Nauru

Niue

Niue is made up of one island and is home to a population of 1611 people. It is located in the South Pacific and has a land area of 298 km² with 75 km of coastline. Approximately 44.5% of the population of Niue resides within approximately 500 m of the coastline (Table 8.2). Unlike other countries in this case study, no people live within 50 m of the coast, and only 1.2% of the population lives within 100 m. Figure 8.6 shows the overall distribution of the population in Niue. While the proportion of people living close to the coast is relatively smaller than that of many of the countries discussed in this study, the settlements are still relatively closer to the coast. The central part of the island has very few houses.

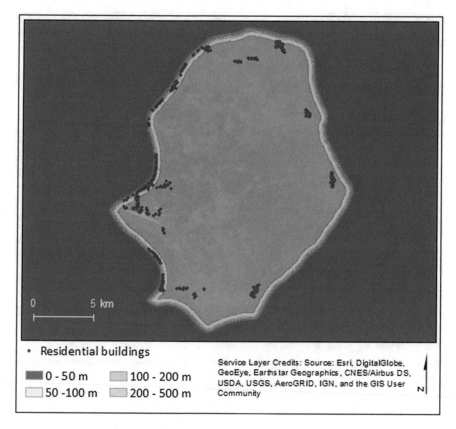

Fig. 8.6 Population distribution on Niue

Palau

Palau is an island country in the western Pacific that is made up of 250 islands. The population of Palau is 17,661, with Koror being the most populous with a total land area of 495 km² and 514 km of coastline. The capital of Palau is Ngerulmud, which is located on the island of Babeldaob; the international airport is also located on this island. Approximately 59.7% of the population of Palau is located within approximately 500 m of the coastline (Table 8.2); 7% of the population lives within 50 m of the coastline, while 16% lives within 100 m of the coastline. Figure 8.7 shows the population distribution on Meyungs Island. Here, the entire population is within 500 m of the coast, with a large majority within 200 m.

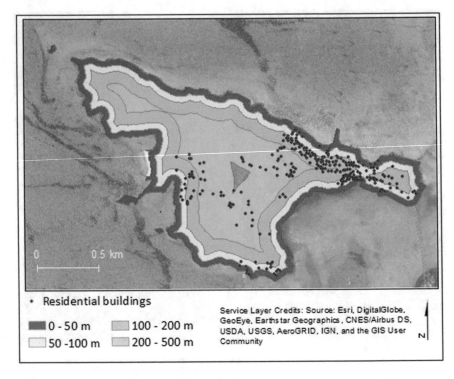

Fig. 8.7 Population distribution on Meyungs in Palau

Samoa

Samoa, which has a population of 187,820, is made up of seven islands, two of which are large and the rest of which are relatively small. It has a land area of 3046 km^2 with 482 km of coastline. The main population centres of Samoa are Upolu and Savai'i. The capital Apia and the international airport are located on the island of Upolu. Around 48.5% of the population lives within approximately 500 m of the coastline. Around 5.2% of the population lives within 50 m of the coastline, while 14.7% live within 100 m (Table 8.2). Figure 8.8 shows the population distribution on Savai'i in Samoa. While the people on this island are generally not within 500 m of the coast, they still live closer to the coast than inland and are scattered all around the island.

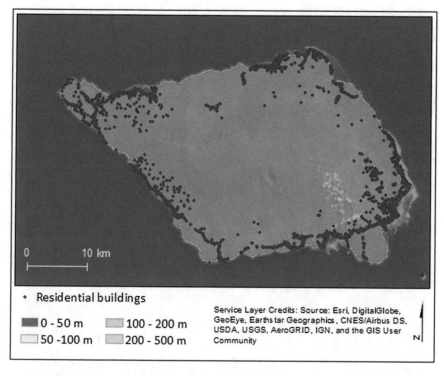

Fig. 8.8 Population distribution on Savai'i in Samoa

The Solomon Islands

The Solomon Islands is made up of 413 islands with an estimated population of 515,870. It is located in the South Pacific and has a land area of 29,672 km² and a coastline of 8848 km. The Solomon Islands' main population centre is on the island of Guadalcanal, which is where the capital city of Honiara is also situated. About 39.4% of the population of the Solomon Islands lives within approximately 500 m of the coastline (Table 8.2); 3.2% of the population lives within 50 m of the coastline while 8.4% lives within 100 m. Figure 8.9 shows the population distribution on two islands in the Solomon Islands, namely, Ghizo and Kolombangara. On both the islands, the population is primarily located on the coast, with Ghizo Island having almost all of its population very close to the coast in the eastern part of the island. Kolombangara Island is somewhat similar, but is a much larger island. The population is still close to the coast and on the south and eastern part of the island.

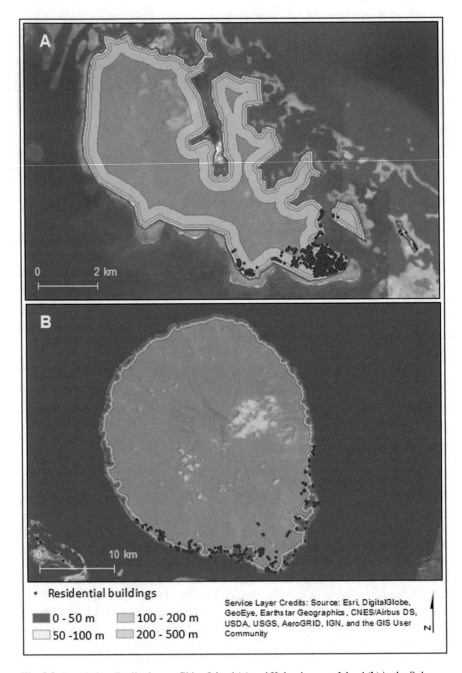

Fig. 8.9 Population distribution on Ghizo Island (**a**) and Kolombangara Island (**b**) in the Solomon Islands

Tonga

Tonga is made up of 176 small islands and atolls with an estimated population of 103,252. It has a land area of 847 km² and 929 km of coastline. Tonga's main population centre is on the island of Tongatapu, which is where the capital city Nuku'alofa is situated. About 49.9% of the population of Tonga resides within approximately 500 m of the coastline, while 3% live within 50 m and 8.8% live within 100 m of the coastline (Table 8.2). Figure 8.10 shows the population distribution on Tongatapu in Tonga. The majority of the population lives close to the coast on the northern side of the island.

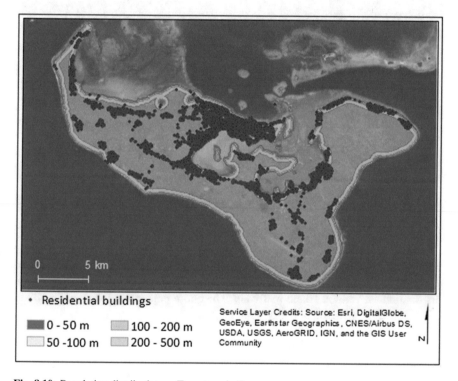

Fig. 8.10 Population distribution on Tongatapu in Tonga

Tuvalu

Tuvalu is made up of 10 islands with an estimated population of 10,782. It is located in the South Pacific and has a much smaller land area of 44.5 km² with 233 km of coastline. Tuvalu's main population centre and capital is Funafuti, which has an international airport and main harbours. More than 99% of the population of Tuvalu lives within approximately 500 m of the coastline; 33.6% lives within 50 m of the

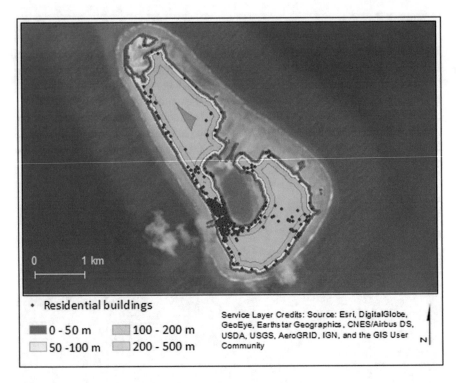

Fig. 8.11 Population distribution on Vaitupu Island in Tuvalu

coastline, while 64.3% lives within 100 m (Table 8.2). The large proportion of the population of Tuvalu living in close proximity to the sea makes it very vulnerable to sea-level rise and other effects of climate change. Figure 8.11 shows the population distribution on Vaitupu Island in Tuvalu. The entire population lives within 500 m of the coastline, with the majority within 200 m.

Vanuatu

Vanuatu is made up of 82 small islands with an estimated population of 234,023. It has a land area of 13,526 km^2 with 3234 km of coastline. Vanuatu's main population centre is on the island of Efate, which also holds the capital city of Port Vila. About 43.8% of the population of Vanuatu lives within approximately 500 m of the coastline; 2.6% of the population lives within 50 m of the coastline, while 8.4% lives within 100 m (Table 8.2). Figure 8.12 shows the distribution of the population on Gaua and Espiritu Santo in Vanuatu. For both of these islands, the bulk of the population resides close to the coast. The inland areas of both islands are devoid of settlements.

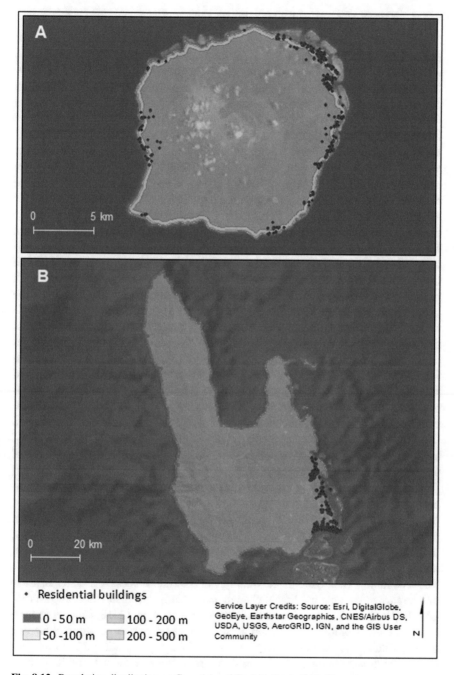

Fig. 8.12 Population distribution on Gaua (**a**) and Espiritu Santo (**b**) in Vanuatu

8.2.4 Discussion and Conclusions

For all 12 countries combined, 8% of the population is located within 50 m, 19.6% within 100 m, 34.8% within 200 m, and 54.6% within 500 m of the coastline. Only 45% of the population lives beyond 500 m of the coastline (Fig. 8.13). Overall, a very large proportion of the population of these 12 Pacific island countries lives in close proximity to the sea. Of the 12 countries analysed, Niue, Samoa, the Solomon Islands, Tonga, and Vanuatu are the only countries with more than 50% of the population living beyond 500 m of the coastline. Figure 8.13 shows that for Kiribati, the Marshall Islands, and Tuvalu, 67% of the population lives within 100 m of the coastline, over 90% lives within 200 m, and around 98% lives within 500 m of the coastline. Thus, only 2% of the population lives beyond 500 m of the coastline for these three countries. Kiribati, Tuvalu, the Marshall Islands, and Nauru are some of the smallest countries in the South Pacific Ocean (Kumar and Taylor 2015). Except for Nauru, the land area of the other three countries is split into small islands, and very little area, if any, is beyond 500 m from the coastline. In terms of elevation, the Cook Islands, Kiribati, the Marshall Islands, and Tuvalu are atoll islands with mostly low-lying areas; for Kiribati, the Marshall Islands, and Tuvalu, almost the entire islands are below 10 m in elevation (Nunn et al. 2016).

Reef islands and atolls are perceived as particularly fragile and one of the most threatened coastal systems by sea-level rise (White et al. 2007); thus, even slight rises in sea level can have substantial impacts (McLean et al. 2001). The three main inferred immediate effects are shoreline recession, inundation of low lands, and seawater intrusion into freshwater bodies (Woodroffe 2008). Population aggregation along the coast in low atoll islands poses a threat to the shallow, fresh groundwater. The limited land areas already restrict the quantities of freshwater. Overextraction along with storm surge and sea-level rise increases the intensity of seawater intrusion into fresh groundwater resources, which is the major source of water in many atolls (White et al. 2007).

Population distribution and its associated risks are better estimated when combined with other factors for island countries. Elevation is a very important factor because people living on higher grounds close to the coast are less vulnerable than those living within the same distance of the coast but on lower grounds. Unfortunately, high-quality elevation data, such as LiDAR-generated digital elevation models, are not available for any of the Pacific island countries with whole country coverage. The second important factor is the lithology of the islands because a sandy island is much more vulnerable to climate change, particularly storm surges, than volcanic islands (Nunn et al. 2016; Kumar et al. 2018). Kumar et al. (2018; Chapter 4) have reported the vulnerabilities of different island types based on four physical factors, namely, lithology, area, maximum elevation, and shape (circularity). The lithologies of the islands and countries can be used to further examine the vulnerability of the populations of these Pacific island countries.

Table 8.3 shows the number of islands in each of the 12 countries analysed in this study, with only islands greater than 1 ha used for the analysis. Figure 8.14 shows

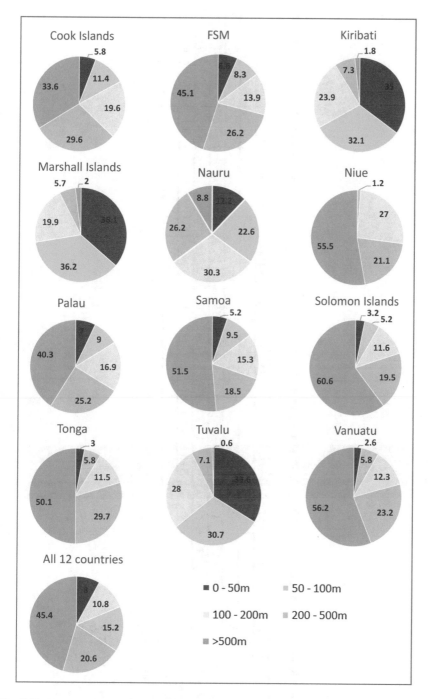

Fig. 8.13 Percentage of population within each interval from the coast for each of the countries and overall total for all 12 countries

Table 8.3 Number of islands in each country (only islands with an area greater than 1 ha are included) and their lithology (see Nunn et al. (2016) for details on categorisation)

	Cook Islands	FSM	Kiribati	Marshall Islands	Nauru	Niue	Palau	Samoa	Solomon Islands	Tonga	Tuvalu	Vanuatu
Composite high	4	2					2		28	3		11
Composite low	2								3			3
Limestone high			1		1	1	15		13	24		11
Limestone low		3	5				6		47	65		4
Reef island	9	91	27	34			8		129	21	10	6
Volcanic high		27					2	7	126	9		41
Volcanic low		4							67	2		5
Total	15	127	33	34	1	1	33	7	413	124	10	81

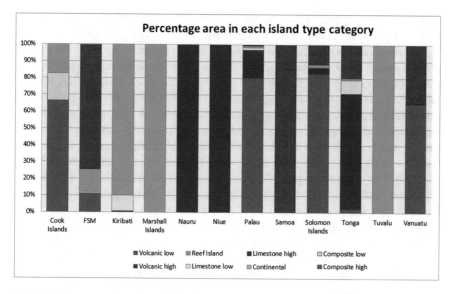

Fig. 8.14 The percentage composition of each of the 12 countries based on lithology (see Nunn et al. (2016) for descriptions of lithology and justification on selection of high and low values)

the percentage area of each country in the different lithology groups (see Nunn et al. (2016) and Chap. 2 for detailed descriptions of the lithology groups and cut-off values used to differentiate low and high islands). Figure 8.14 shows that the 12 countries are generally very different lithologically. Kiribati, the Marshall Islands, and Tuvalu have islands that are primarily reefal in origin. The Marshall Islands and Tuvalu are composed of only reef islands, while 90% of Kiribati (by area) is reefal in lithology. Nauru and Niue are 100% limestone high, while 70% of Tonga (by area) is limestone high. The Cook Islands, Palau, the Solomon Islands, and Vanuatu are primarily composite high island countries, with areas of composite high ranging from 65% to 82%. Samoa is entirely of volcanic high in origin.

Combining the above information with the population distribution data, the countries of Kiribati, the Marshall Islands, and Tuvalu were distinct. These countries are composed of reef islands ranging from an area of 90% for Kiribati to 100% for the Marshall Islands and Tuvalu. They have the highest proportion of population living within close proximity to the coast, with 68.8%, 74.3%, and 64.9% of the population living within 200 m of the coast for Kiribati, the Marshall Islands, and Tuvalu, respectively. The combination of reefal lithology, which has been ranked as the most vulnerable to physical change by Kumar et al. (2018), and the high percentage of the population residing within close proximity to the coast makes these three countries and their populations very vulnerable to sea-level rise, storm surges, and other coastal effects of climate change. The maximum elevations of Kiribati, the Marshall Islands, and Tuvalu are 81 m, 10 m, and 4.6 m, respectively, while the average elevations are 1.8 m, less than 1 m, and 2 m, respectively. Thus, these countries have very low elevations, are primarily of sandy or reefal origins, and also have a very large proportion of their population living in close proximity to the coast.

The implications of these results emphasise the importance of prioritising the populations of the small islands of the Pacific for future adaptation to coastal hazards. Climate change and its associated impacts, particularly sea-level rise, are critical issues for islands in the coming century. Small islands are particularly vulnerable owing to their smaller size, insularity, and remoteness. The land areas of the Pacific island region are comparatively small and isolated from one another with 10.8 million people, which comprises only 0.15% of the world's total population. Overall, the Pacific islands have a high ratio of shoreline to land area and are naturally highly susceptible to negative impacts from rising sea levels. Additionally, because the economic base is fairly narrow, generally focused on primary production and tourism, and scattered along the coastal fringe with a lower carrying capacity, the Pacific island countries are especially vulnerable to sea-level rise (Barnett 2001).

While some countries in the Pacific have already started community relocation, this is not a viable solution for all Pacific island countries. This is because of the physical nature of many of the countries, the fact that the total area is split into many small islands, and their dispersed nature. Many of the islands have a maximum elevation of only 3 m above sea level, so relocation within an island or from one island to another is not possible.

References

Air Worldwide (2011) Pacific Catastrophe Risk Assessment and Financing Initiative (PCRAFI) Component 2: Exposure Data Collection and Management. In: Bank TW (ed) Belgium: Technical Report Submitted to the World Bank

Albert S, Grinham A, Gibbes B, Leon J, Church J (2016) Sea-level rise has claimed five whole islands in the Pacific: first scientific evidence. Environ Res Lett 11(5):1–9. https://doi.org/10.1088/1748-9326/11/5/054011/pdf

Barnett J (2001) Adapting to climate change in Pacific Island countries: the problem of uncertainty. World Dev 29(6):977–993. https://minerva-access.unimelb.edu.au/handle/11343/34507

Barnett J (2011) Dangerous climate change in the Pacific Islands: food production and food security. Reg Environ Chang 11(1):229–237. https://doi.org/10.1007/s10113-010-0160-2

Barnett J, Adger WN (2003) Climate dangers and atoll countries. Clim Chang 61(3):321–337

Barragán JM, de Andrés M (2015) Analysis and trends of the world's coastal cities and agglomerations. Ocean Coast Manage 114:11–20. https://doi.org/10.1016/j.ocecoaman.2015.06.004

Birk T (2012) Relocation of reef and atoll island communities as an adaptation to climate change: learning from experience in Solomon Islands. In: Climate Change and Human Mobility: Global Challenges to the Social Sciences. Cambridge University Press, Cambridge, pp 81–109

Campbell J, Barnett J (2010) Climate change and small island states: power, knowledge and the South Pacific. Routledge, London

Caramel L (2014) Besieged by the rising tides of climate change. http://www.theguardian.com/environment/2014/jul/01/kiribati-climate-change-fiji-vanua-levu

DESA (2015) World population prospects: The 2015 revision, key findings and advance tables. United Nations, New York

Duvat V, Magnan A, Pouget F (2013) Exposure of atoll population to coastal erosion and flooding: a South Tarawa assessment, Kiribati. Sustain Sci 8(3):423–440. https://doi.org/10.1007/s11625-013-0215-7

Factsheet (2017) People and Oceans. Paper presented at The Ocean Conference, New York

Haberkorn G (2008) Pacific Islands' population and development: Facts fictions and follies. NZ Popul Rev 33(34):95–127

Kumar L, Taylor S (2015) Exposure of coastal built assets in the South Pacific to climate risks. Nat Clim Chang 5(11):992

Kumar L, Eliot I, Nunn PD, Stul T, McLean R (2018) An indicative index of physical susceptibility of small islands to coastal erosion induced by climate change: an application to the Pacific islands. Geomat Nat Haz Risk 9(1):691–702. https://doi.org/10.1080/19475705.2018.1455749

Leatherman SP, Beller-Simms N (1997) Sea-level rise and small island states: an overview. J Coast Res 24:1–16. https://www.jstor.org/stable/pdf/25736084.pdf

Martínez M, Intralawan A, Vázquez G, Pérez-Maqueo O, Sutton P, Landgrave R (2007) The coasts of our world: Ecological economic and social importance. Ecol Econ 63(2–3):254–272. https://doi.org/10.1016/j.ecolecon.2006.10.022

Matsuoka Y, Kainuma M, Morita T (1995) Scenario analysis of global warming using the Asian Pacific Integrated Model (AIM). Energy Policy 23(4–5):357–371

McCubbin S, Smit B, Pearce T (2015) Where does climate fit? Vulnerability to climate change in the context of multiple stressors in Funafuti, Tuvalu. Glob Environ Chang 30:43–55. https://doi.org/10.1016/j.gloenvcha.2014.10.007

McLean R, Tsyban A, Burkett V, Codignotto J, Forbes D, Mimura N et al (2001) Coastal zones and marine ecosystems. Retrieved from Working Group II to the Third Assessment Report of the Intergovernmental Panel on Climate Change. http://papers.risingsea.net/IPCC.html

Mertz O, Halsnæs K, Olesen JE, Rasmussen K (2009) Adaptation to climate change in developing countries. Environ Manag 43(5):743–752. https://doi.org/10.1007/s00267-008-9259-3

Mimura N (1999) Vulnerability of island countries in the South Pacific to sea level rise and climate change. Clim Res 12(2–3):137–143

Mohan P, Strobl E (2017) The short-term economic impact of tropical Cyclone Pam: an analysis using VIIRS nightlight satellite imagery. Int J Remote Sens 38(21):5992–6006. https://doi.org/10.1080/01431161.2017.1323288

Mote PW, Salathe EP (2010) Future climate in the Pacific Northwest. Clim Chang 102(1–2):29–50. https://doi.org/10.1007/s10584-010-9848-z

Nicholls RJ, Wong PP, Burkett V, Codignotto J, Hay J, McLean R et al (2007) Coastal systems and low-lying areas. In: Parry M, Canziani O, Palutikof J, van der Linden P, Hanson C (eds) Climate change 2007: impacts, adaptation and vulnerability. Cambridge University Press, Cambridge, p 44

Nunn PD (2013) The end of the Pacific? Effects of sea level rise on Pacific Island livelihoods. Singap J Trop Geogr 34(2):143–171. https://doi.org/10.1111/sjtg.12021

Nunn PD, Mimura N (1997) Vulnerability of South Pacific island nations to sea-level rise. J Coast Res 24:133–151

Nunn PD, Kumar L, Eliot I, McLean RF (2016) Classifying Pacific islands. Geosci Lett 3(1):7. https://doi.org/10.1186/s40562-016-0041-8

Olsthoorn AA, Maunder WJ, Tol RS (2002) Tropical cyclones in the southwest Pacific: impacts on Pacific island countries with particular reference to Fiji. In: Climate, change and risk. Routledge, London, pp 245–268

Parker R (2018) Unregulated population migration and other future drivers of instability in the Pacific. https://think-asia.org/bitstream/handle/11540/8562/Parker_Unregulated_population_migration.pdf?sequence=1:http://hdl.handle.net/11540/8562

Post JC, Lundin CG (1996) Guidelines for integrated coastal zone management. The World Bank, Washington, DC

PRISM (2019) Statistics of Pacific Island Countries and Territories. Pacific Community (SPC). https://prism.spc.int. Accessed 2 June 2019

Ramsay D (2011) Coastal erosion and inundation due to climate change in the Pacific and East Timor. In: Stephens S, Bell R (eds) Pacific Australia Climate Change Science and Adaptation Planning (PACCSAP) program. Department of Climate Change and Energy Efficiency, Australia, p 78

Seto KC, Fragkias M, Güneralp B, Reilly MK (2011) A meta-analysis of global urban land expansion. PLoS One 6(8):e23777. https://doi.org/10.1371/journal.pone.0023777

Small C, Nicholls RJ (2003) A global analysis of human settlement in coastal zones. J Coast Res 19(3):584–599. https://www.jstor.org/stable/4299200

Terry JP, Lau AA (2018) Magnitudes of nearshore waves generated by tropical cyclone Winston, the strongest landfalling cyclone in South Pacific records. Unprecedented or unremarkable? Sediment Geol 364:276–285. https://doi.org/10.1016/j.sedgeo.2017.10.009

Van Aalst MK (2006) The impacts of climate change on the risk of natural disasters. Disasters 30(1):5–18. https://doi.org/10.1111/j.1467-9523.2006.00303.x

Von Glasow R, Jickells TD, Baklanov A, Carmichael GR, Church TM, Gallardo L, Raine R (2013) Megacities and large urban agglomerations in the coastal zone: interactions between atmosphere, land, and marine ecosystems. Ambio 42(1):13–28. https://doi.org/10.1007/s13280-012-0343-9

Warrick R, Oerlemans J (1990) Sea level rise. In: Climate change: The IPCC scientific assessment. Cambridge University Press, Cambridge

Watson RT, Zinyowera MC, Moss RH, Dokken DJ (1998) The regional impacts of climate change. IPCC, Geneva. http://www.tucson.ars.ag.gov/unit/publications/PDFfiles/1244.pdf

White I, Falkland T, Perez P, Dray A, Metutera T, Metai E, Overmars M (2007) Challenges in freshwater management in low coral atolls. J Clean Prod 15(16):1522–1528. https://doi.org/10.1016/j.jclepro.2006.07.051

Woodroffe CD (2008) Reef-island topography and the vulnerability of atolls to sea-level rise. Glob Planet Chang 62(1–2):77–96

Chapter 9
Agriculture Under a Changing Climate

Viliamu Iese, Siosiua Halavatau, Antoine De Ramon N'Yeurt,
Morgan Wairiu, Elisabeth Holland, Annika Dean, Filipe Veisa,
Soane Patolo, Robin Havea, Sairusi Bosenaqali, and Otto Navunicagi

9.1 Introduction

Agriculture plays an important role in Pacific Island Countries (PICs) as a source of livelihood and food for communities. Agriculture provides 70–80% of food for people in Papua New Guinea (PNG), Solomon Islands, and Vanuatu; 40–60% of food for Polynesian countries such as Tonga, Samoa, and Cook Islands; and about 30–40% in rural atoll islands in Tuvalu and Kiribati (Bourke 2005; Allen 2015; NMDI 2018; Iese et al. 2018).

The role of agriculture in PICs is diverse, as are the challenges it faces. Agriculture plays a more significant role in higher islands with larger land areas such as PNG, Fiji, Solomon Islands, Vanuatu, Samoa, and Tonga. In these countries, commercial agriculture and involvement of the private sector is more visible and active. The major commercial crops in these countries include taro in Samoa; sugarcane in Fiji; oil palms and coffee in PNG; oil palms, copra, and cocoa in Solomon Islands; kava

V. Iese (✉) · A. D. R. N'Yeurt · M. Wairiu · E. Holland · F. Veisa · S. Bosenaqali
O. Navunicagi
Pacific Centre for Environment and Sustainable Development, University of the South
Pacific, Suva, Fiji
e-mail: viliamu.iese@usp.ac.fj

S. Halavatau
Land Resource Division—Pacific Community (SPC), Suva, Fiji

A. Dean
University of New South Wales, Sydney, NSW, Australia

S. Patolo
Mainstreaming of Rural Development Innovations, Tonga Trust (MORDI TT),
Nuku'alofa, Tonga

R. Havea
School of Computing, Information and Mathematical Sciences, The University of the South
Pacific, Suva, Fiji

© Springer Nature Switzerland AG 2020
L. Kumar (ed.), *Climate Change and Impacts in the Pacific*, Springer Climate,
https://doi.org/10.1007/978-3-030-32878-8_9

(*Piper methysticum*) and coffee in Vanuatu; and squash, vanilla, watermelons, and root crops in Tonga. There is also higher institutional and functional capacity in these countries in terms of agriculture officers, agriculture-focused NGOs, and development partners' participation in the agriculture sector (Sisifa et al. 2016).

In smaller and low-lying islands (Tuvalu, Kiribati, Cook Islands, Republic of the Marshall Islands (RMI), and Federated States of Micronesia (FSM)), agriculture plays a significant role but mainly for in-country use and semicommercial production. These countries are challenged with limited land areas, poor soils, and large distances between islands (making the islands difficult and expensive to reach and challenging outer-island transportation). Farming in these countries is mainly for household consumption and cultural practices (Sisifa et al. 2016).

About 70% of agricultural systems in the Pacific are rain fed, making them highly vulnerable to variations in rainfall (FAO 2010). Only 2% of the entire 30 million km^2 of the Pacific area is land (Sisifa et al. 2016). Larger Melanesian countries such as PNG, New Caledonia, Solomon Islands, Vanuatu, and Fiji represent about 90% of the land area in the Pacific Islands. The small land area available for agriculture purposes is a major challenge in PICs, especially in low-lying and atoll islands (Sisifa et al. 2016). Most of the good agricultural lands are located near rivers and coastal plains, making them highly exposed to floods and saltwater inundation and intrusion. The atolls and low-lying islands are particularly exposed to coastal flooding and saltwater intrusion and inundation, as most have a highest point at or below 5 m above sea level. Ongoing pressure on land from population growth, urbanization, and infrastructure development reduces the land available for agriculture. The continuous reduction of soil fertility and increasing incidences of pests, diseases, and invasive species also contributes to the vulnerability of Pacific agricultural systems (Sisifa et al. 2016).

The percentage of arable land in different PICs is highly variable. About 30–60% of the land in atolls such as RMI, Tuvalu, and Kiribati is arable land (flat lands that could be plowed). The percentage of arable land looks high, but as the average elevation of these atolls is about 3–5 m above mean sea level, arable lands are highly exposed to sea level rise, saltwater inundation, intrusion, droughts, and salt spray (Sisifa et al. 2016). On the other hand, larger landmasses such as PNG, Solomon Islands, and Fiji have low percentages of arable land, as most land areas have steep slopes and are difficult to cultivate for large-scale agriculture. Agricultural lands in higher islands are vulnerable to heavy rainfall leading to flooding and landslides. Arable lands are mostly located close to coastal areas where they are vulnerable to saltwater inundation and intrusion. Some of the flat lands that would be suitable for agriculture have been taken for developments such as roads, industrial establishments, and residential dwellings (Taylor et al. 2016).

Climate change is seriously impacting the agriculture sector in PICs through increasing the severity of extreme weather events and changing rainfall patterns, sea levels, and increasing average temperatures (IPCC 2019). These changes have had direct impacts on crops and livestock and crucial infrastructure such as roads and outer-island jetties. PICs are committed to reducing the negative impacts of climate change on the agriculture sector. Adaptation actions and risk reduction measures have been employed at different levels to assist farmers to adapt to climate change and climate variability (Iese et al. 2016; McGregor et al. 2016). Many agricultural resilience projects supported by development partners and agencies are building the adaptive capacity of farmers. Farmers are adjusting planting times, shifting to resilient varieties of crops and livestock, improving soil organic matter, adopting agroforestry and low-carbon farming, relocating farms, and integrating climate change and disaster risk management into agriculture policies at national and sub-national levels (Iese et al. 2015, 2016; Taylor et al. 2016; Wairiu et al. 2012).

Despite efforts to reduce the impacts of climate change on agriculture in PICs, the sector will continue to remain threatened by climate change for two reasons. First, climate change-induced hazards such as floods, droughts, and tropical cyclones are expected to increase in severity in the future (IPCC 2019). Sea level and temperatures will also continue to rise. Second, the efforts of PICs to adapt have not been sufficient to prevent damages, resulting in compounding residual impacts that give communities the experience of being in constant "recovery mode." Given the great importance of the agriculture sector for livelihoods, food security, and culture, PICs must transform the sector to be more resilient in the face of the ever-increasing risks associated with climate change.

This chapter provides insights on strategies to transform the agriculture sector to build resilience and productivity in a changing climate. In PICs, the term agriculture is sometimes used to refer to crops, livestock, fisheries, aquaculture, and forestry. This chapter focuses on land-based agriculture, which means crops, livestock, and forestry. We will first provide an overview of the role of agriculture in PICs followed by short summaries of impacts of climate change and climate variability on the agriculture sector. Additional projected future impacts of climate change on crops such as taro, cassava, and potato will be presented based on Pacific-led crop simulation modelling research. Specific case studies are also presented to illustrate key drivers needed to transform agricultural production in PICs. The case studies illustrate systems-oriented approaches that recognize the foundational importance of healthy soils in building the resilience of the agricultural sector. The final part of the chapter provides an overview of key principles that need to be addressed going forward in order to transform Pacific agriculture to be productive and resilient in a changing climate.

9.2 The Role of Agriculture in PICs

Agricultural production in PICs has been growing very slowly over recent decades. Annual production rates grew steadily in the Pacific region since the 1960s but have slowed down since the 1990s in most countries for which there are data (Halavatau 2016). The annual growth rate of the agriculture sector between 2000 and 2008 has varied between countries. Solomon Islands and Cook Islands recorded 4.2 and 3.2% growth, respectively. The growth rate ranges between 1.3 and 1.9% in Vanuatu, Tonga, PNG, and Kiribati. Samoa and Fiji recorded negative growth rates of −0.9

and −2.4%, respectively (Halavatau 2016). There has been a continuous decline in the contribution of agriculture to GDP in PICs. For example, Solomon Islands recorded a decrease in the contribution of agriculture to GDP from 24.2% in 2012 (NMDI 2018) to 16.2% in 2013 (Government of Solomon Islands 2015). Samoa recorded a decline in the contribution of agriculture to GDP from 13.6% in 2006 (Government of Samoa 2016) to 6.4% in 2014 (NMDI 2018). Tonga experienced a decline in the contribution of agriculture to GDP from 26.3% in 2004–2005 to 19.2% in 2009–2010 (Government of Tonga 2016) and 14% in 2015 (NMDI 2018). Fiji also recorded a decrease in the contribution of agriculture to GDP from 16% in 1990 to 10.4% in 2014 (Ministry of Agriculture 2014; NMDI 2018). The declining growth rate of production is due to loss of soil fertility, increases in pests and diseases, demographic changes, and the impacts of climate change and climate variability. The declining contribution of agriculture to GDP is due to the declining growth rate in production, as well as changes in trade agreements, price fluctuations due to the vagaries of markets, and reliance on a few export crops. The growing contribution to GDP of other sectors, such as tourism, may also be a factor in some countries.

9.2.1 Significance in GDP

Despite its decline, agriculture is still a critically important economic and livelihood sector in PICs (Halavatau 2016). Larger PICs, especially in Melanesia, have a higher contribution from agriculture to GDP as shown in Table 9.1. PNG had a contribution from agriculture, forestry, and fisheries to GDP of 18.8% in 2014. Vanuatu and Solomon Islands have agriculture sectors that contribute 24.4 and 24.2% of GDP, respectively, in 2012. Most of the agriculture contributions in the three mentioned countries come from export of commodities such as copra, palm oil, cocoa, kava, and coffee. Tonga recorded a 14% agriculture contribution to GDP in 2015, while Samoa recorded 6.4% in 2014 and Fiji recorded 10.4% agriculture contribution in 2013. Most of the contribution of agriculture to GDP is attributed to exports of fish, root crops, kava, fruits, coconut products, and sugar (only for Fiji).

The contribution of agriculture to GDP in smaller PICs including atolls varies widely. The FSM recorded a 14.5% contribution in 2013. FSM's main export products are betel nut, kava, some cooked root crops, bananas, and vegetables. Kiribati's agriculture sector contributed 25.6% to GDP in 2011. Copra is the main export product from Kiribati. Tuvalu (11.4% in 2011) and Niue (17.4% in 2012) do not export agriculture products, but agriculture contributes enormously to local food consumption and provides other social-economic benefits. The agriculture sectors in Cook Islands (2.1% in 2012), Nauru (1.2% in 2004), and Palau (1.4% in 2014) have small contributions to GDP, but agriculture plays an important role in food and domestic markets and social activities.

Table 9.1 Summary of the role of agriculture in PICs

Country	Agricultural trade		Total land area (2003) (ha)	% of arable and perm. cropland	% of household income from agriculture	% of labor force engaged in agriculture and forestry	% of GDP from agriculture
	% of total exported	% of total imported					
Fiji	15.8 (2013)	9.1 (2013)	1,827,000	13.7 (2009)	15.8 (2008)	16.9 (2008)	10.4 (2013)
Papua New Guinea	19.2 (2013)	2.2 (2013)	45,286,000	2.6 (2012)	72.4 (2005)	10.8 (2010)	18.8 (2014)[a]
Solomon Islands	60 (2013)	14.1 (2013)	2,799,000	3.8 (2012)	42 (2006)	51.7 (2013)	24.2 (2012)
Vanuatu	74 (2014)	11 (2014)	1,219,000	41 (2007)	28 (2012)	53.2 (2009)	24.4 (2013)
Samoa	5.5 (2013)	19.9 (2013)	283,000	13 (2009)	19.5 (2008)	33 (2011)	6.4 (2014)
Tonga	44.4 (2014)	19.4 (2014)	72,000	41.3 (2011)	23.4 (2014)	23 (2016)	14 (2015)
Cook Islands	0.1 (2013)	12.8 (2012)	24,000	1.6 (2009)	8.3 (2011)	2.2 (2011)	2.1 (2012)
Tuvalu	0.1 (2013)	3.1 (2013)	3000	60 (2012)	15.2 (2016)	16.9 (2016)	11.4 (2011)
FSM	14.7 (2013)	15.1 (2013)	70,000	31.4 (2012)	19.2 (2013)	16.2 (2013)	14.5 (2013)
Kiribati	54.1 (2012)	18.9 (2012)	73,000	42 (2012)	37.8 (2006)	9.1 (2015)	25.6 (2011)
RMI	1.2 (2014)	0.2 (2014)	18,000	64 (2013)	Fisheries employed 10% of population	0.3 (2006)	In 2012, Fisheries contributed 1.2% to GDP (Republic of the Marshall Islands 2013)
Nauru	0 (2013)	6.2 (2013)	2000	4 square km	3.9 (2013)	4.8 (2011)	1.2 (2004)
Palau	0.3 (2013)	9.1 (2013)	46,000	10.9 (2012)	2.5 (2006)	9.8 (2015)	1.4 (2014)
Niue	2.9 (2013)	10.5 (2013)	26,000	19.2 (2009)	7.1 (2016)	10 (2016)	17.4 (2012)

The data have been compiled from NMDI (2018), FAO (2018a, b), Republic of the Marshall Islands (2013), and Government of Nauru (2005)
[a]PNG National Statistics Office. https://www.nso.gov.pg/images/NationalAccounts2007-2014.pdf

9.2.2 Significance of Agriculture at the Household Level

The most important role of agriculture is providing food and sustaining incomes for households especially in rural areas in PICs. The percentage of households that name agriculture as the main source of income varies widely between countries. For example, in Palau and Nauru, agriculture is the main source of income for just 3% of households, whereas in PNG 75% of households rely on agriculture as the primary source of income (see Table 9.1). In terms of labor force employment, the agriculture and forestry sector employs about 17% in Fiji, 50% in Vanuatu and Solomon Islands, 23% in Tonga, and 33% in Samoa (NMDI 2018).

Indicators for measuring the contribution of agriculture in PICs tend to focus on production aspects that underestimate the crucial role of the sector. In reality, most people in PICs are involved in agriculture in one form or another, either as producers or consumers. In Solomon Islands, PNG, and Vanuatu, 80% of the population is involved in subsistence farming and depends entirely on subsistence production for daily sustenance (this is almost the entire rural population of these countries) (Government of Solomon Islands 2015; Government of Vanuatu 2015b; Independent State of PNG Ministry of Agriculture 2007). In Vanuatu, the people living in urban areas (about 20% of the population) also rely on agricultural products bought from local markets for food and cultural activities. In Niue, about 87% of households are actively involved in agriculture (Department of Agriculture, Forestry and Fisheries 2015). Around 65% of Tonga's population live in rural areas and are dependent on agriculture and fisheries as their main source of livelihood (Government of Tonga 2016). In 2011, about 67.5% of the population of Cook Islands were subsistence farmers who relied mostly on agriculture for daily sustenance (Government of Cook Islands 2015). In Nauru, a country with no agriculture exports and with very limited arable land for agriculture, 70% of the average diet is sourced from locally produced food (Government of Nauru 2005).

9.2.3 Significance of Agriculture for Social Cultural Activities

Agriculture is an important sector for sustaining culture and maintaining social bonds and practices in PICs. Most of the items needed for cultural exchange and practices are from agricultural products (Barnett 2011). The welcoming ceremonies in Samoa and in Fiji are practiced with kava. High-value cultural items are exchanged during weddings, funerals, ceremonies for forgiveness, and marriage proposals. These items are raw and processed agriculture products such as fine mats, tapa, root crops, pigs, betel nuts, and handicrafts (Allen 2015; Ministry of Agriculture 2014). For example, in the Federated States of Micronesia, yams, sakau (kava), breadfruit, taro, and pigs are crucial for ceremonies and gifting to cement culture and social bonds (FSM Department of Resources and Development 2012). Subsistence and household gardens allow the participation of everyone in the family where men, women, and children work together.

9.2.4 Resourcing Agriculture

The limited budget allocated for the agriculture sector by national governments is one of the major constraints in developing and building the resilience of agriculture in PICs. Although agriculture is considered to be a very important livelihood and economic sector, the annual budget allocations are small. For example, as reported by the Solomon Islands Agriculture and Livestock Sector Policy 2015–2019, the budget for the agriculture department is less than 2% of the national budget, which does not reflect the importance of agriculture in the national economy. The policy states that this lack of resourcing is a "major constraint" to providing essential services to rural populations (Government of Solomon Islands 2015). Samoa's annual budget for the agriculture sector was only 2.4% of the national budget in 2015–2016 (NMDI 2018). The government emphasized the need for development partners and the private sector to contribute toward implementing the Samoa Agriculture Sector Plan 2016–2020 (Government of Samoa 2016). The atoll nation of Tuvalu allocated an average of about 0.3% of the national budget to the Ministry of Natural Resources between 2012 and 2014, and agriculture received only about 2.6% of the limited allocation to the ministry (Government of Tuvalu 2016). In many PICs, the largest expense for most ministries is salaries. For example, in Tonga the Ministry of Agriculture, Forestry, Food and Fisheries (MAFFF) uses about 60–70% of their budget allocation for salaries and 30% on operational costs such as fuel, electricity bills, and others (Government of Tonga 2016). The limited national budget allocation to the agriculture sectors of PICs restricts the provision of services to farmers in urban and rural areas.

The government agriculture departments rely on projects funded by donors and contributions from the private sector to supplement the services and enhance coverage of services. Donor funds also provide support for agricultural research and other technologies needed to improve productivity. As highlighted by Fig. 9.1, Pacific Island Countries have been receiving an average of USD 67.6 million per year between 2007 and 2016 in official development assistance for the agriculture sector. Assistance for agriculture increased between 2007 and 2011 but decreased from 2012 to 2016. Overall, there was a slight decline over the 10 years from 2007 to 2016 (OECD 2018a).

Bilateral commitments in support of the Rio markers (biodiversity, climate change mitigation, climate change adaptation, and desertification) are tagged in the Development Assistance Committee (DAC) Creditor Reporting System (CRS) database. This data shows that aid activities targeting climate change adaptation in the agriculture sector have fluctuated from year to year but have increased slightly

Official development assistance to Oceania for the agriculture, forestry and fisheries sector, including for adaptation and mitigation activities (2007 - 2016)

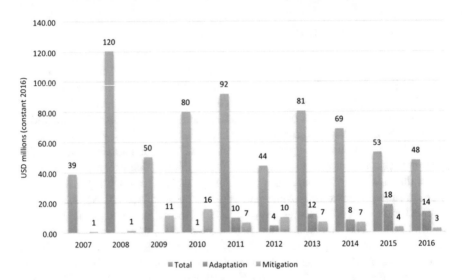

Fig. 9.1 Total development assistance for the agriculture sector in PICs (Oceania) between 2007 and 2016. Note: to avoid double counting, mitigation and adaptation activities should not be added together. Adaptation and mitigation funds are part of the overall development assistance for the agriculture sector. Assistance is from OECD countries so China is not included in this analysis. (Source: OECD 2018a, b)

between 2009 and 2016. On the other hand, bilateral commitments for activities targeting mitigation in the agriculture sector increased rapidly between 2008 and 2010 and have steadily decreased since then. It is important to note that in the OECD-DAC database the same activity can be marked for multiple environmental objectives (e.g., both adaptation and mitigation) so adding up the adaptation-related finance and mitigation-related finance could result in double counting. Bilateral commitments for adaptation and mitigation represent a potentially overlapping subset of total bilateral commitments to the agriculture sector (OECD 2018b).

The Green Climate Fund (GCF) and other multilateral climate funds are also providing assistance for adaptation and mitigation in PICs. Unfortunately, out of the eight countries with approved projects by the GCF, only Vanuatu's project will provide benefits to the agriculture sector. Vanuatu's project will focus on improving climate services, which will assist farmers to prepare to reduce the risks induced by climate hazards. Vanuatu has also been benefiting from a GCF readiness program on Reducing Emissions from Deforestation and Forest Degradation (REDD+) (GCF 2018).

9.3   Impacts of Extreme Events and Rising Sea Levels on Agriculture

Agricultural production in PICs is very vulnerable to the negative impacts of climate hazards. Both sudden-onset climatic hazards such as tropical cyclones, droughts, floods, and storm surges and slow-onset hazards such as sea level rise and rising average temperature are affecting the agriculture sector in PICs. This section provides a brief summary of the impacts of climate change on the agriculture sector in PICs, according to large-scale post-disaster needs assessments and other research. More details on the impacts of climate change on agriculture in PICs can be found in Taylor et al. (2016), Barnett (2011), Iese et al. (2015), Bourke (1999), Allen and Bourke (2001), Wairiu et al. (2012), Iese et al. (2016), and McGregor et al. (2016).

9.3.1   Observed Impacts of Climate Change and Climate Variability on Agriculture

Climate change is impacting agricultural production in PICs through slow-onset stressors such as rising average temperatures, shifting rainfall patterns, and sea level rise. At the same time, agricultural production is being impacted by the influence of climate change in increasing the frequency and/or severity of extreme weather events. There is a greater understanding of the agricultural losses and damages caused by extreme weather events compared to those caused by slow-onset events. This is because the costs of extreme weather events have been quantified fairly precisely by post-disaster needs assessments. Tropical Cyclone (TC) Evan in 2012 caused WST 62 million (USD[1] 23.8 million) in losses and damages to the agriculture sector in Samoa and FJD 37.7 million (USD (see Footnote 1) 11.7 million) in Fiji (Government of Samoa 2013; Government of Fiji 2013). TC Pam, a Category 5 cyclone, devastated Vanuatu in 2015 and caused losses and damages to the agriculture sector valued at USD 56.5 million (Government of Vanuatu 2015a). TC Pam also generated waves that caused severe damages to food sources, water and infrastructure in Tuvalu and Kiribati. The waves and strong winds destroyed about 30–90% of crops on many islands of Tuvalu. The economic impacts of TC Pam were estimated to be 25% of Tuvalu's projected GDP in 2015 (Katea 2016). Severe Tropical Cyclone Winston in 2016 incurred losses and damages on the agriculture sector valued at FJD 542 million (USD (see Footnote 1) 254.7 million) (Government of Fiji 2016). TC Gita, which affected the agriculture sector in Tonga in 2018, resulted in losses and damages valued at TOP 97.5 million (USD (see Footnote 1) 42.8 million) (Government of Tonga 2018). Flooding in Honiara, Solomon Islands, in 2014 affected over 9000 households in Guadalcanal Island, destroying more than

---

[1] https://www.xe.com/currencyconverter/. Accessed 25 Mar 2019.

75% of household food gardens in these areas. The total loss and damage to agriculture from the flooding in Honiara was an estimated USD 18 million (Reliefweb 2018). In the RMI, a 16% fall in copra production occurred in 1996 due to heavy rainfall during an El Niño year (Republic of Marshall Islands 2013). According to a group of farmers in Ranadivi, Tavua in Fiji, the ongoing 2018 drought caused a loss of sugarcane (Fiji Times 2018). Farmers have not been able to cultivate the land for the following year's planting season as the soil has been too dry to plow.

Shifts in the normal rainfall patterns have also been affecting local farmers in PICs. A crop modelling simulation on impacts of rainfall on taro production in Santa Isabel, Solomon Islands, revealed that the likely cause for losses in taro yield was nitrogen stress caused by an increase in daily rainfall on Santa Isabel in 2011, leading to increased runoff. The steepness of the slopes where taro is cultivated, combined with the practice of removing all weeds and leaving no ground cover, has also contributed to this impact (as some farmers believe that a clean farm with exposed soil is a sign of a hardworking farmer) (Quity 2012). In Nabukelevu, Serua province in Fiji, rainfall has increased compared to the past 20 years according to the villagers, causing a loss of cash crops. Mandarin trees, which are the main source of income in the community, are not fruiting anymore, and duruka (Fiji asparagus) is not budding, causing a loss of income and a sudden change in sources of livelihood in the villages. The villagers are now spending more time hunting wild pigs than gathering wild fruits for food and income (UNESCO 2017).

Rising sea levels and storm surges causing saltwater inundation, intrusion and salt spray are affecting coastal agricultural production, especially in atolls and low-lying areas in PICs. Sea level rise in combination with other climate hazards decreases the availability of fresh groundwater for agriculture. Most groundwater in Pacific Islands exists beneath the surface of mostly limestone permeable islands as freshwater lenses. Sea level rise will eventually lead to a reduction in freshwater in Pacific Islands, especially on atolls and along the coastal areas of larger islands as freshwater lenses are contaminated by salinity intrusion from below or inundation from above. Freshwater is being contaminated as salinity intrusion is causing a reduction in freshwater. Water salinity as well as soil salinity in atolls and along coastal areas will increase with the rise in sea levels. For example, assessments conducted in 2009 of wells in the outer islands in Chuuk State in the Federated States of Micronesia found, in most cases, that well water had increased in salinity since 1984 (Shigetani 2009). On Tarawa, Kiribati, it has been found that by 2050 rainfall could decline by 10% and sea levels could rise by 0.4 m, shrinking the thickness of the freshwater lens by as much as 38% (The World Bank 2000). In Funafuti, Tuvalu, salinization of the groundwater lens due to sea level rise has impacted giant swamp taro production, worsening prior salinization of the water lens caused by the construction of the Funafuti airstrip (Lewis 1989). Freshwater lens observation in Laura, Majuro in the Marshall Islands showed that the 1998 up-coning of the freshwater lens as a result of a long drought still exists and has changed little in shape (Koda et al. 2017). Interestingly, the results indicated that saltwater intrusion did not affect crops despite the increase in groundwater withdrawals. Studies in Tonga showed increasing salinity of wells located on the low-lying coastal areas because of saltwater intrusion (Government of Tonga 2012). Groundwater reserves in the northwest of Savai'i, Samoa, are becoming more saline due to saltwater intrusion, resulting in the abandonment of some boreholes (Berthe et al. 2014). Increased salinity of groundwater is influenced by factors such as sea level rise resulting in seawater intrusion (from below) or inundation (from above), rising temperatures resulting in higher evaporation rates, and decreased rainfall.

The impacts of salinity on soil and crops have been observed by farmers especially in low-lying islands. Saltwater intrusion into inland gardens (primarily for taro production) has already begun to affect some atoll islands of Solomon Islands like Ontong Java, making tubers yellow and bitter and unsuitable for consumption (Maeke 2013). Furthermore, increases in salinity are reportedly impacting the growth of giant swamp taro in Tuvalu (Tekinene 2014). Saltwater intrusion during strong westerly winds and rough spring tides destroyed giant swamp taro pits in Funafuti, Tuvalu. On Nanumaga Island, the giant swamp taro pit on the northern part of the island is frequently flooded during high tides, affecting the growth and reducing the yield. Impacts of salinity on crops are more visible during droughts as was observed in the La Niña event of 2011 (Tekinene 2014). In Nataleira village in Fiji, farmers stopped planting rice at the lowland areas because of saline soils from saltwater inundation during storm surges. Loss of yield due to the salinization of groundwater and soils is expected to continue to affect food security and income security, increasing hardship and poverty in Pacific households (UNESCO 2017).

9.3.2   Future Impacts of Climate Change on Crops in PICs

Studies investigating the future impacts of climate change on agricultural crops in the Pacific are limited. Nevertheless, a small number of researchers have begun to explore this topic by applying crop models to explore future projections for specific sites and crop varieties. Unfortunately, the results of crop models have to date remained within the domain of researchers and have not been widely adopted by policy-makers to inform agricultural adaptation and mitigation strategies.

The impacts of climate change on future taro production in the Solomon Islands were investigated by applying the Decision Support System for Agrotechnology Transfer (DSSAT version 4.5) model package. Maeke (2013) simulated the yield of *Tango Sua*, a cultivar of Taro (*Colocasia esculenta*), using the soil profile from Bellona Island, historical daily weather data from Honiara, and future climate change projections for Solomon Islands. The future climate change projections came from Pacific Climate Futures version 2.0, a web-based tool built upon extensive analysis of global models from climate change in the Pacific, produced by the Pacific Climate Change Science Program in 2011. Projected changes in temperature and rainfall were considered, based on 20-year time periods around 2030, 2055, and 2090. By 2030, temperature is projected to increase by 1 °C and annual rainfall is projected to increase by 8%, leading to a simulated reduction in yield of between 13.1 and 23.2%. Temperature and rainfall are both projected to continue to increase to 2090, leading to further reductions in simulated yield of up to 36% by 2055 and up to 57.7% by 2090. The projected yield reductions are likely due to nitrogen leaching from runoff associated with excess rainfall. The most vulnerable site in Bellona Island is Sa'aiho which has the highest projected yield reduction due to its limited soil fertility and loosely packed soil structure that is susceptible to leaching. When future carbon dioxide concentrations are added to the simulations, the projected yield for *Tango Sua* still declined for two sites by 2030 by between 2.8 and 5.9% (but by a lesser degree than for simulations that only considered temperature and rainfall). Simulations showed an increase in yields of between 0.5 and 1.4% by 2055 and 4.6 and 6.7% by 2090. However, for Sa'aiho (a site with poor soils), yields continued to decline by 12.5%, 10.5%, and 12%, respectively, in 2030, 2055, and 2090. The major variables contributing to this change in projected yield are soil type and quality, increase of temperature and rainfall, as well as carbon dioxide fertilization. There is more work needed to validate DSSAT taro model in PICs.

The decline in taro yield under projected changes in rainfall and temperature for 2030, 2055, and 2090 is also in line with decline in yield of taro varieties such as *Tausala-Samoa* and *Lehua* varieties simulated for high volcanic island (Santa Isabel) in Solomon Islands (Quity 2012).

Nand (2013) simulated the impacts of El Niño Southern Oscillation (ENSO) and future climate scenarios on future yields of the *Desiree* potato variety in Rakiraki and Koronivia in Fiji. Simulations showed that during seven El Niño years (between 1960 and 2012), the average yield in Rakiraki declined between 30 and 60%, while the average yield in Koronivia increased by 31%. The varying yield responses at each site were due to the effect of El Niño on the amount of rainfall received. The western side of Viti Levu (Rakiraki) received below normal rainfall during El Niño years. On the other hand Koronivia, which is located on the windward side of Viti Levu Island, received a fairly normal amount of rainfall during El Niño years (Nand et al. 2016). When simulating future climate change using the Pacific Climate Futures for Fiji, the *Desiree* variety produced zero yield in 2055 for Rakiraki site only and in 2099 for both Rakiraki and Koronivia sites. The increase of temperature and rainfall variability are the two main variables that affect the simulated future yield of the *Desiree* variety. The *Desiree* variety is sensitive to slight increases of temperature as it reduces tuber formation (Nand 2013). The use of crop models to assess the impacts of climate change on crops revealed that the impacts are very specific depending on the type of variety, crop, and soil or location of the farm.

The impacts of future climate change projections were also simulated on cassava varieties in Fiji (branching and non-branching) using the Agricultural Production Systems Simulator (APSIM) model. Cassava yields were projected to decline by up to 9% by 2030 and up to 18% by 2050. In addition to declines in yield, the year-to-year variability was shown to increase by up to 19% by 2030 and up to 28% by 2050 (the increase in variability is driven by more frequent lower yielding years) (McGregor et al. 2016). According to Crimp et al. (2012), the severity of impacts on cassava yields depends mainly on soil types and soil characteristics at different locations.

9.4   Case Studies: Reducing the Impacts of Climate Change on Agriculture

Pacific Island Countries have been working hard in adapting and coping to reduce the impacts of climate change on agriculture. Most national and community-based adaptation initiatives for the agriculture sector are donor-funded projects implemented by national governments, regional organizations, NGOs, private sector actors, and farmers. However, most approaches to date have not taken a systems-oriented approach and have treated components of the agriculture system separately. For instance, projects have supported vulnerability assessments, the introduction of new early maturing crop varieties and resilient varieties of crops and livestock, farm diversification and agroforestry, and reducing the risks of flood hazards through developing drainage systems and floodgates. While many of these activities have been useful in reducing vulnerability, more can be done to take a systems-oriented approach, recognizing the health of soil as the foundation of resilience. This section presents selected case studies that illustrate useful approaches that—if adopted more widely—could support the transformation of the agriculture sector in Pacific Island Countries. Adopting a holistic, systems-oriented approach is crucial to building the resilience of the agriculture sector. Addressing the health of soil is fundamental and is a key aspect of the system that is typically neglected in vulnerability assessments. Each of the case study illustrates how building the resilience of the agriculture sector to climate change starts with the soil. The first example is a policy case study (of the Tonga Agriculture Sector Plan—TASP) that shows how systems-oriented approaches to resilience that focus on household food security and improving the health of the soil first can be effectively embedded into the overarching plan for the agriculture sector as a whole. The second case study illustrates the benefits of broadening beyond the usual household and community vulnerability assessments to instead consider the water catchment as a whole, considering land use, landforms, biodiversity, water, and agricultural practices. The last three case studies focus on improving soil health over the long term using organic sources such as compost, mucuna, and seaweed fertilizer. The case study on the atoll islands looks at how to increase agricultural resilience on a small landmass and with poor soils and limited water using a combination of applied research on soils, crop diversification, and technologies to increase water use efficiency.

9.4.1   Tonga Agriculture Sector Plan for a Resilient and Sustainable Agriculture System

A good example of a country's leadership to improve agricultural production in a changing climate is the Tonga Agriculture Sector Plan (TASP) 2016–2020. Agriculture is very important for Tonga as 75% of the population live in rural areas and depend predominantly on agriculture and fisheries for their livelihoods.

Although only 10% of farmers are categorized as commercial farmers, the contribution of the agriculture sector to GDP has averaged around 20% in the last 10 years. Climate change and natural hazards have severely impacted the agriculture sector in Tonga in the past, and the country has consistently been in the top three most risk-prone countries in the world from natural hazards (climate and non-climate hazards), according to the World Risk Index (United Nations University 2014, 2016; Government of Tonga 2016).

Given this, it is understandable that the TASP places heavy emphasis on climate change adaptation and resilient agriculture. Climate change is integrated in all aspects of the sector plan. The term "climate change" is mentioned 72 times, while "climate" is mentioned 141 times in the plan. The TASP states:

> Future agricultural development initiatives will need to heed the importance of including climate change adaptation (CCA) and disaster risk reduction (DRR) into programmes and projects that target the sector. The best way to achieve this is to focus on building resilience, with traditional production systems forming a strong foundation. (Government of Tonga 2016)

Two of the TASP's four strategic objectives are particularly relevant to building resilience to climate change. These are "to develop a climate-resilient environment" and "to develop diverse, climate-resilient farming systems for the Kingdom's islands." The TASP also has three specific objectives (for the whole plan), which are to (1) develop baseline knowledge for sustainable management of soil and water (for agriculture), (2) develop climate-resilient guidelines and indicators for diverse farming systems, and (3) build capacity for climate-resilient agriculture (diverse farming systems and adaptive communities).

Activities under the TASP include upgrading and equipping soil laboratories, upgrading soil profiles, conducting a national soil survey, and updating national soil maps. These outputs will facilitate improved understanding among Tongan farmers of organic matter and soil carbon content as well as soil nutrient availability and how current farming practices and changing climate parameters affect these. From these activities, it is hoped that knowledge will be generated on ideal combinations of organic fertilizers and green manures for improving soil health, increasing soil water retention capacity, and sustaining crop yield with minimum impacts on environment.

Another important activity in the TASP is to improve meteorological information and data availability for agriculture. This is being achieved through a partnership project between the Tongan Government and the APEC Climate Center (APCC), from South Korea. From this partnership, soil-moisture balance maps will be developed that link relevant soil parameters from soil profile data (in particular available water-holding capacity) with relevant meteorological data such as derived potential evapotranspiration (PET). These maps will highlight areas with moisture deficits and drought risks. Also proposed is the further development and refinement of crop models, which can be linked to current and future analyses of climate risks and adaptation options. Targeted research will also be undertaken to support improved agrometeorological advice for decision-making. Climate forecasts and potential impacts have been translated into the Tongan language and made available on the Tonga Meteorological Division's website.

The TASP focuses on a "systems approach" to agriculture rather than a crop-based approach. In a changing climate, where multiple variables are changing simultaneously, viewing agriculture from a systems-oriented perspective is more likely to ensure resilience. The systems approach is centered on soil health and water availability and how farming management techniques, climate parameters, and other biological and human impacts affect the balance of the system.

The TASP is focused on the creation of knowledge and usable information about the current status of the system and how climate change will impact the system going forward. This has included the development of decision support tools such as crop models and maps (using Geographic Information Systems) to frame information so that it can be used to inform decision-making for resilient agriculture. In addition, the TASP has facilitated the transfer of information between different stakeholders through active and participatory research projects at research stations and farmer field schools.

The TASP has a community-focused approach founded on improving food security for communities before commercialization of products. This is based on the "community readiness" approach that is at the heart of the community planning processes led by the Ministry of Internal Affairs, the NGO Mainstreaming of Rural Development Innovation Tonga Trust (MORDI TT), and funded by the International Fund for Agriculture Development (IFAD). The community readiness approach focuses on empowering communities to take control of their own development process through building community capacity and transforming the surrounding enabling environment (through developing new markets and giving communities the skills to find funding to develop critical infrastructure).

The TASP was funded by multiple donors with interests in agricultural development, climate risk resilience, and sustainable livelihoods of communities. There is a very clear call in the TASP for NGOs and the private sector to support the upscaling of agricultural research and advisory services to extend their reach to every farmer and community in Tonga.

9.4.2 USAID Food Security Adaptation Project: Sabeto District, Nadi, Fiji Islands[2]

The food security project *"Enhanced Climate Change Resilience of Food Production Systems for Selected PICTs (Fiji, Kiribati, Samoa, Solomon Islands, Tonga, and Vanuatu)"* used a holistic assessment approach to develop a plan for increasing resilience of food systems. The Pacific Community (SPC) implemented the

---

[2] This case study was summarized from a field assessment report "Community based climate change vulnerability assessment of the Sabeto Catchment" by SPC (2013).

USAID-funded project in all six countries. In Fiji, the Sabeto District at Nadi, Viti Levu, was chosen as the project site. The Sabeto catchment covers 13,819 ha and is located halfway between Nadi and Lautoka. Almost 96% of the catchment is native land and 4% is freehold land. Landowners' consultation is vital before any development takes place in the catchment. Four villages from the Sabeto District were selected, namely, Korobebe, Nagado, Naboutini, and Nakorokoroyawa. The intervention was in two parts: (1) the analysis of the vulnerability of food and agriculture systems in the villages and (2) implementing agriculture interventions to improve food production, food security, and livelihoods in the villages.

Three different community-based methods were used to assess the vulnerability of the agriculture systems in the villages. These methods were land use surveys, participatory rural appraisal (hazard mapping, climate change impact mapping), and household income and expenditures surveys aimed at assessing the sensitivity and coping capacity of the communities.

The land use survey was conducted by the Government of Fiji's Ministry of Primary Industry and the Land Resources Division within SPC. The survey provided a description of land resources, soil types and structures, land availability, limitations, and potential uses. This was done through collection and preparation of soil maps and land use capability maps for the Sabeto catchment area. Satellite images (1:10,000) were used to identify land use types. The assessment also covered food sources and consumption patterns of villagers. The information collected was used to determine the vulnerability of the selected communities as well as their agricultural production environment. Field findings were integrated into GIS, and maps were drawn to show soil types, land use, land capability, and boundaries.

As part of the participatory rural appraisal, SPC developed their own methodology to assess the vulnerability of food security systems at the community level. The climate change vulnerability assessments found that the vulnerability index (out of 5) for all four communities ranges from high (3.75) to very high (4.35). A food availability assessment found that communities' reliance on processed foods is high, ranging from 51 to 54.4% or carbohydrates and 56.4–66.8% for proteins. The high reliance on imported and processed foods increases food insecurity, as these foods are expensive relative to household incomes and prices of imported and processed foods are volatile, as they are influenced by international markets. This detracts from the ability of households to pay for essential costs such as education, health, and building resilient houses. Overall, the assessment of soil, land use, land capability, food security, and climate change vulnerability showed a need to devise adaptation measures to reduce impacts of climate change on villages' food systems.

Agricultural adaptation measures were put in place to diversify agriculture systems, improve soil fertility, and ultimately increase the resilience of food security and livelihoods at all villages. Crop nurseries were established in all villages to raise and multiply climate-resilient crops and trees before distributing to farmers. The crop nurseries included traditional crops and varieties, such as wild yams, cultivars that are commonly planted in the area, and varieties from the climate-resilient collection from the Centre for Pacific Crops and Trees (CePaCT) at the Pacific Community. The climate-resilient collection included taro, cassava, and drought-tolerant, yellow-fleshed sweet potato varieties, which are rich in pro-vitamin A. Resilient crops were distributed to farmers for field evaluation, with the aim of increasing the stability of production during droughts, floods, and cyclones. Demonstration farms were also established to show the use of mucuna bean (*Mucuna pruriens*) in maintaining and improving soil fertility and how contour farming on slopes can expand farming areas. Piggery and chicken demonstration farms were established around the villages to encourage farmers to raise their own livestock to improve protein sources and reduce reliance on purchased foods. Women's groups were also trained in beekeeping and honey production as an additional source of income.

This case study is unique because of the approach used to assess the vulnerability of agriculture and food security for the communities. The team used an integrated vulnerability method where scientific and technological analysis of soil types, land use, and land availabilities was combined with participatory methods through community consultation. Through the participatory methods, the communities mapped their risk areas (in relation to the impacts of hazards), ranked the impacts and vulnerability, and discussed solutions to adapt the agriculture system in each village. The spatial scale of the assessment covered the whole catchment rather than village boundaries and encompassed the whole agriculture system including land, water, crops, livestock, and sociocultural factors.

9.4.3 Improving Soil Health and Land Availability: Atolls and High Islands

Soil is the foundation of crop production, and as such, the health of soil is very important for the resilience of agriculture systems. Crops grow more vigorously when all the important plant nutrients are available for plant growth; yield increases, and crops are more tolerant to climate-influenced hazards such as droughts, sea level rise, and pests and diseases. Most large-scale, commercial farmers and agriculture departments in Pacific Island Countries use chemical fertilizers to "speed up" the growth of crops. Unfortunately, the long-term use of chemical fertilizers changes the biological composition of soil (making it more acidic), and excess nutrient runoff has detrimental impacts on waterways and oceans. Despite this, few research and adaptation projects have focused on improving soil health through techniques such as organic composting, green manure, and diversification of crops.

Atoll Soil Health Project

The project "*Improving soil health, agricultural productivity and food security on atolls*" focused on the development of sustainable soil health technologies for atolls in order to improve production of staple root crops and nutritious crops. The project was funded by the Australian Centre for International Agricultural Research (ACIAR) and was implemented in partnership with the Government of Kiribati, the Government of Tuvalu, the Government of Marshall Islands, the Pacific Community (SPC), and many universities in Australia (ACIAR 2018).

In Kiribati, one cassava variety, six varieties of sweet potato, and three varieties of taro were collected from farmers in Kiribati (from the atolls of Abemama and Butaritari and Banaba Island) and evaluated in the atoll of Tabiteuea North. The taro varieties were evaluated against three introduced varieties from the CePaCT climate-ready collection. The evaluation revealed that yields of sweet potato are better when compost is applied in the subsoil and planting materials are inserted into flat soil which is mounded after a few weeks. In Tuvalu, the evaluation revealed that certain varieties of sweet potato performed better than others (e.g., the PNG variety performed better than Banaba and PRAP varieties). The better-performing varieties were distributed to outer islands to help communities recover after Tropical Cyclone Pam.

The project also developed a participatory productivity index in each country based on seven factors. The factors were (1) income ($/week and percentage to buy food), (2) decision-making on land use, (3) farming skills (farmers, agriculture department), (4) farm productivity of current production, (5) farm resilience (above and below ground), (6) quality of land (from participatory method and soil test), and (7) greenhouse gas emission reduction (biodiversity and soil carbon). After the assessments, spider-web diagrams were drawn to show the relationships between each of the factors and production. Farming skills and household incomes were consistently the most limiting factors for increasing production. The next most limiting factors were lack of farm resilience and low productivity of current production. The spider-web diagrams display baseline information for communities, against which progress can be measured both during and after the completion of the project.

Soil testing was also conducted in Kiribati, Tuvalu, and the Marshall Islands using the *"Palintest SKW 500 Quick Soil Test Kit."* It was found that most test sites had limited potassium and all had limited iron, copper, and manganese. The project also analyzed more than 100 soil and compost samples from Kiribati, Tuvalu, and the Marshall Islands. The analysis consistently showed high pH and very low levels of potassium and manganese. Even in improved soils (that had had additions of compost), levels of potassium and manganese were found to be only marginal to adequate. The analysis showed low levels of copper and iron in some of the soils, but levels of available and total potassium and available zinc were surprisingly adequate to good. Nutrient omission trials have been discussed to determine whether soil tests are accurately predicting potassium levels for high pH sandy soils. Recipes for improved compost were created using different combinations of manure, ash, sea cucumber, and green leaves. The application of compost did not improve each nutrient significantly, but the combined effect of small increases in multiple nutrients significantly improved growth and yield.

The use of irrigation was also trialled in the field during this project. Initially, field assistants manually irrigated crops using buckets, but water was not reaching the rooting zone of plants. The water issue was resolved when a well was dug and

Fig. 9.2 Terraced pits with "rooms" for different crops such as giant swamp taro, sweet potato, and taro

water was pumped to an overhead water tank, providing a consistent source of gravity-fed water for irrigation. Soil water meters were also installed to gauge how much water should be applied and detect water distribution in the soil. Using water meters to fine-tune how much water should be applied proved effective in conserving water while ensuring plants received enough to thrive. This tool could be applied on other atolls, where water is generally a limited resource.

Another achievement of this project was increasing the diversity of root crops and vegetables grown in Kiribati by encouraging households to plant a variety of crops in their giant swamp taro pits. Both women and men applied compost in the pits, created terraces, and planted giant swamp taro and other vegetables in different "rooms" in the pits (Fig. 9.2).

Use of Invasive Seaweed as an Organic Fertilizer

The Pacific Small Island Developing States (PSIDS) consists of a variety of both high and low islands, each with its own characteristics of habitats, complexity of land features, as well as agricultural fertility (Thaman et al. 2002). With growing populations and the ever-increasing need for more intensive and effective food security efforts, farmers in PSIDS increasingly rely on soil additives, mainly in the form of chemical NPK (nitrogen, phosphorus, potassium) fertilizers. However these are both expensive and detrimental to the environment as sources of coastal eutrophication. Since the PSIDS are in a marine environment, marine-based substitute solutions to soil enhancement would be a natural alternative. Pacific nations such as Fiji and Tuvalu in the Central Pacific have recently been faced with algal bloom issues (N'Yeurt and Iese 2015a, b), and these events have been linked to excessive inputs of nutrients into the coastal environment. Notably, a high percentage of these anthropogenic nutrient inputs come from runoff of chemical fertilizers and manure from terrestrial agriculture.

Ongoing research at the University of the South Pacific (N'Yeurt and Iese 2015a) has demonstrated the effectiveness of seaweed fertilizer as a substitute for their chemical counterparts. Liquid fertilizers made from the brown seaweed *Sargassum polycystum* and the red seaweed *Gracilaria edulis* have been found to contain high values of the essential nutrients phosphorus and potassium, as high as 14–15 mg/L for phosphorus and up to 6 mg/L for potassium (Soreh et al. in prep.). Conversely, plant-derived fertilizers are usually low in nitrogen, and seaweed extracts fall into that category. In atolls such as Tuvalu, *Sargassum* seaweed is currently used as a fertilizer on vegetables such as tomato and cucumber, being harvested from the beaches and applied with minimal processing (usually just washing in freshwater). On the culturally important Island of Beqa in Fiji, other marine plants such as the seagrass *Syringodium isoetifolium* are used in a similar manner on tomato and kava plantations.

While not yet widespread in use, seaweed-derived fertilizers in both unprocessed and processed (liquid, composted) form have a very high potential to replace chemical fertilizers, with a much lesser impact on the environment. The deficiency in nitrogen found in marine-plant-based fertilizers could be easily addressed by blending with protein-rich compost produced from animal sources, such as fishmeal or even the coral-eating crown-of-thorn starfish (COTS). Work in progress by the authors and colleagues on COTS-based fertilizer is very encouraging, and the production of cost-effective, nutritionally balanced organic fertilizers based on marine pest species is now an attractive reality for farmers in SIDS. Such initiatives, in addition to contributing to the food security of local communities, can have significant impacts on economic security through the sale of excess fertilizer to other farmers and the community at large. An overarching additional benefit is the reduction of marine pest species and a decrease in the input of excess nutrients in coastal waters, promoting healthy marine ecosystems such as coral reefs, seagrass beds, and coastal mangroves which are each in their own right bountiful sources of food security for local PSIDS communities.

Use of Mucuna to Improve Soil Fertility and Resilience

The important role of mucuna in improving soil fertility was proven during field trials conducted in Taveuni, Fiji (Lal 2013), and four locations in Samoa (Anand 2016). The continuous cultivation of taro in Taveuni for 30 years led to reductions of yield due to low soil fertility. Farmers applied large amounts of chemical fertilizers that negatively affected the environment. The taro yields continued to decline. As a response, a research project focused on improving taro yield through the use of mucuna beans (*Mucuna pruriens*) as a fallow crop. A comparison of mucuna and a typical grass fallow with and without lime and rock phosphate applications recorded a 100% increase of Olsen available P (a measure of plant-available soil phosphorus) with mucuna fallow for both 6 and 12 months. The mucuna fallow plot also showed a 50% increase of nitrogen in 6 months and a 100% increase in 12 months of trials. For total organic carbon, a slight decrease was recorded after 6 months, but there was a significant increase of total organic carbon between 6- and 12-month mucuna fallow duration. Mucuna fallow had significantly higher biomass production and accumulated higher levels of nitrogen, phosphorus, potassium, and calcium in its foliage. Furthermore, as the durations of fallow increased from 6 to 12 months, total soil organic carbon, nitrogen, and potassium bulk density and earthworm numbers increased significantly. Overall, there was a 33.5% increase of yield of taro under mucuna fallow compared to grass fallow at Taveuni, Fiji (Lal 2013).

In a multi-agroecological site study in Samoa, fallow with mucuna and grass significantly improved soil active carbon stocks upon decomposition. Mucuna fallow contributed to the largest addition of biomass across all the agroecological sites in Samoa. Mucuna was also the most superior cover crop for improving soil active carbon and soil biological activities. The yield of taro was comparatively higher under mucuna fallow than grass fallow. Comparative economic analysis of mucuna fallow technology showed a 98% increase in gross profit for Salani and Safaatoa sites in Samoa (Anand 2016). Mucuna as a green farming technology not only increases soil health and overall yield and its associated economic benefits but also reduces labor requirements and chemical inputs. This technology is suitable to increase yield of crops in a changing climate and overcultivated lands. Both studies were funded by ACIAR through scholarships to the School of Agriculture and Food Technology, at the University of the South Pacific.

The nutritional value of crops is rarely considered in agricultural adaptation activities in PICs. This is unfortunate given the alarming rate of nutritional disorders such as anemia, vitamin A deficiency, obesity, and other nutrition related noncommunicable diseases in PICs (Thaman 1995). Studies have also implied that most protein in the highlands of PNG comes from sweet potato and other plant products. The food supplies committee in the Solomon Islands conducted a study on the production of sweet potato at household level, and the study stressed the need to increase production primarily of sweet potato in home gardening since there was a correlation between vitamin deficiency and those that do not have garden plots. In Kiribati, to combat vitamin A deficiency and anemia in heavily populated areas of South Tarawa, campaigns and competitions have been started to promote growing vegetables and fruit trees. The Foundation of the Peoples of the South Pacific International (FSPI) has been a strong advocate in Kiribati of increasing local production of vegetables that are rich in vitamin A (FAO (2018a, b)).

The case studies discussed above highlight the use of effective partnerships between national governments, regional and international institutions, universities, and local farmers to improve crop production through research, technology transfer, and climate-smart agriculture approaches. The success of agriculture in a changing climate depends on applied research and capacity building of farmers. The use of basic tools to understand the current status of soil, water, and community practices helped to provide targeted technologies, which addressed factors limiting agricultural production.

9.5 Looking Ahead: Strategic Directions for Transforming Agriculture Under a Changing Climate

This section discusses strategic directions for building resilient agriculture in PICs. It first analyzes national agriculture sector plans of PICs and distills key priorities in relation to climate change that—if effectively realized—could facilitate in the transformation of the agriculture sector going forward. The section then discusses key opportunities for realizing these priorities, including integrated vulnerability assessments, applied research, genuine partnerships, and resourcing arrangements.

9.5.1 Priorities of Agriculture Sector Plans in Relation to Climate Change

PICs have developed many strategies, policies, and plans to improve the productivity of the agriculture sector in the future, both in relation to agricultural commodities and increasing food security. Key themes/priorities in relation to building the resilience of the agriculture sector in a changing climate include healthy soils and access to land, secure and sustainable water supplies, diversification of farming systems (including agroforestry), embracing climate-smart agriculture (including reducing greenhouse gas emissions and adopting resilient crop and livestock varieties), and resourcing (see Table 9.2).

Healthy Soils, Sustainable Land Management, and Access to Land

Almost all PICs have agricultural plans that emphasize the need to maintain and enhance soil health in order to improve agricultural productivity. The case studies described in Sect. 9.4 highlight interventions that have made soil health the center of sustainable agricultural production. When soil health is maintained, organic carbon content and water-holding capacity are enhanced, and important nutrients and micronutrients are made more accessible. Healthy soils also promote pH balance and biological activity of soils, which in turn leads to increased crop resilience and yields. In a changing climate with more frequent and intense extreme events and increasing temperatures and soil salinity, PICs need to focus on improving and maintaining soil health to increase and sustain agricultural production. The agricultural plans of PICs also discuss more broadly the need to improve sustainable resource and land management.

Many lands that are suitable for farming in PICs remain unused due to communities having migrated to the urban areas or overseas. Mechanisms to facilitate access to land in these circumstances should be included in agriculture strategies at the community level. Some PICs are changing their laws to facilitate access to idle lands. For example, Fiji and Samoa have passed laws to allow the lease of custom-

Table 9.2 Key priorities relating to climate change in the agriculture and food security plans of Pacific Island Countries

Country (years of plan)	Key themes	Distillation of key themes
Samoa (2016–2020)	• Strengthen the capacity and resilience of farmers and fishers to prepare and recover from climate threats and disasters affecting agriculture and rural livelihoods • Sustainable agricultural and fisheries resource management practices in place and climate resilience and disaster relief efforts strengthened	• Build resilience and capacity of farmers and fishers • Sustainable resource/land management • Improved disaster relief
Cook Islands (2015)	• Healthy soil and conduct agroecological adaptive research • Support sustainable agriculture and its gradual adaptation to agroecological production models • An agriculture insurance package is developed in accordance with the Pacific Disaster Risk Financing and Insurance (PDRFI) program (supported by World Bank)	• Healthy soils • Sustainable agriculture • Agricultural insurance
FSM (2012–2016)	• Enhanced environmental services and sector resilience to natural disasters and climate change • Appropriate and well-managed use of trees in agricultural systems (agroforestry and home gardens) • Land management practice has much to offer in terms of nutrient cycling and ecosystem services required in organic farming	• Enhanced environmental services • Farm diversification (agroforestry and home gardens) • Sustainable land management
Fiji (2020)	• Adoption of climate change agriculture in Fiji—including agroforestry and new drainage techniques • Effective management of soil carbon • To be prepared and ensure better productivity when facing climate change, thus ensuring food security • Farmers are also contributing to the broad goal of climate change agriculture, which is to decrease greenhouse gas (GHG) emissions	• Farm diversification (agroforestry) • Healthy soils/enhanced soil carbon • Climate-smart agriculture

(continued)

Table 9.2 (continued)

Country (years of plan)	Key themes	Distillation of key themes
Kiribati (2013–2016)	• When organic matter levels in soils are improved, the soils will hold more water for plant use • Selecting varieties that are more adaptable to harsh atoll conditions and potential climate change impacts of increased temperature, drought, and seawater intrusion • Develop integrated and holistic food production systems resilient to impacts of climate change and contributing to food security • Potential development of a new pest and disease regime addressing impacts of climate change will be undertaken with capacity	• Healthy soils • Resilient varieties • Food security
Nauru (2005–2025)	• Develop locally tailored approaches and initiatives to mitigate the causes of climate change and adapt to its impacts • Practical and relevant climate change adaptation measures and initiatives implemented and sustained	• Mitigation/reduced greenhouse gas emissions
RMI (2013)	• Promote climate "smart" farming systems, and evaluate new crop cultivars to identify those which are more tolerant of drought and saline soil and water conditions • Undertake enhanced planning and interventions to address climate vulnerabilities in food security and nutrition	• Climate-smart agriculture • Food security
Palau (2015)	• Local farmers to adapt production methods to the impacts of climate change • Real-time extension services can help prevent potential losses from investments • The recovery period from the impacts of extreme storm surge and sea level rise should be researched and integrated into business plans as well as grant and loan agreements • Uses climate proofed state-of-the-art sustainable land management practices and organic practices and maximizes local inputs and renewable energy to produce 80% of Palau's food needs by 2025	• Enhanced agricultural extension services • Sustainable land management and organic agriculture • Food security
PNG (2007–2016)	• To increase land resource (soils, water, climate, geomorphology) surveys • Improve availability and quality of land resources (soils, water, climate, and geomorphology) information—acquire soil, land, and climate survey equipment	• Conduct more applied research • Increase applied research capacity and equipment

(continued)

Table 9.2 (continued)

Country (years of plan)	Key themes	Distillation of key themes
Solomon Islands (2015–2019)	• Nutrients extracted from the soils should be replaced, forests replanted, soil degraded, and overgrazing reversed • Conservation farming techniques such as agroforestry, fallow, cover crops, intercrop, and contour planting • Promote agroforestry with the use of intercropping to reduce vulnerability • Develop crops that are resilient to natural disasters • Crop insurance schemes where possible or disaster funds	• Healthy soils • Conservation farming • Farm diversification (agroforestry) • Resilient varieties • Agricultural insurance
Tonga (2016–2020)	• Healthy soils, secure and sustainable water, diverse farming systems, and adaptive communities. • Develop a climate-resilient environment—low-carbon, climate-resilient farming systems • "Future Farmers Programme"—return to family farms and to become climate-resilient subsistence and/or commercial farmers and change agents within their communities • Soil fertility tools, agrometeorology, crop models, and technology transfer for food processing	• Healthy soils • Food security • Climate-smart agriculture
Vanuatu (2016)	• Mainstream climate variability, climate change and disaster risk reduction using adaptation and mitigation strategies in all agriculture initiatives and developments • Build risk reduction capacity of farming communities through training and awareness to adapt and mitigate effects of climate variability, climate change, and natural disasters • Promote mitigation strategies in all farming practices • Prioritize the introduction of climate- and risk-resilient crops for cultivation by farmers • Provide adequate funding for activities to address climate variability, climate change, and disaster risk reduction	• Mainstream adaptation and mitigation strategies in agriculture • Risk reduction and capacity building at the community level • Reduce greenhouse gas emissions • Resilient crops • Resourcing

ary lands for longer periods of time for agriculture and development purposes (iTaukei Land Trust Board 2019; Government of Samoa 2008).

Sustainable Water Supply for Agriculture

Sustainable water supply in Pacific agriculture is important, but Tonga is the only country that has included the need for sustainable water supply in their agriculture sector plan. The impacts of drought on agriculture are becoming more prominent in PICs. Droughts can last for months or years, and farmers and governments are finding it hard to respond and recover. Lack of water destroys plants and livestock, while droughts also tend to be associated with high temperatures that dry out soils, making them hard to cultivate for the following season. Availability of water for irrigation during droughts is a major problem for farmers. Atolls and low-lying islands with no rivers rely heavily on rainfall for growth of crops. For high islands with rivers, long distances between farms and water sources and lack of finances and equipment for irrigation are the main issues with setting up irrigation systems. Measures for increasing the availability of water in PICs should be considered in agriculture sector plans. Solutions for increasing sustainable water supply will be different for atoll islands and higher islands with rivers, but measures could include water storage infrastructure, water distribution infrastructure, and technologies that increase water use efficiency such as smart water meters.

Diverse Farming Systems

Traditionally, the agriculture systems of PICs are organic and highly diverse including livestock, vegetables, spices, and culturally important plants. The diversity of traditional agriculture systems has enabled Pacific communities to provide food for themselves and survive many disasters. Households cultivate home gardens with different types of crops and plants that are used for food, medicine, adornment during dances, and decorating buildings during special occasions. Home gardens should be encouraged in urban areas where land is limited. Agroforestry farming is another technology that has been widely practiced in PICs to increase diversification. With this type of farming, the combination of crops, trees, and livestock can promote productivity while maintaining the balance of the ecosystem, including soil, air, water, and environmental health. Many PICs promote farm diversification in their agriculture sector plans, including Tonga, Solomon Islands, Fiji, and FSM.

Climate-Smart Agriculture

Climate-smart agriculture is an approach that aims to sustainably increase agricultural productivity by building the resilience of the agriculture sector to climate change impacts while reducing greenhouse gas emissions from the sector where

Table 9.3 Mitigation sectors in the Nationally Determined Contributions of PICs submitted to the UNFCCC

Country	Electricity	Energy	Transportation	Waste	Agriculture
Samoa	√				
Cook Islands	√				
FSM	√	√	√		
Fiji	√	√			
Kiribati	√	√			
Nauru	√				
RMI	√	√	√	√	
Palau	√		√	√	
PNG	√				
Solomon Islands	√	√	√		
Tonga	√	√	√	√	√
Vanuatu	√				

Source: Redrawn from Lloyd (2018)

possible (FAO (2018a, b)). The term encompasses many activities, some of which have been practiced in PICs for a long time. Activities that qualify as illustrations of climate-smart agriculture include the promotion of agroforestry, soil organic carbon sequestration, organic agriculture, and the use of early warning systems and agro-meteorology in farming systems.

Climate-smart agriculture also includes activities to reduce greenhouse gas emissions from agriculture. About six countries include mitigating greenhouse gas emissions and practicing low-carbon agriculture in their agriculture policies. However, only Tonga has included the agriculture sector as a mitigation sector in their Nationally Determined Contributions (NDCs) report to the UNFCCC (see Table 9.3).

Focus on Food Security

There is a need to increase and sustain agricultural production for food security and livelihoods in Pacific communities. Many PICs mention food security as a priority in their agriculture sector plans; however, agriculture ministries and departments tend to focus more on supporting large-scale farmers to increase yields of commodity crops, rather than a "food first" approach based on agriculture as a system rather than a commodity. In national agriculture planning and development, commercial farmers or representatives of farmers associations (who are mostly commercial farmers) are represented, but rural farmers or subsistent/semi-subsistent farmers are rarely included. Support should be given to subsistence farmers to improve production and help them become semicommercial farmers. Pathways should be initiated and supported to facilitate the "readiness" of communities to invest in value adding to agricultural products.

9.5.2 Opportunities for Transforming the Agriculture Sector

The priorities for future agriculture development in a changing climate are very clear in national agriculture policies and in many ways mirror the good practices illustrated in the case studies discussed in this chapter. The key challenge going forward is realizing these priorities. The discussion below focuses on opportunities for realizing the priorities discussed above. These opportunities relate to integrated vulnerability and risk assessments, applied research, building and utilizing effective partnerships, and securing resources.

Vulnerability and Risk Assessments of Food Security and the Agriculture Sector (Before, During, and After Hazards)

The use of integrated vulnerability assessment tools is useful for understanding the current and future exposure of agriculture systems to climatic hazards and how they will impact crop growth, yield, food nutrition values, soil nutrients and carbon, water availability, and markets. A production index should also be included in the assessments to identify the limiting factors to agricultural production. Baseline information of current soil profiles, land use, landforms and availability maps, short-term projections of rainfall, cyclones, sea level rise, temperature, and historical impacts should be established in each country. Proper agroecological zones should be established based on soil types, landscape, and weather patterns to map vulnerability at specific sites. Tools and technologies such as participatory rural appraisal, soil testing kits, water collection and distribution systems, soil water meters, and GIS technologies should be supported. Agriculture decision support tools such as crop simulation models or impact models should be applied to provide important information to support communities to understand current, short-term, and long-term risks of climate change on agriculture.

Applied and Accessible Research

Ongoing agriculture and climate change research supported by national governments, private sector, NGOs, research organizations, universities, and farmers is generating important knowledge, skills, and information farmers need in order to build the resilience of PICs' agriculture systems. Previous research has ranged from evaluation of resilient crops and livestock, soil health and water availability, agroforestry, ecosystem-based adaptation, traditional coping mechanisms to reduce risks, development and application of crop models and impact models to assess impacts of climate change on crops and livestock, and how to reduce losses and damages in the agriculture sector. Unfortunately, these research projects have been driven by external donors and have usually only involved a small number of local stakeholders such as government officers in research stations or commercial farmers.

This has meant that the application and adoption of research findings by farmers and communities in PICs has generally been limited. There are also issues with the accessibility of many research products, which are generally written in English using highly technical language and are kept in hard copy in libraries or government offices. Many of these publications have never reached the hands of farmers or policy-makers and have therefore colloquially been named "dusty technologies." Publications should be simplified, translated into the local language, and digitized so that they can be shared more widely.

Crops that are deemed to be resilient (which are kept in national and regional germplasm centers) should be properly evaluated both before and after distribution to farmers. For example, crop simulation models can be used to simulate potential yields of crops such as taro, sweet potato, and cassava before distribution and to develop crop profiles on how, when, and where to plant. Farmers should also be supported to document the performance of resilient varieties in the field, as there is a lack of evaluation in how resilient varieties—many of which have come from overseas—perform in PICs.

Standardized methods should be developed to monitor the effectiveness and efficiency of adaptation and mitigation actions in the agriculture sector. To date, monitoring and evaluation tends to be donor driven and project-focused and usually does not extend beyond the lifetime of projects. There is a need for cross-project evaluation methods that measure progress toward the goals and objectives of agriculture sector plans. The best practices generated from evaluation of adaptation and mitigation initiatives will help inform where limits to adaptation exist and when transformational, "step-change" approaches are needed. This would involve determining what level of risks (arising from droughts, cyclones, saltwater, floods, poor soil) an adaptation option can tolerate in different countries and communities before residual losses occur. Developing risk tolerance and adaptation limits will assist decision-making on what type of adaptation option to adopt, what types of technology transfer are needed, and at what point new, innovative, or "transformational" approaches need to be devised. Such approaches will enable PICs to document residual loss and damages.

Genuine Partnerships

Agriculture is a sector that stretches from the ridge to the reefs—involving different formal and informal sectors, different land tenure systems, diverse ecosystems, and different regional and international development partners. Continuously engaging different partners and stakeholders in decision-making and in sharing resources and capacity when implementing activities is crucial. This reduces competition for roles and improves complementary actions for growth of the agriculture sector. The involvement of the private sector, NGOs, farmer associations, and subsistence farmers in agriculture research, innovation, and information sharing should be supported.

Resourcing

All Pacific Island governments allocate less than 5% of their national budgets to the agriculture sector. Resources for the implementation of national agriculture policies mainly come from development partners. The overreliance on development partners for funding tends to result in an overabundance of pilot projects and a lack of follow-through to scale up best practices. The lack of capital and investment in technologies such as irrigation, crop models, soil testing kits, GIS and remote sensing tools, and resilient cropping systems is continuing to expose crops and trees to adverse impacts of extreme events and climate change, therefore affecting farmers and communities. All PIC governments should increase budget allocations to the agriculture sector if they sincerely plan to increase the resilience of the sector.

Cook Islands and Solomon Islands have prioritized development of an insurance scheme for the agricultural sector to transfer risks associated with climate change. This will help farmers recover faster after disasters (Table 9.2). Insurance for farmers has been successfully introduced in other regions of the world (Linnerooth-Bayer and Mechler 2009), and it should be considered as an important risk transfer mechanism to support farmers to cover losses from disasters in PICs. A major challenge is setting insurance premiums at a level that is affordable to households across the Pacific and convincing households that insurance is a worthwhile investment. To help reduce insurance premiums, there may be a case for the establishment of public-private partnerships between the insurance industry and different levels of government to invest in risk reduction measures. Governments of PICs could also consider providing subsidies on insurance premiums to incentivize the uptake of insurance products.

Another opportunity for securing resourcing to transform the agriculture sector is the Green Climate Fund and other climate change and disaster risk reduction funds. The Green Climate Fund supports both adaptation and mitigation. Despite the fact that all countries of the Pacific prioritized agriculture or food security resilience in their national adaptation plans, out of the eight PICs that have had project proposals approved by the Green Climate Fund, only Vanuatu included climate services for farmers and forestry in their activities, and only Tonga included agriculture as a mitigation sector in their Nationally Determined Contributions (NDCs) submitted to the UNFCCC. Many of the practices that increase the resilience of the agriculture sector in PICS also have mitigation co-benefits. For example, practicing agroforestry and increasing soil health and carbon sequestration increase the resilience of agricultural production while reducing greenhouse gas emissions. PICs should consider including the agriculture sector in project proposals to the Green Climate Fund and other climate change and disaster risk reduction funds. There are many small-scale projects that are producing best practices and increasing agricultural productivity at the community or sub-national levels. The good practices from these projects should be upscaled and shared more widely. The upscaling of good agricultural practices can be funded by the communities themselves, governments, private sector, NGOs, and development partners.

The priorities of the agriculture plans discussed above are closely aligned with priority areas for intervention at the international level with the Koronivia Joint Work on Agriculture Initiative (hereafter the Koronivia Initiative), a decision that came out of the UNFCCC COP 23 meeting in Bonn. The Koronivia Initiative recognizes methods for assessing adaptation co-benefits, embracing integrated systems for soil carbon, health and fertility and water management, improved manure and nutrient management, improved livestock management systems, and the importance of food security dimensions (ECBI 2018). The Koronivia Initiative aims to support implementation; facilitate knowledge sharing, capacity building, and technology transfer; and aid in the mobilization of finance (CCAFS 2018; UNFCCC 2017). The close alignment between the key priorities of the national agriculture sector plans of PICs and the Koronivia Initiative presents opportunities to seek resourcing support.

9.6 Conclusion

The agriculture sector in PICs is critically important for food security and livelihoods at the household, community, national, and regional levels. Climate change is already impacting and will continue to impact agriculture in both the short and long term. PICs need to transform the agriculture sector for it to remain prominent and relevant in Pacific communities. There is a need for the agriculture sector in PICs to become resilient to the negative impacts of climate change while simultaneously increasing production to feed a growing population. In addition, there is a need to reduce the negative environmental impacts of unsustainable agriculture on soil, waterways, and the atmosphere (through the release of greenhouse gas emissions). Agricultural transformation can be achieved in PICs through focusing on a systems-oriented perspective that recognizes the foundational importance of healthy soils. Opportunities exist to strengthen existing partnerships and forge new ones to address information and resourcing constraints.

References

ACIAR (2018) ACIAR SMCN-2014-089: improving soil health, agricultural productivity and food security on atolls. Annual Report, 2018

Allen MG (2015) Framing food security in the Pacific Islands: empirical evidence from an island in the Western Pacific. In: Regional environmental change, vol 15. Springer, Heidelberg, pp 1341–1353

Allen BJ, Bourke, RM (2001) The 1997 drought and frost in PNG: overview and policy implications', food security for Papua New Guinea. In: Proceedings of the Papua New Guinea Food and Nutrition 2000 Conference, Papua New Guinea University of Technology, Lae, pp 155–163

Anand S (2016) Developing a taro (Colocasia esculenta) production system based on genotype and fallow system for economic and environmental sustainability under local conditions in Samoa. Ph.D. Thesis. School of Agriculture and Food Technology, Faculty of Business and Economics, University of the South Pacific, Apia, Samoa

Barnett J (2011) Dangerous climate change in the Pacific Islands: food production and food security. Reg Environ Change 11(Suppl. 1):229–237. https://doi.org/10.1007/s10113-010-0160-2

Berthe L, Seng DC, Asora L. (2014) Multiple stresses, veiled threat: saltwater intrusion in Samoa. Presented at the Samoa Conference III: opportunities and challenges for a sustainable cultural and natural environment, 25–29 Aug, National University of Samoa, Apia, Samoa

Bourke RM (1999) Vanuatu Agriculture System Survey—April–May, Canberra, Australia

Bourke RM (2005) Sweet potato in Papua New Guinea: the plant and people. In: Ballard C (ed) The sweet potato in Oceania: a reappraisal. University of Pittsburgh and University of Sydney, Sydney, pp 15–24

CCAFS (Climate Change, Agriculture and Food Security (2018). Koronivia: setting the stage for an agricultural transformation. https://ccafs.cgiar.org/blog/koronivia-setting-stage-agricultural-transformation#.W6NDzGgzaUk. Accessed 20 Sept 2018.

Crimp S, Taylor M, Naululvula P, Hargreaves J et al (2012) Understanding the implications of climate change for Pacific staple food production—a cassava case study. 10th International Conference on Southern Hemisphere Meteorology and Oceanography. 23–27 April 2012, Noumea, New Caledonia

Department of Agriculture, Forestry and Fisheries (2015) Niue Agriculture Sector Plan 2015–2019. http://pafpnet.spc.int/policy-bank. Accessed 18 Sept 2018

ECBI (European Capacity Building Initiative) (2018) COP23 adopts Koronivia Joint Work on Agriculture. https://www.ecbi.org/news/cop23-adopts-koronivia-joint-work-agriculture. Accessed 20 Sept 2018

FAO (2010) Pacific Food Security Toolkit: building resilience to climate change; root crop and fishery production. Rome, Italy. Available at http://www.fao.org/docrep/013/am014e/am014e04.pdf.

FAO (2018a) Climate-smart-agriculture. http://www.fao.org/climate-smart-agriculture/en/. Accessed 20 Sept 2018

FAO (Food and Agriculture Organization of the United Nations) (2018b) Country profiles. http://www.fao.org/countryprofiles/en/. Accessed 19 Sept 2018

Fiji Times (2018) Cane production drops in face of prolonged dry weather. 8 Sept 2018

FSM (Federated States of Micronesia) (2012) Federated States of Micronesia Agriculture Policy 2012–2016. Department of Resources and Development. http://pafpnet.spc.int/policy-bank. Accessed 18 Sept 2018

GCF (Green Climate Fund) (2018) Country Profiles. https://www.greenclimate.fund/countries. Accessed 19 Sept 2018

Government of Cook Islands (2015) Agriculture and food sector—After MDG vision: healthy soils, healthy foods—sustaining our common livelihoods. http://pafpnet.spc.int/policy-bank. Accessed 18 Sept 2018

Government of Fiji (2013) Fiji post-disaster needs assessment: Tropical Cyclone Evan, 17 Dec 2012. https://www.gfdrr.org/sites/default/files/publication/Fiji_Cyclone_Evan_2012.pdf

Government of Fiji (2016) Fiji post-disaster needs assessment. Tropical Cyclone Winston, 20 Feb 2016. https://www.gfdrr.org/sites/default/files/publication/Post%20Disaster%20Needs%20Assessments%20CYCLONE%20WINSTON%20Fiji%202016%20(Online%20Version).pdf

Government of Nauru (2005) National Sustainable Development Strategy 2005–2025. Development Planning and Policy Division, Ministry of Finance and Economic Planning. http://pafpnet.spc.int/policy-bank. Accessed 18 Sept 2018

Government of Samoa (2008) Land Title Registration Act 2008. http://extwprlegs1.fao.org/docs/pdf/sam88408.pdf. Accessed 10 Mar 2019

Government of Samoa (2013) Samoa post-disaster needs assessment Cyclone Evan 2012. http://www.gfdrr.org/sites/gfdrr/files/SAMOA_PDNA_Cyclone_Evan_2012.pdf.

Government of Samoa (2016) Agriculture Sector Plan 2016–2020. Governance, Institutional and Strategic Frameworks, vol 1. http://pafpnet.spc.int/policy-bank. Accessed 18 Sept 2018

Government of Solomon Islands (2015) Solomon Islands Agriculture and Livestock Sector Policy 2015–2019. Pacific Agriculture Policy Bank. http://pafpnet.spc.int/policy-bank. Accessed 18 Sept 2018

Government of Tonga (2012) Second National Communication. In response to its obligations under the United Nations Framework Convention on Climate Change. Second National Communication on Climate Change Project, Ministry of Environment and Climate Change, Nuku'alofa, Tonga

Government of Tonga (2016) The Kingdom of Tonga: Tonga Agriculture Sector Plan 2016–2020. http://pafpnet.spc.int/policy-bank. Accessed 18 Sept 2018

Government of Tonga (2018) Post disaster rapid assessment. Tropical Cyclone Gita, 12 Feb 2018. https://reliefweb.int/sites/reliefweb.int/files/resources/tonga-pdna-tc-gita-2018.pdf

Government of Tuvalu (2016) Tuvalu Agriculture Strategic Marketing Plan 2016–2025. http://pafpnet.spc.int/policy-bank. Accessed 18 Sept 2018

Government of Vanuatu (2015a) Vanuatu Agriculture Sector Policy 2015–2030. http://pafpnet.spc.int/policy-bank. Accessed 18.09.2018

Government of Vanuatu (2015b) Vanuatu post-disaster needs assessment. Tropical Cyclone Pam, Mar, 2015. https://reliefweb.int/sites/reliefweb.int/files/resources/vanuatu_pdna_cyclone_pam_2015.pdf

Halavatau SM (2016) Regional partnership to address food production crisis in the Pacific Islands. Bangkok, Thailand

Iese V, Maeke J, Holland E, Wairiu M, Naidu S (2015) Farming adaptations to the impacts of climate change and extreme events in Pacific Island countries. In: Ganpat WG, Isaac W-AP (eds) Impacts of climate change on food security in Small Island Developing States. IGI Global, Hershey, pp 166–194. https://doi.org/10.4018/978-1-4666-6501-9.ch006

Iese V, Paeniu L, Pouvalu SIF, Tuisavusavu A, Bosenaqali S, Wairiu M, Nand M, Maitava K, Jacot Des Combes H, Chute K, Apis-Overhoff L, Veisa F, Devi A (2016) Food security: best practices for the Pacific. Suva, Fiji: Pacific Centre for Environment and Sustainable Development (PaCE-SD). The University of the South Pacific (USP), Suva. https://doi.org/10.1080/103577 18.2012.658620

Iese V, Holland E, Wairiu M, Havea R, Patolo S, Nishi M, Hoponoa T, Bourke RM, Dean A, Waqainabete L (2018) Facing food security risks: the rise and rise of the sweet potato in the Pacific Islands. Glob Food Sec 18:48–56

Independent State of Papua New Guinea Ministry of Agriculture and Livestock (2007) National Agriculture Development Plan 2007–2016. Implementation plan: growing the economy through agriculture, vol 2. http://pafpnet.spc.int/policy-bank. Accessed 18 Sept 2018

IPCC (2019) Summary for policymakers. In: Pörtner H-O, Roberts DC, Masson-Delmotte V, Zhai P, Tignor M, Poloczanska E, Mintenbeck K, Nicolai M, Okem A, Petzold J, Rama B, Weyer N (eds) IPCC special report on the ocean and cryosphere in a changing climate (in press)

iTaukei Land Trust Board (2019) Land ownership in Fiji Booklet. https://www.tltb.com.fj/getattachment/Media/Brochures/Land-Ownership-in-Fiji-Booklet-(1).pdf.aspx?lang=en-US. Accessed 15 Feb 2019

Katea T (2016) Tuvalu Progress Report. 16th RAV Tropical Cyclone Committee meeting. Honiara, Solomon Islands. 29 Aug–2 Sept, 2016. https://www.wmo.int/pages/prog/www/tcp/documents/RAV_TCC-16_Tuvalu_ProgressReport.pdf.

Koda K, Tsutomu K, Lorennji R, Robert A, DeBrum H, Lucky J, Paul P (2017) Freshwater lens observation: case study of Laura Island, Majuro atoll, Republic of the Marshall Islands. World Academy of Science, Engineering and Technology. Int J Agric Biosyst Eng 11(2):82–85

Lal R (2013) Influence of Mucuna (Mucuna pruriens) fallow crop on selected soil properties and taro yield in Taveuni, Fiji. Master of Agriculture Thesis. School of Agriculture and Food Technology, University of the South Pacific, Apia, Samoa

Lewis J (1989) Sea level rise: some implications for Tuvalu. Environmentalist 9(4):269–275. https://doi.org/10.1007/BF02241827

Linnerooth-Bayer J, Mechler R (2009) Insurance against losses from natural disasters in developing countries. DESA Working Paper No. 85. ST/ESA/2009/DWP/85, Oct, Austria

Lloyd B (2018) Climate change in Pacific Island countries: a review. Policy Brief No. 20. Toda Peace Institute, Japan

Maeke J (2013) Vulnerability and impacts of climate change on food crops in raised atoll communities: a case study of Bellona community in Solomon Islands. Master of Science Thesis. University of the South Pacific, Fiji

McGregor A, Taylor M, Bourke RM, Lebot V (2016) Vulnerability of staple food crops to climate change. In: Taylor M, McGregor A, Dawson B (eds) Vulnerability of Pacific Island agriculture and forestry to climate change. Pacific Community (SPC), Noumea, pp 161–238

Ministry of Agriculture (2014) Fiji 2020 Agriculture sector policy agenda: modernizing agriculture. Pacific Agriculture Policy Bank. http://pafpnet.spc.int/policy-bank. Accessed 18 Sept 2018

N'Yeurt ADR, Iese V (2015a) Marine plants as a sustainable source of agrifertilizers for Small Island Developing States (SIDS). In: Ganpat WG, Isaac W-A (eds) Impacts of climate change on food security in Small Island Developing States. IGI Global, Hershey, pp 280–311. ISBN: 978-1-4666-6501-9

N'Yeurt ADR, Iese V (2015b) The invasive brown alga *Sargassum polycystum* in Tuvalu, South Pacific: assessment of the bloom and applications to local agriculture and sustainable energy. J Appl Phycol 27(5):2037–2045. https://doi.org/10.1007/s10811-014-0435-y

Nand M (2013) Evaluating the impacts of climate change and climate variability on potato production in Banisogosogo, Fiji. Master of Science Thesis, University of the South Pacific, Fiji

Nand MM, Iese V, Singh U, Wairiu M, Jokhan A, Prakash R (2016) Evaluation of decision support system for agrotechnology transfer SUBSTOR potato model (v4.5) under tropical conditions. South Pac J Nat Appl Sci 34(1):1. https://doi.org/10.1071/SP16001

NMDI (National Minimum Development Indicators) (2018) Pacific Regional Information System. Statistics for Development Division, Secretariat of the Pacific Community. http://www.spc.int/nmdi/agriculture_households. Accessed 19 Sept 2018

OECD (2018a) GeoBook: ODA by sector—bilateral commitments by donor and recipient. https://stats.oecd.org/Index.aspx?DataSetCode=DACGEO. Accessed 15 Sept 2018

OECD (2018b) Aid activities targeting global environment objectives. https://stats.oecd.org/Index.aspx?DataSetCode=RIOMARKERS. Accessed 15 Sept 2018

Quity G (2012) Assessing the ecological impacts of climate change on root crop production in high islands: a case study in Santa Isabel, Solomon Islands, Master of Science Thesis, University of the South Pacific, Fiji

Reliefweb (2018) Solomon Islands: flash floods—April 2014. https://reliefweb.int/disaster/fl-2014-000045-slb. Accessed 22 Sept 2018

Republic of the Marshall Islands (2013) Republic of the Marshall Islands Food Security Policy: for a food secure Marshall Islands. http://pafpnet.spc.int/policy-bank. Accessed 18 Sept 2018

Shigetani M (2009) Federated States of Micronesia: Preliminary Damage Assessment (PDA) High Tide Event, Dec 7–12, 2008

Sisifa A et al (2016) Pacific communities, agriculture and climate change. In: Taylor M, McGregor A, Dawson B (eds) Vulnerability of Pacific Island agriculture and forestry to climate change. Pacific Community (SPC), Noumea, pp 5–46

SPC (2013) Community based climate change vulnerability assessment of the Sabeto catchment: enhancing community adaptation to climate change. In: The field PRA report. Pacific Community, Suva

Taylor M, Lal P, Solofa D, Amit S, Fereti A, Nichol N, Scott G, Starz C (2016) Agriculture and climate change: an overview. In: Taylor M, McGregor A, Dawson B (eds) Vulnerability of Pacific Island agriculture and forestry to climate change. Pacific Community (SPC), Noumea, pp 103–160

Tekinene M (2014) An assessment of the impacts of climate change on cultivated pulaka (Cyrtosperma chamissonis) in Tuvalu, Master of Science Thesis, University of the South Pacific, Fiji

Thaman RR (1995) Urban food gardening in the Pacific Islands: a basis for food security in rapidly urbanizing small-island states. Habitat Intl 19(2):209–224

Thaman RR, Meleisea M, Makasiale J (2002) Agricultural diversity and traditional knowledge as insurance against natural disasters. Pac Health Dialog 9:76–85

The World Bank (2000) Cities, seas, and storms: managing change in Pacific Island economies, Adapting to climate change, vol IV. World Bank, Washington

UNESCO (2017) Towards climate change resilience: minimising loss and damage in Pacific SIDS communities. United Nations Educational, Scientific and Cultural Organization, Paris 07 SP, France

UNFCCC (United Nations Framework Convention on Climate Change) (2017) Issues relating to agriculture. Draft conclusions proposed by the Chair. Addendum. FCCC/SBSTA/2017/L.24/Add.1. Subsidiary Body for Scientific and Technological Advice, 47th session, Bonn, 6–15 Nov

United Nations University (2014) World Risk Report. https://i.unu.edu/media/ehs.unu.edu/news/4070/11895.pdf. Accessed 10 Mar 2019

United Nations University (2016) World Risk Report. https://collections.unu.edu/eserv/UNU:5763/WorldRiskReport2016_small_meta.pdf. Accessed 10 Mar 2019

Wairiu M, Lal M, Iese V (2012) Climate change implications for crop production in Pacific Islands region. In: Aladjadjiyan A (ed) Food production—approaches, challenges and tasks. InTech, London, pp 67–86. https://doi.org/10.5772/1870

Chapter 10
Impacts of Climate Change on Marine Resources in the Pacific Island Region

Johanna E. Johnson, Valerie Allain, Britt Basel, Johann D. Bell, Andrew Chin, Leo X. C. Dutra, Eryn Hooper, David Loubser, Janice Lough, Bradley R. Moore, and Simon Nicol

10.1 Introduction

10.1.1 Physical and Biological Features of the Region

The Pacific Island region encompasses 22 Pacific Island countries and territories (PICTs) that span much of the tropical and subtropical Pacific Ocean. The combined exclusive economic zones (EEZs) of PICTs cover an area of >27 million km^2, but only 2% of their combined jurisdictions is land (see Chaps. 1 and 2). PICTs are therefore often referred to as 'large ocean states'.

J. E. Johnson (✉)
C2O Pacific, Port Vila, Vanuatu

College of Science and Engineering, James Cook University, Townsville, QLD, Australia
e-mail: j.johnson@c2o.net.au

V. Allain
Pacific Community (SPC), Noumea, New Caledonia

B. Basel
C2O Pacific, Port Vila, Vanuatu

Ecothropic, San Cristóbal de las Casas, Chiapas, Mexico

J. D. Bell
Australian National Centre for Ocean Resources and Security, University of Wollongong, Wollongong, NSW, Australia

Conservation International, Arlington, VA, USA

A. Chin
College of Science and Engineering, James Cook University, Townsville, QLD, Australia

L. X. C. Dutra
CSIRO Oceans and Atmosphere Business Unit, Queensland BioSciences Precinct, St Lucia, Brisbane, QLD, Australia

School of Marine Studies, Faculty of Science, Technology and Environment, The University of the South Pacific, Suva, Fiji

© Springer Nature Switzerland AG 2020
L. Kumar (ed.), *Climate Change and Impacts in the Pacific*, Springer Climate, https://doi.org/10.1007/978-3-030-32878-8_10

From an oceanographic and management perspective, the region is dominated by the Western and Central Pacific Ocean (WCPO), which supports vast areas of coastal and oceanic habitats and many species of fish, invertebrates and other animals. Given the scale of this ocean area, it is unsurprising that the region has the greatest dependence on marine resources in the world. These resources, particularly fish and invertebrates, provide a significant source of animal protein, income from artisanal fishing and tourism livelihoods and hold important cultural values for communities in the Pacific Islands region (Sect. 10.2). For example, fish consumption in the Pacific Islands has been a cornerstone of food security; per capita fish consumption in many PICTs is 3–5 times the global average, and, in rural areas, fish often supplies 50–90% of dietary animal protein (Bell et al. 2009; Bell et al. 2018a) The oceanic fisheries resources (mainly tuna) in the EEZs of PICTs also make major contributions to economic development in the region.

10.1.2   Ethnic and Cultural Diversity and Demography

People began to populate the Pacific Islands region approximately 50,000 years ago. Through successive migrations, these early wayfarers eventually established communities across the vast Pacific Ocean (Kayser 2010). Today, these communities exhibit immense cultural and linguistic diversity (Crowley 1999), are commonly governed by customary belief systems and laws that structure society, are the foundation of people's identity and underpin the use and management of natural resources (Hviding 1996).

In recognition of the variation in the physical nature, biogeography, ethnic origin and cultural differences among PICTs, the region is divided into three subregions—Melanesia, Micronesia and Polynesia.

———————————————

E. Hooper
C₂O Pacific, Port Vila, Vanuatu

D. Loubser
Ecosystem Services Ltd, Wellington, New Zealand

J. Lough
Australian Institute of Marine Science (AIMS), Townsville, QLD, Australia

B. R. Moore
Institute for Marine and Antarctic Studies (IMAS), University of Tasmania, Hobart, TAS, Australia

National Institute of Water and Atmospheric Research (NIWA), Nelson, New Zealand

S. Nicol
Pacific Community (SPC), Noumea, New Caledonia

10.2 Importance of Marine Resources to the People in the Pacific

10.2.1 Cultural and Social Importance of Marine Resources

Given that 98% of the jurisdictions of PICTs is ocean, the majority of Pacific Island people share a high dependence on marine resources for food security (as a critical protein source), travel, economic development and as source of income to support livelihoods. This dependence has become deeply intertwined with cultural identity, religious beliefs and social structures (Hviding 1996). For example, in the Trobriand Islands of Papua New Guinea (PNG), origin stories, family relationships and ceremonies revolve around marine life (Ruddle 1998). More generally, there are examples of traditional fisheries management, customary restrictions (taboos), ceremonial fishing practices, the use of teeth and bones of marine animals as power symbols and sacred relationships between individuals and specific marine species that are guided by customary laws (Ruddle and Panayotou 1989; Hyndman 1993; Cinner and Aswani 2007; Whimp 2008; Friedlander et al. 2013; Veitayaki et al. 2014).

However, Pacific Island cultures are increasingly experiencing radical social, economical and political changes. Previously subsistence-based economies are being monetised. Building on the legacy of colonialism, leadership structures and religious authorities are changing as a result of urban migration and the presence of Abrahamic religions. Education systems are being transformed, population pressure on limited resources is rising and the forces of globalisation and technology are increasingly present (Ruddle 1993, 1998). These complex, and largely foreign, drivers of change are continually altering the traditional cultural relationships between communities and marine resources, even while communities continue to be heavily dependent upon the ocean for their wellbeing and livelihoods.

Traditionally, there is a wide range of capacity that communities have developed to cope with natural disasters, particularly cyclones. For example, surveys of women from Tanna Island in Vanuatu after tropical cyclone Pam said that they had known to start fastening down house roofing materials several hours before the cyclone, and there were a number of signs based on 'traditional knowledge' that the cyclone would significantly damage housing and resources (RESCCUE 2015). Many Pacific communities rely on traditional knowledge and other institutions of social capital, local governance, customary marine tenure and self-enforcement capacity to adapt to climate impacts and provide a resilient way forward (Heenan et al. 2015; Johnson et al. 2018).

10.2.2   Importance of Fisheries and Aquaculture for Food Security, Livelihoods and Economic Development

Fisheries are critical for food security, livelihoods and economic development in most PICTs (Bell et al. 2011; Gillett 2016; Gillett and Tauati 2018) (Table 10.1). Across the region, a range of small-scale fisheries target demersal fish and invertebrates associated with coastal habitats, and tuna and other large pelagic fish in nearshore waters, for food security and livelihoods (Pratchett et al. 2011; Johnson et al. 2017; Bell et al. 2018a). In some PICTs (e.g. Fiji, New Caledonia, Tonga and Vanuatu), these fisheries are supplemented by catches of deepwater snapper, grouper and emperors from reef slopes and seamounts. Recent estimates suggest that the total catch of small-scale coastal fisheries in the region is around 164,000 tonnes, worth approximately USD 453 million. These fisheries provide the primary or secondary source of income for an average of 50% of households in many coastal communities (Pinca et al. 2010; Gillett 2016; Bell et al. 2018a).

Industrial purse-seine and longline tuna fisheries make substantial contributions to economies and societies across the region. These fisheries target four species: skipjack tuna, yellowfin tuna, bigeye tuna and South Pacific albacore. Collectively, these industrial fisheries harvest an average of around 1.5 million tonnes of tuna each year from the exclusive economic zones (EEZs) of Pacific Island countries and territories (PICTs) and supply ~30% of the world's tuna (Williams et al. 2017). The licence fees obtained from distant fishing water operating within the EEZs of PICTs make significant contributions to economic development—eight countries and territories receive 10% to >90% of all their government revenue from these licence fees, with five of these countries receiving 45–60% of their revenue in this way (FFA 2016). Approximately 23,000 jobs have also been created through processing tuna onshore, crewing on tuna-fishing vessels and placing observers on purse-seine vessels (FFA 2016). PICTs have also been alerted to the need for tuna to make greater contributions to the supply of fish for local food security (Bell et al. 2015a).

To sustain and enhance the socio-economic benefits of tuna to the region, Pacific Island leaders endorsed a Regional Roadmap for Sustainable Pacific Fisheries in 2015. The Roadmap aims to improve the sustainability of industrial tuna fisheries, increase employment in the sector, add value to the catch and allocate more tuna for domestic food security (FFA and SPC 2015).

10.2.3   Importance of Marine Resources for Tourism

Coral reefs in many PICTs have strong potential to support the development and growth of sustainable tourism, particularly for small-scale ecotourism ventures that can provide income for local communities. This potential is based on the appeal of and access to clear water locations with diverse fish life and a high percentage of coral cover favoured by many tourists (Uyarra 2005).

Table 10.1 Estimates of average fresh fish consumption (kg/person/year) within coastal communities (from Pinca et al. 2010) and annual (2014) catches in tonnes and as percentage of total catch, for categories of coastal fisheries in Pacific Island countries and territories (PICTs)

PICT	Fish consumption (kg/person/year)	Demersal fish						Nearshore pelagic fish						Targeted invertebrates		Subsistence invertebrates		Total catch
		Subsistence		Commercial		Total		Subsistence		Commercial		Total		tonnes	%	tonnes	%	
		tonnes	%	tonnes	%	tonnes	%	tonnes	%	tonnes	%	tonnes	%					
Melanesia																		
Fiji	83.6	10,336	38.3	7070	26.2	17,406	64.5	2400	8.9	3030	11.2	5430	20.1	900	3.3	3264	12.1	27,000
New Caledonia	34.3	1827	37.7	840	17.3	2667	55.0	350	7.2	210	4.3	560	11.5	300	6.2	1323	27.3	4850
PNG	37.7	13,860	33.4	3124	7.5	16,984	40.9	14,000	33.7	2082	5.0	16,082	38.8	1294	3.1	7140	17.2	41,500
Solomon Islands	115.1	10,640	40.2	4317	16.3	14,957	56.5	6000	22.7	1850	7.0	7850	29.7	302	1.1	3360	12.7	26,468
Vanuatu	20.4	1434	36.7	685	17.5	2119	54.2	560	14.3	369	9.4	929	23.8	52	1.3	806	20.6	3906
Micronesia																		
FSM	81.4	2696	51.1	1136	21.5	3832	72.6	605	11.5	561	10.6	1166	22.1	28	0.5	254	4.8	5280
Guam	na	11	10.0	22	18.9	33	28.9	28	24.7	50	43.5	78	68.2	1	0.8	2	2.2	114
Kiribati	111.9	8026	42.2	5917	31.1	13,942	73.4	2280	12.0	1669	8.8	3949	20.8	14	0.1	1094	5.8	19,000
Marshall Islands	111.5	1980	44.0	895	19.9	2875	63.9	750	16.7	596	13.3	1346	29.9	9	0.2	270	6.0	4500
Nauru	52.8	71	19.0	44	11.9	115	30.9	134	36.0	118	31.6	252	67.6	1	0.3	5	1.3	373
CNMI	na	230	46.7	24	4.9	254	51.6	70	14.2	115	23.3	185	37.5	3	0.6	50	10.2	492
Palau	69.5	455	21.5	518	24.5	973	46.0	438	20.7	346	16.3	783	37.0	1	0.0	358	16.9	2115
Polynesia																		
American Samoa	na	68	42.0	29	18.1	97	60.1	36	22.2	13	7.8	49	30.0	0	0.0	16	9.9	162
Cook Islands	78.9	96	22.5	58	13.7	154	36.2	166	38.9	88	20.6	253	59.4	4	1.0	14	3.4	426

(continued)

Table 10.1 (continued)

PICT	Fish consumption (kg/person/year)	Demersal fish						Nearshore pelagic fish						Targeted invertebrates		Subsistence invertebrates		Total catch
		Subsistence		Commercial		Total		Subsistence		Commercial		Total						
		tonnes	%	tonnes	%	tonnes	%	tonnes	%	tonnes	%	tonnes	%	tonnes	%	tonnes	%	
French Polynesia	59.2	1711	21.3	2203	27.5	3914	48.8	212	2.6	3042	38.0	3254	40.6	421	5.2	428	5.3	8016
Niue	33.5	20	12.0	2	1.3	22	13.4	123	74.7	9	5.3	132	80.0	0	0.0	11	6.6	165
Pitcairn Islands	na	4	45.9	1	9.6	5	55.5	1	10.0	2	19.3	3	29.3	0	4.4	1	10.8	9
Samoa	73.8	2160	21.6	1657	16.6	3817	38.2	1000	10.0	2723	27.2	3723	37.2	621	6.2	1840	18.4	10,000
Tokelau	na	126	31.5	14	3.5	140	35.0	198	49.5	22	5.5	220	55.0	4	1.0	36	9.0	400
Tonga	81.2	2052	29.7	2840	41.2	4892	70.9	300	4.3	368	5.3	668	9.7	692	10.0	648	9.4	6900
Tuvalu	142.4	792	55.2	126	8.8	918	64.0	284	19.8	171	11.9	455	31.7	3	0.2	60	4.2	1435
Wallis and Futuna	53.8	510	57.1	135	15.1	645	72.2	68	7.6	15	1.7	83	9.2	69	7.7	97	10.9	894
Total	–	59,104	36.0	31,658	19.3	90,762	55.3	30,001	18.3	17,447	10.6	47,448	28.9	4718	2.9	21,077	12.9	164,005

Adapted from Bell et al. (2018a)

Estimates for subsistence and commercial catches are shown for demersal fish and nearshore pelagic fish; targeted invertebrate fisheries are considered to be 100% commercial

na estimate not available

For PICTs with tourism capacity, healthy reefs with abundant fish life, sharks and rays act as a tourism attraction and provide significant income at national and local levels. For example, in the Great Barrier Reef, sharks are important tourism icons and key attractions (Stoeckl et al. 2010). Shark and ray tourism is likewise important in many PICTs, such as French Polynesia and Fiji (Brunnschweiler and McKenzie 2010; Clua et al. 2011), and could make a substantial contribution to GDP (Vianna et al. 2012) as they offer opportunities for snorkelers and non-swimmers. The non-extractive values of sharks and rays are instrumental in driving shark protection and promoting local ecotourism opportunities. For example, the entire EEZ of Palau has been declared a shark 'sanctuary', and large 'shark parks' have been established in many areas of the Pacific.

Reef tourism is expected to contribute to GDP on the macroscale in the form of employment and foreign exchange inflows (Beeton 2006) and help address food insecurity issues and environmental degradation on the microscale through flow-on benefits from local economic development (Dutra et al. 2011).

However, unless managed well, tourism may have unwanted consequences, such as habitat degradation through construction of infrastructure (e.g. airports, hotels and roads), overfishing to provide meals for tourists, unequal distribution of wealth and changes to local traditions and lifestyle (Dutra et al. 2011; Movono et al. 2018). Such outcomes would undermine the very features on which tourism depends (Hough 1990). Therefore, developing and promoting coral reefs as tourism destinations in PICTs strongly depends on sustainable management to ensure healthy coral reefs and communities, especially given the intricate relationship between resource dependence for food and income and healthy coastal ecosystems.

10.3   Vulnerability of Marine Resources to Climate Change

10.3.1   Impacts of Local Changes in Climate

Marine resources in the tropical Pacific Ocean are strongly influenced by their oceanic environment. This 'marine climate' encompasses water temperatures, salinity, nutrient availability, dissolved oxygen concentrations, aragonite saturation state (a measure of how easily calcifying marine organisms, such as corals, can extract calcium and carbonate ions from seawater to build their skeletons and shells), large- and small-scale water circulation patterns, waves and sea level. The spatial patterns of these features of the ocean vary seasonally and inter-annually. The inter-annual variations are driven primarily by El Niño-Southern Oscillation (ENSO; McPhaden et al. 2006) events (the major source of global inter-annual tropical climate variability with its epicentre in the ocean-atmosphere circulation system of the tropical Pacific), modulated on decadal time scales by the Inter-decadal Pacific Oscillation (IPO) (Power et al. 1999). Due to human interference in the global energy budget, the tropical Pacific marine climate has already warmed by ~1 °C since pre-industrial times. The magnitude of future climate-related changes is strongly dependent on global actions to constrain further additional warming in the oceans to 0.5–1.0 °C.

The most recent Intergovernmental Panel on Climate Change (IPCC) Assessment Report—IPCC-AR5 (2013)—provides strong evidence from observational records that the global ocean surface and subsurface climate is changing (Rhein et al. 2013). For the tropical Pacific, these changes include widespread warming of subsurface waters and sea surface temperatures (SST), especially in the western Pacific, freshening of surface waters in the western equatorial Pacific and under the major atmospheric convergence zones, expansion of the Western Pacific Warm Pool, strengthening of the South Pacific Gyre, increased stratification of the water column which limits vertical supply of nutrients to the surface, decreasing concentrations of dissolved oxygen, lowering of the aragonite saturation state and rising sea level (Cravatte et al. 2009; Durack and Wijffels 2010; Ganachaud et al. 2011; Heron et al. 2016; Lough et al. 2018). These changes are consistent with those expected in a warming climate system with an intensified hydrological cycle.

Based on numerical climate models, substantial changes in tropical Pacific Ocean circulation patterns, SST and vertical temperature structure are projected to continue over the twenty-first century with the magnitude of change related to the global trajectory of greenhouse gas emissions. Likely changes in major currents include weakening of the South Equatorial Current and South Equatorial Counter Current and strengthening of the Equatorial UnderCurrent in the western Pacific. The surface and subsurface oceans will continue to warm, and the Western Pacific Warm Pool will expand further eastwards and become fresher. Nutrient supply to surface waters through upwelling will decrease as stratification increases and the upper ocean mixed layer becomes shallower. Dissolved oxygen concentrations may increase near the equator but are likely to decrease at higher latitudes of the tropical Pacific. There will be continued decline of the aragonite saturation state and increase in sea levels (Ganachaud et al. 2011; BOM and CSIRO 2011, 2014; Lough et al. 2011, 2016; van Hooidonk et al. 2014).

Superimposed on these trends in average tropical Pacific marine climate over the twenty-first century, the region will continue to be impacted by severe weather events and inter-annual modulation of climate. Tropical cyclones are a major source of physical destruction to coastal marine ecosystems, such as coral reefs, seagrass and mangroves. Although there may be fewer tropical cyclones in the future (Christensen et al. 2013), those that do occur are likely to be more intense and destructive (such as Severe TC Pam which impacted Vanuatu in March 2015 and Severe TC Winston which impacted Fiji in February 2016). ENSO events will continue to cause substantial inter-annual disruptions of tropical Pacific climate, modulating SST, tropical cyclone, ocean circulation and rainfall patterns (Lough et al. 2011, 2016). There is mounting evidence that the more extreme El Niño and La Niña events (such as 1997–1998 and 2015–2016) will occur more frequently in the future (Power et al. 2013; Cai et al. 2015).

IPCC predictions expect global mean sea-level rise during the twenty-first century to exceed the rate observed during 1971–2010 for all Representative Concentration Pathway (RCP) scenarios (Table 10.2). This is principally because of

RCP	Minimum (m)	Maximum (m)
RCP2.6	0.26	0.55
RCP4.5	0.32	0.63
RCP6.0	0.33	0.63
RCP8.5	0.45	0.82

Table 10.2 RCP sea-level rise predictions

Source: Church et al. (2013)

increases in ocean warming and loss of mass from glaciers and ice sheets. Local sea-level rises in the Pacific follow the global IPCC predictions (i.e. 70 cm by 2070 and 100 cm by 2100) (Church et al. 2006).

Overall, the marine resources of the tropical Pacific are entering an era of potentially profound changes to the ocean that supports them, and will be increasingly subject to more extreme, and often more destructive, weather events, as described in earlier chapters.

10.4   Impacts of Climate Change on Marine Habitats

10.4.1   Oceanic Habitats and Food Webs

The projected changes in ocean circulation are expected to alter the timing, location and extent of the upwelling processes on which most oceanic primary productivity depends. Changes in the vertical structure of the water masses and in the depth and strength of the thermocline will also impact the availability of nutrients. The production of phytoplankton at the base of oceanic food webs is primarily constrained by the availability of nutrients (e.g. nitrogen), and/or micronutrients (e.g. iron). Because phytoplankton rapidly exhaust the limited nutrients of surface waters, substantial primary production occurs only where deep, nutrient-rich waters are brought to the surface by upwelling and eddies, or when the thermocline becomes shallower and/or weaker, allowing the diffusion of nutrients from the deep nutrient-rich water masses towards the surface (Le Borgne et al. 2011).

In turn, production of organisms at higher trophic levels in the food web (e.g. zooplankton, micronekton, mid-level and top predators) is constrained by variations in phytoplankton production, size structure and composition (Woodworth-Jefcoats et al. 2013) and directly by environmental factors such as temperature and ocean acidification.

A range of studies point to reduced phytoplankton production as the ocean warms in relation to nutrient supply. A 9–33% decrease in phytoplankton in the tropical western Pacific ecological provinces is projected under a high emissions scenario (Le Borgne et al. 2011) with subsequent decrease in zooplankton. Reduction of the phytoplankton biomass is also projected in the north Pacific by 2100, causing a decline in the biomass of zooplankton and all higher trophic level groups. At the global level, a 2–20% decrease in mean primary productivity is projected by 2100 under a high emissions scenario, including in the tropics (Henson et al. 2013; Steinacher et al. 2010).

Important differences in phytoplankton production are expected to occur within the tropical Pacific region (Sarmiento et al. 2004). However, there is still much uncertainty about the nature and extent of these changes (Chavez et al. 2011). For example, the low primary production subtropical gyres in the Pacific could expand by ~30% by 2100, and the productive Pacific Equatorial Divergence could contract by 28% (Polovina et al. 2011). In contrast, other models project increases in subsurface phytoplankton concentrations down to 100 m depth that could offset the decline in surface phytoplankton (Matear et al. 2015), and an increase in primary production as temperature increases (Taucher and Oschlies 2011). The latter projection is based on the positive impact of increased temperature on the microbial loop activity that degrades dying phytoplankton and increases the availability of recycled nutrients for new phytoplankton growth (Behrenfeld 2011). Other studies also demonstrate that the direct effect of temperature on phytoplankton might induce a greater demand for nitrogen which is already in limited supply (Toseland et al. 2013).

Knowledge about the impact of climate change on phytoplankton, zooplankton and micronekton is limited by several factors, including our poor understanding of marine system responses to multifactorial physicochemical climate drivers, the complexity of life cycles and species' interactions, the difficulties in representing all ecosystem functions at multiple levels (organism physiology, populations, communities) and the unknown potential of marine organisms to adapt behaviourally, physiologically, genetically and phenotypically to the unprecedented pace of current climate change (Behrenfeld et al. 2016; Hallegraeff 2010; Petitgas et al. 2012).

10.4.2 Coral Reefs

There is more than 160,000 km^2 of coral reef habitat in the Pacific Island region (Hoegh-Guldberg et al. 2011). Although coral reefs provide valuable ecosystem services in their own right, when they form a habitat mosaic with mangroves and seagrasses, they sustain a greater diversity of organisms and higher fisheries production and protect the coastline against erosion and storms more effectively (Veitayaki et al. 2017; Guannel et al. 2016; Moberg and Folke 1999; Zann 1994). Since the 1970s, the deterioration of coral reefs and associated coastal habitats such as mangroves and seagrasses has been accelerating (Albert et al. 2017; Guannel et al. 2016; Hassenruck et al. 2015). Climate change is currently the strongest driver affecting coral reef dynamics (Aronson and Precht 2016) through higher ocean temperatures, sea-level rise, ocean acidification, more intense storms and cyclones and the synergistic effects that climate drivers have with each other and with non-climate drivers, such as overfishing, sediment runoff and pollution. These impacts are expected to increase as the climate continues to change (Dutra et al. 2018).

In particular, increases in Pacific Ocean sea surface temperatures (SST) are causing widespread impacts on coral reefs due to thermal coral bleaching (Adjeroud
et al. 2009; Cumming et al. 2000; Davies et al. 1997; Hughes et al. 2017; Kleypas
et al. 2015; Lough 2012; Lovell et al. 2004; Mangubhai 2016; Obura and Mangubhai
2011; Rotmann 2001). There are several negative effects associated with mass
bleaching events, including increased reef bioerosion (Chaves-Fonnegra et al.
2018), reduction of coral calcification rates (De'ath et al. 2009; Nurse et al. 2014)
and changes in coral spawning events (Keith et al. 2016). Bleaching also affects
colony size and timing of coral spawning (Paxton et al. 2016), slows swimming
speed of coral larvae and reduces the number of viable recruits (Singh 2018).
Increased SST acts synergistically with increased nutrients and sediment loads to
amplify bleaching effects (Wiedenmann et al. 2012), affecting coral recovery period
after bleaching.

Tropical cyclones are becoming more intense in the Pacific (Elsner et al. 2008),
causing the loss of coral reef and mangrove areas (Guillemot et al. 2010; Johnson
et al. 2016a; Mangubhai 2016; Singh 2018). Flood events associated with cyclones
cause substantial soil erosion, leading to increased sediment and nutrient runoff
onto coral reefs (Guillemot et al. 2010; Levin et al. 2015; Mangubhai 2016; Terry
2007; Veitayaki 2018). Pacific coral reefs stressed by pollution, overfishing and
other climate change stressors may be slower to recover after cyclones (Zann 1994).

Oceans are becoming more acidic as they absorb the excess carbon dioxide from
the atmosphere (Barros and Field 2014; Bates et al. 2014; Dore et al. 2009; IPCC
2014; Johnson et al. 2016b). Ocean acidification has been shown to weaken coral
skeletons, slow coral growth, change the abundance and structure of coral communities (Enochs et al. 2015, 2016; Fabricius et al. 2011), decrease coral diversity and
recruitment (Fabricius et al. 2011) and reduce the abundance of crustose coralline
algae, an inducer of coral larval settlement (Fabricius et al. 2015). These negative
effects on corals most likely facilitate macroalgae growth, causing a shift from a
coral-dominated to algal-dominated state (Enochs et al. 2015).

Weaker reef systems will be far more susceptible to damage from other pressures, including bioerosion, eutrophication, coral disease, intense storms and thermal bleaching as corals become more fragile (Meissner et al. 2012; Nuttall and
Veitayaki 2015; Szmant and Gassman 1990; van Hooidonk et al. 2014). Weakened
coral skeletons due to ocean acidification can trigger stress-response mechanisms,
which affect the rates of tissue repair, skeletal density, feeding rate, reproduction
and early life stage survival (Albright and Mason 2013; Cooper et al. 2008;
D'Angelo et al. 2012; Enochs et al. 2015; Fabry et al. 2008; Kroeker et al. 2010;
Szmant and Gassman 1990).

Sea level has been gradually rising globally and accelerating in the last decades
in some Pacific Islands (Dean and Houston 2013; Jevrejeva et al. 2009; Kench et al.
2018). Sea-level rise effects on Pacific Ocean coral reefs are complex and challenging to predict. In principle, this extra depth provides space for growth that may be
beneficial to corals (van Woesik et al. 2015; Woodroffe and Webster 2014; Saunders
et al. 2016). However, under all climate change scenarios, it is more likely that
there will be more coastal erosion due to rising sea levels (Barros and Field 2014;

Kench et al. 2018), thus increasing turbidity and sedimentation in coastal waters, negatively affecting corals and other reef organisms, at least in the early inundation period until sea level stabilises (Brown et al. 2017a, b; De'ath and Fabricius 2010). Without sustained ecological recovery, very few reefs would be able to keep pace with projected sea-level rise, especially under RCP4.5 or RCP8.5, potentially resulting in reef submergence. This will lead to changes in wave energy regimes, increasing sediment mobility, shoreline change and island overtopping (Kench et al. 2015, 2018; Saunders et al. 2014).

There is also concern that the progressive degradation of coral reefs could increase the incidence of ciguatera fish poisoning and other problems related to toxic algae. The organisms responsible for ciguatera and ciguatera-like symptoms are dinoflagellate microalgae in the genera *Gambierdiscus*, *Prorocentrum* and *Ostreopsis*. These microalgae live as epiphytes on dead coral, turf algae and macroalgae and are ingested by grazing herbivorous fish. The microalgae produce a range of toxins that bioaccumulate through the food chain (Dalzell 1993; Roué et al. 2013). Greater availability of the preferred substrata of these microalgae—dead coral and macroalgae—resulting from increased coral bleaching events and cyclones of greater intensity will likely increase the incidence of ciguatera in the region (Pratchett et al. 2011; Rongo and van Woesik 2013).

10.4.3   Seagrass Meadows

Globally there are over 60 species of seagrasses, with 14 species and 1 subspecies recorded from the tropical Pacific (Ellison 2009). Species richness decreases from west to east with the greatest species richness being found in PNG. Seagrasses are absent or unreported from the Cook Islands, Nauru, Niue, Pitcairn Islands, Tokelau and Tuvalu (Waycott et al. 2011).

Evidence suggests that seagrasses are declining globally, mainly due to anthropogenic impacts (Short and Wyllie-Echeverria 1996; Orth et al. 2006; Waycott et al. 2009). Threats include sediment runoff affecting water quality, construction, dredging and landfill activities. In addition to these direct threats, there are increasing threats from climate change from increasing SST, ocean acidification, more intense storms and cyclones and sea-level rise.

Increasing concentrations of atmospheric CO_2, resulting in equivalent increases in seawater CO_2 levels, have the potential to cause seagrass production to increase. However, increasing CO_2 levels are also likely to increase the production of epiphytic algae on seagrass leaves, which may negatively impact seagrasses through shading and competition (Beer and Koch 1996).

Increases in SST are expected to stress seagrasses, resulting in distribution shifts (Hyndes et al. 2016) and changes in reproduction, growth rates and carbon balance (Short and Coles 2001). When temperatures approach the upper thermal limit for a species, productivity is reduced and eventually the plant will die (Coles et al. 2004). High temperatures can also increase the growth of epiphytes, which can outcompete seagrasses (see above). The thermal tolerance of the different species and their optimum temperature for photosynthesis will influence how they will cope with increased ocean temperatures.

As a result of increasing dissolved CO_2, the pH of seawaters will decrease due to the dissolved CO_2 forming an equilibrium with carbonic acid, which dissociates to add protons to the water which makes the water more acidic. Under expected future increased CO_2 concentrations, ocean acidification could be buffered locally by photosynthesis in seagrass stands (Beer et al. 2006). In turn, this could increase seagrass photosynthesis and productivity (Palacios and Zimmerman 2007).

Increased runoff, nutrient levels and wave power from more severe cyclones are expected to impact seagrasses, reducing photosynthesis and damaging the plants. Shallow-rooted and smaller seagrass species, such as *Halophila* spp., are more likely to be damaged by the increased wave action than deeper-rooted species such as *Enhalus acoroides*. Post cyclone, seagrasses do have the ability to re-establish quickly (Short et al. 2006).

Rising sea levels can adversely impact seagrasses due to increases in water depths, thereby reducing light and reducing photosynthesis and growth. However, increasing turbidity and seawater intrusions on land or into estuaries and rivers could favour landward migration providing there are no barriers to migration (Short and Neckles 1999; Waycott et al. 2011; Saunders et al. 2013).

10.4.4 Mangroves

Typically, mangroves are located along the shore and have a number of ecological roles, including buffering waves to create sheltered environments that support many species of fish and invertebrates (McLeod and Salm 2006). Mangroves also provide feeding areas for the adults of many species of demersal fish, some of which reside on reefs during the day and forage over this range of habitats at night (Nagelkerken and Van der Velde 2004).

Mangrove habitats are currently impacted due to careless land management in coastal catchments and through direct removal to construct coastal infrastructure (Lotze et al. 2006; Waycott et al. 2009). While poorly managed forestry, agriculture and mining operations can deliver toxic pollutants that damage mangrove areas, increased sedimentation from activities in catchments can increase the area of mangrove habitat and reduce the vulnerability of mangroves to SLR (Lovelock et al. 2015).

Although increased CO_2 emissions could enhance the growth of mangroves, sea-level rise is expected to cause significant reductions in their area because the trees cannot tolerate extended immersion in seawater. While some common mangroves, such as *Avicennia marina*, can display a high level of tolerance to waterlogging, responses are extremely variable as they relate to length of time immersed, depth of immersion, salinity, temperature and other environmental variables (Alongi 2015).

Mangroves and other coastal ecosystems trap and vertically deposit sediment, allowing them to raise substrate levels, reduce inundation and maintain conditions for plant growth (Kirwan and Megonigal 2013). This ability to adapt is being compromised however by increasing anthropogenic impacts (e.g. dam building and water extraction for irrigation) on upstream river systems, reducing the delivery of sediments to mangrove habitats (Lovelock et al. 2015) and their ability to deposit substrate. In addition, the pace of SLR is expected to be greater than the ability of mangroves to 'keep up' under mid to high RCPs (Lovelock et al. 2015).

While the physiological impacts of climate change on mangroves are significant, the ecological impacts are also marked. Steep terrain of volcanic islands in the western Pacific, where most mangroves occur (Waycott et al. 2011), will prevent landward migration with sea-level rise. Where the terrain is suitable, rapid sea-level rise could outstrip the capacity of mangroves to migrate. By 2050, the area of mangroves across all PICTs could be reduced by 50% under a high emissions scenario (Waycott et al. 2011). The ecological benefits of maintaining mangroves as a buffer against climate change-related impacts are significant. The protection of coastlines by mangroves against storm surges was well documented during the 2005 boxing day tsunami (Rabinovich et al. 2015) and the same holds true of the protection of coastlines by mangroves during cyclonic events (Marois and Mitsch 2015).

While the global distribution of mangroves is decreasing, due to a variety of anthropogenic causes, there is evidence that climate change is causing a poleward shift in the distribution of mangroves. Poleward migration is extending the latitudinal limits of mangroves due to warmer winters and decreasing frequency of extreme low temperatures in subtropical areas. This may, however, result in a decline in mangrove area, structural complexity and/or in functionality in the tropical Pacific as warming conditions exceed the physiological limits of some mangrove species (Godoy and de Lacerda 2015; Alongi 2015).

The value of mangrove forests goes beyond their value for mitigating the effects of climate change through their ability to sequester large volumes of carbon. Mangroves have a value in supporting the adaptation of coastal communities to the already visible impacts of climate change. Mangroves and, to a lesser extent, seagrass have long played a role in the subsistence economy of many Pacific Island communities. The value of these services has been estimated in a number of studies (Pascal 2014; Warren-Rhodes et al. 2011). In Vanuatu, estimates are that subsistence fisheries and wood collection make up around 14% of the total value of the services provided by mangrove ecosystems (Pascal 2014). And the minimum annual subsistence value from mangroves in the Solomon Islands is estimated to be SBD $2500–10,718/household/year, which represented 38–160% of annual cash incomes (Warren-Rhodes et al. 2011).

Lal (2003) does question the validity of economic valuation of mangroves, particularly in light of ecosystem fragmentation and paucity in understanding of these ecosystems across the Pacific. What is clear, however, in those island states that have mangrove forests, that there is a significant value to the subsistence livelihoods of communities who utilise the goods and services provided by these ecosystems.

The loss of these systems, either by climate change or through development and other pressures, will compromise the resilience of communities to a wider range of environmental stressors than just climate change.

10.5   Impacts of Climate Change on Fish and Invertebrates

10.5.1   Coastal Fisheries

Climate change and ocean acidification are expected to have a range of indirect and direct effects on coastal fish and invertebrates in the region and the fisheries they support. Overall, recent modelling of expected changes in abundance and distribution of demersal fish in the tropical Pacific indicates that significant decreases in production may occur, with declines exceeding 50% under RCP8.5 projected by 2100 (Asch et al. 2017). The effects of increased CO_2 emissions on the productivity of invertebrates in the region have received less attention, but one analysis projects decreases of 5% by 2050 and 10% by 2100 and reductions in quality and size due to reduced aragonite saturation levels (Pratchett et al. 2011).

Indirect Effects

Indirect effects of climate change will occur through the changes to coastal fish habitats (see Sect. 10.3.1). Declines of coral-dependent species are expected, following declines in live coral cover. Increased cover of turf and macroalgae may provide more favourable conditions for some herbivorous species, leading to increases in their abundance, at least in the short term (Johnson et al. 2017). Accordingly, these species are expected to become even more dominant in catches and will become the primary focus for many fishers. Declines in live coral cover and increases in cover of dead coral and algae are expected to result in increased occurrence of ciguatera fish poisoning, furthering localised shortfalls in fish supply (Pratchett et al. 2011; Johnson et al. 2017). Changes in the strength of ocean currents are likely to result in changes to spatial and temporal patterns of larval dispersal and settlement, with impacts on recruitment.

Direct Effects

Increases in SST and associated changes have decreased global fisheries production (Free et al. 2019) and are expected to have significant effects on demersal fish and invertebrates in the Pacific Island region (Munday et al. 2008; Pratchett et al. 2011; Asch et al. 2017; Johnson et al. 2017). For example, projected increases in SST are likely to exceed the optimum thermal levels of many species, as well as alter individual performance, leading to changes in abundance, distribution, growth, reproduction and mortality (Pratchett et al. 2011). Reductions in pelagic larval duration due to increases in SST, combined with altered ocean currents, may reduce connectivity of fish populations among islands. Seasonal changes in temperature may lead to alterations in spawning time for certain species (Munday et al. 2008), potentially resulting in a mismatch between timing of spawning and optimal conditions for larval survival and dispersal.

Ocean acidification is likely to have significant negative impacts on coastal fisheries resources of the Pacific Islands. For demersal fish, increased boldness and activity (Munday et al. 2013), altered auditory preferences (Simpson et al. 2011), loss of lateral movement (Domenici et al. 2012) and impaired olfactory function (Munday et al. 2009; Cripps et al. 2011; Devine et al. 2012) have been observed in individuals raised in elevated CO_2. These changes are expected to alter the homing and settlement success of juveniles and their ability to detect and avoid predators, with implications for connectivity, survivorship and population replenishment. For gastropod and bivalve molluscs and echinoderms, lower aragonite saturation levels are expected to reduce calcification rates, making individuals more vulnerable to predation (Pratchett et al. 2011), leading to declines in the abundance of bivalves and gastropods gleaned for local consumption and the size and quality of products for export.

10.5.2   Oceanic (Tuna) Fisheries

The direct and indirect effects of climate change on tuna are expected to be more difficult to observe than for coral reef fish because climatic variability has strong effects on the distributions of these species (Hobday and Evans 2013). This is particularly the case for skipjack tuna—the locations where the best catches of this species are made in the Western and Central Pacific Ocean (WCPO) can vary by up to 4000 km due to ENSO events (Lehodey et al. 1997). The projected effects of climate change on all four species of tuna are being assessed with a spatial ecosystem and population dynamics model (SEAPODYM) (Lehodey et al. 2008). Preliminary analyses indicate that there will be an eastward and poleward shift in distribution and a reduction in total biomass, for both skipjack and yellowfin tuna under the RCP8.5 emissions scenario (Fig. 10.1) (Lehodey et al. 2013, 2017). At the scale of the EEZs of PICTs, abundances are generally expected to decrease west of 170 °E and increase east of 170 °E. By 2050, the greatest decreases in abundances of skipjack and yellowfin tuna, relative to 2005, are projected to occur for PNG, the Federated States of Micronesia, Nauru and Palau (Bell et al. 2018b). These patterns are expected to persist and intensify by 2100. East of 170 °E, substantial increases in biomass relative to 2005 are projected to occur for skipjack tuna in Vanuatu, New

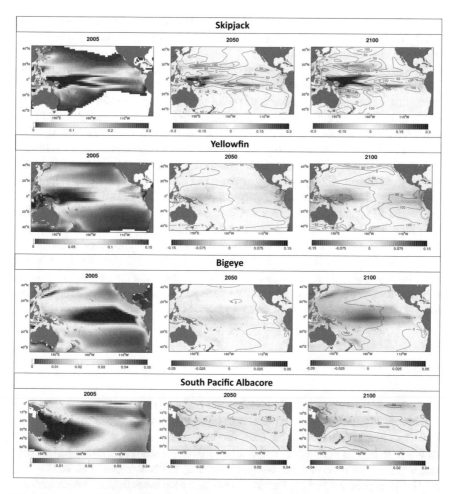

Fig. 10.1 Average historical (2005) distributions of skipjack, yellowfin and bigeye tuna and South Pacific albacore (t/km²) in the tropical Pacific Ocean, and projected changes in biomass of each species relative to 2005 under the RCP8.5 emissions scenario for 2050 and 2100, simulated using SEAPODYM. Isopleths in the projections for 2050 and 2100 represent the relative percentage change in biomass caused by climate change (*Source*: Bell et al. 2018b)

Caledonia, Pitcairn Islands and French Polynesia and for yellowfin tuna in French Polynesia (Bell et al. 2018b).

The responses of bigeye tuna and South Pacific albacore to climate change are expected to differ from the responses by skipjack and yellowfin tuna. Strong decreases in biomass of bigeye tuna are projected to occur in the EEZs of all PICTs, with the declines exceeding 60% in the waters of several PICTs by 2100 (Fig. 10.1). For South Pacific albacore, larvae and juveniles are expected to move poleward towards the Tasman Sea after 2050 (Fig. 10.1), resulting in a decrease in the Coral Sea by 2050. As a consequence, the biomass of adult albacore is projected to decline by ~30%, relative to the year 2000. However, this decline could be reversed after 2080, when the north Tasman Sea is expected to become a spawning area for this species (Lehodey et al. 2015).

The impacts of ocean acidification on tuna and other large pelagic fish species are not yet clearly understood. Preliminary behavioural experiments and simulation modelling indicate that the effects of ocean acidification on mortality of tuna are likely to be lower than the impacts of temperature increases for the expected average increase in ocean acidity by 2100 (Lehodey et al. 2017; Laubenstein et al. 2018; Watson et al. 2018; WCPFC 2018). However, recent projections of ocean acidification suggest that seasonal and spatial variability in acidity levels may have been underestimated previously and that some areas are likely to be more acidic than expected (McNeil and Sasse 2016; Kwiatkowski and Orr 2018). The extremes of the projected variability in ocean acidification (McNeil and Sasse 2016) fall within the range that resulted in mortalities and deformities in experimental trials on yellowfin tuna (Bromhead et al. 2015; Frommel et al. 2016). The areas to the east of 170° where tuna are projected to be less impacted by increasing ocean temperatures coincide with areas where seasonal extremes in ocean acidity are expected to occur. It is possible, therefore, that the quality of areas expected to provide a refuge from temperature stress could be compromised by ocean acidification.

10.5.3 Sharks and Rays

More than 130 species of sharks and rays are believed to occur in the Pacific (Lack and Meere 2009) although this is likely to be an underestimate. Sharks and rays are important in Pacific fisheries, but they also have a wide range of social and cultural values (e.g. Hylton et al. 2017) and tourism values (Vianna et al. 2012) and are crucial for the livelihoods of some communities (e.g. in PNG; Vieira et al. 2017).

The main pressure facing sharks and rays in the Pacific is fishing, which includes large-scale industrial fishing and more localised impacts from small-scale fisheries. However, long-term, accurate data on the catch of sharks in these fisheries are limited, especially from coastal fisheries, and there is great uncertainty about species-specific catch rates, catch fate and trade (Clarke et al. 2006). Furthermore, some species may be subject to additional pressures from habitat loss and degradation, especially when important habitats such as nursery grounds are threatened. There are several indications that sharks are under significant pressure in the Pacific (Lack and Meere 2009; Clarke et al. 2013). As a result, several species of pelagic sharks have been listed in Appendix II of the Convention on International Trade in Endangered Species of Wild Fauna and Flora (CITES).

The diversity of shark and ray species and their associated biological and ecological traits mean that predicted climate change impacts vary greatly among species. A comprehensive risk assessment for tropical sharks and rays identified both direct and indirect pressures and a spectrum of vulnerability across species (Chin et al. 2010). In general, sharks and rays are relatively large and mobile species that are fairly adaptable and resilient. This means they can respond to changing conditions by moving to more favourable conditions and exploit new resources (Chin and Kyne 2007).

Pelagic sharks are highly dependent on food resources and associated oceanographic factors. As such, oceanographic changes and shifts in prey distribution will likely affect pelagic sharks (Chin et al. 2010). Furthermore, increased unpredictability in the timing and location of upwellings could affect provisioning and survival. However, pelagic sharks are highly adaptable and may be able to respond to changing conditions and prey distribution.

Coastal sharks occur on mud and sand flats, reef flats, seagrass beds, coral reefs and reef lagoons. These habitats are highly exposed to climate change impacts such as storms, sea-level rise and significant changes in environmental envelopes (e.g. temperature, salinity, pH). Climate impacts that reduce the availability of suitable habitats and prey may have indirect effects on coastal sharks and rays (Chin et al. 2010). Increased rainfall and runoff and increasing temperatures may cause direct physiological stress that affect shark movement, development and behaviour (Schlaff et al. 2014; Heinrich et al. 2014). Overall, some coastal species may be able to adapt to changing conditions, but changes in local abundance and distribution are likely. Other species with specific habitat requirements may experience declines if those habitats are degraded or lost, especially in spatially small and/or isolated locations that have limited habitat availability. Deepwater sharks are extremely poorly understood, and climate change effects on these species cannot be reliably predicted.

A central factor in predicting climate change effects on sharks and rays in the Pacific Island region is the assumption that many species are able to physiologically tolerate a range of conditions, and when limits are exceeded, relocate to more favourable conditions. Certainly these adaptive behaviours are already widely evident (Schlaff et al. 2014; Chin et al. 2010). Nevertheless, these assessments are largely theoretical and many unknowns remain, such as effects of increased temperature and ocean acidification on shark and ray physiology. Recent experimental data suggest that these changes may have significant impacts on some sharks and rays, affecting body condition, growth, pigmentation and the ability to detect and pursue prey (Gervais et al. 2016; Rosa et al. 2017). However, some species may be tolerant of extreme conditions (Heinrich et al. 2014). The limited available information suggests that sharks and rays exposed to extreme environmental ranges, such as a reef flat specialist like the epaulette shark *Hemiscyllium ocellatum*, may be able to physically tolerate increased temperature and acidity much better than species such as pelagic sharks (Heinrich et al. 2014), but effects on other key traits such as hunting ability are still largely unexplored.

10.6   Impacts of Changes to Marine Resources on Communities and Culture

10.6.1   Impacts of Changes to Marine Resources on Communities and Culture

Given the social and cultural importance of marine resources in the Pacific Island region (Friedlander et al. 2013; Kittinger 2013), the changes described above are causing a spectrum of impacts on communities and cultures.

Coastal communities are highly dependent on marine resources for local subsistence (Bell et al. 2009). A reduction in local catch due to climate impacts results in a reduction of protein sources for local populations, especially those that are not integrated into market economies. For these subsistence-based and traditional communities, a decrease in traditional protein sources leads to the need to acquire other dietary protein (Bell et al. 2009). One strategy to meet this need may be using aquaculture systems or raising poultry and/or livestock for local consumption. In some communities, this change could constitute a disruption in local culture and identity. For example, in some communities on the northern coast of Efate, Vanuatu, community members were resistant to developing these alternative protein sources because of a conflict with their identity as fishermen. An alternative strategy for acquiring protein sources is purchasing protein. For subsistence communities, this strategy signifies entering the market economy and generating cash income. Income generation may come in the form of monetising local resources, which can create added pressure on already vulnerable local marine resources, or emigration to urban centres for gainful employment (Craven 2015).

These impacts create a cascade of other effects on communities and culture. In many parts of the Pacific, men and community leaders are often the individuals who emigrate. In tight-knit and small communities that are common in much of the Pacific, the temporary or permanent absence of these individuals leaves a gap in leadership structures and social networks, leading to a breakdown or dilution of local governance, traditional knowledge and social cohesion (Craven 2015). Emigration also results in greater external influences in communities (Maron and Connell 2008), which can displace traditional leadership and affect customary beliefs, culture and marine resource management.

The shift to cash-based rather than subsistence-based economies alters traditional diets, which are linked to cultural identity. Changes in diet are often accompanied by negative health impacts, including diabetes, resulting from increased consumption of sugar and processed foods (Evans et al. 2001). The increased use, or introduction, of alcohol is also common and can result in further social impacts including possible increases in domestic violence (Leonard 2001; WHO 2005; Livingston 2011).

Consequentially, impacts of changes to marine resources reverberate throughout Pacific Island communities, including the loss or dilution of traditional knowledge, local practices, social cohesion and values. In the context of climate adaptation, this is especially significant because of the important role of traditional and local knowledge in assisting communities in adapting to climate-related challenges (Adger et al. 2013).

The result is a reduction in community resilience. In light of climate-related challenges, actively working to increase community resilience will be of escalating importance. Additionally, actively increasing the resilience of communities that depend on and manage marine resources is an important component for increasing the resilience of the marine ecosystems that they locally manage (Berkes et al. 2000; Kittinger 2013). If approached appropriately, implementing adaptation actions to increase community resilience can also contribute to the resilience of marine resources to climate-related impacts (Adger et al. 2013).

10.6.2 Impacts of Changes to Marine Resources on Tourism and Aquaculture

Climate change will affect many tourism sectors from reef-based activities to cruise ships, large resorts, ecotourism ventures and sports fishing. This will be in the form of direct impacts of sea-level rise and more intense storms damaging coastal infrastructure including resorts and maritime facilities (Guannel et al. 2016) and reducing visitation indirectly due to degradation of coastlines and reef habitats (War Sajjad et al. 2014). In addition, climate change is expected to negatively affect aesthetics, cultural connections between traditional communities and their marine environments, spiritual values and other ecosystem services that contribute to industries and human wellbeing (Hughes et al. 2017; Johnson et al. 2016b).

Aquaculture is an important industry in many PICTs and includes farming marine and coastal species to support livelihoods, growing freshwater species for local food security and the hatchery rearing of juveniles for restocking programmes to replenish depleted fisheries for high-value marine invertebrates (Bell et al. 2005; Pickering et al. 2011). The main marine and coastal species grown to produce commodities for export or sale to lucrative local markets are black-lipped pearl oysters (*Pinctada margaritifera*) in French Polynesia, Cook Islands and Fiji, and penaeid shrimps in New Caledonia. The culture of seaweed (*Kappaphycus alvarezii*) has also been established for the benefit of marginalised coastal communities in the outer-island provinces of Fiji, Kiribati, PNG and Solomon Islands. Nile tilapia (*Oreochromis niloticus*) is grown for local food in inland and coastal locations, mainly in PNG, Fiji and Vanuatu and to a lesser extent in the Cook Islands, Samoa, American Samoa, Guam, Saipan and Northern Marianas. Milkfish (*Chanos chanos*) is also produced for this purpose but at a much smaller scale (Johnson et al. 2017).

In 2010, the value of aquaculture production in the Pacific Island region was estimated to be USD 200–250 million (Ponia 2010). Farming of black pearls, shrimp and seaweed provides thousands of people with full-time or part-time work (Bell et al. 2011; Pickering et al. 2011). Both coastal and freshwater aquaculture in the tropical Pacific are expected to be affected by climate change, particularly commodities with calcareous shells that will be impacted by ocean acidification. Under a high emission scenario, coastal aquaculture is projected to be directly vulnerable to increasing rainfall and cyclone intensity, higher SST, ocean acidification and sea-level rise and indirectly vulnerable to changes in supporting habitats (Pickering et al. 2011). Climate change is also expected to affect the viability of coastal aquaculture enterprises due to (1) greater temperature stress likely to increase the vulnerability of several species to pathogens and parasites (Yukihira et al. 2000; Pouvreau and Prasil 2001), or harmful algal blooms, and (2) effects of global warming on the production of fishmeal elsewhere in the world that is likely to increase the cost of high-protein formulated feeds for carnivorous species, affecting the economic viability of fish and shrimp farming.

Climate change may also affect future opportunities for development of the non-extractive industries, e.g. catch and release sports fishing and ecotourism, being planned to take pressure off coastal fisheries. For PICTs with tourism capacity, healthy marine ecosystems are necessary to deliver tourism attractions and provide income at national and local levels. Climate change impacts on marine habitats and species will affect the opportunities for new tourism businesses and the success of current ventures.

10.6.3   Impacts on Economic Development and Government Revenue

Redistribution of skipjack and yellowfin tuna (Fig. 10.1) is expected to result in lower catches across the prime fishing grounds by 2050. The reduced catches could affect licence revenues from purse-seine fishing and potentially the plans to increase employment based on fish processing if it becomes more difficult to deliver the tuna required by canneries in PNG and Solomon Islands to operate efficiently (Bell et al. 2018b). Other possible negative impacts on economic development may occur from the eastward redistribution of bigeye tuna and poleward movement of South Pacific albacore (Fig. 10.1) if a greater proportion of longline fishing eventually occurs outside the EEZs of PICTs.

On the other hand, the projected eastward redistribution of skipjack and yellowfin tuna due to climate change could result in opportunities for PICTs in the eastern WCPO, e.g. French Polynesia, and PICTs in the subtropics, e.g. Vanuatu and Fiji, to obtain increased economic benefits. However, although the percentage increases in catch could be substantial in these EEZs, the scale of benefits is likely to be modest because present-day catches are low (Bell et al. 2018b).

10.7   Adaptation Options

10.7.1   Community-Based Adaptation and Resilience

Given the impacts previously discussed, the present and future wellbeing of many Pacific Island communities depends on reducing community vulnerability and increasing resilience. Resilience is the capacity of a human or natural system to endure the impacts of a stress (e.g. a cyclone or an economic crisis) and adapt, with the potential of the system transforming into something new or stronger (Béné et al. 2012; Folke et al. 2002; Frankenberger et al. 2014). Resilience can be fortified through adaptation planning and implementation in accordance with the specific characteristics (including geography, governance and culture) of the community in question (Adger et al. 2013).

Community-based adaptation (CBA) refers to participatory processes that involve the local population in all levels of adaptation planning and implementation. As a result, adaptation planning is directed by the community members according to local knowledge, priorities and cultural values while being complemented by climate and natural resource management science (Reid 2015; Forsyth 2017).

Consequentially, CBA processes also reinforce local capacity, traditional knowledge and governance structures. Simultaneously CBA increases the likelihood of positive adaptation outcomes by empowering resource managers rather than increasing community dependence on external aid (Aalst et al. 2008; Berkes et al. 2000). CBA often incorporates the low-cost and low-tech adaptation strategy of ecosystem-based adaption (EBA), which increases resilience by protecting and strengthening ecosystem services and biodiversity (CBD 2009; Reid 2015). The combination results in a place-specific and locally driven socio-ecological approach to adaptation.

In recent years, the scientific and development community has progressed a variety of community-based vulnerability assessment and adaptation planning tools (McLeod et al. 2015). The CBA tool *Adapting to a Changing Climate: Guide to Local Early Action Planning* (LEAP) *and Management Planning* was developed by the Micronesia Conservation Trust in a collaborative process among scientific experts, resource managers, conservation practitioners and community members. It has been adapted over time, based on the experience of practitioners and field tests (Gombos et al. 2013).

The LEAP is particularly appropriate for small communities that have control over the governance of their local natural resources, high dependence on natural resources and limited economic opportunities. Combined with a low cost for implementation per community (McLeod et al. 2015), the LEAP is a suitable CBA tool for the socio-economic and cultural characteristics of many communities in the Pacific Island region. The tool has been successfully implemented in Solomon Islands, PNG, Timor-Leste, the Philippines, Indonesia and Malaysia (Wongbusarakum et al. 2015; Jolis et al. 2014), among other locations including the Caribbean nation of Cuba (Basel et al. 2018).

The LEAP builds awareness about climate change impacts by explaining climate science within the context of local experience. The process also engages communities in a participative planning process based on local knowledge, skills and resources, with the support of additional technical expertise as necessary. The four stages of the LEAP process are: (1) getting ready for raising awareness and planning, (2) understanding climate change and your climate story, (3) carrying out a field-based threat and vulnerability assessment and (4) finalising your Local Early Action Planning (Gombos et al. 2013).

The cross-sectorial approach of the LEAP addresses both social and ecological factors, allowing for integrated adaptation actions. Instead of only assessing marine resources and management, the tool includes all social and natural resources the community depends upon. These include terrestrial resources whose management has direct impacts (e.g. contamination) and indirect impacts (e.g. resource pressure and over-extraction) on marine resources. The result is a comprehensive planning process designed to maximise benefits and avoid maladaptation or unintended consequences that could otherwise result from adaptation actions (Wongbusarakum et al. 2015).

For any CBA tool to be effective, including the LEAP, it is important that programmes and facilitators account for inherent structural inequalities in communities, including but not limited to gender, land tenure, sustainable livelihood options and possible positive or negative repercussions of adaptation actions (Forsyth 2017). It is also important for adaptation actions to be supported by regional and national policies and programmes (Reid 2015) and to take into account potential political, economic and social drivers of vulnerability (Forsyth 2017). For the LEAP, this signifies the need for an appropriately experienced facilitation team and the inclusion of these considerations in programme planning and implementation. Field experience indicates that the tool is most effective when (1) adapted to the specific community educational and cultural context, (2) enhanced by the ample use of experiential learning activities and (3) used in conjunction with 'Semi-Quantitative Assessment of Vulnerability to Climate Change' (Johnson et al. 2016c; Basel et al. 2018).

10.7.2 Adaptations for Food Security

A range of practical adaptations have been proposed for maintaining the important role fish plays in food security in the Pacific Island region (SPC 2008; Bell et al. 2009, 2011, 2018a). Summarised below, these adaptations can be categorised into two broad groups: those that focus on minimising the gap between sustainable harvests from coastal fish habitats and the quantities of fish recommended for good nutrition of growing human populations and those that focus on filling the gap (for a full description, see Bell et al. 2018a).

Adaptation Options to Minimise the Gap

1. *Manage and restore catchment vegetation*, through maintaining good coverage of vegetation on slopes and wide riparian buffer zones to reduce damage to coastal habitats through increased turbidity, sedimentation and eutrophication from runoff and erosion.
2. *Minimise other degradation of coastal habitats*, including through maintaining water quality by controlling pollution from sewage, chemicals and waste and eliminating activities that reduce the structural complexity and extent of habitats, such as destructive fishing practices or excessive harvesting of mangrove timber for firewood.
3. *Provide for landward migration of mangrove habitats*, by preventing infrastructure development in low-lying areas, allowing for inundation of low-lying areas suitable for mangrove colonisation, and planting young trees to fast-track mangrove establishment.
4. *Sustain production of coastal demersal fish and invertebrates*, by strengthening of community-based ecosystem approaches founded on 'primary' fisheries management practices and simple harvest controls, such as size limits, spatial and temporal closures and gear restrictions (Cochrane et al. 2011).
5. *Maximise the efficiency of spatial management*, by ensuring areas protected from fishing are designed to account for the ecology of target fish species and that habitat mosaics and migration corridors important for connectivity are preserved.
6. *Diversify catches of coastal demersal fish*, by transferring effort to species expected to increase in local abundance with climate change while ensuring production is maintained within sustainable bounds.

Adaptations to Fill the Gap

1. *Transfer coastal fishing effort from demersal fish to nearshore pelagic fish*, particularly tuna, through the deployment of nearshore fish aggregation devices (FADs).
2. *Expand fisheries for small pelagic species,* such as mackerel, anchovies, pilchards, sardines, scads and fusiliers, through use of alternative fishing technologies (e.g. the 'bagan' fishing platform used in Southeast Asia) at sustainable levels.
3. *Extend the shelf life of fish harvests*, by training communities in how to improve traditional methods for smoke curing, salting and drying of small pelagic fish.
4. Increase access to small tuna and by-catch offloaded by industrial fleets during transhipping operations.

10.7.3   Adaptations for Livelihoods

Coral reef degradation directly affects the livelihoods of Pacific Islanders through reduced local income and inflow of foreign exchange due to the decline of fisheries and associated increase in food insecurity (Zeller et al. 2015; Bell et al. 2018a). Habitat degradation negatively affects reef-based tourism (War Sajjad et al. 2014), increases the risk of property damage due to reduced coastal protection provided by coral reefs (Guannel et al. 2016) and promotes substantial negative changes on aesthetics, cultural connections between Pacific Islanders and their marine environments, spiritual values and other ecosystem services that contribute to human wellbeing (Johnson et al. 2016b; Hughes et al. 2017).

Many of the adaptations described in Sect. 10.4.2 will also assist coastal communities to continue to engage in catching and selling fish at local and urban markets to support their needs for cash income. As human populations continue to grow, the only marine resources capable of supporting more livelihoods will be the oceanic fish species. Expanding the use of nearshore FADs to make them part of national infrastructure for both food security and income generation will be an essential adaptation for harnessing the potential to create more livelihoods for coastal communities (Bell et al. 2015a, b).

There is scope for marine aquaculture to provide more livelihoods, for example, through the culture of wild-caught milkfish in atoll nations and the coastal areas of high islands (Pickering et al. 2011). However, on high islands, the availability of freshwater and the simplicity of farming Nile tilapia mean that the most practical way of generating income through aquaculture in Melanesia will be in small ponds (Johnson et al. 2017).

While traditional systems of resources management implemented in many PICTs have historically been effective at protecting habitats and maintaining fisheries because the harvest of marine resources is strongly controlled, accelerated population growth and increasing resource demand are undermining traditional management systems. Therefore, traditional management needs to be supported by more formal sustainable management to address these existing pressures and the projected effects of climate change in the future. For example, marine protected areas (MPAs) are commonly applied but are rarely compared to other management options or assessed for their cost-effectiveness or feasibility. The Noumea Strategy (SPC 2015) clearly states that over-reliance on site-based approaches, such as MPAs, is unlikely to achieve widespread goals of fisheries management and hence proposes other ecosystem-based fundamental approaches. This creates positive flow-on effects on reef-related livelihoods, such as fisheries and tourism, which can be greatly improved when local communities are engaged in the development and adoption of conservation measures in line with traditional governance structures and values.

Increased tourism can benefit communities under stress from climate change but only if managed appropriately. The economic benefits expected from tourism may not reach local communities as anticipated where the benefits rely on intermediaries, such as travel agents and airlines. In such situations, people from remote communities have little control over the industry from which they hope to benefit (Rugendyke and Connell 2008).Therefore, tourism initiatives in PICTs should be planned to enhance not only national economic and employment benefits but also the wellbeing of the communities that host the activities.

Any intervention in PICTs looking into improving livelihoods needs to consider local development aspirations, potential social, economic and environmental costs and benefits, local dynamics of village governance, social rules and protocols, and traditional forms of knowledge that can inform long-term solutions (Remling and Veitayaki 2016).

Another important factor is that tenure and associated political systems differ substantially across PICTs (Aswani et al. 2017). As a result, it is not possible to apply measures and strategies uniformly across the Pacific. Local environmental, social and governance contexts must be considered when implementing adaptation programmes (Dutra et al. 2018).

Building capacity is essential for improving local understanding about the complexities involved in climate change and adaptation, as well as for helping communities prepare for the future. Such capacity should be built around practical discussions, for example, how to manage receding shorelines and processes to rehabilitate coastal habitats and protect local forests, water catchment areas and food sources. Coastal communities in PICTs understand that only a healthy environment surrounding coral reef systems can support their basic needs for food, income and clean water in the long term, and understand the benefits of healthy ecosystems as a buffer for climate change (Remling and Veitayaki 2016; Veitayaki in press; Heenan et al. 2015).

10.7.4 Economic Development and Government Revenue

Two main types of adaptations could assist PICTs in the central and eastern areas of the WCPO to harness greater economic benefits from the projected eastward migration of skipjack and yellowfin tuna, and PICTs in the western areas of the WCPO to reduce the potentially negative implications for their economies. The first involves development of flexible management measures to allow fishing effort to shift east, while ensuring that large quantities of tuna can still be channelled through the established and proposed fish processing operations in the west. The second is optimising the productivity and value of tuna resources across the region.

The vessel day scheme (VDS) (Havice 2013) for managing purse-seine fishing effort across the EEZs of the eight Pacific Islands countries—the Parties to the Nauru Agreement (PNA)—that yield ~30% of the world's tuna already provides much of the flexibility to maintain the socio-economic benefits that PICTs receive from skipjack tuna as it responds to climate variability and climate change (Bell et al. 2011). The VDS allows licence revenues to be shared among PNA member countries regardless of the effects of ENSO phase on the best locations for catching this species. The VDS will also adjust the number of fishing days that PICTs can sell to foreign fleets as climate change alters the distribution of skipjack tuna. The global sourcing provisions of the Interim Economic Partnership Agreement that PNG has with the European Commission also allows fish to be delivered to the nation's canneries, regardless of where they are caught.

Finding ways to add more value to skipjack tuna would allow PICTs to offset the consequences of lower projected catches caused by climate change. Value-adding would create the opportunity to increase licence fees, helping PICTs to obtain more government revenue in the short term, and to maintain the present-day contributions of licence fees to economies when abundances of skipjack tuna decline due to climate change.

Similarly, investments in seasonal forecasting tools may assist industry and fisheries administrations with balancing the immediate consequences of climate variability and change (Hobday et al. 2018). Such tools would allow industry to plan and prepare for reduced access to resources in poor years (e.g. diversify processing operations to add more value to reduced catches) and maximise access and opportunities in good years. Forecasting used in this manner has the potential to increase the resilience of industries, and enable them to remain economically viable when operational circumstances are affected by climate change.

At the broader level of regional management, arrangements are needed to optimise sustainable catches both now and as the distributions of tuna stocks change. For this to happen, the Western and Central Pacific Commission (WCPFC) needs to negotiate the adoption of adequate conservation and management measures that reduce fishing mortality to sustainable levels across all the EEZs and high seas (Hanich and Tsamenyi 2014). New approaches, based on implementing decision-rules that transparently and equitably distribute the conservation burden in accordance with pre-agreed principles are required (Hanich and Ota 2013). It will also be important to limit investments in vessels operated by distant water nations fishing on the high seas (which are not covered by the VDS), and effort creep by such vessels (McIlgorm 2010).

Ultimately, the eastward redistribution of tuna may make it necessary to consider an amalgamation of WCPFC and the Inter-American Tropical Tuna Commission, which is the regional fisheries management organisation for the Eastern Pacific Ocean.

10.8   Future Research

Priority knowledge gaps to better increase our understanding of the effects of climate change on *coastal resources*, particularly coastal fisheries, and the consequences for food security and livelihoods include:

1. Monitoring shifts in the distributions of species, and the effects of climate-induced changes in species composition on ecosystems, by establishing long-

term monitoring at specific sites to measure distributional shifts and biological impacts as well as environmental conditions, such as pH, SST and habitat condition.

2. Further examination of the synergistic effects of increasing ocean acidification, SST and other anthropogenic stressors, on the biology and ecology of demersal fish and invertebrates, and the ability of target fisheries species to adapt to these changes.

3. Undertaking a cost-benefit analysis of the key adaptation options for food security to inform sustainable and adaptive management, noting that a holistic approach needs to incorporate a range of adaptation tools.

Dedicated sampling programmes will be needed to monitor the effects of climate change on key coastal habitats, coral reefs and target species, and the fisheries they support, to provide information for adaptive management. These programmes will require an experimental approach that controls for the effects of other stressors, such as fishing pressure, poor land management practices and pollution (SPC 2013). Importantly, the status of coastal resources in PICTs is either uncertain or impacted due to rapidly growing human populations and other pressures, such as coastal development. Therefore, there is an urgent need to better understand the most effective options for sustainable management of coastal resources to address these existing pressures recognising that the expected effects of climate change will potentially exacerbate these impacts. For example, marine protected areas (MPAs) are commonly applied but are rarely compared to other management options or assessed for their cost-effectiveness or feasibility. The Noumea Strategy (SPC 2015) clearly states that over-reliance on site-based approaches, such as MPAs, is unlikely to achieve widespread goals of fisheries management and hence proposes other ecosystem-based fundamental approaches.

Much of the research needed to improve knowledge of the projected effects of climate change on *tuna fisheries* in the WCPO centres around strengthening SEAPODYM. In particular, research is needed to improve the biogeochemical component of SEAPODYM and estimates of future fishing effort, which need to be coupled to projections from an ensemble of global climate models to estimate catches of the four species of tuna under various climate change scenarios (Lehodey et al. 2011). Increased access to operational fisheries data is also needed to validate SEAPODYM—the more closely the resolution of the industrial fisheries catch data matches the resolution of the environmental data, the better the predictive performance of the SEAPODYM model.

Another important gap in knowledge is the spatial structure of stocks (i.e. number of separate self-replenishing populations) within the ranges of the four tropical tuna species. Recent analysis of conventional tagging data indicates that there could be at least three separate stocks of bigeye tuna across the tropical Pacific Ocean, with the possibility of up to another six stocks (Schaefer et al. 2015). In addition, recent genetic analyses of the population structure of yellowfin tuna indicate the potential for separate stocks between the western Pacific (Australia) and central Pacific (Tokelau) (Grewe et al. 2015). Furthermore, archival tagging studies indicate that the maximum displacement of an individual yellowfin tuna was ~1350 km (Schaefer et al. 2011). Once the spatial stock structure of each tuna species has been identified, SEAPODYM can be used to model the response of each separate tuna stock to climate change and ocean acidification. The finer-scale understanding of tuna stock structure will improve stock assessment and enable regional fisheries managers to identify which countries share each stock and how much of each stock occurs in high seas areas.

Sharks and rays in many areas of the Pacific are already heavily impacted by human activities, and it is difficult to disentangle climate change impacts from existing threats and impacts. Nevertheless, securing the future of the Pacific's sharks and rays in a changing climate would be aided by the following targeted research:

1. A systematic review of the diversity and status of sharks and rays in each PICT to describe biodiversity and identify key threats
2. For highest-risk species, targeted interdisciplinary research to identify, trial and evaluate management options
3. Improved data on the extent of the most significant existing threats, i.e. fishing, specifically, improved data on catch composition and fate, risk assessment and improved stock assessment, especially for small-scale fisheries that are not well studied and generally undervalued

The challenges facing sharks and rays are complex due to their biological and ecological diversity, the dearth of information for many species, the range of uses and values and the interactions they have with fisheries and communities. This means that case-by-case research and conservation programmes need to be devised to tailor management and conservation actions to specific contexts (Dulvy et al. 2017).

10.9 Conclusions

The greatest challenges for Pacific people are likely to be from sea-level rise, loss of coastal habitats, declining productivity of demersal fish and invertebrates and an eastward shift in distribution of some tuna species as a result of climate change. Changes in ocean temperature and water circulation are expected to impact coastal ecosystems, reducing fish and invertebrate productivity. Warming temperatures and ocean acidification could also affect dispersal and settlement of larvae, affecting colonisation and connections between coral reef and seagrass areas, fish behaviour and growth. Habitat declines, particularly of coral reefs, are already occurring and will impact on the species, communities and industries that depend on these ecosystems. Impacts will be complex, widespread and difficult to accurately predict.

Invertebrates such as trochus, green snails and pearl oysters are likely to decline as lower pH weakens their shells, reduces growth and causes higher mortality. Climate change effects on coastal fisheries will largely be due to the indirect impacts of changes in the extent and condition of coastal fish habitats. Resulting declines in coastal fish and invertebrate populations will widen the gap between fish needed by growing human populations and sustainable harvests, with shortages expected in some Pacific nations (e.g. Papua New Guinea, Solomon Islands) by 2035 and ecosystem-based fisheries management to support sustainable fisheries stocks. Alternative incomes will be needed where fishing operations are negatively affected.

The projected slow-onset declines in coastal fisheries productivity due to climate change have important implications for the food security of PICTs. The magnitude of the loss and damage and whether it can be mitigated depend on the severity of coral reef decline (that provides much of the coastal fisheries production across the region), as well as population growth, the area of coral reef per capita and the distance of reefs from population centres.

Adaptations for maintaining the important role of fish for food security in the region (SPC 2008; Bell et al. 2011) centre on minimising the gap between the quantities of fish required for good nutrition and the fish available from coastal habitats due to population growth and productivity declines (1) using appropriate best management of coastal fish habitats and stocks, (2) increasing access to tuna for rural and urban populations and (3) boosting pond aquaculture. The implications of the projected changes in production of coastal fisheries and aquaculture for sustainable livelihoods are that (1) livelihoods may need to switch from one resource to another and (2) more flexible arrangements may be needed for operating fishing and aquaculture ventures.

The four species of tropical tuna are expected to have relatively low vulnerability to the projected physical and chemical changes to the WCPO and to alterations in oceanic food webs, because they can move to areas with their preferred temperature conditions (Bell et al. 2013, 2018b). This conclusion is tempered, however, by the unknown effects of ocean acidification, which may comprise the more favourable temperature conditions for tuna expected to occur further to the east. The projected changes in distribution and decreased productivity of tuna underscore the need for effective management. The small national economies with a high dependence on licence fees are likely to be vulnerable to these changes by 2050. It is possible, however, that the plans to improve the value of tuna in the Regional Roadmap for Sustainable Pacific Fisheries could maintain existing levels of government revenue from licence fees even though catches decline.

Addressing the implications of climate change for Pacific Island nations requires resourcing as well as financial commitment for effective implementation. Inadequate resourcing has been an ongoing issue for local PICT capacity to implement climate change adaptation and mitigation actions, something that is needed for resilient coastal ecosystems and communities. In many cases, this will require the development and implementation of basic but robust management systems and will also require significant education and awareness-raising as well as enforcement at all levels—from government to individual communities.

References

Aalst MK, Cannon T, Burton I (2008) Community level adaptation to climate change: the potential role of participatory community risk assessment. Glob Environ Change 18(1):165–179. https://doi.org/10.1016/j.gloenvcha.2007.06.002

Adger WN, Barnett J, Brown K, Marshall N, O'Brien K (2013) Cultural dimensions of climate change impacts and adaptation. Nat Clim Change 3(2):112

Adjeroud M, Michonneau F, Edmunds PJ, Chancerelle Y, de Loma TL, Penin L, Thibaut L, Vidal-Dupiol J, Salvat B, Galzin R (2009) Recurrent disturbances, recovery trajectories, and resilience of coral assemblages on a South Central Pacific reef. Coral Reefs 28:775–780

Albert S, Saunders MI, Roelfsema CM, Leon JX, Johnstone E, Mackenzie JR, Hoegh-Guldberg O, Grinham AR, Phinn SR, Duke NC, Mumby PJ, Kovacs E, Woodroffe CD (2017) Winners and losers as mangrove, coral and seagrass ecosystems respond to sea-level rise in Solomon Islands. Environ Res Lett 12:094009

Albright R, Mason B (2013) Projected near-future levels of temperature and pCO_2 reduce coral fertilization success. PLoS One 8:e56468

Alongi DM (2015) The impact of climate change on mangrove forests. Curr Clim Change Rep 1(1):30–39

Aronson RB, Precht WF (2016) Physical and biological drivers of coral-reef dynamics. Coral Reefs World 6:261–275

Asch RG, Cheung WWL, Reygondeau G (2017) Future marine ecosystem drivers, biodiversity, and fisheries maximum catch potential in Pacific Island countries and territories under climate change. Mar Policy 88:285–294

Aswani S, Albert S, Love M (2017) One size does not fit all: critical insights for effective community-based resource management in Melanesia. Mar Policy 81:381–391

Barros V, Field C (2014) Climate change 2014 impacts, adaptation, and vulnerability part A: global and sectoral aspects working group II contribution to the fifth assessment report of the Intergovernmental Panel on Climate Change Preface. Climate change 2014: impacts, adaptation, and vulnerability, pt A: global and sectoral aspects, pp Ix–Xi

Basel B, Fernández M, Correa López P (2018) Informe: Fuertes y Preparados Frente al Cambio Climático en el Naranjo del Toa, 15 al 17 de enero del 2018. Unidad de Servicios Ambientales Alejandro de Humboldt, Departamento de Conservación, Ministerio de Ciencia, Tecnología, y Medio Ambiente, Baracoa

Bates NR, Astor YM, Church MJ, Currie K, Dore JE, Gonzalez-Davila M, Lorenzoni L, Muller-Karger F, Olafsson J, Santana-Casiano JM (2014) A time-series view of changing surface ocean chemistry due to ocean uptake of anthropogenic CO_2 and ocean acidification. Oceanography 27:126–141

Beer S, Koch E (1996) Photosynthesis of seagrasses vs. marine macroalgae in globally changing CO_2 environments. Mar Ecol Prog Ser 141:199–204

Beer S, Mtolera M, Lyimo T, Björk M (2006) The photosynthetic performance of the tropical seagrass Halophila ovalis in the upper intertidal. Aquat Bot 84:367–371

Beeton S (2006) Community development through tourism. Landlinks Press, Canberra

Behrenfeld MJ, O'Malley RT, Boss ES, Westberry TK, Graff JR, Halsey KH, Milligan AJ, Siegel DA, Brown MB (2016) Revaluating ocean warming impacts on global phytoplankton. Nat Clim Change 6:323–330. https://doi.org/10.1038/nclimate2838

Behrenfeld M (2011) Biology: uncertain future for ocean algae. Nat Clim Chang 1:33–34. https://doi.org/10.1038/nclimate1069

Bell JD, Rothlisberg PC, Munro JL, Loneragan NR, Nash WJ, Ward RD, Andrew NL (2005) Restocking and stock enhancement of marine invertebrate fisheries. Adv Mar Biol 49:1–370

Bell JD, Kronen M, Vunisea A, Nash WJ, Keeble G, Demmke A, Pontifex S, Andréfouët S (2009) Planning the use of fish for food security in the Pacific. Mar Policy 33:64–76

Bell JD, Johnson JE, Hobday AJ (eds) (2011) Vulnerability of tropical Pacific fisheries and aquaculture to climate change. Secretariat of the Pacific Community, Noumea, 925pp

Bell JD, Ganachaud A, Gehrke PC, Griffiths SP, Hobday AJ, Hoegh-Guldberg O, Johnson JE, Le Borgne R, Lehodey P, Lough JM, Matear RJ, Pickering TD, Pratchett MS, Sen Gupta A, Senina I, Waycott M (2013) Mixed responses of tropical Pacific fisheries and aquaculture to climate change. Nat Clim Change 3:591–599

Bell JD, Allain A, Allison EH, Andréfouët S, Andrew NL et al (2015a) Diversifying the use of tuna to improve food security and public health in Pacific Island countries and territories. Mar Policy 51:584–591

Bell JD, Albert A, Andréfouët S, Andrew NL, Blanc M et al (2015b) Optimising the use of near-shore fish aggregating devices for food security in the Pacific Islands. Mar Policy 56:98–105

Bell JD, Cisneros-Montemayor A, Hanich Q, Johnson JE, Lehodey P, Moore BR et al (2018a) Adaptations to maintain the contributions of small-scale fisheries to food security in the Pacific Islands. Mar Policy 88:303–314

Bell JD, Allain A, Sen Gupta A, Johnson JE, Hampton J, Hobday AJ, Lehodey P, Lenton A, Moore BR, Pratchett MS, Senina I, Smith N, Williams P (2018b) Chapter 14: Climate change impacts, vulnerabilities and adaptations: Western and Central Pacific Ocean marine fisheries. In: Barange M, Bahri T, Beveridge M, Cochrane K, Funge-Smith S, Poulain F (eds) Impacts of climate change on fisheries and aquaculture: synthesis of current knowledge, adaptation and mitigation options. FAO fisheries technical paper 627

Béné C, Wood RG, Newsham A, Davies M (2012) Resilience: new utopia or new tyranny? Reflection about the potentials and limits of the concept of resilience in relation to vulnerability reduction programmes. IDS working paper #405

Berkes F, Colding J, Folke C (2000) Rediscovery of traditional ecological knowledge as adaptive management. Ecol Appl 10(5):1251–1262. https://doi.org/10.2307/2641280

BOM and CSIRO (2011) Climate change in the Pacific: scientific assessment and new research, Regional overview, vol 1. Australian Bureau of Meteorology and Commonwealth Scientific and Industrial Research Organisation, Melbourne, 257pp

BOM and CSIRO (2014) Climate variability, extremes and change in the Western Tropical Pacific: new science and updated country reports. Pacific-Australia climate change science and adaptation planning program technical report. Australian Bureau of Meteorology and Commonwealth Scientific and Industrial Research Organisation, Melbourne, 358pp

Bromhead D, Scholey V, Nicol S, Margulies D, Wexler J, Stein M, Hoyle S, Lennert-Cody C, Williamson J, Havenhand J, Ilyina T, Lehodey P (2015) Ocean acidification impacts on tropical tuna populations. Deep Sea Res II 113:268–279

Brown CJ, Jupiter SD, Albert S, Klein CJ, Mangubhai S, Maina JM, Mumby P, Olley J, Stewart-Koster B, Tulloch V, Wenger A (2017a) Tracing the influence of land-use change on water quality and coral reefs using a Bayesian model. Sci Rep-UK 7:4740

Brown CJ, Jupiter SD, Lin HY, Albert S, Klein C, Maina JM, Tulloch VJD, Wenger AS, Mumby PJ (2017b) Habitat change mediates the response of coral reef fish populations to terrestrial run-off. Mar Ecol Prog Ser 576:55–68

Brunnschweiler JM, McKenzie J (2010) Baiting sharks for marine tourism: comment on Clua et al. (2010). Mar Ecol Prog Ser 420:283–284

Cai W, Santoso A, Wang G, Yeh S-W, An S-I, Cobb KM et al (2015) ENSO and greenhouse warming. Nat Clim Change 5:849–859

Chaves-Fonnegra A, Riegl B, Zea S, Lopez JV, Brandt M, Smith T, Gilliam DS (2018) Bleaching events regulate shifts from coral to excavating sponges in algae-dominated reefs. Glob Change Biol 24:773–785

Chavez FP, Messié M, Pennington JT (2011) Marine primary production in relation to climate variability and change. Annu Rev Mar Sci 3:227–260. https://doi.org/10.1146/annurev.marine.010908.163917

Chin A, Kyne PM (2007) Chapter 13: vulnerability of Chondrichthyan fishes of the Great Barrier Reef to climate change. In: Johnson JE, Marshall PA (eds) Climate change & the Great Barrier Reef. Great Barrier Reef Marine Park Authority, Townsville, pp 393–425

Chin A, Kyne PM, Walker TI, McAuley RB (2010) An integrated risk assessment for climate change: analysing the vulnerability of sharks and rays on Australia's Great Barrier Reef. Glob Chang Biol 16(7):1936–1953

Christensen JH, Krishna KK, Aldrian E, An S-I, Cavalcanti IFA, de Castro M et al (2013) Climate phenomena and their relevance for future regional climate change. In: Stocker TF, Qin D, Plattner G-K, Tignor M, Allen SK, Boschung J et al (eds) Climate change 2013: the physical science basis. Contribution of working group I to the fifth assessment report of the Intergovernmental Panel on Climate Change. Cambridge University Press, Cambridge

Church JA, White NJ, Hunter JR (2006) Sea-level rise at tropical Pacific and Indian Ocean islands. Glob Planet Change 53(3):155–168

Church JA, Clark PU, Cazenave A, Gregory JM, Jevrejeva S, Levermann A, Merrifield MA, Milne GA, Nerem RS, Nunn PD, Payne AJ, Pfeffer WT, Stammer D, Unnikrishnan AS (2013) Sea level change. In: Stocker TF, Qin D, Plattner G-K, Tignor M, Allen SK, Boschung J, Nauels A, Xia Y, Bex V, Midgley PM (eds) Climate change 2013: the physical science basis. Contribution of working group I to the fifth assessment report of the Intergovernmental Panel on Climate Change. Cambridge University Press, Cambridge

Cinner JE, Aswani S (2007) Integrating customary management into marine conservation. Biol Conserv 140(3–4):201–216

Cochrane KL, Andrew NL, Parma AM (2011) Primary fisheries management: a minimum requirement for provision of sustainable human benefits in small-scale fisheries. Fish Fish 12(3):275–288

Coles R, McKenzie L, Campbell S, Mellors J, Waycott M, Goggin L (2004) Seagrasses in Queensland waters. CRC Reef Research Centre Brochure, Townsville, 6pp

Convention on Biological Diversity (CBD) (2009) Connecting biodiversity and climate change mitigation and adaptation: report of the second ad hoc technical expert group on biodiversity and climate change (CBD technical series no. 41). Secretariat of the Convention on Biological Diversity, Montreal

Clarke SC, McAllister MK, Milner-Gulland EJ, Kirkwood GP, Michielsens CG, Agnew DJ, ... Shivji MS (2006) Global estimates of shark catches using trade records from commercial markets. Ecol Lett 9(10):1115–1126

Clarke SC, Harley SJ, Hoyle SD, Rice JS (2013) Population trends in Pacific Oceanic sharks and the utility of regulations on shark finning. Conserv Biol 27(1):197–209

Clua E, Buray N, Legendre P, Mourier J, Planes S (2011) Business partner or simple catch? The economic value of the sicklefin lemon shark in French Polynesia. Mar Freshw Res 62(6):764–770

Cooper TF, De'ath G, Fabricius KE, Lough JM (2008) Declining coral calcification in massive Porites in two nearshore regions of the northern Great Barrier Reef. Glob Change Biol 14:529–538

Cravatte S, Delcroix T, Zhang D, McPhaden M, Leloup J (2009) Observed freshening and warming of the western Pacific Warm Pool. Clim Dyn 33:565–589

Craven LK (2015) Migration-affected change and vulnerability in rural Vanuatu. Asia Pac Viewpoint 56:223–236. https://doi.org/10.1111/apv.12066

Cripps IL, Munday PL, McCormick MI (2011) Ocean acidification affects prey detection by a predatory reef fish. PLoS One 6:e22736

Crowley T (1999) Linguistic diversity in the Pacific. J Socioling 3(1):81–103. https://doi.org/10.1111/1467-9481.00064

Cumming RL, Toscano MA, Lovell ER, Carlson BA, Dulvy NK, Hughes A, Koven JF, Quinn NJ, Sykes HR, Taylor OJS, Vaughn D (2000) Mass bleaching in the Fiji Islands. In: Moosa MK, Soemodihardjo S, Soegiarto A, Rominihtarto K, Nontji A, Soekarmo, Surharsono (eds) Proceedings of the ninth international coral reef symposium, Bali, 23–29 Oct 2000, pp 1161–1169

D'Angelo C, Smith EG, Oswald F, Burt J, Tchernov D, Wiedenmann J (2012) Locally accelerated growth is part of the innate immune response and repair mechanisms in reef-building corals as detected by green fluorescent protein (GFP)-like pigments. Coral Reefs 31:1045–1056

Dalzell P (1993) Management of ciguatera fish poisoning in the South Pacific. South Pacific Commission, Noumea, New Caledonia

Davies JM, Dunne RP, Brown BE (1997) Coral bleaching and elevated sea-water temperature in Milne Bay Province, Papua New Guinea, 1996. Mar Freshwater Res 48:513–516

De'ath G, Fabricius K (2010) Water quality as a regional driver of coral biodiversity and macroalgae on the Great Barrier Reef. Ecol Appl 20:840–850

De'ath G, Lough JM, Fabricius KE (2009) Declining coral calcification on the Great Barrier Reef. Science 323:116–119

Dean RG, Houston JR (2013) Recent sea level trends and accelerations: comparison of tide gauge and satellite results. Coast Eng 75:4–9

Devine B, Munday PL, Jones GP (2012) Homing ability of adult cardinalfish is affected by elevated carbon dioxide. Oecologia 168:269–276

Domenici P, Allan B, McCormick MI, Munday PL (2012) Elevated carbon dioxide affects behavioural lateralization in a coral reef fish. Biol Lett 8:78–81

Dore JE, Lukas R, Sadler DW, Church MJ, Karl DM (2009) Physical and biogeochemical modulation of ocean acidification in the central North Pacific. Proc Natl Acad Sci U S A 106:12235–12240

Durack PJ, Wijffels SE (2010) Fifty-year trends in global ocean salinities and their relationship to broad-scale warming. J Clim 23:4342–4362

Dutra LXC, Haworth RJ, Taboada MB (2011) An integrated approach to tourism planning in a developing nation: a case study from Beloi (Timor-Leste). In: Dredge D, Jenkins J (eds) Stories of practice: tourism policy and planning. Ashgate, Farnham, pp 269–293

Dutra LXC, Haywood MDE, Singh SS, Piovano S, Ferreira M, Johnson JE, Veitayaki J, Kininmonth S, Morris CW (2018) Impacts of climate change on corals relevant to the Pacific Islands. In: Townhill B, Buckley P (eds) Pacific marine climate change report card. Commonwealth Marine Economies Programme, London, pp 132–258

Dulvy NK, Simpfendorfer CA, Davidson LN, Fordham SV, Bräutigam A, Sant G, Welch DJ (2017) Challenges and priorities in shark and ray conservation. Curr Biol 27(11):R565–R572

Ellison JC (2009) Wetlands of the Pacific Island region. Wetl Ecol Manag 17:169–206

Elsner JB, Kossin JP, Jagger TH (2008) The increasing intensity of the strongest tropical cyclones. Nature 455:92–95

Enochs IC, Manzello DP, Donham EM, Kolodziej G, Okano R, Johnston L, Young C, Iguel J, Edwards CB, Fox MD, Valentino L, Johnson S, Benavente D, Clark SJ, Carlton R, Burton T, Eynaud Y, Price NN (2015) Shift from coral to macroalgae dominance on a volcanically acidified reef. Nat Clim Change 5:1083

Enochs IC, Manzello DP, Kolodziej G, Noonan SHC, Valentino L, Fabricius KE (2016) Enhanced macroboring and depressed calcification drive net dissolution at high-CO_2 coral reefs. Proc R Soc B Biol Sci 283. https://doi.org/10.1098/rspb.2016.1742

Evans M, Sinclair RC, Fusimalohi C, Liava'a V (2001) Globalization, diet, and health: an example from Tonga. Bull World Health Organ 79(9):856–862

Fabricius KE, Langdon C, Uthicke S, Humphrey C, Noonan S, De'ath G, Okazaki R, Muehllehner N, Glas MS, Lough JM (2011) Losers and winners in coral reefs acclimatized to elevated carbon dioxide concentrations. Nat Clim Change 1:165–169

Fabricius KE, Kluibenschedl A, Harrington L, Noonan S, De'ath G (2015) In situ changes of tropical crustose coralline algae along carbon dioxide gradients. Sci Rep-UK 5:9537

Fabry VJ, Seibel BA, Feely RA, Orr JC (2008) Impacts of ocean acidification on marine fauna and ecosystem processes. ICES J Mar Sci 65:414–432

FFA [Forum Fisheries Agency] (2016) Tuna development indicators 2016. Pacific Islands Forum Fisheries Agency, Honiara

FFA and SPC (2015) Future of Pacific fisheries: a regional roadmap for sustainable Pacific fisheries. Pacific Islands Forum Fisheries Agency, Pacific Community, Honiara, Noumea

Folke C, Carpenter S, Elmqvist T, Gunderson L, Holling CS, Walker B (2002) Resilience and sustainable development: building adaptive capacity in a world of transformations. Ambio 31(5):437–440

Forsyth T (2017) Community-based adaptation to climate change. Oxford research encyclopedia of climate science. https://doi.org/10.1093/acrefore/9780190228620.013.602

Frankenberger TR, Constas MA, Nelson S, Starr L (2014) Nongovernmental organizations' approaches to resilience programming. http://www.ifpri.org/publication/nongovernmental-organizations-approachesresilienc-programming

Free CM, Thorson JT, Pinsky ML, Oken KL, Wiedenmann J, Jensen OP (2019) Impacts of historical warming on marine fisheries production. Science 363:979–983

Friedlander AM, Shackeroff JM, Kittinger JN (2013) Customary marine resource knowledge and use in contemporary Hawai'i. Pac Sci 67(3):441–460

Frommel AY, Margulies D, Wexler JB, Stein MS, Scholey VP, Williamson JE, Bromhead D, Nicol S, Havenhand J (2016) Ocean acidification has lethal and sub-lethal effects on larval development of yellowfin tuna, Thunnus albacares. J Exp Mar Biol Ecol 482:18–24

Ganachaud AS, Sen Gupta A, Orr JC, Wijffels SE, Ridgway KR, Hemer MA et al (2011) Observed and expected changes to the tropical Pacific Ocean. In: Bell JD, Johnson JE, Hobday AJ (eds) Vulnerability of tropical Pacific fisheries and aquaculture to climate change. Secretariat of the Pacific Community, Noumea, pp 101–187

Gillett R (2016) Fisheries in the economies of Pacific Island countries and territories. Pacific Community, Noumea

Gillett R, Tauati MI (2018) Fisheries of the Pacific Islands: regional and national information. Fisheries and aquaculture technical paper no. 625. Food and Agriculture Organization of the United Nations, Apia

Godoy MD, de Lacerda LD (2015) Mangroves response to climate change: a review of recent findings on mangrove extension and distribution. An Acad Bras Cienc 87(2):651–667

Gombos M, Atkinson S, Wongbusarakum S (2013) Adapting to a changing climate: guide to local early action planning (LEAP) and management planning. Micronesia Conservation Trust, Pohnpei

Grewe PMP, Feutry PPL, Hill PL et al (2015) Evidence of discrete yellowfin tuna (Thunnus albacares) populations demands rethink of management for this globally important resource. Sci Rep 5:16916. https://doi.org/10.1038/srep16916

Gervais C, Mourier J, Rummer J (2016) Developing in warm water: irregular colouration and patterns of a neonate elasmobranch. Mar Biodivers 4:743–744

Guannel G, Arkema K, Ruggiero P, Verutes G (2016) The power of three: coral reefs, seagrasses and Mangroves protect coastal regions and increase their resilience. PLoS One 11:e0158094

Guillemot N, Chabanet P, Le Pape O (2010) Cyclone effects on coral reef habitats in New Caledonia (South Pacific). Coral Reefs 29:445–453

Hanich Q, Ota Y (2013) Moving beyond rights based management: a transparent approach to distributing the conservation burden in tuna fisheries. Int J Mar Coast Law 28:135

Hanich Q, Tsamenyi M (2014) Progress in the implementation of conservation and management measures for bigeye and yellowfin tunas in the western and central pacific: sharing the conservation burden and benefit. In: Lodge MW, Nordquist MH (eds) Law of the sea: Liber amicorum Satya Nandan. Brill/Nijhoff, Leiden/Boston, pp 358–380

Hallegraeff GM (2010) Ocean climate change, phytoplankton community responses, and harmful algal blooms: a formidable predictive challenge1. J Phycol 46:220–235. https://doi.org/10.1111/j.1529-8817.2010.00815

Hassenruck C, Hofmann LC, Bischof K, Ramette A (2015) Seagrass biofilm communities at a naturally CO_2-rich vent. Environ Microbiol Rep 7:516–525

Havice E (2013) Rights-based management in the Western and Central Pacific Ocean tuna fishery: economic and environmental change under the Vessel Day Scheme. Mar Policy 42:259–267

Heenan A, Pomeroy R, Bell J, Munday P, Cheung W, Logan C, Brainard R, Amri AY, Alino P, Armada N, David L, Guieb R, Green S, Jompa J, Leonardo T, Mamauag S, Parker B, Shackeroff J, Yasin Z (2015) A climate-informed, ecosystem approach to fisheries management. Mar Policy 57:182–192

Henson SA, Cole H, Beaulieu C, Yool A (2013) The impact of global warming on the seasonality of ocean primary productivity. Biogeosciences 10:4357–4369. https://doi.org/10.5194/bg-10-4357-2013

Heinrich DDU, Rummer JL, Morash AJ, Watson S-A, Simpfendorfer CA, Heupel MR, Munday PL (2014) A product of its environment: the epaulette shark (Hemiscyllium ocellatum) exhibits physiological tolerance to elevated environmental CO_2. Conserv Physiol 2:cou047

Heron SF, Maynard J, van Hooidonk R, Eakin CM (2016) Warming trends and bleaching stress of the world's coral reefs 1985–2012. Sci Rep 6:38402. https://doi.org/10.1038/srep38402

Hobday AJ, Evans K (2013) Detecting climate impacts with oceanic fish and fisheries data. Clim Change 119:49–62

Hobday A, Spillman CM, Eveson JP, Hartog JR, Zhang X, Brodie S (2018) A framework for combining seasonal forecasts and climate projections to aid risk management for fisheries and aquaculture. Front Mar Sci 5(137):1–9

Hoegh-Guldberg O, Andrefouet S, Fabricius KE, Diaz-Pulido G, Lough JM, Marshall PA, Pratchett MS (2011) Vulnerability of coral reefs in the tropical Pacific to climate change. In: Bell J, Johnson JE, Hobday AJ (eds) Vulnerability of tropical Pacific fisheries and aquaculture to climate change. Secretariat of the Pacific Community, Noumea

Hough M (1990) Out of place: restoring identity to the regional landscape. Yale university press, New Haven

Hughes TP, Kerry JT, Alvarez-Noriega M, Alvarez-Romero JG, Anderson KD, Baird AH, Babcock RC, Beger M, Bellwood DR, Berkelmans R, Bridge TC, Butler IR, Byrne M, Cantin NE, Comeau S, Connolly SR, Cumming GS, Dalton SJ, Diaz-Pulido G, Eakin CM, Figueira WF, Gilmour JP, Harrison HB, Heron SF, Hoey AS, Hobbs JPA, Hoogenboom MO, Kennedy EV, Kuo CY, Lough JM, Lowe RJ, Liu G, McCulloch MTM, Malcolm HA, McWilliam MJ, Pandolfi JM, Pears RJ, Pratchett MS, Schoepf V, Simpson T, Skirving WJ, Sommer B, Torda G, Wachenfeld DR, Willis BL, Wilson SK (2017) Global warming and recurrent mass bleaching of corals. Nature 543:373

Hviding E (1996) Guardians of Marovo Lagoon: practice, place, and politics in maritime Melanesia, Pacific Islands monograph series, 14. University of Hawaii Press, Honolulu, HI

Hylton S, White WT, Chin A (2017) The sharks and rays of the Solomon Islands: a synthesis of their biological diversity, values and conservation status. Pac Conserv Biol 23:324–334

Hyndes GA, Heck KL Jr, Vergés A, Harvey ES, Kendrick GA, Lavery PS, Wernberg T (2016) Accelerating tropicalization and the transformation of temperate seagrass meadows. Bioscience 66(11):938–948

Hyndman D (1993) Sea tenure and the management of living marine resources in Papua New Guinea, 4th edn., vol 16, Dec. Pacific Studies

IPCC [Intergovernmental Panel on Climate Change] (2014) In: Team CW, Pachauri RK, Meyer LA (eds) Climate change 2014: synthesis report. Contribution of working groups I, II and III to the fifth assessment report of the Intergovernmental Panel on Climate Change. IPCC, Geneva, p 151

IPCC [Intergovernmental Panel on Climate Change] AR5 (2013) In: Stocker TF, Qin D, Plattner G-K, Tignor M, Allen SK, Boschung J, Nauels A, Xia Y, Bex V, Midgley PM (eds) Climate change 2013: the physical science basis. Contribution of working group I to the fifth assessment report of the Intergovernmental Panel on Climate Change. Cambridge University Press, Cambridge, 1535pp. https://doi.org/10.1017/CBO9781107415324

Jevrejeva S, Grinsted A, Moore JC (2009) Anthropogenic forcing dominates sea level rise since 1850. Geophys Res Lett 36:L20706

Johnson J, Welch D, Fraser A (2016a) Climate change impacts in North Efate, Vanuatu. Vanuatu RESCCUE project, Noumea

Johnson JE, Bell JD, Sen Gupta A (2016b) Pacific Islands ocean acidification vulnerability assessment. SPREP, Apia

Johnson JE, Welch DJ, Maynard JA, Bell JD, Pecl G, Tobin A, Robins J, Saunders T (2016c) Assessing and reducing vulnerability to climate change: moving from theory to practical decision-support. Mar Policy 74:220–229. https://doi.org/10.1016/j.marpol.2016.09.024

Johnson JE, Bell JD, Allain V, Hanich Q, Lehodey P, Moore B, Nicol S, Pickering T (2017) The Pacific Islands: fisheries and aquaculture and climate change. In: Philips B, Ramirez M (eds) Implications of climate change for fisheries and aquaculture: a global analysis. Wiley Publications, New York, pp 333–379

Johnson JE, Monnereau I, Welch DJ, McConney P, Szuster B, Gasalla MA (2018) Climate change adaptation: vulnerability and challenges facing small-scale fisheries in small islands. In: FishAdapt conference proceedings, Food and Agriculture Organization of the United Nations, FAO Fisheries and Aquaculture proceedings no. x. FAO, Rome

Jolis G, Sumampouw M, Saleh E, Razak FRA (2014) Vulnerability assessment and local early action planning (Va-Leap) for climate change adaptation in the Malaysian coral triangle area: a case study from Selakan Island, Semporna, Sabah, Malaysia. In: International conference on marine science & aquaculture (ICOMSA), 2015, Kota Kinabalu, Sabah, Malaysia

Kayser M (2010) The human genetic history of oceania: near and remote views of dispersal. Curr Biol 20(4):R109–R201. https://doi.org/10.1016/j.cub.2009.12.004

Keith SA, Maynard JA, Edwards AJ, Guest JR, Bauman AG, van Hooidonk R, Heron SF, Berumen ML, Bouwmeester J, Piromvaragorn S, Rahbek C, Baird AH (2016) Coral mass spawning predicted by rapid seasonal rise in ocean temperature. Proc Biol Sci 283:20160011–20160019

Kench PS, Thompson D, Ford MR, Ogawa H, McLean RF (2015) Coral islands defy sea-level rise over the past century: records from a central Pacific atoll. Geology 43:515–518

Kench PS, Ford MR, Owen SD (2018) Patterns of island change and persistence offer alternate adaptation pathways for atoll nations. Nat Commun 9:605

Kirwan ML, Megonigal JP (2013) Tidal wetland stability in the face of human impacts and sea-level rise. Nature 504(7478):53

Kittinger JN (2013) Human dimensions of small-scale and traditional fisheries in the Asia-Pacific region. Pac Sci 67(3):315–325

Kleypas JA, Castruccio FS, Curchitser EN, McLeod E (2015) The impact of ENSO on coral heat stress in the western equatorial Pacific. Glob Change Biol 21:2525–2539

Kroeker KJ, Kordas RL, Crim RN, Singh GG (2010) Meta-analysis reveals negative yet variable effects of ocean acidification on marine organisms. Ecol Lett 13:1419–1434

Kwiatkowski L, Orr JC (2018) Diverging seasonal extremes for ocean acidification during the twenty-first century. Nat Clim Change 8:141–145. https://doi.org/10.1038/s41558-017-0054-0.

Lack M, Meere F (2009) Regional action plan for sharks: guidance for Pacific Island Countries and Territories on the conservation and management of sharks

Lal P (2003) Economic valuation of mangroves and decision-making in the Pacific. Ocean Coast Manag 46(9–10):823–844

Laubenstein TD, Rummer JL, Nicol S, Parsons DM, Pether SMJ, Stephen Pope S, Smith N, Munday PL (2018) Correlated effects of ocean acidification and warming on behavioral and metabolic traits of a large Pelagic fish. Diversity 10:35. https://doi.org/10.3390/d10020035.

Le Borgne R, Allain V, Griffiths SP, Matear RJ, McKinnon AD, Richardson AJ, Young JW (2011) Vulnerability of open ocean food webs in the tropical Pacific to climate change. In: Bell JD, Johnson JE, Hobday AJ (eds) Vulnerability of tropical Pacific fisheries and aquaculture to climate change. Secretariat of the Pacific Community (SPC), Noumea, pp 189–249

Lehodey P, Bertignac M, Hampton J, Lewis A, Picaut J (1997) El Niño-Southern Oscillation and tuna in the western Pacific. Nature 389:715–718

Lehodey P, Senina I, Murtugudde R (2008) A spatial ecosystem and populations dynamics model (SEAPODYM)—modeling of tuna and tuna-like populations. Prog Oceanogr 78:304–318

Lehodey P, Hampton J, Brill RW, Nicol S, Senina I, Calmettes B, Pörtner HO, Bopp L, Ilyina T, Bell JD, Sibert J (2011) Vulnerability of oceanic fisheries in the tropical Pacific to climate change. In: Bell JD, Johnson JE, Hobday AJ (eds) Vulnerability of tropical Pacific fisheries and aquaculture to climate change. Secretariat of the Pacific Community, Noumea, pp 433–492

Lehodey P, Senina I, Calmettes B, Hampton J, Nicol S (2013) Modelling the impact of climate change on Pacific skipjack tuna population and fisheries. Clim Change 119:95–109

Lehodey P, Senina I, Nicol S, Hampton J (2015) Modelling the impact of climate change on South Pacific albacore tuna. Deep-Sea Res II 113:246–259

Lehodey P, Senina I, Calmettes B, M. Dessert, S. Nicol, J. Hampton, N. Smith, T. Gorgues, O. Aumont, M. Lengaigne, C. Menkes, M. Gehlen. (2017) Modelling the impact of climate change including ocean acidification on Pacific yellowfin tuna. Western and Central Pacific fisheries commission scientific committee, thirteenth regular session, Rarotonga, Cook Islands, 9–17 Aug 2017

Leonard K (2001) Domestic violence and alcohol: what is known and what do we need to know to encourage environmental interventions? J Subst Use 6(4):235–247. https://doi.org/10.1080/146598901753325075

Levin LA, Liu KK, Emeis KC, Breitburg DL, Cloern J, Deutsch C, Giani M, Goffart A, Hofmann EE, Lachkar Z, Limburg K, Liu SM, Montes E, Naqvi W, Ragueneau O, Rabouille C, Sarkar SK, Swaney DP, Wassman P, Wishner KF (2015) Comparative biogeochemistry-ecosystem-human interactions on dynamic continental margins. J Mar Syst 141:3–17

Livingston M (2011) A longitudinal analysis of alcohol outlet density and domestic violence. Addiction 106(5):919–925. https://doi.org/10.1111/j.1360-0443.2010.03333.x

Lough JM (2012) Small change, big difference: sea surface temperature distributions for tropical coral reef ecosystems, 1950–2011. J Geophys Res Oceans 117. https://doi.org/10.1029/2012JC008199

Lough JM, Meehl GA, Salinger MJ (2011) Observed and projected changes in surface climate of the tropical Pacific. In: Bell JD, Johnson JE, Hobday AJ (eds) Vulnerability of tropical pacific fisheries and aquaculture to climate change. Secretariat of the Pacific Community, Noumea, pp 49–99

Lough JM, Sen Gupta A, Power SB, Grose MR, McGree S (2016) Observed and projected changes in surface climate of tropical Pacific Islands. In: Taylor M, McGregor A, Dawson B (eds) Vulnerability of Pacific Island agriculture and forestry to climate change. Secretariat of the Pacific Community, Noumea, pp 47–101

Lough JM, Anderson KD, Hughes TP (2018) Increasing thermal stress for tropical coral reefs: 1871–2017. Sci Rep 8:6079. https://doi.org/10.1038/s41598-018-24530-9

Lovell E, Sykes H, Deiye M, Wantiez L, Carrigue C, Virly S, Samuelu J, Solofa A, Poulasi T, Pakoa K, Sabetian A, Afzal D, Hughes A, Sulu R (2004) Status of coral reefs in the South West Pacific: Fiji, Nauru, New Caledonia, Samoa, Solomon Islands, Tuvalu and Vanuatu. In: Wilkinson C (ed) Status GCRMN, Townsville

Lovelock CE, Cahoon DR, Friess DA, Guntenspergen GR, Krauss KW, Reef R, Saintilan N (2015) The vulnerability of Indo-Pacific mangrove forests to sea-level rise. Nature 526(7574):559

Lotze HK, Lenihan HS, Bourque BJ, Bradbury RH et al (2006) Depletion, degradation, and recovery potential of estuaries and coastal seas. Science 312:1806–1809

Mangubhai S (2016) Impact of tropical cyclone Winston on coral reefs in the Vatu-i-Ra Seascape, report no. 01/16, Suva, Fiji, 27pp

Marois DE, Mitsch WJ (2015) Coastal protection from tsunamis and cyclones provided by mangrove wetlands—a review. Int J Biodiver Sci Ecosyst Serv Manag 11(1):71–83

Maron N, Connell J (2008) Back to Nukunuku: employment, identity and return migration in Tonga. Asia Pac Viewpoint 49(2):168–184

Matear RJ, Chamberlain MA, Sun C, Feng M (2015) Climate change projection for the western tropical Pacific Ocean using a high-resolution ocean model: implications for tuna fisheries. Deep-Sea Res II Top Stud Oceanogr 113:22–46. https://doi.org/10.1016/j.dsr2.2014.07.003

McIlgorm A (2010) Economic impacts of climate change on sustainable tuna and billfish management: insights from the Western Pacific. Prog Oceanogr 86(1–2):187–191

Mcleod E, Margles Weis SW, Wongbusarakum S, Gombos M, Dazé A, Otzelberger A, Hammill A, Agostini V, Urena Cot D, Wiggins M (2015) Community-based climate vulnerability and adaptation tools: a review of tools and their applications. Coast Manag 43(4):439–458

McNeil BI, Sasse TP (2016) Future ocean hypercapnia driven by anthropogenic amplification of the natural CO_2 cycle. Nature 529:383–386

McLeod E, Salm RV (2006) Managing mangroves for resilience to climate change. International Union for Conservation of Nature, Gland, Switzerland

McPhaden MJ, Zebiak SE, Glantz MH (2006) ENSO as an integrating concept in earth science. Science 314:1740–1745

Meissner KJ, Lippmann T, Sen Gupta A (2012) Large-scale stress factors affecting coral reefs: open ocean sea surface temperature and surface seawater aragonite saturation over the next 400 years. Coral Reefs 31:309–319

Moberg F, Folke C (1999) Ecological goods and services of coral reef ecosystems. Ecol Econ 29(2):215–233

Movono A, Dahles H, Becken S (2018) Fijian culture and the environment: a focus on the ecological and social interconnectedness of tourism development. J Sustain Tour 26:451–469

Munday PL, Jones GP, Pratchett MS, Williams A (2008) Climate change and the future of coral reef fisheries. Fish Fish 9:261–285

Munday PL, Dixson DL, Donelson JM, Jones GP, Pratchett MS, Devitsina GV, Doving KB (2009) Ocean acidification impairs olfactory discrimination and homing ability of a marine fish. Proc Natl Acad Sci U S A 106:1848–1852

Munday PL, Pratchett MS, Dixson DL, Donelson JM, Endo GGK, Reynolds AD, Knuckey R (2013) Elevated CO_2 levels affects the behaviour of an ecologically and economically important coral reef fish. Mar Biol 160:2137–2144

Nagelkerken I, Van der Velde G (2004) Relative importance of interlinked mangroves and seagrass beds as feeding habitats for juvenile reef fish on a Caribbean island, Marine Ecology Progress Series, vol 274, pp 153–159

Nurse LA, McLean RF, Agard J, Briguglio LP, Duvat-Magnan V, Pelesikoti N, Tompkins E, Webb A (2014) Small Islands. In: Barros VR, Field CB, Dokken DJ, Mastrandrea MD, Mach KJ, Bilir TE, Chatterjee M, Ebi KL, Estrada YO, Genova RC, Girma B, Kissel ES, Levy AN, MacCracken S, Mastrandrea PR, White LL (eds) Climate change 2014: impacts, adaptation, and vulnerability. Part B: regional aspects. Contribution of working group II to the fifth assessment report of the Intergovernmental Panel on Climate Change. Cambridge University Press, Cambridge, pp 16133–16654

Nuttall P, Veitayaki J (2015) Oceania is vast, Canoe is centre, Village is anchor, Continent is margin. In: Smith HD, De Vivero JLS, Agardy TS (eds) Routledge handbook of ocean resources and management. Routledge, London, pp 560–575

Obura D, Mangubhai S (2011) Coral mortality associated with thermal fluctuations in the Phoenix Islands, 2002–2005. Coral Reefs 30:607–619

Orth RJ, Curruthers TJB, Dennison WC, Duarte CM, Fourqurean JW, Heck KL, Hughes AR, Kendrick GA, Kenworthy WJ, Olyarnik S, Short FT, Waycott M, Williams SL (2006) A global crisis for seagrass ecosystems. Bioscience 56:987–996

Palacios SL, Zimmerman RC (2007) Response of eelgrass Zostera marina to CO_2 enrichment: possible impacts of climate change and potential for remediation of coastal habitats. Mar Ecol Prog Ser 344:1–13

Pascal N (2014) Economic valuation of mangrove ecosystem services in Vanuatu: case study of Crab Bay (Malekula Is.) and Eratap (Efate Is.)—summary report. IUCN, Suva, Fiji, 18pp

Paxton CW, Baria MVB, Weis VM, Harii S (2016) Effect of elevated temperature on fecundity and reproductive timing in the coral Acropora digitifera. Zygote 24:511–516

Pickering T et al (2011) Vulnerability of aquaculture in the tropical Pacific to climate change. In: Bell JD, Johnson JE, Hobday AJ (eds) Vulnerability of tropical Pacific fisheries and aquaculture to climate change. Secretariat of the Pacific Community, Noumea

Pinca S, Kronen M, Friedman K, Magron F, Chapman L, Tardy E, Pakoa K, Awira R, Boblin P, Lasi F (2010) Regional assessment report: profiles and results from survey work at 63 sites across 17 Pacific island countries and territories. Secretariat of the Pacific Community, Noumea, 512 pp

Petitgas P, Rijnsdorp AD, Dickey-Collas M, Engelhard GH, Peck MA, Pinnegar JK, Drinkwater K, Huret M, Nash RDM (2012) Impacts of climate change on the complex life cycles of fish. In: Fisheries & Oceanography

Polovina JJ, Dunne JP, Woodworth PA, Howell EA (2011) Projected expansion of the subtropical biome and contraction of the temperate and equatorial upwelling biomes in the North Pacific under global warming. ICES J Mar Sci. J, Cons

Ponia B (2010) A review of Pacific aquaculture in the Pacific Islands 1998–2007: tracking a decade of progress through official and provisional statistics. Secretariat of the Pacific Community Aquaculture technical report, Noumea

Pouvreau S, Prasil V (2001) Growth of the black-lip pearl oyster, Pinctada margaritifera, at nine culture sites of French Polynesia: synthesis of several sampling designs conducted between 1994 and 1999. Aquat Living Resour 14:155–163

Power S, Casey T, Folland C, Colman C, Mehta V (1999) Interdecadal modulation of the impact of ENSO on Australia. Clim Dyn 15:319–324

Power S, Delage F, Chung C, Kociuba G, Keay K (2013) Robust twenty-first-century projections of El Niño and related precipitation variability. Nature 502:541–545

Pratchett MS, Munday PL, Graham NAJ, Kronen M, Pinca S, Friedman K, Brewer TD, Bell JD, Wilson SK, Cinner JE, Kinch JP, Lawton RJ, Williams AJ, Chapman L, Magron F, Webb A (2011) In: Bell JD, Johnson JE, Hobday AJ (eds) Vulnerability of tropical Pacific fisheries and aquaculture to climate change. Secretariat of the Pacific Community, Noumea, pp 493–576

Rabinovich AB, Geist EL, Fritz HM, Borrero JC (2015) Introduction to "Tsunami science: ten years after the 2004 Indian Ocean tsunami. Volume I". Pure Appl Geophys 172(3–4):615–619

Reid H (2015) Ecosystem- and community-based adaptation: learning from community-based natural resource management. Clim Dev 8(1):4–9. https://doi.org/10.1080/17565529.2015.1034233

Remling E, Veitayaki J (2016) Community-based action in Fiji's Gau Island: a model for the Pacific. IJCCSM 8(3):375–398

RESCCUE [Restoration of Ecosystem Services and Climate Change Adaptation] (2015) Initial diagnosis document: Vanuatu. Report to the Pacific Community (SPC), Noumea, 120pp

Rhein M, Rintoul SR, Aoki S, Campos E, Chambers D, Feely RA (2013) Observations: ocean. In: Stocker TF, Qin D, Plattner G-K, Tignor M, Allen SK, Boschung J et al (eds) Climate change 2013: the physical science basis. Contribution of working group I to the fifth assessment report of the Intergovernmental Panel on Climate Change. Cambridge University Press, Cambridge

Rotmann S (2001) Coral bleaching event on Lihir Island, Feb–Mar 2001

Rongo T, van Woesik R (2013) The effects of natural disturbances, reef state, and herbivorous fish densities on ciguatera poisoning in Rarotonga, southern Cook Islands. Toxicon 64:87–95

Rosa R, Rummer JL, Munday PL (2017) Biological responses of sharks to ocean acidification. Biol Lett 13(3):20160796

Roué M, Cruchet P, Ung A, Darius T, Chinain M, Laurent D (2013) Giant clams: new vectors of ciguatera. Toxicon 75(SI):206–206

Ruddle K (1993) External forces and change in traditional community-based fishery management systems in the Asia-Pacific region. Maritime Anthropol Stud 6(1–2):1–37. Accessed 26 Apr 2018

Ruddle K (1998) The context of policy design for existing community-based fisheries management systems in the Pacific Islands. Ocean Coast Manag 40(2–3):105–126. https://doi.org/10.1016/s0964-5691(98)00040-4

Ruddle K, Panayotou T (1989) The organization of traditional inshore fishery management systems in the Pacific. In: Rights based fishing. Springer, Dordrecht, pp 73–93. https://doi.org/10.1007/978-94-009-2372-0_4

Rugendyke B, Connell J (2008) 15 marginal people and marginal places. In: Tourism at the grass-roots: Villagers and visitors in the Asia-Pacific, vol 274

Saunders MI, Leon J, Phinn SR, Callaghan DP, O'Brien KR, Roelfsema CM, Lovelock CE, Lyons MB, Mumby PJ (2013) Coastal retreat and improved water quality mitigate losses of seagrass from sea level rise. Glob Change Biol 19:2569–2583. https://doi.org/10.1111/gcb.12218

Saunders MI, Leon JX, Callaghan DP, Roelfsema CM, Hamylton S, Brown CJ, Baldock T, Golshani A, Phinn SR, Lovelock CE, Hoegh-Guldberg O, Woodroffe CD, Mumby PJ (2014) Interdependency of tropical marine ecosystems in response to climate change. Nat Clim Change 4:724

Saunders MI, Albert S, Roelfsema CM et al (2016) Coral Reefs 35:155. https://doi.org/10.1007/s00338-015-1365-0

Sarmiento JL, Gruber N, Brzezinski MA, Dunne JP (2004) High-latitude controls of thermocline nutrients and low latitude biological productivity. Nature 427(6969):56

Schaefer K, Fuller D, Block B (2011) Movements, behavior, and habitat utilization of yellowfin tuna (Thunnus albacares) in the Pacific Ocean off Baja California, Mexico, determined from archival tag data analyses, including unscented Kalman filtering. Fish Res 112:22–37

Schaefer K, Fuller D, Hampton J, Caillot S, Leroy B, Itano D (2015) Movements, dispersion, and mixing of bigeye tuna (Thunnus obesus) tagged and released in the equatorial Central Pacific Ocean, with conventional and archival tags. Fish Res 161:336–355

Schlaff AM, Heupel MR, Simpfendorfer CA (2014) Influence of environmental factors on shark and ray movement, behaviour and habitat use: a review. Rev Fish Biol Fish 24(4):1089–1103

Short FT, Coles R (eds) (2001) Global seagrass research methods. Elsevier, Amsterdam, 482pp

Short FT, Neckles HA (1999) The effects of global climate change on seagrasses. Aquat Bot 63(3–4):169–196

Short FT, Wyllie-Echeverria S (1996) Natural and human-induced disturbance of seagrasses. Environ Conserv 23:17–27

Short FT, Koch E, Creed JC, Magalhaes KM, Fernandez E, Gaeckle JL (2006) SeagrassNet monitoring across the Americas: case studies of seagrass decline. Mar Ecol 27:277–289

Simpson SD, Munday PL, Wittenrich ML, Manassa R, Dixson DL, Gagliano M, Yan HY (2011) Ocean acidification erodes crucial auditory behaviour in a marine fish. Biol Lett 7:917–920

Singh S (2018) Contribution of Symbiodinium clades to Pocillopora's response to thermal stress in Fiji and French Polynesia. MSc thesis, The University of the South Pacific, Suva

SPC [Secretariat of the Pacific Community] (2008) Fish and food security. Policy Brief 1/2008. Secretariat of the Pacific Community, Noumea

SPC [Pacific Community] (2008) Fish and food security: Policy Brief 1/2008. Secretariat of the Pacific Community, Noumea, New Caledonia

SPC [Pacific Community] (2013) Status report: Pacific Islands Reef and nearshore fisheries and aquaculture. Secretariat of the Pacific Community, Noumea, New Caledonia

SPC [Pacific Community] (2015) A new song for coastal fisheries – Pathways to change: the Noumea strategy. Pacific Community, Noumea, New Caledonia

Stoeckl N, Birtles A, Farr M, Mangott A, Curnock M, Valentine P (2010) Live-aboard dive boats in the Great Barrier Reef: regional economic impact and the relative values of their target marine species. Tour Econ 16:995–1018

Steinacher M, Joos F, Frölicher TL, Bopp L, Cadule P, Cocco V, Doney SC, Gehlen M, Lindsay K, Moore JK, Schneider B (2010) Projected 21st century decrease in marine productivity: a multi-model analysis. Biogeosciences 7(3):979–1005

Szmant AM, Gassman NJ (1990) The effects of prolonged bleaching on the tissue biomass and reproduction of the reef coral Montastrea annularis. Coral Reefs 8:217–224

Taucher J, Oschlies A (2011) Can we predict the direction of marine primary production change under global warming? Geophys Res Lett 38:L02603. https://doi.org/10.1029/2010GL045934

Terry JP (2007) Tropical cyclones: climatology and impacts in the South Pacific. Springer, New York

Toseland ADSJ, Daines SJ, Clark JR, Kirkham A, Strauss J, Uhlig C, Lenton TM, Valentin K, Pearson GA, Moulton V, Mock T (2013) The impact of temperature on marine phytoplankton resource allocation and metabolism. Nat Clim Chang 3(11):979

van Hooidonk R, Maynard JA, Manzello D, Planes S (2014) Opposite latitudinal gradients in projected ocean acidification and bleaching impacts on coral reefs. Glob Change Biol 20:103–112

van Woesik R, Golbuu Y, Roff G (2015) Keep up or drown: adjustment of western Pacific coral reefs to sea-level rise in the 21st century. R Soc Open Sci 2:150181

Veitayaki J (2018) Ocean in US: security of life in the world's largest ocean. University of Hawaii Press, Honolulu, HI

Veitayaki J, Breckwoldt A, Sigarua T, Bulai NAR (2014) Living from the sea: culture and marine conservation in Fiji. Taukei Trust Fund Board, Suva

Veitayaki J, Waqalevu V, Varea R, Rollings N (2017) Mangroves in small island development states in the Pacific: an overview of a highly important and seriously threatened resource. In: Participatory mangrove management in a changing climate. Springer, Tokyo, pp 303–327

Vianna GMS, Meekan MG, Pannell DJ, Marsh SP, Meeuwig JJ (2012) Socio-economic value and community benefits from shark-diving tourism in Palau: a sustainable use of reef shark populations. Biol Conserv 145:267–277

Vieira S, Kinch J, White W, Yaman L (2017) Artisanal shark fishing in the Louisiade Archipelago, Papua New Guinea: socio-economic characteristics and management options. Ocean Coast Manag 137:43–56

Uyarra E (2005) Knowledge, diversity and regional innovation policies: theoretical issues and empirical evidence of regional innovation strategies. Manchester, Prest, Discussion Paper Series, 05–16

War Sajjad A, Kumar A, Vyas V (2014) Assessment of riparian buffer zone of Chandni Nalla-A stream in Narmada Basin, India. Adv Appl Sci Res 5(2):102–110

Warren-Rhodes K, Schwarz AM, Boyle LN, Albert J, Agalo SS, Warren R et al (2011) Mangrove ecosystem services and the potential for carbon revenue programmes in Solomon Islands. Environ Conserv 38(4):485–496

Watson SA, Allan BJM, McQueen DE, Nicol S, Parsons DM, Pether SMJ, Pope S, Setiawan AN, Smith N, Wilson C, Munday PL (2018) Ocean warming has a greater effect than acidification on the early life history development and swimming performance of a large circumglobal pelagic fish. Glob Change Biol 24:4368–4385. https://doi.org/10.1111/gcb.14290

Waycott M, Duarte CM, Carruthers TJB, Orth RJ, Dennison WC, Olyarnik S, Calladine A, Fourqurean JW, Heck KL Jr, Hughes AR, Kendrick GA, Kenworthy WJ, Short FT, William SL (2009) Accelerating loss of seagrasses across the globe threatens coastal ecosystems. Proc Nat Acad Sci U S A 106:12377–12381

Waycott M, McKenzie LJ, Mellors JE, Ellison JC, Sheaves MT, Collier C, Schwarz A-M, Webb A, Johnson JE, Payri CE (2011) Vulnerability of mangroves, seagrasses and intertidal flats. In: Bell JD, Johnson JE, Hobday AJ (eds) Vulnerability of tropical Pacific fisheries and aquaculture to climate change in the tropical. Secretariat of the Pacific Community, Noumea

WCPFC (2018) Western and Central Pacific Fisheries Commission,13th Regular Session of the Scientific Committee. Ecosystems and Bycatch Mitigation Theme. Working Paper: Lehody, P., ISenina, I., Calmettes, B., Dessert, M., Nicol, S., Hampton, J., Smith, N., Gorgues, T., Aumont, O., Lengaigne, M., Menkes, C., Gehlen, M. Modelling the impact of climate change including ocean acidification on Pacific yellowfin tuna

Whimp G (2008) Cetaceans and citations: a survey of the English literature on the role of cetaceans in South Pacific island cultures. Tuhinga 19:169–184. Accessed 26 Apr 2018

WHO [World Health Organization] (2005) Multi-country study on women's health and domestic violence against women: summary report of initial results on prevalence, health outcomes and women's responses. World Health Organization, Geneva

Wiedenmann J, D'Angelo C, Smith EG, Hunt AN, Legiret F-E, Postle AD, Achterberg EP (2012) Nutrient enrichment can increase the susceptibility of reef corals to bleaching. Nat Clim Change 3:160–164

Williams P, Terawasi P Reid C (2017) Overview of tuna fisheries in the Western and Central Pacific Ocean, including economic conditions—2016. In: 13th Scientific committee regular session, WCPFC-SC13-2017/GN-WP-01, 66pp (also available at https://www.wcpfc.int/node/29889).

Wongbusarakum S, Gombos M, Parker B, Courtney C, Atkinson S, Kostka W (2015) The local early action planning (LEAP) tool: enhancing community-based planning for a changing climate. Coast Manag 43(4):383–393. https://doi.org/10.1080/08920753.2015.1046805

Woodroffe CD, Webster JM (2014) Coral reefs and sea-level change. Mar Geol 352:248–267

Woodworth-Jefcoats PA, Polovina JJ, Dunne JP, Blanchard JL (2013) Ecosystem size structure response to 21st century climate projection: large fish abundance decreases in the central North Pacific and increases in the California Current. Glob Change Biol 19:724–733. https://doi.org/10.1111/gcb.12076

Yukihira H, Lucas JS, Klump DW (2000) Comparative effects of temperature on suspension feeding and energy budgets of the pearl oysters Pinctada margaritifera and P. maxima. Mar Ecol Prog Ser 195:179–188

Zann LP (1994) The status of coral reefs in South Western Pacific Islands. Mar Pollut Bull 29:52–61

Zeller D, Harper S, Zylich K, Pauly D (2015) Synthesis of underreported small-scale fisheries catch in Pacific island waters. Coral Reefs 34(1):25–39

Chapter 11
Freshwater Availability Under Climate Change

Tony Falkland and Ian White

11.1 Introduction

The 2014 UN Conference on Small Island Developing States (SIDS) in Samoa re-emphasised the particular vulnerability of small island states to natural and human-driven global changes and extreme events. The SIDS statement detailed numerous challenges in islands for the vital freshwater sector including pollution; the overexploitation of surface, ground and coastal waters; saline intrusion; drought and water scarcity; soil erosion; inadequate water and wastewater treatment; and lack of access to sanitation and hygiene, and projected changes in rainfall patterns and sea level related to climate change may have significant impacts on freshwater supply (UNGA 2014). In order to assess those impacts, it is necessary to consider the diverse sources of freshwater and their use by island communities, the hydrological processes influencing those sources, the current issues surrounding management of freshwater sources and the projected changes in climate.

T. Falkland (✉)
Island Hydrology Services, Canberra, ACT, Australia
e-mail: tony.falkland@netspeed.com.au

I. White
Fenner School of Environment and Society, Australian National University,
Canberra, ACT, Australia
e-mail: Ian.White@anu.edu.au

© Springer Nature Switzerland AG 2020
L. Kumar (ed.), *Climate Change and Impacts in the Pacific*, Springer Climate,
https://doi.org/10.1007/978-3-030-32878-8_11

11.2   Freshwater Resources Availability

Sources of freshwater on Pacific islands vary due to their diversity in physiographic, geological, hydrological and demographic characteristics. This chapter focuses mainly on hydrological characteristics and their influences on the availability of water resources though others are also considered.

11.2.1   Diversity of Islands and Influences on Water Resources

Island area, shape, topography, soils and lithology greatly influence both the occurrence and distribution of natural freshwater sources. Large, mountainous volcanic islands have considerable surface water and groundwater resources while small, low-lying coral sand and limestone islands have and no surface water and very limited groundwater. Details of physiographic and geological characteristics of Pacific islands are discussed, respectively, in Chaps. 1 and 2.

Population densities in Pacific islands range from less than 20 to more than 30,000 people/km^2 (see Chap. 8). These greatly influence demand for water and the degree of contamination of often scarce water resources. In small islands, non-sustainable extraction of groundwater can lead to saline intrusion. Human settlements and agriculture, forestry and mining can lead to biological and chemical contamination of both surface and groundwater. Soil erosion due to inappropriate land clearing and consequent sedimentation of surface water and near-shore reefs, pollution of surface and groundwater from sanitation systems and agricultural chemicals are continuing challenges. Because of the diversity of island communities, water supply systems range from household or small community systems such as rainwater harvesting systems in rural areas to medium and large public water supply systems using surface water, groundwater or, in some cases, desalinated water. The demographic characteristics of Pacific islands are considered in Chap. 8.

11.2.2   Types of Freshwater Resources

Freshwater resources in Pacific islands fall into two categories: those that are naturally occurring and require a relatively low level of technology to develop them and those that involve a higher level of technology, which sometimes are referred to as 'non-conventional' water resources.

Naturally occurring freshwater resources include surface water, groundwater and rainwater, collected mainly from roofs. The second category includes the less common desalination, importation and wastewater reuse. Naturally occurring water resources are inevitably more economical to develop than 'non-conventional' water resources. In addition to freshwater resources, brackish groundwater and seawater are used mainly for washing and toilet flushing where freshwater is scarce.

Table 11.1 summarises the main freshwater resources and uses for most Pacific islands, including the 14 independent Pacific island countries (PICs), Hawaiian Islands (State of the USA) and a number of territories or provinces of other countries. Not all countries in the Pacific region are included in Table 11.1.

Table 11.1 Summary of main freshwater resources and uses in Pacific islands

Location	Main freshwater resources[a]	Main freshwater uses[b]
Independent countries		
Cook Islands	SW, GW, RW	WS, T
Federated States of Micronesia (FSM)	SW, GW, RW	WS
Fiji	SW, GW, RW, D (tourist resort only)	WS, T, H, I
Kiribati	GW, RW, D (limited)	WS
Marshall Islands	RW (from airport catchment and buildings), GW, D (regular use on Ebeye, emergency use in others)	WS
Nauru	D, RW, GW (limited)	WS
Niue	GW, RW	WS
Palau	SW, GW, RW	WS
Papua New Guinea (PNG)	SW, GW, RW	WS, M, H
Samoa	SW, GW, RW	WS, T, H
Solomon Islands	SW, GW, RW	WS
Tonga	GW, RW, SW (limited)	WS
Tuvalu	RW, GW (limited), D (emergency)	WS
Vanuatu	SW, GW, RW, D (emergency)	WS, T, H
State and territories of USA		
Hawaiian Islands	SW, GW, RW	WS, T, I
American Samoa	SW, GW, RW	WS
Guam	SW, GW, RW, D	WS, T
Commonwealth of the Northern Mariana Islands (CNMI)	SW, GW, RW	WS
Territories of France		
French Polynesia	SW, GW, RW, D	WS, T, H
New Caledonia	SW, GW, RW, D	WS, T, M, H
Wallis and Futuna	SW, GW, RW	WS
Others		
Tokelau (territory of New Zealand)	RW	WS
Pitcairn Islands (territory of UK)	GW, RW	WS
Rapa Nui (Easter Island) (territory of Chile)	GW, RW	WS
Galápagos Islands (province of Ecuador)	SW, GW, RW, D (limited)	WS, T

[a]*SW* surface water, *GW* groundwater, *RW* rainwater, *D* desalination
[b]*WS* water supply to communities, *T* tourism, *H* hydroelectricity, *M* mining, *I* irrigation

Not shown in Table 11.1 is bottled water which is increasingly being used in many Pacific islands for drinking water, particularly in urban centres. Bottled water is produced locally in some PICs including Cook Islands, Fiji, Tonga and Samoa.

The main water resources and uses are described in more detail below.

Surface Water

Surface water occurs on volcanic islands in the form of rivers, ephemeral and perennial streams and springs, and as freshwater lagoons, lakes and swamps. Rivers only occur in the larger islands where rainfall is abundant, for example, PNG and the larger islands of Fiji and Solomon Islands. Perennial streams and springs occur mainly in volcanic islands where the permeability of the rock is low. Many streams in small, steep catchments are ephemeral and only flow for several hours or days after rainfall. On some islands, dams have been constructed to store surface runoff for water supply such as Vaturu Dam on the Nadi River, Viti Levu, Fiji, and the Fena Valley Dam, Guam, and for hydroelectricity generation in French Polynesia, Fiji, New Caledonia, Samoa and PNG.

Freshwater lakes are found on some larger volcanic islands but generally not on smaller volcanic islands. These occur in the craters of extinct volcanoes such as on Upolu, Samoa, or depressions in the topography. The small volcanic island of Niuafo'ou in the northern part of Tonga has a number of fresh and brackish lakes within its crater.

Low-lying coral islands and limestone islands, for example, the atoll islands of Kiribati, Marshall Islands, Tokelau and Tuvalu and the largely limestone islands of Nauru and Niue, rarely have surface fresh water resources. Small lakes and swamps on these islands are normally brackish. One exception is a central freshwater lagoon on the small atoll island Teraina (Washington Island), Kiribati, which occurs due to high rainfall and the lagoon having just one regulated passage to the sea.

Groundwater

Groundwater occurs on most islands as either perched (high-level) or basal (low-level) aquifers. In some islands, such as Nauru, fresh groundwater is found only in very limited parts of the island owing to the highly permeable karst limestone which promotes mixing of fresh and underlying saline water. In very small limestone islands, this mixing is so rapid that there is no or very little fresh groundwater as in some outer islands of Tonga and in Nauru.

Perched aquifers commonly occur over horizontal confining layers (aquitards) or, less commonly, as dyke-confined aquifers which are formed when vertical volcanic dykes trap water as in some of the islands of Hawaii and French Polynesia. Springs are sourced from perched aquifers, often well above sea level.

Basal aquifers occur at or below sea level and are found on many low islands of sufficient size and rainfall and in the coastal margins of high volcanic islands. On many small coral and limestone islands, unconfined basal aquifers take the form of 'freshwater lenses' (or 'groundwater lenses'), which underlie part or all of these islands. An example of a relatively thick freshwater lens is shown in Fig. 11.1.

Basal aquifers tend to be more important than perched aquifers because of their generally larger storage volumes. Basal aquifers are, however, vulnerable to saline intrusion from seawater, and over-pumping can cause salinisation of water supplies. The use of infiltration galleries (Fig. 11.1) for groundwater pumping rather than conventional vertical boreholes can minimise saline intrusion.

The term 'freshwater lens' is somewhat misleading as it implies a distinct freshwater aquifer. There is no distinct boundary between freshwater and seawater but rather a transition zone as shown in Fig. 11.1. The base of the freshwater zone can be defined on the basis of a salinity criterion such as electrical conductivity or chloride ion concentration. Diagrams such as that in Fig. 11.1 exaggerate the vertical scale relative to the horizontal scale. In reality, the vertical scale is very small in comparison to the horizontal scale, often by a factor of 50–100 so that the fresh groundwater zone is a thin veneer of freshwater over saline water.

Freshwater lenses often have asymmetric shapes with the deepest portions displaced towards the less permeable side (often the lagoon side) of the island, as

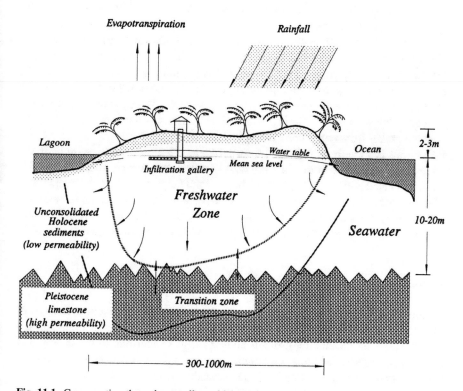

Fig. 11.1 Cross section through a small coral island showing main features of a freshwater lens (exaggerated vertical scale) and location of an infiltration gallery

shown in Fig. 11.1. Typically, the freshwater zone on a small coral island is about 5–10 m thick, with a transition zone of a similar thickness. In islands where the freshwater zone is less than about 5 m thick, the transition zone is often much thicker than the freshwater zone. The freshwater and transition zone thicknesses are dynamic and vary with changing groundwater recharge and, where it occurs, groundwater pumping.

Rainwater

Rainwater harvesting systems are common in most Pacific islands. In islands with moderate to high rainfall and limited or no fresh groundwater such as Tokelau, Tuvalu and some islands in Tonga, rainwater harvested from the roofs of houses and community buildings and stored in rainwater tanks is the primary source of freshwater. In other Pacific islands, rainwater is used as a supplementary source to other water sources, especially groundwater. In dry periods, rainwater use is limited to potable water needs (drinking, cooking and hand-washing). Common materials for rainwater tanks are ferrocement, steel, timber, fibreglass and polyethylene. In recent years, polyethylene tanks have become popular for household rainwater collection, and in some islands these tanks, with typical capacities of 3–6 kL but as large as 25 kL, are locally manufactured.

In some islands such as the Cook Islands, large open buildings have been constructed solely to harvest rainwater and store it in two 45-kL concrete storage tanks. On the island of Fongafale, Funafuti atoll, Tuvalu, rainwater is collected from the roofs of houses, community buildings and a concrete basketball court and stored in above-ground tanks and underground concrete cisterns. Where shortages are experienced at household tanks during extended dry periods, water is delivered by government tanker from community storages for a fee.

Rainfall is sometimes collected from ground surfaces. A prime example is the concrete runway on Majuro atoll, Marshall Islands, where surface runoff from rainfall is collected and pumped to large above-ground storages which are major contributors to the water supply on Majuro (USAID 2009). In contrast, simple rainwater collection systems consisting of plastic barrels located under the crown of coconut trees where rainfall runoff is concentrated are used in some outer islands of PNG.

Desalination

Desalination is a relatively expensive and complex method of obtaining freshwater for remote, small islands. The cost of producing desalinated water is almost invariably higher than developing groundwater or surface water due to the energy and operating expenses. Desalination systems also require trained operators to ensure the necessary operation and maintenance procedures are implemented.

The reverse osmosis (RO) method of desalination is used as the primary source of freshwater for Ebeye island, Kwajalein Atoll, Marshall Islands which has a population of about 9600 and the highest population density of any Pacific island. Desalinated water using RO is also the primary freshwater supply source for approximately 11,500 people on Nauru during frequent droughts and as a supplementary source to rainwater and limited groundwater when rainfall is plentiful. Vanuatu has installed solar powered RO units for Ambae and Aniwa islands for supplementary use for populations of 10,500 and 350, respectively (PRDR 2018).

Other countries have installed RO units for emergency use during droughts in the past decade such as on South Tarawa, Tarawa atoll and Banaba Island, Kiribati; Majuro atoll and some outer islands of the Marshall Islands; and Funafuti atoll, Tuvalu. RO units are also used for producing bottled water in Tongatapu, Tonga, and on a small number of tourist islands such as in Cook Islands and Fiji. Desalination will become a significant additional source of freshwater to groundwater and rainwater for South Tarawa in the next few years due to limited groundwater and increasing demand.

Production of freshwater from RO units using electricity from solar panels is an attractive option for remote locations or to reduce reliance on fossil fuels. Its disadvantage is that unless battery-backed, water production is limited to daylight hours. Batteries tend to have a short operational life in small islands. Direct solar distillation of freshwater from seawater is also possible. Limited production rates and the area needed for solar collection mean that it is mostly restricted to emergency supply of drinking water.

The long-term operational performance of RO units in Pacific islands has been generally poor, except on islands where they have become the major source of freshwater such as Nauru and Ebeye. Reasons for failures on other islands include poor operation and maintenance, high operation costs, lack of technical expertise and lack of manufacturer support (Freshwater and Talagi 2010; Duncan 2011).

Importation

Imported bottled water has become an alternative source of drinking water in some islands. This is normally imported, but in some islands such as Tongatapu, Tonga and Majuro atoll, Marshall Islands, bottled water has been produced from local desalination (RO) units. The cost of bottled water is invariably much higher than for water supplied by local water authorities. In addition, disposal of plastic bottles is an increasing problem in small islands, such as Tokelau and Funafuti, Tuvalu.

Water is imported between islands in some countries, especially as an emergency measure during droughts. Water has been imported by sea transport (boats or barges) during droughts, for instance, to outer islands of Fiji, PNG and Tonga. Bottled water was also imported to Tokelau from Samoa following a 6-month dry period in 2011 (Anderson et al. 2017).

During the period of major phosphate production, Nauru relied on imported water for much of its supply. In 1988, the estimated daily water consumption on the island was 1300 kL, of which about 30% was supplied as back-cargo from Melbourne, Australia, and the rest was mainly supplied by rainwater (55%) and a small amount (15%) from groundwater (Jacobson and Hill 1988). Despite above-average rainfall conditions on Nauru in 2001–2002, expensive potable water was shipped from Kosrae, FSM, as the main desalination plant ceased to operate and the centralised rainwater collection system had badly deteriorated. The cost of one shipment of 1.5 million L was reported to be AUD 87,000, equivalent to AUD 58 per kL, or approximately ten times greater than the cost of desalinated water (AusAID 2002).

During water shortages on some islands, people travel by boat or canoe to collect water from nearby islands with more plentiful water sources. Water is also piped to islands close to larger islands with more plentiful water resources. In Samoa, for example, freshwater is supplied by submarine pipeline over a distance of approximately 4 km from the western end of Upolu to Manono Island.

Other Sources

The extraction of water from humid air is also a possible source of freshwater and battery-backed, solar powered units are being trialled around the world, including in Vanuatu (Zero Mass Water 2019). The production rate is, however, small and limited to drinking water only, the initial costs are high, and maintaining the units in harsh Pacific environments is a challenge.

Other sources of water, for non-potable uses, include brackish groundwater, seawater and treated wastewater. There are many examples of brackish groundwater use in order to conserve valuable freshwater resources on small islands. Some atoll islands have wells with brackish water which is used for all purposes including drinking and cooking.

Seawater is used for toilet flushing and as a potential source for fire-fighting in densely populated parts of Tarawa and Majuro atolls and all of Ebeye island, Kwajalein Atoll, Marshall Islands. Dual pipe systems are used to distribute freshwater and seawater to houses and other connections. Seawater is also used on some islands for cooling electric power generation plants and for ice making. Treated wastewater is not a common source of non-potable water in small islands, but is used for irrigation of garden and recreational areas at some tourist resorts and hotels such as in Fiji.

During severe drought conditions, or after natural disasters, coconuts have been used as a substitute for drinking water in some small remote areas. People on some of the smaller outer islands in Fiji, Kiribati and PNG, for instance, have survived on coconuts during drought periods. The coconut tree is very salt-tolerant and can continue to produce coconuts even when groundwater is brackish.

Further Information

Further information on the water resources of the selected countries are presented in many publications including specific country water resources reports and others covering the wider region. Publications covering the 14 PICs and other islands in the Pacific Ocean include Carpenter et al. (2002), Falkland (2002, 2011), Scott et al.

(2003) and Duncan (2011). WMO et al. (2012) provide surface water resource information for Cook Islands, FSM, Fiji, Palau, PNG, Samoa, Solomon Islands and Vanuatu. Other publications covering the water resources of many islands within the Pacific Ocean as well as other parts of the world include UNESCO (1991), IETC (1998) and Tribble (2008). Vacher and Quinn (1997) provide extensive information about groundwater on many islands including 16 case studies of Pacific islands. Groundwater in Pacific islands is also extensively covered in Dixon-Jain et al. (2014).

Water resources publications specific to the most vulnerable islands in the Pacific (i.e. those with coral sand or limestone islands with only limited groundwater resources and rainwater) include White et al. (2002, 2007), USAID (2009), White and Falkland (2010), Bailey et al. (2016), Werner et al. (2017) and Post et al. (2018).

11.2.3 Freshwater Uses

Water Supply

The main consumptive use of freshwater in many Pacific islands is water supply for urban and rural communities and agriculture. The conjunctive use of groundwater and rainwater, when available, for potable and non-potable purposes, respectively, is common in some islands.

Per capita freshwater usage varies considerably between PICs and within islands of these countries depending on availability, quality, type and age of water distribution systems, cultural and socio-economic factors and system management. Freshwater usage varies from relatively low values of approximately 20–50 litres per person per day (Lpd), where water is very limited to much higher values. The minimum safe water requirement to satisfy essential health and hygiene needs in an emergency situation is 20–30 Lpd as shown in Fig. 11.2 (WHO and WEDC 2013)

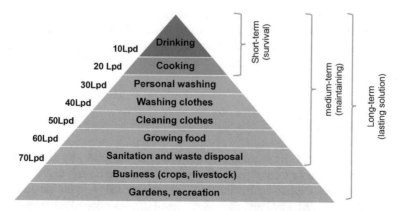

Fig. 11.2 Hierarchy of water requirements. (Modified from WHO and WEDC 2013)

while approximately 50 Lpd is recommended as a basic water requirement for domestic water supply (Howard and Bartram 2013; WHO 2017).

Water usage tends to be higher in urban than in rural areas for a number of reasons, including the use of water consuming devices such as washing machines and leakage from pipe distribution systems. Typical per capita water usage in well-managed water supply systems is in the order of 100–200 Lpd but can be as high as 1000 Lpd. For Rarotonga, the estimated water demand plus leakage in 2013 was estimated as 14,300 kL/day (AECOM 2013). This is equivalent to about 1100 Lpd based on the population of about 13,000 in both 2011 and 2016 (MFEM 2018).

Leakage from water supply pipelines and other losses including illegal connections and uncontrolled overflows at community or household tanks in urban centres and larger rural villages are a major issue. Water losses equal to or greater than 50% have been measured or estimated in a number of urban water supply systems in PICs and territories including Koror-Airai, Palau (Gerber 2010); Majuro, Marshall Islands (SOPAC 2007); and South Tarawa, Kiribati (White 2011; GHD 2017). The estimated water losses (or 'non-revenue water') in 2016 from the public water supply system on South Tarawa was 92% of the groundwater pumped (GHD 2017). Of 25 urban water utilities and other water supply agencies in Pacific islands, 13 had equal to or greater than 50% water losses (PWWA 2016).

Typical community water supplies in rural areas have a distribution pipe network using water from surface or groundwater sources. Surface water systems normally use gravity flow pipelines from springs or streams to tanks or standpipes in the village. Groundwater systems often use petrol, diesel or solar pumps, which may be operated for a number of hours each day, to supply water to a storage tank feeding standpipes within the village. Rural water supply systems are often managed by village or island councils or community 'water committees'. In some cases such as village water supply schemes in Tonga, a small fee is charged to households in order to cover operational expenses.

Urban water supply systems commonly consist of source works, groundwater pumping areas and/or surface water collection and storage, transmission pipelines and networks of distribution pipes to consumers. Urban water supply systems are generally operated by either a water authority or a water division within a government department. In a limited number of cases such as Port Vila, Vanuatu, a private water company is responsible for water supply. Cost recovery by fixed fee or metered usage has been implemented in urban areas of some Pacific island countries and territories.

Other Freshwater Uses

Freshwater is also used for subsistence agriculture, limited industry and mining in larger islands such as those in Fiji, New Caledonia, Solomon Islands and PNG. Non-consumptive uses include hydroelectricity generation using dams on rivers (in French Polynesia, Fiji, PNG and Samoa, as previously mentioned) or river intakes (Hawai'i and Pohnpei (FSM), Samoa and Solomon Islands) as well as navigation on larger rivers (e.g. in PNG). The percentages of national electricity generation from hydroelectric schemes were 40% in 2017 for PNG (IHA 2018) and 55% in 2013 for Fiji (IRENA 2015). Further hydroelectricity generation plants are planned for several PICs including Fiji, PNG, Samoa, Solomon Islands and Vanuatu.

11.3   Main Hydrological Influences on Freshwater Availability

The most important hydrological processes which impact on water resources are precipitation and evapotranspiration. In the Pacific region, precipitation most commonly occurs as rainfall. The magnitude and variability of rainfall and, to a lesser extent, evapotranspiration have a major influence on surface water, groundwater and rainwater resources.

11.3.1   Rainfall

Figure 11.3 shows a map of the Pacific region with the wide spatial distribution of mean annual rainfall. A summary of key rainfall characteristics for 14 PICs is presented in Table 11.2. These characteristics include mean annual rainfall and the coefficient of variation (Cv) of annual rainfall, a measure of annual rainfall variability, for the capital in each country and also the range of mean annual rainfalls in each country.

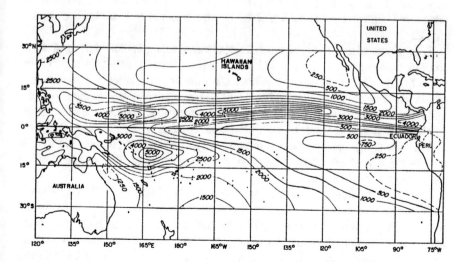

Fig. 11.3 Mean annual rainfall contours (mm/year) in the Pacific region. (From UNESCO 1991; modified from Taylor 1973)

Table 11.2 Summary of rainfall characteristics for the 14 Pacific island countries

Country	Mean annual rainfall (mm) at the capital	Coefficient of variation (Cv) of annual rainfall at the capital	Range of mean annual rainfall in country (mm)
Cook Islands	2000	0.20	1800–4500
FSM	4700	0.12	2600–8200
Fiji	3000	0.19	1500–6000
Kiribati	2000	0.47	900–3100
Marshall Islands	3300	0.15	2200–3300
Nauru	2100	0.54	2100
Niue	2100	0.24	2100
Palau	3700	0.13	3200–4300
PNG	1100	0.24	900–9000
Samoa	2900	0.20	2500–7500
Solomon Islands	2000	0.20	1800–9000
Tonga	1700	0.24	1700–2500
Tuvalu	3500	0.20	2400–4000
Vanuatu	2100	0.27	2000–4000

Notes: Rainfall data was obtained from several sources including water resources reports and national meteorological services in some PICs. Mean annual rainfalls are shown to the nearest 100 mm. The coefficient of variation (Cv) of annual rainfall (=standard deviation/mean) is a measure of temporal variability of annual rainfall with higher values indicating greater variability. The range of mean annual rainfall, obtained from all available rain gauges, illustrates the spatial variation of rainfall in each country

Figure 11.4 shows the mean annual rainfall for the capital in each country and the lowest and highest mean annual rainfall in each country as well as the Cv of annual rainfall.

For the countries with high islands, rainfall varies topographically from coasts to mountains where the lifting of cloud masses over mountains results in higher rainfall at higher elevations and on windward sides and lower rainfall on leeward sides of islands. This orographic effect is evident on larger islands such as those in PNG and Solomon Islands but is also apparent on much smaller islands such as Pohnpei, FSM, and Rarotonga, Cook Islands. Orographic effects in the island of Viti Levu, Fiji, results in higher mean annual rainfall of approximately 3000 mm in Suva on the eastern side of the islands compared with approximately 1900 mm in Nadi on the western side of the island.

There is considerable temporal variability of rainfall across the Pacific. The Cv of annual rainfall at the capitals of the 14 Pacific islands varies from a low value of 0.12 for FSM to a high value of 0.54 for Nauru (Table 11.2). The highest Cv's are associated with countries showing relatively lower mean annual rainfall such as Kiribati and Nauru. These high temporal variabilities are mainly due to El Niño and La Niña events most evident close to the equator (see Chap. 3).

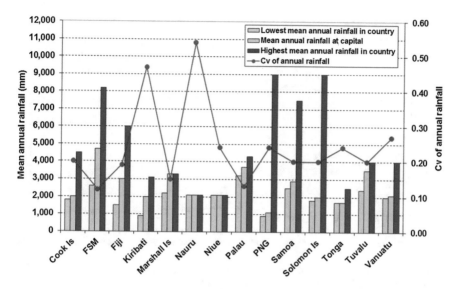

Fig. 11.4 Rainfall characteristics for 14 Pacific island countries

11.3.2   Evapotranspiration

Evapotranspiration is also a very important component of the hydrological cycle and of significance to water resources availability. It comprises the processes of evaporation from open surfaces and transpiration from vegetation and is dependent on many factors including solar radiation, humidity, wind speed, temperature, air pressure and atmospheric carbon dioxide levels. Potential evaporation occurs when there is an adequate supply of moisture available at all times.

Mean annual potential evaporation over the western and central Pacific region, summarised in Fig. 11.5, varies between about 1600 and 1800 mm (4.4–4.9 mm/day). The contours in Fig. 11.5, based on estimates for selected low-lying islands to remove the effects of altitude on evaporation in high islands, were made by Nullet (1987) using the Priestley-Taylor method to estimate potential evaporation (Priestley and Taylor 1972).

Estimates of potential evaporation in other studies using the Penman equation show values which are reasonably similar or lower. For example, Thompson (1986a) provides a potential evaporation estimate of 1460 mm/year for Tongatapu which is about 10% less than the approximate 1600 mm/year from Fig. 11.5. For the Cook Islands, Thompson (1986b, c) estimated potential evaporation in the range from about 1800 mm/year, averaged over four northern islands, to about 1500 mm/year, averaged over four southern islands. The estimate for the northern Cook Islands matches very well with Fig. 11.5, while the estimate for the southern Cook Islands is about 7% less than the approximate 1600 mm/year in Fig. 11.5. Mean annual potential evaporation rates for nine locations on the island of Tutuila, American Samoa (Izuka et al. 2005), were estimated from climate station data using the

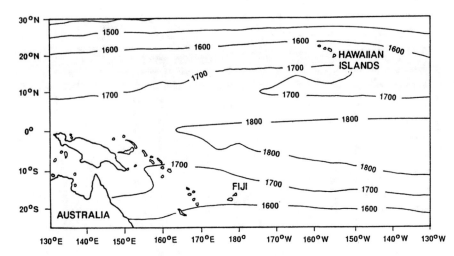

Fig. 11.5 Mean annual potential evaporation contours (mm/year) in the western and central Pacific region. (From UNESCO 1991; modified from Nullet 1987)

Penman equation (Penman 1948). These estimates varied from about 1580 mm (4.3 mm/day) near sea level to about 1100 mm (2.7 mm/day) at an elevation close to 500 m above sea level.

For Kiritimati (previously called Christmas Island) in eastern Kiribati, Porteous and Thompson (1996) estimated potential evaporation from pan evaporation data (using a pan factor of 0.7) to be approximately 1900 mm/year which is reasonably consistent with Fig. 11.5. For Tarawa in western Kiribati, a potential evaporation estimate of about 1800 mm/year was made in Falkland (1992) based on pan evaporation data for 1981–1991 and a pan factor of 0.8. Measurements on Tarawa using a climate station and the assumption that equilibrium evaporation is more applicable in tropical climates than the Priestley-Taylor method used by Nullet (1987) suggest a lower annual potential evaporation rate of 1420 mm or 3.9 mm/day (White et al. 1999), which is about 25% lower than in Fig. 11.5.

As shown in Fig. 11.5, the spatial variation of potential evaporation in the Pacific region is reasonably low, because of its dependence primarily on solar radiation. In addition, the temporal variation of potential evaporation at most islands, where estimates have been made, is relatively low compared with the temporal variations in rainfall.

On a monthly basis, evapotranspiration does not vary greatly, especially near the equator. For locations near sea level in the northern Cook Islands and American Samoa, the monthly estimated potential evaporation values using the Penman equation are 166 and 167 mm in October compared with 126 and 121 mm in June, respectively (Thompson 1986b; Izuka et al. 2005). These months correspond to the highest and lowest solar radiation at these locations.

Actual evapotranspiration from a land mass, which is of most importance in water resources studies, is dependent not only on the potential evaporation but also on the availability of moisture in the soil and the type and leaf area index of vegetation. Actual evapotranspiration can be less than, equal to or, in highly advective situations, greater than potential evaporation. Over annual periods, actual evapotranspiration in islands tends to be between about 40% and 80% of potential evapotranspiration.

11.3.3   The Water Balance

The difference between rainfall and actual evapotranspiration determines the water available for surface runoff and groundwater recharge. For islands with no surface runoff, such as atoll islands with high permeability sandy soils, mean recharge equals mean rainfall minus mean actual evapotranspiration. Based on a number of atoll island water balance studies, mean recharge and actual evapotranspiration vary in the ranges 30–50% and 70–50% of mean rainfall, respectively. Further information and case examples are available in Nullet (1987), UNESCO (1991) and Vacher and Quinn (1997).

11.3.4   Characteristics of Five Selected Islands

Details of hydrological and other characteristics are provided in Table 11.3 for five selected Pacific islands comprising three atoll islands and two volcanic islands spread across the Pacific. These islands have been selected to emphasise the large diversity in their characteristics and the impacts that these have on the availability of freshwater resources.

Table 11.3 shows the wide variation of rainfall between the five islands. The mean annual rainfall varies from just over 1000 mm (Kiritimati) to approximately 3500 mm (Fongafale). The Cv of annual rainfall shows a very high variability for Kiritimati (0.74) and a relatively low variability for the other four islands (0.19–0.24). Despite the relatively low and highly variable rainfall on Kiritimati, the island has significant groundwater resources partly because of the large size of the atoll and partly due to the relatively low permeability of the sediments compared with some other atoll islands. By comparison, the atoll islands of Fongafale and Ebeye, which have higher and less variable rainfall (both with Cv's of 0.19), show very little and no fresh groundwater, respectively, due to their narrower widths and higher permeability of the sediments. The latter two islands have been built with larger (up to boulder) size sediments due to cyclonic events that impact the islands, particularly Fongafale, compared with the relatively tranquil non-cyclonic conditions that have built Kiritimati with mainly coral sands and gravels. On Fongafale and Ebeye, the freshwater sources are rainwater and desalinated water. On the two larger and volcanic islands Rarotonga and Efate, the main freshwater resources are surface water and a combination of surface water and groundwater, respectively. The island size,

Table 11.3 Summary of characteristics and rainfall for five selected islands

Island	Kiritimati	Ebeye	Fongafale	Rarotonga	Efate
Country	Kiribati	Marshall Islands	Tuvalu	Cook Islands	Vanuatu
Island type	Atoll	Atoll island	Atoll island	Volcanic	Volcanic
Geology	Coral sand and gravel, with underlying karst limestone			Volcanic	
Land area (km²)	388	0.31	1.4	67	900
Maximum elevation (m)	13	3	5	653	647
Population	6500	9600	4500	13,000	84,000
Year of census	2015	2011	2012	2016	2016
Population density (people/km²)	17	31,000	3200	190	90
Annual rainfall (mm) (1951–2017)					
Mean	1036	2519	3478	1882	2271
Maximum	3686	3540	5141	3021	4104
Minimum	177	1507	2055	1090	1236
Cv	0.74	0.19	0.19	0.20	0.24
Rainfall trend (mm/year)					
Linear regression	+11.6**	−4.7^{ns}	−5.1*	−6.4*	+0.7^{ns}
Non-parametric	+6.7**	−4.6^{ns}	−5.7*	−6.2**	+0.5^{ns}
Estimated annual evapotranspiration (mm)	1800	1750	1700	1600	1600
Freshwater resources					
Main	Groundwater	Desalinated water	Rainwater	Surface water	Groundwater and surface water
Supplementary	Rainwater	Rainwater	Desalinated water	Rainwater	Rainwater

Notes: Populations are rounded to nearest 100. Population densities rounded to nearest 100 for Ebeye and Fongafale, and nearest 10 for Rarotonga and Efate. Significance of trends: ^{ns} indicates not significant, ** indicates significant at 0.05 level, * indicates significant at 0.1 level

geology, hydrogeology and topography are all factors in addition to rainfall as to whether groundwater is a significant resource.

Rainfall trends were calculated for each of the five islands using the annual rainfall records for the common period 1951–2017. The trends were calculated using both linear regression and a non-parametric method. Kiritimati shows a statistically significant increasing annual rainfall trend of about 12 or 7 mm depending on the method used. Similar increasing trends in rainfall are shown for other equatorial islands including Nauru and Tarawa atoll, Kiribati. Rarotonga shows a statistically significant decreasing annual rainfall trend of about 6 mm. The other three islands show lesser negative or slightly positive trends, but the results are not statistically significant (at 0.1 level).

11.4 Current Freshwater Issues in Pacific Islands

11.4.1 Water Security

Freshwater availability and security in the Pacific islands are affected to various degrees by current climate variability, non-climate-related natural hazards and human activities and impacts.

Water security can be defined as 'the capacity of a population to safeguard sustainable access to adequate quantities of acceptable quality water for sustaining livelihoods, human well-being, and socio-economic development, for ensuring protection against water-borne pollution and water-related disasters, and for preserving ecosystems in a climate of peace and political stability' (UN-Water 2013). A simpler definition is 'the availability of an acceptable quantity and quality of water for health, livelihoods, ecosystems and production, coupled with an acceptable level of water-related risks to people, environments and economies' (Grey and Sadoff 2007).

On small islands, water security depends on both the availability and quality of water resources as well as the condition of water supply facilities that provide or deliver the water to communities.

The most vulnerable islands and parts of islands in terms of water security from both climate and non-climate-related factors are the following:

- Small, low-lying islands particularly atoll islands with limited land area and fresh groundwater resources. Countries and territories with islands in this category include FSM, Kiribati, Marshall Islands, Tokelau and Tuvalu while other countries have some atoll islands including Cook Islands and French Polynesia.
- Small, mainly limestone islands with little or no fresh groundwater which rely on rainwater or desalinated water for freshwater needs such as Nauru and some islands in Fiji and Tonga. In some similar islands, water is obtained or imported from nearby islands during droughts when rainwater is depleted.
- Crowded urban and peri-urban areas, which are at major risk because of lack of adequate water supply and often polluted groundwater or surface water sources. Examples are South Tarawa, Tarawa atoll, Kiribati where local groundwater, polluted from sanitation systems and animals, is often used for non-potable uses and sometimes for potable uses, and Honiara, Solomon Islands.
- Remote parts of larger islands, which are at risk during droughts if water resources are depleted and food crops fail, due to the difficulty of access for emergency assistance and the time taken to regrow crops once rainfall returns to normal. Examples are found in the higher parts of larger Pacific islands including PNG.
- Very low level parts of islands, which are at risk of sea overtopping, erosion and temporary salinisation of groundwater from waves caused by storms, cyclones or tsunamis in addition to potential inundation from projected sea level rise. Examples are the atoll islands in the northern Cook Islands, Tokelau and Tuvalu and the low-lying parts of islands in Samoa, Solomon Islands and Tonga.

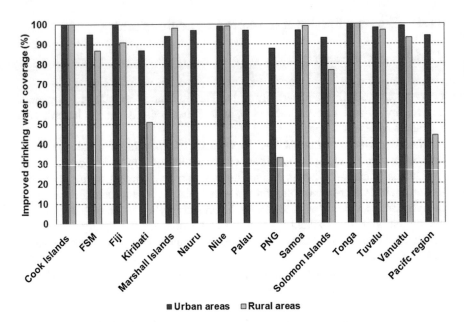

Fig. 11.6 Percentage of populations using improved drinking water sources in urban and rural areas of Pacific island countries, 2015. (Using data from UNICEF and WHO 2015)

11.4.2   Access to Safe Water

Although Pacific island populations have access to freshwater, the water is often not safe to drink or not available in adequate quantities to meet basic health needs.

Safe drinking water was defined by the United Nations Millennium Project Task Force on Water and Sanitation as 'water that is safe to drink and available in sufficient quantities for hygienic purposes' (UN Millennium Project 2005). Safety means that the water should be free from contamination by pathogens, hazardous chemicals or radiological hazards. Drinking water includes water used for drinking, cooking, personal hygiene and similar uses (WHO 2016).

The status of water supply and sanitation in 2015 for the Pacific region consisting of 14 PICs and six territories is provided in UNICEF and WHO (2015). The term 'improved drinking water' is used to describe safe water sources such as piped household water connections and other sources including public taps and stand-pipes, groundwater boreholes, protected dug wells, protected springs and rainwater collection. Not all improved drinking water sources are considered safe (Shaheed et al. 2014). Reasons for this include pipelines not always supplying safe water especially where water supply is intermittent, 'protected' dug wells which may be contaminated from nearby pollution sources and water stored in containers within households which can become contaminated after safe water is provided via pipeline.

Figure 11.6 shows the percentage of populations in the 14 PICs and the combined Pacific region (Oceania) using improved drinking water sources in urban and rural areas of PICs in 2015 using data from UNICEF and WHO (2015). The percentages for the six territories (American Samoa, Guam, Northern Mariana Islands, New Caledonia, French Polynesia and Tokelau) were all at or very close to 100%.

From Fig. 11.6, it is evident that the urban populations in PICs have generally better access to improved drinking water sources than rural populations. Kiribati (87%) and PNG (88%) have the lowest urban improved drinking water percentages. For rural populations, access to improved drinking water sources is lowest for PNG (33%), Kiribati (51%) and Solomon Islands (77%). Since 1990, the situation has not changed significantly for urban populations, but it has changed significantly for rural populations. Piped water supply systems cover only 25% of the population in the Pacific region, the lowest percentage of any region in the world except sub-Saharan Africa (UNICEF and WHO 2015). The percentages of improved water sources for the combined urban and rural populations of the 14 PICs in 2015 were, respectively, 94% and 44%. The low combined rural percentage is due to the large rural PNG population relative to rural populations in the other PICs.

Overall, 52% of the people in the Pacific region had access to improved drinking water in 2015 compared with the world average of 91% (UNICEF and WHO 2015). This low percentage for the Pacific region, which has increased from only 50% in 1990 compared with a population increase of about 70% in the same period, is influenced by the large PNG population which was about 70% of the total population in the Pacific region in 2015. If PNG is removed, the percentage of improved water sources in the remaining Pacific region is 91% equal to the world average (WHO 2016). For individual PICs and territories, the percentage of the population with access to improved water supplies varied from about 40% for PNG to 100% claimed for American Samoa, Cook Islands, Guam, French Polynesia, Tonga and Tokelau (UNICEF and WHO 2015). The PICs with 52% of people having access to improved drinking water fell well short of the Millennium Development Goal 7C of 73% by 2015 (UNICEF and WHO 2015).

ADB (2014) conservatively estimated the annual economic costs of poor water and sanitation conditions in South Tarawa, Kiribati, at between approximately AUD 550 and AUD 1080 per household which was equivalent to 2% to 4% of the country's 2013 Gross Domestic Product. The study shows that vulnerable groups have the most to gain from improvements, as they are more likely to suffer from illnesses including diarrhoeal diseases.

11.4.3   Impacts of Current Climate Variability

Current climate variability already imposes major challenges for freshwater availability in the Pacific islands through frequent droughts, floods, landslides and inundation by waves.

Droughts

Across the Pacific, droughts are commonly associated with El Niño or La Niña episodes, depending on geographic location. These often lead to lengthy water supply shortages in many Pacific islands and have resulted in declarations of emergencies such as those in Tuvalu and Tokelau in 2011. Droughts also severely challenge agriculture in many islands and can cause major reduction in hydroelectricity generation as in Fiji and Samoa.

There is a strong relationship between the El Niño Southern Oscillation (ENSO) and inter-annual rainfall which in turn influences surface and groundwater resources. The frequency, severity and duration of droughts on high islands are exemplified by reductions of stream flows and lowering of perched aquifer water tables leading to reductions or cessations in spring flows. On low islands and the low-lying parts of high islands, droughts lead to contraction in the thickness, areal extent and volume of freshwater lenses and coastal aquifers, which can increase the salinity of the groundwater both during and for some months following the end of droughts.

The severe impacts of the 1982–1983 El Niño drought on rainfall, stream flows, groundwater and water supplies in northern Pacific countries (FSM, Marshall Islands, Palau) and territories (Guam and the Commonwealth of the Northern Mariana Islands) are described by van der Brug (1986). The severe impacts of the major 1997–1998 El Niño drought on stream flows, groundwater, water supplies and agriculture have been documented for a number of PICs including PNG and Solomon Islands (Barr 1999), Fiji (Terry 2002; Terry and Raj 2002), Majuro atoll, Marshall Islands (Presley 2005), and other countries including Cook Islands, Samoa and Tonga (Scott et al. 2003). The droughts of 1997–1998 and 2015–2016 affected over a million people in the rural parts of PNG with severe impacts on water, food, agriculture, education, health and economic sectors (Barr 1999; PNG National Disaster Centre 2016).

The impacts of the long 1998–2001 La Niña drought in many islands of Kiribati, when total rainfall on Tarawa was only a third of the mean rainfall, are described in White et al. (2007) and White and Falkland (2010). This severe drought resulted in the declaration of a state of emergency followed by supply and installation of several desalination plants, none of which operated for more than a few years. The La Niña drought of 2011 severely impacted parts of Samoa, Tokelau and Tuvalu (Kuleshov et al. 2014). On Funafuti, Tuvalu's main atoll, rainwater supply became critically low when the 3-month rainfall from July to September and the 6-month rainfall from April to October were the lowest on record with only a quarter and a third of the mean rainfalls, respectively, for these periods. A similar shortage of rainwater occurred on the three atolls of Tokelau. States of emergency were declared for both Tokelau and Tuvalu, and bottled water and a number of desalination plants were supplied by donors (Kuleshov et al. 2014; PACC 2015).

The 1997–1998 drought in Majuro led to the installation of five large desalination (RO) plants (Presley 2005). A state of emergency was declared throughout the Marshall Islands in the 2015–2016 El Niño drought. During this drought, the runoff from the Majuro airport catchment was severely depleted in early 2016 (Leenders et al. 2017). The other main water source on Majuro, groundwater from the Laura freshwater lens, maintained reasonable thickness (Sinclair et al. 2016). In outer (rural) islands, rainwater tanks were depleted, and desalination units were installed in 35 communities. The economic impact was estimated at approximately USD 5 million (Leenders et al. 2017). Groundwater was not used extensively owing to a Marshall Islands Environment Protection Agency requirement to discontinue its use when total dissolved salts (TDS) exceed 500 mg/L. As Leenders et al. (2017) have noted, there is scope to use groundwater with a higher salinity up to a 'palatable limit' of 1200 mg/L as this does not present a health risk.

Severe droughts have caused major health impacts including malnutrition from reduced food supply, conjunctivitis, scabies and influenza-like-illnesses (PNG National Disaster Centre 2016; Leenders et al. 2017). A severe diarrhoea outbreak occurred in Funafuti during the 2011 drought. Emont et al. (2017) found that low rainwater tank levels and decreased handwashing frequency were key factors in the incidence of diarrhoea. The outbreak was controlled after implementation of a hygiene promotion campaign despite ongoing drought conditions and limited water availability.

Droughts have also significantly impacted on electricity generation in Viti Levu, Fiji, and Upolu, Samoa. For instance, low rainfalls in the Monasavu catchment resulted in very low water levels in 2013 and 2014 (IRENA 2015). In droughts, diesel generators are required to make up the shortfall in electricity generation with associated greater costs (SOPAC 2003; IRENA 2015).

Tropical Cyclones, Storms, Floods and Landslides

Tropical cyclones and storms often cause severe wind damage, floods and hillside erosion with consequent downstream damage and sedimentation. The highest rainfall intensities and maximum daily rainfalls on small islands are normally associated with tropical cyclones and tropical storms or depressions. Cyclones often cause major damage to infrastructure, including water supply infrastructure and agriculture, and some cause loss of life. Terry (2007) presents a detailed account of the occurrence and impacts of tropical cyclones in the South Pacific region.

Floods caused by tropical cyclones and storms are a major problem in urban centres and villages located near major rivers and frequently cause loss of life, destruction of houses and infrastructure and damage to agricultural land.

Examples of flooding impacts include the larger islands of Fiji, Papua New Guinea, Samoa, Solomon Islands, Tonga and Vanuatu. Terry et al. (2004) and Terry (2007) describe the impacts of major flooding resulting from Cyclone Ami in 2003 on Vanua Levu, Fiji. Flooding was made worse by the simultaneous occurrence of peak discharges and a strong storm surge. Yeo et al. (2007) describe 100 years of flooding on the Ba River in north-western Viti Levu, Fiji. Kuleshov et al. (2014) recorded impacts of floods from tropical depressions in January and March 2012 in the western part of Viti Levu which caused deaths of eight people and displacement of 3500 to temporary shelters. Infrastructure and crop damage was extensive.

Estimated losses due to the January 2012 flood (Kuleshov et al. 2014) were approximately FJD 40 million (about USD 20 million). The March 2012 event resulted in record flood heights at the towns of Nadi and Ba. Cyclone Winston in February 2016, one of the most intense tropical cyclones on record in the Pacific, caused massive damage and destruction in the islands of Fiji. An estimated 31,000 houses were damaged or destroyed, and over 40% of the population were affected in various ways (OCHA 2016). The economic cost was estimated at USD 1.4 billion. Flooding was extensive, water and sanitation systems were badly damaged and more than 240,000 people required emergency water supplies (OCHA 2016).

For the 5-year period 1997–2002, flooding affected approximately half a million people in PNG and, in terms of people affected, was the second worst natural disaster in PNG after droughts (World Bank 2009). Cyclone Guba in November 2007 caused severe flooding and landslides in Oro Province, in the south-eastern part of the main island of PNG during its slow development phase. The flooding resulted in more than 200 fatalities and significant loss of infrastructure and food crops (WMO 2008). Extreme rainfall in 2016 resulted in major floods and landslides in the Morobe Province of PNG (Reliefweb 2016). Cyclone Ita in April 2014 caused extreme rainfall and resulted in some of the worst flash flooding on record in Solomon Islands affecting the capital Honiara and villages across Guadalcanal Province. There were over 20 fatalities and over 50,000 people were affected (OCHA 2014).

In Samoa, Cyclones Ofa (1990), Val (1991) and Evan (2012) caused major flooding and damage to water storage, treatment plants and pipelines in and near Apia, Samoa, with consequent major disruptions to water supply (Samoa Government 2011, 2013). Forested areas within water supply catchments, which act to reduce floods, were also damaged and hydrological measurement infrastructure was destroyed. The estimated damage costs from Cyclone Evan were about USD 200 million in Samoa and USD 100 million in Fiji (Kuleshov et al. 2014).

Cyclone Pam in March 2015 was the most intense tropical cyclone in the Southern Hemisphere in that year and the second most intense tropical cyclone ever in the South Pacific basin (Handmer and Iveson 2017). It impacted on a number of PICs and territories including Fiji, Kiribati, Solomon Islands, Tuvalu, New Caledonia and especially Vanuatu. In Vanuatu, water supplies were damaged or contaminated with salt water leaving nearly half the population in need of clean drinking water. However, while there was extensive damage, populated areas were not subject to significant damaging storm surges, landslides or flooding and deaths were low (Handmer and Iveson 2017).

Cyclone Gita in February 2018 caused extensive damage and flooding in Samoa and in the islands of Tongatapu and 'Eua in Tonga. In Tonga, many houses and buildings and associated roofs and gutters used for rainwater collection were damaged or destroyed, and there was increased risk of groundwater contamination from sanitation systems (IFRC 2018).

Major landslides are also caused by torrential rainfall during tropical cyclones and storms. In 2002, heavy rainfall associated with Typhoon Chata'an caused 250 landslides on several islands in Chuuk State, FSM, with loss of life and destruction or damage to many buildings and infrastructure including water supply systems (Harp et al. 2004). Terry (2007) provides details of landslides triggered by intense rainfall due to cyclones in Fiji, Samoa and Solomon Islands which also have caused loss of life, destruction and damage. Further information regarding cyclones in the Pacific region is presented in Chap. 6.

Wave Overwash and Groundwater Inundation

Wave overwash resulting from storm and cyclone-driven waves has occurred on a number of low atoll islands in the northern Cook Islands, Marshall Islands, Tokelau and Tuvalu. Apart from loss of life and destruction of infrastructure, the freshwater lenses on some of these small low-lying islands have been partially inundated with seawater as a result of overtopping by waves generated by cyclonic storms. Cyclone Martin in November 1997 caused 10-m waves to inundate much of the populated areas of Manihiki atoll, northern Cook Islands. In addition to causing the loss of 19 lives, destroying or damaging 90% of the houses and resulting in an estimated USD 8 million damage cost (de Scally 2008), the seawater intruded the groundwater.

Many months or even years may be required for rainfall recharge to 'flush' the saltwater from freshwater lenses and restore wells to a potable condition. Cyclone Percy in February 2005 caused extensive damage and saline intrusion into freshwater lenses on the three islands of Pukapuka atoll, northern Cook Islands. Measured groundwater salinity profiles at monitoring boreholes showed considerable recovery of the upper part of the freshwater lenses after about 1 year as the more dense saline water moved through the freshwater lens (Terry and Falkland 2010; Terry et al. 2013). The saltwater plume was still evident 2.2 years after inundation in the lower part of the freshwater lens, as shown in the salinity profiles for one of the monitoring boreholes (Fig. 11.7). At 12 years after the saline intrusion, the salinity profile had returned to the pre-saline intrusion condition. It is most likely that this would have occurred earlier than 12 years, but the only other set of data obtained (in September 2012) was not considered reliable for use. The results of more recent groundwater modelling studies of wave overwash and groundwater inundation on atoll and other islands are provided in Sect. 11.5.7.

Severe storms generated by cyclones can also cause wind damage to houses and other buildings and severely impact on rainwater collection systems and even storage tanks. Coastal erosion processes caused by severe storms can modify and even reduce the land area overlying freshwater lenses. On Majuro atoll, Marshall Islands, storm-generated waves have overtopped parts of the island including the airport where surface runoff from rainfall is collected for water supply (Spennemann and Marschner 2000; PACC 2014).

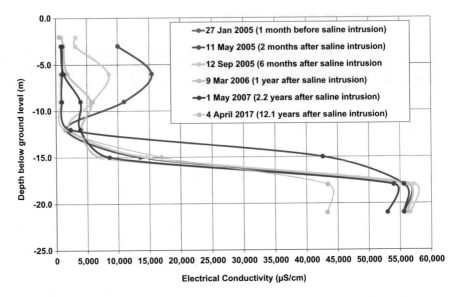

Fig. 11.7 Salinity profiles for monitoring borehole on Motu Ko, the largest of three islands on Pukapuka atoll, before and after Cyclone Percy in February 2005. The groundwater table at the borehole was about 2 m below ground level

11.4.4 Impacts of Non-climate-Related Natural Hazards

Non-climate-related natural hazards including volcanic eruptions, earthquakes and resulting tsunamis and landslides have caused major destruction and damage of infrastructure, including water supply systems on some Pacific islands. Other impacts include temporary salinisation of groundwater resources in small islands and coastal areas of larger islands.

Many Pacific islands are subject to seismic activity. Some countries have active volcanoes including PNG, Solomon Islands, Tonga and Vanuatu, while others have dormant volcanoes. Explosive volcanic activity can overwhelm water supplies including contamination from ash and catastrophic damage from volcanic blast (Scott 2002). The volcanic eruption in 1994 at Rabaul, the provincial capital of New Britain, PNG, caused thick ash deposits through the town and the destruction of infrastructure, including water supply infrastructure. This led to the resettlement of the population and additional water supply development.

Earthquakes and tremors have been experienced in a number of the selected countries especially those near tectonic plate boundaries including Samoa, Solomon Islands, Tonga and Vanuatu. As a consequence, water supply systems have been damaged due to cracking of pipes and rainwater tanks requiring emergency water supplies in the short term. The Vanuatu earthquake in November 1999 (Higgins et al. 1999) caused damage to water supplies in a number of islands.

Destructive tsunamis have been generated from submarine earthquakes as occurred in Solomon Islands in 2007 and Samoa in 2009 or submarine landslides generated by earthquakes such as in PNG in 1998. The July 1998 Sissano Lagoon-Aitape tsunami on the north coast of the main island of PNG caused the loss of more than 2200 lives, the destruction of villages including water supply facilities and the resettlement inland of 10,000 people (McSaveney et al. 2000; Ripper et al. 2001; Davies et al. 2001). The 2007 submarine earthquake and tsunami in western Solomon Islands caused loss of life, destruction and damage to infrastructure including water supply infrastructure on a number of islands (IFRC 2007). Similarly, the tsunami resulting from a major submarine earthquake south of Samoa in 2009 caused destruction and loss of life in Samoa, American Samoa and Tonga. In Upolu, Samoa, 43 villages and associated water supply infrastructure were damaged or destroyed, resulting in the resettlement of some coastal villages to higher areas and associated assessment and planning for the development and use of alternative water resources (UNICEF 2009a, b). The same tsunami impacted villages near the coast on Niuatoputapu island, Tonga, which led to their relocation away from the coastline (World Bank 2014).

Apart from damage to infrastructure, coastal inundation from tsunamis and cyclone-generated waves can cause temporary salinisation and pollution from sewage and petroleum products of wells used to extract groundwater for many coastal villages. Submarine landslides can also cause catastrophic changes to atolls with the loss of whole or parts of islands as, for example, on the atolls in the northern Cook Islands (Hein et al. 1997).

11.4.5   Impacts of Human Activities

A wide range of human activities can decrease the availability and security of water resources, water supplies and impact on water infrastructure.

In urban and peri-urban areas, rapidly increasing populations place increasing demands on the naturally available water resources. Piped water supply systems in Pacific island countries, with a few exceptions, have large losses which decreases access to adequate quantities of safe freshwater. High population densities and inadequate and poorly located sanitation facilities can lead to pollution of groundwater and surface water resources and resultant water quality degradation.

Expanding population pressures in islands with limited land area mean that water source areas are difficult to protect from settlement development. As a result, the incidence of water-borne disease is high in some Pacific islands. The rate of diarrhoeal diseases, linked to contaminated drinking water and poor sanitation conditions, is on average four to five times higher than in developed countries in the region such as Australia and New Zealand (WHO and SOPAC 2008). The high incidence of diarrhoeal diseases (including dysentery, typhoid and cholera) is mainly caused by contaminated drinking water which is linked to poor sanitation and hygiene. Outbreaks of cholera in PICs have been linked to contaminated water such as in Tarawa, Kiribati, in 1977 and in FSM: Chuuk in 1982–1983 and Pohnpei in 2000. The incidence of diarrhoeal diseases in PICs varies with water availability and climate and high incidences are associated with low water availability and higher temperatures (Singh et al. 2001).

Large numbers of animals in and near streams and groundwater sources, especially pigs, add to pollution (Fukumoto et al. 2016). Other sources of pollution include oil and fuel leaks and spills and agricultural chemical on rural land (Crennan and Berry 2002; van der Velde et al. 2007; White and Falkland 2010).

Inappropriate technology applied to rural water supply and sanitation can also lead to problems. Poor operational management exemplified by over-pumping coastal aquifers can result in salinisation of groundwater in some islands. Pumping systems are sometimes installed without the necessary investigations and the continued monitoring required to ensure sustainable pumping rates.

On high islands, clearing of native vegetation and conversion to open land for grazing of animals or planting of crops or open-cut mining increases peak stream flows and turbidity after heavy rainfall with consequent downstream water quality deterioration extending sometimes to near-shore reefs. Clearing can also lead to decreased baseflow (low flow) in streams due to compaction of soils causing decreased infiltration and lower groundwater yields for other purposes such as potable water supply.

Land ownership is central to community structures in many PICs. Inherent in this is the traditional custom that land ownership infers ownership of all contained resources, including water resources. When governments develop water sources on customary or owned land, disputes often arise between landowners and government agencies. Disputes have led to damage of water supply infrastructure including vandalism of groundwater pumping stations and monitoring boreholes in Tarawa, Kiribati (White et al. 1999), and water intakes and pipelines in Solomon Islands (Powell et al. 2006).

Poor water governance and management is also a major factor in decreasing current water security. Problems include lack of adherence to water policy and plans and absence of legislation and ineffective coordination and administration of water sector agencies. With a few exceptions, information on national water resources, their availability and quality as well as their use is limited in PICs. In some countries, even rainfall is no longer routinely monitored or recorded.

The absence or ineffectiveness of water source protection increases the vulnerability of water resources to human-induced contamination. Many water monitoring programs started under development aid projects, including rainfall, surface water and groundwater resource and water supply monitoring, continue for only a few years after project completion owing to lack of government priority and budgetary support. This is especially evident on outer islands of the PICs and some territories. Human and financial resource capacity limitations often prevent even essential operation and maintenance tasks from being undertaken. Insufficient training, education and ongoing development of water sector personnel and loss of personnel to more lucrative positions within or outside the country are ongoing problems. These are compounded by limited community education, awareness and participation in freshwater management, conservation and protection. At the political level, safe water supply does not appear to be a high priority in some countries. This is partly due to other urgent problems faced and partly due to the fact that traditionally water supply was a family responsibility. One potential solution being used by UNICEF (UNICEF and WHO 2015) is to introduce water, sanitation and hygiene education to school students at all levels.

11.5 Future Impacts of Climate Change on Water Security

To assess future impacts of climate change on water security, this section draws on projections of changing climate in the Pacific made under the Pacific Climate Change Science Program (PCCS P) and Pacific-Australia Climate Change Science and Adaptation Planning Program (PACCSAP) by the Australian Bureau of Meteorology and CSIRO.

11.5.1 Climate Change Projections

Table 11.4 summarises climate change projections for rainfall, droughts and tropical cyclone frequencies for each of the 14 PICs based on CMIP5 global climate models (Australian Bureau of Meteorology and CSIRO 2014).

Some of the projections in Table 11.4 are different from those shown in earlier publications based on CMIP3 global climate models (Australian Bureau of Meteorology and CSIRO 2011a, b) and as summarised in Heath et al. (2014). Also, there were minor inconsistencies in the words used to describe the trends in some of the parameters between the various summaries in Australian Bureau of Meteorology and CSIRO (2014). For instance, the summary at the start of Niue section mentions 'little change' in mean rainfall while the summary mentions 'increase slightly'.

The climate change projections in Australian Bureau of Meteorology and CSIRO (2014) which are common to all PICs with a high degree of confidence are increases in: frequency of extreme rainfall and intensity; mean and extreme air temperatures; sea level; and ocean acidification.

There is no consensus from the CMIP5 models about whether El Niño and La Niña events will become more or less frequent, or whether El Niño-driven sea surface temperature variability will become stronger or weaker in a future warmer climate (Australian Bureau of Meteorology and CSIRO 2014). IPCC (2013) states that there is high confidence that El Niño Southern Oscillation (ENSO) events will remain the dominant mode of interannual variability in the tropical Pacific and that the variability of ENSO-related precipitation will likely intensify due to the increase in moisture availability. Further, IPCC (2013) states that the large variations of ENSO amplitude and spatial pattern leads to low confidence in specific projected changes to ENSO and related regional phenomena within the twenty-first century.

Table 11.4 Summary of climate change projections from Australian Bureau of Meteorology and CSIRO (2014)

Country	Mean Rainfall	Drought frequency and duration	Extreme Rainfall frequency & intensity	Tropical Cyclone frequency
Cook Islands	Similar	Similar	Increase	Decrease
FSM	Increase	Decrease	Increase	Decrease
Fiji	Little change	Slight decrease	Increase	Decrease
Kiribati	Increase	Decrease	Increase	No projections
Marshall Islands	Increase	Decrease	Increase	Decrease
Nauru	Increase	Decrease	Increase	No projections
Niue	Slight increase	Slight decrease	Increase	Decrease
Palau	Increase	Decrease	Increase	Decrease
PNG	Increase	Decrease	Increase	Decrease
Samoa	Little change	Similar	Increase	Decrease
Solomon Islands	Slight increase	Slight decrease	Increase	Decrease
Tonga	Little change	Slight decrease	Increase	Decrease
Tuvalu	Little change	Slight decrease	Increase	Decrease
Vanuatu	Little change	Similar or slight decrease	Increase	Decrease

Note: The colours in the table indicate the confidence rating for each projection in Australian Bureau of Meteorology and CSIRO (2014): blue indicates high, yellow indicates medium and pink indicates low confidence

Australian Bureau of Meteorology and CSIRO (2014) point out that when interpreting climate change projections, it is important to consider current climate variability, such as ENSO status and the Interdecadal Pacific Oscillation (IPO) which strongly affect climate over time periods of a few years and decades. It is also noted that while Table 11.4 lists projections for tropical cyclone frequency, there are no projections available for tropical cyclone intensity which strongly impacts water security.

11.5.2 Impacts from Mean Rainfall and Evaporation Changes

Stream flows, groundwater recharge and water availability have significant potential to be impacted by changes in mean rainfall and evaporation, particularly in the drier parts of the year.

The CMIP5 Projections in Table 11.4 reveal projected increases in mean rainfall for three PICs with high confidence (FSM, Kiribati and Marshall Islands) and a further three PICs with medium confidence (Nauru, Palau and PNG). Kiribati and Nauru are very close to the equator while the other four PICs are within 10° of the equator. Both Niue and Solomon Islands have projections of slight increases in mean rainfall (with low confidence) while the projection for the Cook Islands is for similar mean rainfall (with medium confidence). The projections for the other five PICs (Fiji, Samoa, Tonga, Tuvalu and Vanuatu) indicate little change in mean rainfall (with low confidence). There are no PICs with projected decreases in mean rainfall. Of the 14 PICs, only three had high confidence projections while four had medium and seven had low confidence projections.

Australian Bureau of Meteorology and CSIRO (2014) have not provided any projections for evaporation. While temperature, one factor that influences evaporation, is projected to increase, there are no projections for other factors which also influence evaporation, including solar radiation which is influenced by changes in cloudiness, humidity and wind speed. Australian Bureau of Meteorology and CSIRO (2011a) have projected increases in potential evapotranspiration over much of the region occupied by the 14 PICs but based on CMIP3 climate models. The exception is the region near the equator, including Nauru, Kiribati and Tuvalu, where the relatively large projected rainfall increases exceed the projected smaller changes in potential evapotranspiration. The projected multi-model median changes in annual mean potential evaporation to 2090, relative to 1990, for the A2 (high) emissions scenario varied between −0.1 and +0.5 mm/day (equivalent to about −40 to +180 mm/year). These projected annual evapotranspiration changes are relatively small compared with the mean potential annual evaporation values for the Pacific region of 1600–1800 mm/year (refer to Sect. 11.3.2). Given the relatively minor projected changes in potential evaporation in Australian Bureau of Meteorology and CSIRO (2011), the impacts on mean stream flow and recharge compared with projected changes in rainfall are likely to be relatively small.

Given the above projections of mean rainfall and evaporation, islands close to the equator are likely to experience an improvement in water resources availability, while those further from the equator are likely to have little impact on water resources availability as there is only low to medium confidence in the mean rainfall projections (refer to Table 1.4).

The projections of drought frequency and duration in Table 11.4 reflect the projections in mean rainfall. All six PICs with a projected increase in mean rainfall also show a projected decrease in drought frequency and duration. Similar patterns are shown for droughts in the other PICs with slight increases, similar and little changes in mean rainfall. There is low confidence in the drought frequency and duration projections because of the lack of consensus about projected ENSO changes, which directly influence drought frequency and duration. In addition, the low to medium confidence in the mean rainfall projections implies that drought projections also have low to medium confidence, as shown in Table 11.4.

For the 14 PICs and nearby island territories, impacts on water resources are likely to be larger from changes to the variability of rainfall, due mainly to changes in ENSO activity, rather than from changes in mean rainfall. Future rainfall variability is therefore of much greater importance regarding impacts on water resources.

The lack of consensus in projections regarding ENSO means that there is uncertainty in the projected impacts on water resources.

Historical data shows increasing annual rainfall trends for the islands near the equator. For example, Kiritimati in Kiribati (approximately 2° north of the equator) shows a significant increasing trend using a non-parametric method of about 7 mm/ year for 1951–2017 data (Table 11.3). Similar increasing rainfall trends are shown for Tarawa, Kiribati, and Nauru, both of which are within 2° of the equator. This finding is consistent with the projections in Australian Bureau of Meteorology and CSIRO (2014).

Pacific islands away from the equator tend to have decreasing or near-neutral historic annual rainfall trends. For example, the annual rainfall trends for four islands in Table 11.3 (Ebeye, Fongafale, Efate and Rarotonga) show a decreasing but non-significant trend. Only the negative annual rainfall trend of approximately 6 mm for Rarotonga, located 21° south of the equator, is statistically significant.

A study of 14 rainfall stations with long-term records (some with over 90 years of record) in Fiji found there was no significant long-term trend in annual, wet season or dry season rainfall (Kumar et al. 2014). These authors also mention there is little evidence of long-term climate change in the historical record, consistent with the 'little change' mean rainfall projection in Table 11.4. They also confirm that ENSO has a significant influence on rainfall, especially on the drier western sides of the larger islands, Viti Levu and Vanua Levu.

Of seven long-term rainfall stations in Solomon Islands with more than 40 years of record, only one station showed a significant seasonal trend. Taro Island in the north-west part of the country near the Pacific Warm Pool showed a significant decreasing wet season rainfall trend (White 2016a). The direction of this trend is different to the low confidence 'slight increase' in mean rainfall in Table 11.4.

For seven long-term rainfall stations in Vanuatu with records of between 43 and 64 years, no significant annual, seasonal or 3-monthly rainfall trends was found (White 2016b). At three of the stations, significant but small negative rainfall trends were found for a single month of the year. These findings are consistent with the 'little change' mean rainfall projection in Table 11.4.

11.5.3   Impacts from Extreme Rainfall Frequency and Intensity Changes

The projections for extreme rainfall frequency and intensity in Australian Bureau of Meteorology and CSIRO (2014) show increases for all 14 PICs with a high confidence rating, although the magnitude of projected increases has a low confidence rating.

As an indicator of the change in the intensity of extreme rainfall, Australian Bureau of Meteorology and CSIRO (2014) provides estimates of the change in return frequency of the 1-in-20-year daily rainfall event for each of the 14 PICs using a majority of the climate models. For the very low emission scenario or Representative Concentration Pathway (RCP2.6), the 1-in-20-year daily rainfall event is projected to change from a 1-in-7-year daily rainfall event (for FSM) to a 1-in-13-year daily rainfall event (for Vanuatu) in the 20-year period centred on 2090. For the very high emission scenario (RCP8.5), the 1-in-20-year daily rainfall event is projected to change from a 1-in-4-year daily rainfall event (for Fiji, Kiribati, Niue, Palau, PNG and Solomon Islands) to a 1-in-6-year daily rainfall event (for FSM, Samoa and Tuvalu) in the 20-year period centred on 2090. While these projected changes are significant, their low confidence rating means it is not possible to make quantitative estimates of the impacts of increases in extreme rainfall.

From a qualitative viewpoint, the following negative impacts can be expected from an increase in the frequency and intensity of extreme rainfall:

- Increased flash flooding in rivers and streams with consequent problems including landslides, infrastructure damage and destruction, increased landslides and erosion, especially in cleared, steep catchments, and sedimentation of downstream reaches of streams and rivers and the coastal environment
- Increased economic costs of repairs and replacement of damaged and destroyed infrastructure
- Increased risk of death and injury to people impacted by flash floods and landslides

Beneficial impacts can be expected including:

- Enhanced groundwater recharge to freshwater lenses on coral sand and limestone islands and to coastal aquifers in high islands. Groundwater recharge is enhanced during periods of heavy rainfall as rainfall percolates quickly through the highly permeable soils and thus evaporative losses are minimised. Examples are the atolls of Tarawa and Kiritimati in Kiribati where freshwater lens thicknesses are significantly increased by the very high rainfall associated with El Niño events.
- Increased streamflow into water storages on high islands and potentially increased hydropower production.

11.5.4 Impacts from Mean and Extreme Temperature Changes

The projected air temperature increases appear to have a negligible impact on water demand compared with the water demand increase due to economic development, population increases and losses in pipe networks. An increase in water demand of 2% by 2030 due to projected temperature increase was estimated in a water master plan study for Tarawa, Kiribati (White 2011).

11.5.5   Impacts from Tropical Cyclones

While tropical cyclones are one of the largest threats to the sustainability of Pacific islands and their water security, their projected behaviour under climate change is uncertain. If tropical cyclones and storms are more frequent and intense, these would obviously have greater than current impacts on low-lying islands and coastal areas of high islands (refer to Sects. 11.4.3.2 and 11.4.3.3 and Chap. 6).

Tropical cyclone frequencies are projected to decrease for 12 of the 14 PICs, as shown in Table 11.4. Of these, Fiji, Niue, Samoa, Tonga and Tuvalu have a high confidence rating while PNG, Solomon Islands and Vanuatu have a medium confidence rating and Cook Islands, FSM, Marshall Islands and Palau have a low confidence rating. No projections are provided for Kiribati and Nauru, both close to the equator and generally unaffected by cyclones (Australian Bureau of Meteorology and CSIRO 2014). Despite this, some islands of Kiribati were severely affected by waves and wind caused by Cyclone Pam in 2015 (UNICEF 2015).

Widlansky et al. (2018) reviewed the occurrence of cyclones in four north Pacific islands: Guam, Kwajalein Atoll in the Marshall Islands, Okinawa in Japan and Oahu in the Hawaiian Islands. Of these islands, Guam has experienced the most frequent and severe cyclones while Oahu has experienced the least. These authors determined that the frequency of cyclones is likely to decrease for Guam and Kwajalein but will remain about the same near Okinawa and Oahu based on assessments from the climate models that best simulate the tropical Pacific climate. They also suggest that the maximum intensity of the strongest cyclones may increase near these islands. The results of this study are consistent with Zhang and Wang (2017) that projected a decrease in storms in the north-western Pacific.

11.5.6   Impacts from Mean Sea Level Rise

The prospect of sea level rise has been one of the main concerns to small island and coastal communities for many years (Burns 2002; Terry et al. 2013). Low-lying atoll islands are the most vulnerable to sea level rise with potential shoreline inundation and erosion, inundation and saline intrusion into island freshwater lenses and coastal aquifers (Woodroffe 2008; Storlazzi et al. 2018).

Projected Mean Sea Level

Projected mean sea level (MSL) changes for each of the 14 PICs are summarised in Australian Bureau of Meteorology and CSIRO (2014) using data from IPCC (2013) and a baseline of the 20-year period centred on 1995. Increases in MSL are projected by all models.

For RCP2.6, the projected multi-model mean rise varied between 12 cm (or 0.12 m) for the 20-year period centred on 2030 ('2030') and 0.42 m for the 20-year period centred on 2090 ('2090'). The corresponding lowest 5% and highest 95% confidence limits are 0.07 and 0.60 m for 2030 and 2090, respectively. For RCP8.5, the projected multi-model mean rise varied between the same 0.12 m for 2030 and 0.65 m for 2090 with 5% and 95% confidence limits of 0.07 and 0.92 m for 2030 and 2090, respectively. The confidence ratings are medium for all MSL rise values including those for RCP4.5 and RCP6.0 scenarios.

Using the approximate average of the projected mean rise values for the 14 PICs for the RCP2.6, RCP4.5, RCP6.0, and RCP8.5 scenarios of 0.42, 0.46, 0.49 and 0.63 m, the average annual rises in sea level over the 95 years from 1995 to 2090 (centres of the two 20-year periods) are approximately 4.4, 4.8, 5.2 and 6.6 mm, respectively. Further information about projected MSL changes are provided in Chap. 3.

Actual Sea Level Measurements and ENSO Effects

Between 1993 and 2017, satellite measurements show that there has been a rise in sea level of 3–6 mm/year for the Pacific islands, but with some notable differences between islands (CMEP 2018). Some islands in the Western Pacific (Solomon Islands, Papua New Guinea and Marshall Islands) have experienced a higher rate of sea level rise (up to 6 mm/year), compared to other islands further east (Samoa and Kiribati). This difference in sea level rise is mainly attributed to large scale trends in trade winds.

Shorter-term sea level fluctuations associated with ENSO events are similar to the projected multi-model MSL rise for RCP2.6 in 2090. Widlansky et al. (2015) show the observed sea level anomalies for four Pacific islands (Guam, American Samoa, Kiritimati and Galápagos Islands) for the period 1979–2013. During this period, fluctuations in sea level above and below normal levels were between 0.2 m and 0.4 m during the major El Niño events of 1982–1983 and 1997–1998. The sea level near Guam and Samoa, in the northwest and southwest Pacific, respectively, experienced a decrease while Kiribati and Galápagos Islands, in the central and eastern Pacific, experienced an increase owing to weakening of the trade winds. During La Niña events, sea level is up to 0.2 m above normal level in the western Pacific islands (Widlansky et al. 2017). The influence of current ENSO-related climate variability on sea level fluctuations provides an insight into the likely effects of future similar mean sea level increases.

Impact Studies on Fresh Groundwater

A number of impact studies have been conducted for freshwater lenses on atoll islands using groundwater models for a range of projected MSL rises and rainfall changes. These include Enjebi Island, Enewetak atoll, Marshall Islands (Oberdorfer and Buddemeier 1988), and Bonriki island, Tarawa atoll, Kiribati (Alam and Falkland 1997; World Bank 2000). Both studies used the variable density, two-dimensional model SUTRA (Voss 1984; Voss et al. 1997).

The analysis in World Bank (2000) for Bonriki island, which is the main source of freshwater for South Tarawa, Kiribati, included a study of impacts from eight climate change scenarios involving MSL rise, rainfall changes and potential loss of island width relative to a baseline scenario using average freshwater lens thickness and pumping conditions in the late 1990s. The results showed that an MSL rise of 0.2 m and similar rainfall to the prevailing pattern would cause virtually no change to the freshwater lens. An MSL rise of 0.4 m and similar rainfall would cause a slight increase in freshwater thickness. This is due to the fact that the average level of the freshwater lens, which is influenced by MSL, would rise slightly into less permeable Holocene sediments than the highly permeable underlying karst Pleistocene limestone (refer to Fig. 11.1). Alam and Falkland (1997) considered a number of scenarios including a 0.5 and 1.0 m MSL rise with similar rainfall to the prevailing pattern. These resulted in 2% and 9% increases in freshwater lens thickness. In all the above scenarios, it was assumed there was no loss of land at the edges of the island due to MSL rise. If land is lost at the edges of the island due to inundation from rising sea level and/or erosion from cyclone or storm-driven waves, there would be a significant effect on the freshwater lens. One scenario in World Bank (2000) which assumed a 0.4 m MSL rise, similar rainfall and loss of about 20% of island width due to inundation showed a 29% reduction in freshwater lens volume. For the case of 1.0 m rise, similar rainfall and loss of about 20% of island width, the reduction in freshwater lens volume was a very significant 77% (Alam and Falkland 1997).

From the above analyses, a small rise in MSL would have no detrimental impact on island freshwater lenses unless land is inundated due to the MSL rise and possibly the added effects of waves. However, if land is lost, then the impact of MSL rise will be significant.

A study of impacts on Bonriki was conducted by NIWA (2008) for two possible MSL rises of 0.48 and 0.79 m in a 20-year period centred on 2090. These possible MSL rises are approximately equivalent to the projected MSL rise for Kiribati for the mean RCP6 scenario and greater than the mean RCP8.5 scenario (Australian Bureau of Meteorology and CSIRO 2014). From maps in NIWA (2008) which showed some inundation of the island on the lagoon side, White (2011) estimated that the sustainable yield of the Bonriki freshwater lens could be reduced by about 20%.

More recent studies using a three-dimensional groundwater model of the Bonriki freshwater lens, including the impacts of sea level rise, are provided in Bosserelle et al. (2015), Mack (2015), Sinclair et al. (2015) and Galvis-Rodriguez et al. (2017). Bosserelle et al. (2015) found that the salinity of the pumped water from the freshwater lens was similar for both current and sea level rise scenarios (using 0.24 and 0.51 m rises). As in previous studies, it was assumed there was no loss of land.

Modelling studies using an analytical approach by Ketabchi et al. (2015) showed that freshwater lens volume is more sensitive to recharge, aquifer thickness and hydraulic conductivity, using typical parameter values, than sea level rise impacts.

A widely cited concern in the Pacific for low-lying islands and atolls is saline intrusion of fragile freshwater lenses by climate change-induced sea level rise. This is even cited as a concern in atoll islands with high rainfall such as Tokelau and Tuvalu which rely on freshwater sources other than groundwater including rainwater for most of the time and desalination or bottled water in emergencies.

11.5.7   Impacts from Wave Overwash in Addition to Mean Sea Level Rise

Several authors have used groundwater models to investigate the impacts of wave overwash in addition to mean sea level rise.

Terry and Chiu (2012) and Chiu and Terry (2013) investigated cyclone driven wave conditions with 0.1, 0.2 and 0.4 m sea level rise scenarios. These authors found that the saline intrusion due to overwash for the 0.4 m scenario is less than in current conditions. This is due to the higher water table caused by sea level rise reducing the thickness of the unsaturated zone thickness, i.e. the region in which seawater can quickly infiltrate and accumulate during overwash.

For the Bonriki freshwater lens, a scenario which partially inundates the island shows a significant effect on the salinity of individual pumping systems (infiltration galleries) in the affected area for both dry and wet conditions (Bosserelle et al. 2015; Mack 2015). Recovery from an extreme inundation event was shown to take 2–5 years, depending on recharge from rainfall. Under dry conditions, salinity in the pumped water recovers significantly after inundation in about 5 years. These groundwater investigations and modelling studies identified groundwater recharge and pumping conditions to be more critical than inundation for the sustainability of the Bonriki freshwater lens which is vulnerable to prolonged droughts. The recovery from the effects of groundwater pumping was found to take longer than recovery from the effects of possible seawater inundation. The most effective groundwater management strategy after a possible inundation event and during wet conditions is to switch off affected galleries. During dry conditions, additional non-inundated galleries would also need to be switched off. Regarding the impacts of pumping, Post et al. (2018) demonstrate that the Bonriki freshwater volume has not reached a new equilibrium following introduction of major pumping since the mid-1980s. One of their findings was that adjustment to the stresses of pumping takes a period of nearly three decades. This has implications for other atoll islands where groundwater pumping occurs.

Bailey and Jenson (2014) and Bailey (2015) used a groundwater model to investigate overwash on atoll islands using different combinations of island width, hydraulic conductivity of the sediments, timing of the overwash event and whether wells are present or not. One of their findings was that 6–10 months is required for a 1-m freshwater layer to re-develop on top of the affected freshwater lens for all scenarios. Also, variations in overwash depth, duration and time of year did not significantly affect recovery of the freshwater lens (Bailey 2015).

Holding and Allen (2014) modelled the response and recovery of overwash events for six island types (young volcanic, old volcanic, low coralline limestone, recent sedimentary, upland limestone and near continental bedrock). They found the time taken to return to pre-overwash varied from 1 to 19 years, depending on island type and recharge rate. For the recent sedimentary island type (which includes atoll islands), full freshwater lens recovery took 6 years which is similar to the longest period found for the Bonriki freshwater lens.

The impacts of a seawater inundation event in December 2008 on Roi-Namur Island, Kwajalein Atoll, Marshall Islands, were assessed by Gingerich et al. (2017) using a three-dimensional groundwater model. Rainwater collected from the airport runway and stored in adjacent concrete lined basins is used to artificially recharge the groundwater during the wet season. Rainwater is sufficient for the island's needs in the wet season, but groundwater is needed in the dry season. Two days after the saline intrusion, the salinity of the water pumped from the main infiltration gallery (skimming well) had increased to seawater salinity. By 10 days after the event, the salinity was 10–20% of seawater, but it took nearly 2 years for the freshwater lens to fully recover. As expected, it was found that recharge is an important factor in controlling the recovery of the freshwater lens.

Storlazzi et al. (2018) examined the impact of a large wave event in March 2014 on Roi-Namur Island which overwashed the seaward portion of the island, raised groundwater levels and caused saline intrusion into parts of the island's freshwater lens. Detailed information about this overwash event is presented in Oberle et al. (2017). Storlazzi et al. (2018) projected the impact of sea level rise and overwash on atoll infrastructure and freshwater availability under a variety of climate change scenarios. They concluded that the combined effect of sea level rise and annual overwash due to waves will cause atoll islands to become uninhabitable by mid-twenty-first century because of frequent damage to infrastructure and the inability of freshwater aquifers to recover between overwash events.

While the prospect of wave overwash combined with sea level rise is a serious challenge for low-lying islands and coastal areas of high islands, there are options for water supply other than pumping from groundwater. As previously mentioned, rainwater, desalinated water and a combination of these with brackish groundwater for non-potable uses are already or likely to be used on islands with significant water stress. Examples of these islands, all of which have high population densities, are South Tarawa, Kiribati; Nauru; Ebeye, Marshall Islands; and Funafuti, Tuvalu.

11.5.8 Resilience of Atoll Islands

Woodroffe (2008) noted that erosion of shorelines due to extreme events such as major waves from storms or cyclones is more likely to affect small islands and low-lying coastal areas than a gradual change in sea level. Webb and Kench (2010) show that many reef islands have remained largely stable or increased in size over the previous 20–60 years. These results are contrary to the widespread perceptions that all atoll/reef islands are eroding in response to recent sea level rise. Some are likely to erode, as at present, while others are likely to remain stable with the type and magnitude of changes varying (Webb and Kench 2010).

McLean and Kench (2015) studied changes in area for 146 islands on 12 atolls in 5 PICs (FSM, Kiribati, Marshall Islands, PNG and Tuvalu) and one territory (French Polynesia) over several decades to the mid- late 2000s. Of these islands, 73% showed stable areas while 19% and 8% showed increases and decreases, respectively. They noted that sea level had risen in the study area during the period 1950–2009 by approximately 1 mm/year (Guam) to 5.1 mm/year (Funafuti, Tuvalu) compared with the global average of 1.8 mm/year. They showed that atoll islands have persisted with little loss of land over the last several decades, despite high sea level rises in some parts of the Pacific. Island margins change with normal seasonal erosion and accretion processes and to extreme events. They concluded that sea level rise is just one of series of multiple stressors on atoll islands (McLean and Kench 2015).

A study of 709 islands on 30 Pacific and Indian Ocean atolls, including atolls in the five PICs mentioned above and French Polynesia, over past decades and up to a century revealed that no atoll lost land area (Duvat 2018). Also, the study found that 73% of the islands remained stable in land area while 16% increased in size and 11% decreased in size. The 374 smallest islands with areas less than 5 ha, representing 53% of all the islands studied, showed the highest variability in land area change. All of the 234 islands with areas greater than 10 ha, representing 33% of all the islands studied, remained stable or increased in area. The 16 largest islands, with areas greater than 200 ha, were the most stable. These include seven in the Tuamotu group of French Polynesia, two on Tarawa atoll in Kiribati, three in the Marshall Islands and four in Tuvalu. Bonriki island, South Tarawa, with an area of 863 ha in 1998 showed an increase in land area of 23% over the previous 30-year period (Duvat 2018). An earlier estimate by Biribo and Woodroffe (2013) was 19% increase in land area for Bonriki island.

From the above studies, it is evident that, over the past several decades, atoll islands have been resilient to MSL rises and extreme events. Further monitoring will be required to assess the impacts of sea level rise and waves in the coming decades.

11.6 Concluding Comments

In the face of the uncertainties surrounding the magnitude and timing of climate change (Barnett 2001), its impacts and lack of detail of ecosystem functions in PICs, Barnett (2005) concluded that the only rational adaptation strategy is 'to develop a society's general capacity to cope with change by building up its institutional structures and human resources while maintaining and enhancing the integrity of ecosystems'. In his view, any activity towards ecologically sustainable human development constitutes adaptation. He concluded that, despite limited financial, technological and infrastructure resources in PICs, their communities' well-developed local institutions, resilient social systems, sensitivity to environmental change and high degree of equity, together with their kinship-based, transnational networks, are the basis for considerable capacity to adapt to climate change.

Dovers (2009) argued quite generally that challenges faced in adapting to climate change are not new. Humans have attempted to cope with climate variability for a long time, and he cites examples in developed countries covering water management, local and regional economic vulnerability, biodiversity, health and wellbeing in remote communities, energy reform and emergency and disaster management. Some have concluded that the international preoccupation with adaptation to global climate change in PICs has distracted them from addressing the actual, local sustainability problems facing island communities (Connell 2003), particularly in managing vital freshwater resources and ensuring their longer-term security.

Water is the primary medium through which climate variability, climate change and natural hazards influence livelihoods and well-being of the people in the Pacific region. Better management of water and sanitation is key to effective adaptation responses (WHO 2016). Pacific islands need to manage many issues related to population growth, limited water resource availability, inadequate and deteriorating water supply infrastructure, limited institutional capacity and human resources in the water sector and insufficient and irregular funding sources even for ongoing operations, maintenance and management. The uncertainty of climate variability, climate change and natural hazards compound and exacerbate these management challenges to ensure sustainable water and sanitation for their populations (WHO 2016).

In this chapter, we have examined the great diversity in water resources and their use across PICs. We have considered the hydrological drivers that govern the availability of water and their spatial and temporal variability. We have also looked at projections of the impacts of climate change on water availability. The survival of Pacific people in the face of natural disasters in remote isolated islands with limited resources over hundreds to thousands of years is testament to their innate toughness and resilience. Climate change is a global phenomenon with diverse local impacts. While Pacific islands have to deal with the local impacts, the broader community must assist by minimising the global changes.

References

ADB (2014) Economic costs of inadequate water and sanitation, South Tarawa, Kiribati. Pacific studies series. Asian Development Bank, Manila, 74pp
AECOM (2013) Te Mato Vai—water supply master plan for Rarotonga. Prepared for Ministry of Finance and Economic Management, Government of the Cook Islands by AECOM New Zealand Limited, Oct 2013, 109pp

Alam K, Falkland A (1997) Vulnerability to climate change of the Bonriki freshwater lens, Tarawa. Report no HWR97/11, ECOWISE Environmental, ACTEW Corporation, Prepared for Ministry of Environment and Social Development, Republic of Kiribati, Apr 1997, 19pp

Anderson K, Barnes R, Dunaiski M, Raoof A (2017) Situation analysis of women and children in Tokelau. Prepared by Coram International for United Nations Children Fund (UNICEF), Apr 2017, 59pp

Webb AP, Kench PS (2010) The dynamic response of reef islands to sea-level rise: Evidence from multi-decadal analysis of island change in the Central Pacific. Global and Planetary Change 72(3):234–246

AusAID (2002) Review of water supply infrastructure on Nauru. Prepared by Marcus Howard, Infrastructure Adviser. Australian Agency for International Development, Oct 2002, 26pp

Australian Bureau of Meteorology and CSIRO (2011a) In: Hennessy K, Power S, Cambers G (Scientific eds) Climate change in the Pacific: scientific assessment and new research. Regional overview, vol 1. Pacific Climate Change Science Program, 257pp

Australian Bureau of Meteorology and CSIRO (2011b) In: Hennessy K, Power S and Cambers G (Scientific eds) Climate change in the Pacific: scientific assessment and new research. Country reports, vol 1. Pacific Climate Change Science Program, 273pp

Australian Bureau of Meteorology and CSIRO (2014) Climate variability, extremes and change in the Western Tropical Pacific: new science and updated country reports. Pacific-Australia climate change science and adaptation planning program technical report. Australian Bureau of Meteorology and Commonwealth Scientific and Industrial Research Organisation, Melbourne, Australia, 372pp

Bailey RT (2015) Quantifying transient post-overwash aquifer recovery for atoll islands in the Western Pacific. Hydrol Process 29:4470–4482

Bailey RT, Jenson JW (2014) Effects of marine overwash for atoll aquifers: environmental and human factors. Groundwater 52(5):694–604

Bailey RT, Barnes K, Wallace CD (2016) Predicting future groundwater resources of Coral Atoll Islands. Hydrol Process 30(13):2092–2105

Barnett J (2001) Adapting to climate change in Pacific Island countries: the problem of uncertainty. World Dev 29(6):977–993

Barnett J (2005) Titanic states? Impacts and responses to climate change in the Pacific Islands. J Int Aff 59(1):203–219

Barr J (1999) Drought assessment: the 1997-98 El Nino drought in Papua New Guinea and the Solomon Islands. Aust J Emerg Manag 14(2):31–37

Biribo N, Woodroffe CD (2013) Historical area and shoreline change of reef islands around Tarawa Atoll. Kiribati Sustain Sci 8:345–362

Bosserelle A, Jakovovic D, Post V, Rodriguez SG, Werner A, Sinclair P (2015) Groundwater modelling report. Bonriki Inundation Vulnerability Assessment (BIVA) assessment of sea-level rise and inundation effects on Bonriki Freshwater Lens, Tarawa Kiribati, SPC technical report SPC00010, 167pp

Burns WCG (2002) Pacific Island developing country water resources and climate change. In: Gleick P (ed) The World's water, 3rd edn. Island Press, Washington, DC, pp 113–132

Carpenter C, Stubbs J, Overmars M (eds) (2002) Papers and proceedings of the Pacific regional consultation on water in small island countries, co-ordinated by SOPAC and ADB, Sigatoka, Fiji, 29 Jul–3 Aug 2002 (published on CD-Rom)

CMEP (2018) In: Townhill B, Buckley P, Hills J, Moore T, Goyet S, Singh A, Brodie G, Pringle P, Seuseu S, Straza T (eds) Pacific marine climate change report card 2018. Commonwealth Marine Economies Programme, 12pp

Connell J (2003) Losing ground? Tuvalu, the greenhouse effect and the garbage can. Asia Pac Viewpoint 44(2):89–107

Crennan L, Berry G (2002) Review of community-based issues and activities in waste management, pollution prevention and improved sanitation in the Pacific Islands Region. IWP technical report 2002/03. In Wright A, Stacey N (eds) Issues for community based sustainable resource management and conservation: considerations for the Strategic Action Plan for the International Waters of the Pacific Small Island Developing States. International Waters Programme, South Pacific Regional Environment Programme, 108pp

Davies HL, Davies JM, Lus WY, Perembo RCE, Joku N, Gedikile H, Nongkas M (2001). Learning from the Aitape tsunami. In: Proceedings of the international tsunami symposium, Seattle, WA, USA, Aug 2001, pp 415–424

de Scally FA (2008) Historical tropical cyclone activity and impacts in the Cook Islands. Pac Sci 62(4):443–459

Dixon-Jain P, Norman R, Stewart G, Fontaine K, Walker K, Sundaram B, Flannery E, Riddell A, Wallace L (2014. Pacific Island groundwater and future climates: first-pass regional vulnerability assessment. Record 2014/43. Geoscience Australia, Canberra, 200pp

Dovers S (2009) Normalising adaptation. Glob Environ Change 19(1):4–6

Duncan D (2011) Freshwater under threat, Pacific Islands. United Nations Environment Programme and Secretariat of the Pacific Community, 2012, 58pp

Duvat KEV (2018) A global assessment of atoll island planform changes over the past decades. WIREs Clim Change 2018:e557. https://doi.org/10.1002/wcc.557

Emont JP, Ko AI, Homasi-Paelate A, Ituaso-Conway N, Nilles EJ (2017) Epidemiological investigation of a Diarrhea outbreak in the South Pacific Island Nation of Tuvalu during a Severe La Niña—associated drought emergency in 2011. Am J Trop Med Hyg 96(3):576–582

Falkland A (1992) Review of Tarawa Freshwater Lenses, Republic of Kiribati. Report HWR92/682. Hydrology and Water Resources Branch, ACT Electricity and Water. Prepared for Australian International Development Assistance Bureau

Falkland A (2002) A synopsis of information on freshwater and watershed management issues in the Pacific Islands region. IWP technical report 2002/02. In: Wright A, Stacey N (eds) Issues for community based sustainable resource management and conservation: considerations for the Strategic Action Plan for the International Waters of the Pacific Small Island Developing States. International Waters Programme, South Pacific Regional Environment Programme, 132pp

Falkland A (2011) Report on water security & vulnerability to climate change and other impacts in Pacific Island countries and East Timor. Prepared on behalf of GHD Pty Ltd for Department of Climate Change & Energy Efficiency, Pacific Adaptation Strategy Assistance Program, Aug 2011, 134pp

Freshwater A, Talagi D (2010) Desalination in Pacific Island countries, a preliminary overview. Technical report 437, Secretariat of the Pacific Community Applied Geoscience and Technology Division, Suva, 50pp

Fukumoto GK, Deenik J, Hura M, Kostka M (2016) Piggery impacts to water quality of streams in Pohnpei, Federated States of Micronesia. Water Issues WI-3, College of Tropical Agriculture and Human Resources, University of Hawai'i at Manoa, Jun 2016, 9pp

Galvis-Rodriguez S, Post V, Werner A, Sinclair P (2017) Climate and abstraction impacts in atoll environments (CAIA): sustainable management of the Bonriki Water Reserve, Tarawa, Kiribati. Secretariat of the Pacific Community technical report SPC00054, Suva, Fiji, 142pp

Gerber F (2010) An Economic Assessment of Drinking Water Safety Planning Koror-Airai, Palau. SOPAC Technical Report 440, November 2010, 42pp.

GHD (2017) Non-revenue water (NRW) report. South Tarawa water supply project (49453-001)—project preparatory technical assistance. Prepared for Asian Development Bank and Government of Kiribati, Sept 2017, 132pp

Grey D, Sadoff CW (2007) Sink or Swim? Water security for growth and development. Water Policy 9:545–571

Handmer J, Iverson H (2017) Cyclone Pam in Vanuatu: learning from the low death toll. Aust J Emerg Manag 32(2):60–65

Harp EL, Reid ME, Michael JA (2004) Hazard analysis of landslides triggered by Typhoon Chata'an on July 2, 2002, in Chuuk State, Federated States of Micronesia. U.S. Geological Survey open-file report 2004-1348, 22pp

Heath L, Salinger MJ, Falkland T, Hansen J, Jiang K, Kameyama Y, Kishi M, Lebel L, Meinke H, Morton K, Nikitina E, Shukla PR, White I (2014) Climate and security in Asia and the Pacific (food, water and energy). In: Manton M, Stevenson LA (eds) Climate in Asia and the Pacific. Springer, Dordrecht, pp 129–198

Hein JR, Gray SC, Richmond BR (1997) Chapter 16: Geology and hydrogeology of the Cook Islands. In: Vacher HL, Quinn TM (eds) Geology and hydrogeology of carbonate islands, Developments in sedimentology, vol 54. Elsevier, Amsterdam, pp 503–535

Higgins C, Planitz A, Biliki R (1999) Vanuatu earthquake and tsunami, 27 Nov 1999. Mission report, United Nations Disaster Assessment and Coordination Team, Dec 1999, 33pp

Howard G, Bartram J (2013) Domestic water quantity, service level and health. World Health Organization, Geneva, 33pp

Holding S, Allen DM (2014) From days to decades: numerical modelling of freshwater lens response to climate change stressors on small low-lying islands. Hydrol. Earth Syst. Sci. 19, 933–949

IETC (1998) Source book of alternative technologies for freshwater augmentation in Small Island Developing States. International Environmental Technology Centre, UNEP in collaboration with South Pacific Applied Geoscience Commission and the Water Branch of UNEP, Technical publication series no. 8, 223pp

IFRC (2007) Solomon Islands: earthquake and tsunami response; preliminary appeal no. MDRSB001. International Federation of Red Cross and Red Crescent Societies, 3 Apr 2007, 6pp

IFRC (2018) Emergency plan of action operation update, Tonga, Tropical Cyclone Gita. International Federation of Red Cross and Red Crescent Societies, 31 Mar 2018, 17pp

IHA (2018) 2018 Hydropower status report. International Hydropower Association, 104pp

IPCC (2013) In: Stocker TF, Qin D, Plattner G-K, Tignor M, Allen SK, Boschung J, Nauels A, Xia Y, Bex V, Midgley PM (eds) Climate change 2013: the physical science basis. Contribution of working group I to the fifth assessment report of the Intergovernmental Panel on Climate Change. Cambridge University Press, Cambridge, 1535pp

IRENA (2015) Fiji renewable readiness assessment. International Renewable Energy Agency, Abu Dhabi, 41pp

Izuka SK, Giambelluca TW, Nullet MA (2005) Potential evapotranspiration on Tutuila, American Samoa: U.S. Geological Survey Scientific investigations report 2005-5200, 40pp.

Jacobson G, Hill PJ (1988) Hydrogeology and groundwater resources of Nauru Island, central Pacific Ocean. Bureau of Mineral Resources Record 1988/12, 87pp

Ketabchi H, Mahmoodzadeh D, Ataie-Ashtiani B, Werner AD, Simmons CT (2015) Sea-level rise impact on fresh groundwater lenses in two-layer small islands. Hydrol Process 28:5938–5953

Kuleshov Y, McGree S, Jones D, Charles A, Cottril A, Prakash B, Atalifo T, Nihmei S, Lagomauitumua F, Seuseu SK (2014) Extreme weather and climate events and their impacts on island countries in the Western Pacific: cyclones, floods and droughts. Atmos Clim Sci 4:803–818

Kumar R, Stephens M, Weir T (2014) Rainfall trends in Fiji. Int J Climatol 34:1501–1510

Leenders N, Holland P, Taylor P (2017) Post disaster needs assessment of the 2015–2016 drought, Republic of the Marshall Islands, Feb 2017, 129pp

Mack P (2015) Summary Report, Bonriki Inundation Vulnerability Assessment (BIVA) assessment of sea-level rise and inundation effects on Bonriki Freshwater Lens, Tarawa Kiribati. SPC technical report SPC00014, 41pp

McLean RF, Kench PS (2015) Destruction or persistence of coral atoll islands in the face of 20th and 21st century sea-level rise? WIREs Clim Change 6:445–463

McSaveney MJ, Goff JR, Darby DJ, Goldsmith P, Barnett A, Elliott S, Nongkas M (2000) The 17 July 1998 tsunami, Papua New Guinea: evidence and initial interpretation. Mar Geol 170:81–92

MFEM (2018) Demographic table. Ministry of Finance and Economic Management, Government of the Cook Islands. http://www.mfem.gov.ck/statistics/census-and-surveys/census/142-census-2016. Accessed Dec 2018

NIWA (2008) Sea-levels, waves, run-up and overtopping, Information for Climate Risk Management. Kiribati Adaptation Programme Phase II. Report HAM 2008-22, National Institute for Water and Atmosphere Research Ltd., New Zealand, Sept 2008, 143pp

Nullet D (1987) Water balance of Pacific atolls. Water Resour Bull 23(6):1125–1132

Oberdorfer JA, Buddemeier RW (1988) Climate change, effects on reef island resources. In: Sixth internat. coral reef symposium, Townsville, Australia, vol 3, pp 523–527

Oberle FKJ, Swarzenski PW, Storlazzi CD (2017) Atoll groundwater movement and its response to climatic and sea-level fluctuations. Water 9:650

OCHA (2014) Solomon Islands: worst flooding in history. United Nations Office for the Coordination of Humanitarian Affairs. https://www.unocha.org/story/solomon-islands-worst-flooding-history. Accessed Dec 2018

OCHA (2016) Tropical Cyclone Winston, response & flash appeal, final summary. United Nations Office for the Coordination of Humanitarian Affairs, 13 Jun 2016, 16pp

PACC (2014) Vulnerability and adaptation (V&A) assessment for the water sector in Majuro, Republic of the Marshall Islands. Technical report 5, Pacific Adaptation to Climate Change Programme, Secretariat of the Pacific Regional Environment Programme, 27pp

PACC (2015) PACC demonstration guide: improving rainwater harvesting infrastructure in Tokelau. Technical report 15, Pacific Adaptation to Climate Change Programme, Secretariat of the Pacific Regional Environment Programme, Mar 2015, 29pp

Penman HL (1948) Natural evaporation from open water, bare soil, and grass. Proc R Soc Lond A193:120–146

PNG National Disaster Centre (2016) El Niño 2015-2016 post drought assessment, Report of the Interagency Post Drought Assessment in Papua New Guinea, Sept 2016, 50pp

Porteous AS, Thompson CS (1996) The climate and weather of the Rawaki and Northern Line Islands of Eastern Kiribati. NIWA Science and Technology series no. 42, National Institute of Water and Atmospheric Research Ltd., New Zealand, 60pp

Post VEA, Bosserelle AL, Galvis SC, Sinclair PJ, Werner AD (2018) On the resilience of small-island freshwater lenses: evidence of the long-term impacts of groundwater abstraction on Bonriki Island, Kiribati. J Hydrol 564:133–148

Powell B, Davies P, Costin G, Wairiu M, Ross H (2006) Interdisciplinary approaches to catchment risk management: case studies of Timor Leste and the Solomon Islands. In: International river symposium, Sept 2006, 13pp. http://www.watercentre.org/resources/publications/papers-presentations

PRDR (2018) Pacific regional data repository. http://prdrse4all.spc.int/system/files/pec-fund-vanuatu.pdf. Accessed Dec 2018

Presley TK (2005) Effects of the 1998 drought on the freshwater lens in the Laura Area, Majuro Atoll, Republic of the Marshall Islands. Scientific investigations report 2005-5098. U.S. Geological Survey, Reston Virginia, USA, prepared in cooperation with the Majuro Water and Sewer Company, Majuro Atoll, Republic of the Marshall Islands, 40pp

Priestley CHP, Taylor RJ (1972) On the assessment of surface heat flux and evaporation using large-scale parameters. Mon Weather Rev 100:81–92

PWWA (2016) Benchmarking report, 2016—five years of performance assessment (2011-2015). Pacific Water and Wastewater Association, 60pp

Reliefweb (2016) Floods and landslides follow drought in PNG highlands. International Organization for Migration. https://reliefweb.int/report/papua-new-guinea/floods-and-landslides-follow-drought-png-highlands. Accessed Dec 2018

Ripper ID, Anton L, Letz H, Moihoi M (2001) Preliminary account of the scientific investigations of the 17 July 1998 Sissano Lagoon (Aitape) Tsunami. Report 2001/1, Geological Survey of Papua New Guinea, 35pp

Samoa Government (2011) Strategic program for climate resilience. Prepared for the Pilot Programme for Climate Resilience (PPCR). Ministry of Finance, Feb 2011

Samoa Government (2013) Samoa—post-disaster needs assessment, Cyclone Evan 2012. Ministry of Finance, Mar 2013, 148pp

Scott D (2002) Island vulnerability. In: Carpenter C, Stubbs J, Overmars M (eds) Theme 2 paper. Papers and proceedings of the Pacific regional consultation on water in small island countries co-ordinated by SOPAC and ADB, Sigatoka, Fiji, 29 Jul–3 Aug 2002, 25pp

Scott D, Overmars M, Falkland A, Carpenter C (2003) Pacific dialogue on climate and water. Synthesis report. SOPAC, Feb 2003, 36pp

Shaheed A, Orgill J, Montgomery MA, Jeuland MA, Brown J (2014) Why "improved" water sources are not always safe. Bull World Health Organ 92:283–289

Sinclair P, Singh A, Leze J, Bosserelle A, Loco A, Mataio M, Bwatio E, Galvis-Rodriguez S (2015) Groundwater field investigations, Bonriki Water Reserve, South Tarawa, Kiribati. Bonriki Inundation Vulnerability Assessment (BIVA) Project, Secretariat of the Pacific Community Technical Report SPC00009, Suva, Fiji, 267pp

Sinclair P, Galvis-Rodriguez S, Loco A, Kumar A (2016) Assessment of the 2015-2016 drought impacts on the Laura fresh groundwater lens, Majuro Atoll, Republic of Marshall Islands. SPC report technical report SPC00010

Singh RBK, Hales S, de Wet N, Raj R, Hearnden M, Weinstein P (2001) The influence of climate variation and change on diarrheal disease in the Pacific Islands. Environ Health Perspect 109(2):155–159

SOPAC (2003) El Niño, Oil prices put pinch on Fiji Pacific Energy News, No. 1, Mar 2003, 11pp

SOPAC (2007) National integrated water resource management diagnostic report, Republic of the Marshall Islands, Sustainable Integrated Water Resources and Wastewater Management in Pacific Island Countries, SOPAC miscellaneous report 639, Nov 2007, 51pp

Spennemann DHR, Marschner IG (2000) Stormy years: on the association between the El Niño/Southern Oscillation phenomenon and the occurrence of typhoons in the Marshall Islands. http://marshall.csu.edu.au/Marshalls/html/typhoon/Stormy_Years.html. Accessed Dec 2018

Storlazzi CD, Gingerich SB, van Dongeren A, Cheriton OM, Swarzenski PW, Quataert E, Voss CI, Field DW, Annamalai H, Piniak GA, McCall R (2018) Most atolls will be uninhabitable by the mid-21st century because of sea-level rise exacerbating wave-driven flooding. Sci Adv 4(4)., 9pp

Gingerich SB, Voss CI, Johnson AG (2017) Seawater-flooding events and impact on freshwater lenses of low-lying islands: Controlling factors, basic management and mitigation. Journal of Hydrology 551:676–688

Taylor RC (1973) An atlas of Pacific Islands rainfall. Hawaiian Institute of Geophysics. Data report no. 25, HIG-73-9. Univ. Hawaii, Honolulu, HI, USA

Terry JP (2002) Water resources, climate variability and climate change in Fiji. Asia Pac J Environ Devel 9:86–120

Terry JP (2007) Tropical cyclones: climatology and impacts in the South Pacific. Springer, New York, 210pp

Terry JP, Falkland A (2010) Responses of atoll freshwater lenses to storm-surge overwash in the Northern Cook Islands. Hydrogeol J 18:749–759

Terry JP, Raj R (2002) The 1997–98 El Niño and drought in the Fiji Islands. In: Hydrology and water management in the humid tropics, Proceedings of the Second International Colloquium, 22–26 Mar 1999, Panama, Republic of Panama, IHP-V technical documents in hydrology no. 52. UNESCO, Paris, 2002, pp 80–93

Terry JP, Chui TFM, Falkland A (2013) Atoll groundwater resources at risk: combining field observations and model simulations of saline intrusion following storm-generated sea flooding. In: Wetzelhuetter C (ed) Groundwater in the coastal zones of Asia-Pacific. Springer, Dordrecht, pp 247–270. ISBN: 978-94-007-5648-9

Terry JP, Chui TFM (2012) Evaluating the fate of freshwater lenses on atoll islands after eustatic sea-level rise and cyclone-driven inundation: a modelling approach. Global and Planetary Change 88–89, 76–84.

Terry JP, McGree S, Raj R (2004) The Exceptional Flooding on Vanua Levu Island, Fiji, during Tropical Cyclone Ami in January 2003. Journal of Natural Disaster Science 26(1):27–36

Chui TFM, Terry JP (2013) Influence of sea-level rise on freshwater lenses of different atoll island sizes and lens resilience to storm-induced salinization. Journal of Hydrology 502:18–26

Thompson CS (1986a) The climate and weather of Tonga. New Zealand Meteorological Service, Misc. Publication 188(5), 60pp

Thompson C.S. (1986b) The climate and weather of the Northern Cook Islands. New Zealand Meteorological Service, Misc. Publication 188(3), 45pp

Thompson CS (1986c) The climate and weather of the Southern Cook Islands. New Zealand Meteorological Service, Misc. Publication 188(2), 69pp

Tribble G (2008) Ground water on tropical Pacific Islands—understanding a vital resource: U.S. Geological Survey Circular 1312, 35pp

UN Millennium Project (2005) UN millennium project task force on water and sanitation—health, dignity and development: what will it take? Earthscan, London, 228pp

UNESCO (1991) Hydrology and water resources of small islands, a practical guide. Studies and reports on hydrology no. 49. Prepared by A. Falkland (ed) and E. Custodio with contributions from A. Diaz Arenas and L. Similar and case studies submitted by others. United Nations Educational, Scientific and Cultural Organization, Paris, France, 435pp

UNGA (2014) Report of the third international conference on small island developing states, Apia, Samoa, 1–4 Sept 2014, UN General Assembly. United Nations General Assembly, 70pp. http://www.un.org/ga/search/view_doc.asp?symbol=A/CONF.223/10&Lang=E. Accessed Dec 2018

UNICEF (2009a) Intermediate post-disaster water supply, South eastern region Upolu, Samoa. Final report. Submitted to Samoa Water Authority by RedR on behalf of United Nations International Children's Emergency Fund, Dec 2009, 27pp

UNICEF (2009b) Long term water supply, South eastern region Upolu and Manono Island, Samoa. Preliminary concept report. Submitted to Samoa Water Authority by RedR on behalf of United Nations International Children's Emergency Fund, Dec 2009, 20pp

UNICEF (2015) Cyclone pam humanitarian situation report 13. United Nations International Children's Emergency Fund, 2–4 Apr 2015, 10pp

UNICEF, WHO (2015) Progress on sanitation and drinking-water—2015 update and MDG assessment. United Nations International Children's Emergency Fund and World Health Organization, New York, 90pp

UN-Water (2013) Water security & the global water agenda—a UN-water analytical brief. Prepared by United Nations University—Institute for Water, Environment and Health on behalf of the UN-Water Task Force on Water Security, 47pp

USAID (2009) Adaptation to climate change: case study—freshwater resources in Majuro, RMI, United States Agency for International Development, Aug 2009, 76pp

Vacher HL, Quinn TM (eds) (1997) Geology and hydrogeology of carbonate islands, developments in sedimentology no. 54. Elsevier, Amsterdam, 948pp

van der Brug O (1986) The 1983 drought in the Western Pacific. U.S. Geol. Survey open-file rpt. 85-418, Honolulu, HI, USA, 89pp

van der Velde M, Green SR, Vanclooster M, Clothier BE (2007) Sustainable development in small island developing states: agricultural intensification, economic development, and freshwater resources management on the coral atoll of Tongatapu. Ecol Econ 61:456–468

Voss CI (1984) SUTRA, a finite-element simulation model for saturated-unsaturated, fluid-density-dependent ground-water flow with energy transport or chemically-reactive single-species solute transport. USGS Water Resources investigations report 84-4389, 409pp

Voss CI, Boldt D, Sharpiro AM (1997) A graphical-user interface for the US Geological Survey's SUTRA code using Argus ONE (for simulation of variable–density saturated-unsaturated ground-water flow with solute or energy transport), U S Geological Survey, open-file report 97-421, Reston, VA, 106pp

Werner AD, Sharp HK, Galvis SC, Post VEA, Sinclair P (2017) Hydrogeology and management of freshwater lenses on atoll islands: review of current knowledge and research needs. J Hydrol 551:819–844

White I (2011) Tarawa water master plan, 2010–2030. Prepared for the Kiribati Adaptation Programme Phase II Water Component 3.2.1. Australian National University, Feb 2011, 49pp

White I (2016a) Multi country drought preparedness and response plan design. Solomon Islands—2, seasonal rainfall. Australian National University, Report prepared for UNICEF Pacific, Suva, Fiji, Feb 2016, 39pp

White I (2016b) Multi country drought preparedness and response plan design. Summary of hot spot analysis based on historic rainfall data, Vanuatu. Australian National University, report prepared for UNICEF Pacific, Suva, Fiji, Apr 2016, 26pp

White I, Falkland A (2010) Management of freshwater lenses on small Pacific islands. Hydrogeol J 18:227–246

White I, Falkland A, Crennan L, Jones P, Etuati B, Metai E, Metutera T (1999) Groundwater recharge in low coral islands, Bonriki, South Tarawa, Kiribati: Issues, traditions, and conflicts in groundwater use and management. UNESCO IHP-V, technical documents in hydrology no. 25. UNESCO, Paris, France, 37pp

White I, Falkland A, Etuati B, Metai E, Metutera T (2002) Recharge of fresh groundwater lenses: field study, Tarawa Atoll, Kiribati. In: Hydrology and water resources management in the humid tropics, Proceedings of the second international colloquium, 22–26 Mar 1999, Panama, Republic of Panama, IHP-V technical documents in hydrology no. 52. UNESCO, Paris, 2002, 299–332

White I, Falkland A, Metutera T, Metai E, Overmars M, Perez P, Dray A (2007) Climatic and human influences on groundwater in low atolls. Vadose Zone J 6:581–590

WHO (2015) Human health and climate change in Pacific island countries. World Health Organization Western Pacific Region, 145pp

WHO (2016) Sanitation, drinking-water and health in Pacific island countries: 2015 update and future outlook. World Health Organization Western Pacific Region, 86pp

WHO (2017) Guidelines for drinking-water quality. Fourth edition incorporating the first addendum. World Health Organisation, Geneva, 541pp

WHO, SOPAC (2008) Sanitation, hygiene and drinking-water in the Pacific Island countries—converting commitment into action. World Health Organization and Pacific Islands Applied Geoscience Commission, 2008, 58pp

WHO, WEDC (2013) Technical notes on drinking water, sanitation and hygiene in emergencies, 2nd edn. Prepared for World Health Organization by Water, Engineering and Development Centre, Loughborough University, Leicestershire, UK, Sept 2013

Widlansky MJ, Timmermann A, Cai W (2015) Future extreme sea level seesaws in the tropical Pacific. Clim Change 1:e1500560

Widlansky MJ, Marra JJ, Chowdhury MR, Stephens SA, Miles ER, Fauchereau N, Spillman CM, Smith G, Beard G (2017) Future extreme sea level seesaws in the tropical Pacific. Clim Change J Appl Meteorol Climatol 56(4):849–862

Widlansky MJ, Annamalai H, Gingerich SB, Curt D, Storlazzi CD, Marra JJ, Hodges KI, Choy B, Kitoh A (2018) Tropical cyclone projections: changing climate threats for Pacific Island defense installations. Weather Clim Soc 11, 15pp

WMO (2008) Report of RA V Tropical Cyclone Committee for the South Pacific and Southern East Indian Ocean, twelfth session, Alofi, Niue, 11–17 Jul 2008

WMO, EU, UNESCO, SPC (2012) Catalogue of rivers for Pacific Islands. World Meteorological Organization, European Union, United Nations Educational, Scientific and Cultural Organisation and Secretariat of the Pacific Community, 151pp

Woodroffe CD (2008) Reef-island topography and the vulnerability of atolls to sea-level rise. Glob Planet Change 62:77–96

World Bank (2000) Adapting to climate change. Vol IV in Cities, seas and storms, managing change in Pacific Island economies. Papua New Guinea and Pacific Island Country Unit. The World Bank, 135pp

World Bank (2009) Disaster risk management programs for priority countries. Global facility for disaster reduction and recovery, World Bank and International strategy for disaster reduction, 306pp

World Bank (2014) Implementation completion and results report, Tonga Post Tsunami reconstruction project, Kingdom of Tonga, report no.: ICR00003036, Jun 2014, 50pp

Yeo SW, Blong RJ, McAneney KJ (2007) Flooding in Fiji: findings from a 100-year historical series. Hydrol Sci J 52(5):1004–1015

Zero Mass Water (2019) Meet the technology that's perfecting your water. https://www.zeromasswater.com/ap/source/. Accessed Jan 2019

Zhang C, Wang Y (2017) Projected future changes of tropical cyclone activity over the western North and South Pacific in a 20-km-mesh regional climate model. J Clim 30(15):5923–5941

Chapter 12
Climate Change and Impacts on Biodiversity on Small Islands

Lalit Kumar, Sadeeka Jayasinghe, and Tharani Gopalakrishnan

12.1 Introduction

Biodiversity is the fundamental component of the life support system on Earth and is an integral part of human survival. Biodiversity goes beyond the provisioning of materials for human welfare and livelihoods to include core values such as cultural, recreational, security, social relations, health, and ecosystem services. Food production/agriculture depends on biodiversity for the genetic diversity of trees and crops, pollination, fertilisation of the soil, nutrient recycling, erosion control, pest regulation, and disease control and prevention (Heller and Zavaleta 2009). Globally, 3.2 billion individuals rely on seafood for 20% of their average per capita intake of animal protein (FAO 2016). The majority of the poor in the world reside in rural areas and are dependent on forests, water, wetlands, and pastures for their livelihood. About one-third of the global population rely on wood as their primary source of energy, and forest products provide a significant contribution to the shelter of at least 1.3 billion or 18% of the world's population (Rametsteiner and Whiteman 2014). Climate change impacts on biodiversity, and the ecosystem services they provide have been a key topic of investigation in recent years. Various components of climate change are expected to affect all levels of biodiversity. Climate change can have different effects on people, populations, species, environmental networks, and ecosystems. Over the last two decades, anthropogenic climate change has been a dominant factor that has caused significant changes in biodiversity in different

L. Kumar (✉) · T. Gopalakrishnan
School of Environmental and Rural Science, University of New England,
Armidale, NSW, Australia
e-mail: lkumar@une.edu.au; tgopalak@myune.edu.au

S. Jayasinghe
Department of Export Agriculture, Faculty of Animal Science and Export Agriculture,
Uva Wellassa University, Badulla, Sri Lanka
e-mail: ljayasi2@myune.edu.au

© Springer Nature Switzerland AG 2020
L. Kumar (ed.), *Climate Change and Impacts in the Pacific*, Springer Climate,
https://doi.org/10.1007/978-3-030-32878-8_12

areas and habitat types (Folke et al. 1996). Climate change directly affects biodiversity by altering basic habitat variables such as rainfall and temperature and poses risks through extreme events such as storm surges, drought, hurricanes, flooding, lightning, and fires.

Based on projections from the most recent Intergovernmental Panel on Climate Change report (IPCC 2013), temperatures may increase by 1.6–4.3 °C by 2100 compared to the 1850–1900 baseline. The majority of terrestrial biodiversity is likely to be impacted by climate change since their basic physiological functions are strongly influenced by environmental temperature. Current global warming trends predict that global climate change alone could result in the extinction of over half of the known species on Earth (Parmesan et al. 2005). Temperatures exceeding the physiological limits of species have been reported to have caused mortality in Australian flying-fox species (Welbergen and Davies 2011) and the golden toad (*Bufo periglenes*) (Pounds and Crump 1994). Floods have caused disastrous, species-specific mortality in desert rodents, leading to population declines (Thibault and Brown 2008). Climate, especially seasonality, is likely the main driver of beta-diversity among the rainforest trees in Western Ghats (Davidar et al. 2008). Previous studies have reported climate change impacts on plant phenology, such as earlier leafing, flowering, and fruiting (3–5 days/°C rise in temperature), and delays in autumn activities (Cleland et al. 2007). As plants are fine-tuned to seasonality, the modifications in the timing of plant activities provide the most compelling evidence of the impacts of climate change on them (Cleland et al. 2007). A significant proportion of the variability seen in plant physiology can be linked to changes in climate (Van Vliet et al. 2008).

Shifts in climatic envelopes is a common phenomenon which will affect the biodiversity of particular ecosystems (Chapin et al. 2000; Drielsma et al. 2016; Kogo et al. 2019; Lamsal et al. 2018; Taylor and Kumar 2016). With temperature rises, the geographical location of climate envelopes will change considerably, potentially to the extent that species will not be able to survive in their existing locations any longer. With shifting climatic envelopes, these species will need to either adapt or migrate to cooler and moister environments. Land use change and the associated loss of habitat are significant threats to global biodiversity (Cleland et al. 2007) since they reduce the ability of organisms to adjust their habitat in reaction to climate change. It will also be necessary for marine species to adapt to warmer ocean temperatures. Many ecosystems are physiologically vulnerable to spikes in temperature. Polar regions are now witnessing some of world's fastest as well as most severe climate change, contributing to changes in biodiversity. Since the early twentieth century, Arctic air temperatures have risen by approximately 5 °C; an extra warming of about 4–7° in the Arctic is projected over the next 100 years. Drylands are especially susceptible to a changing climate since small temperature changes can have severe effects on the biodiversity of arid and semi-arid regions. It can significantly affect populations and economies as many individuals depend on dryland biodiversity (Folke et al. 1996). Climate change will also result in hotter and drier deserts. Forest growth may also be affected by small changes in temperature and precipitation. It has been shown that a temperature rise of 1 °C can alter the functioning and structure of forests, while roughly 9% of all known tree species are already in danger of extinction (Willis and Bhagwat 2009). Research has also shown that, in the past few decades, over 20% of freshwater fish species in the world have either become threatened or endangered or totally extinct (Willis and Bhagwat 2009).

Increased atmospheric CO_2 concentrations leads to higher growth rates in many plant species. This is favourable for human beings, but only if this CO_2 'fertilisation' effect is matched by an adequate amount of other resources such as soil moisture and nutrients. The oceans absorb a large amount of CO_2 emissions. This has led to a reduction in the pH of the oceans, which in turn impacts the rate at which many marine organisms construct their skeletons, resulting in a slower recovery of reefs damaged by bleaching or other agents (Staudinger et al. 2013). Increments of CO_2 are also likely to amplify the many impacts on flora and fauna in ecosystems, leading to phenological changes in plants and animals. Studies have shown the effects of forecast climate change on the health of plants, including research that demonstrates that high concentrations of CO_2 in the atmosphere increase rice blast (*Magnaporthe oryzae*) infection risk and the percentage of rice (*Oryza sativa*) plants impacted by sheath blight (Kobayashi et al. 2006). There is increasing evidence that high levels of CO_2 can have a profound impact on host-pathogen dynamics for both plant and animal diseases, including human diseases (Haines et al. 2006). The increase in atmospheric CO_2 and global temperatures influence the amount and diversity of insect damage to plants (Currano et al. 2008). A reduction in the ecological overlap between crops and pollinators can also reduce the dietary breadth of pollinators (Memmott et al. 2007). This can result in extinction or interference of pollinators and/or plants. Studies in the Mediterranean demonstrate more rapid progress in insect phenology compared to plant phenology, indicating the potential for enhanced decoupling of pollinator-flower interactions (Cleland et al. 2007). In combination with a sudden increase in atmospheric CO_2 and temperature, the quantity and variety of insect damage to crops has increased (Currano et al. 2008).

IPCC models have shown that there will be changes in annual rainfall and evapotranspiration in the global context (IPCC 2007; Staudinger et al. 2013). These changes will influence biodiversity both directly and indirectly. Climate models project higher rainfall in Southeast Asia and the temperate regions and lower rainfall in Central Asia, the Mediterranean region, Africa, parts of Australia, and New Zealand. Regions receiving higher rainfall will have a higher flood probability while those receiving lower rainfall will have a higher drought probability, with the likelihood of a higher frequency of extreme climatic events. According to past research, precipitation has a significant impact on biodiversity, its productivity, and species richness. The predicted reduction in rainfall and subsequent reduction in river flow will have a significant effect on aquatic flora and fauna (Folke et al. 1996). Rainfall and its coefficient of variation, and drought in particular, have been reported to decrease the populations of birds and mammals. For example, rainfall in Australian tropical rainforests has been shown to explain bird abundance spatial patterns (Hill et al. 2010), influence the population dynamics of desert bighorn sheep (*Oris canadensis*) in the hills of California (Hill et al. 2010), and abundance of swamp antechinus in Urquhart Bluff in Victoria, Australia (Magnusdottir et al. 2008).

Rainfall and abundance relationships indicate that rainfall drives the dynamics of ungulates in African savannas and that global warming-related changes in rainfall has the potential to affect the spatial distribution, abundance, and richness of these animals (Bellard et al. 2012). The location, extent, and biological diversity of rain-forests are mainly determined by rainfall and its variation (Di Falco and Chavas 2008); however, rainforest type and the kinds of species living there are more temperature-dependent. Increased rainfall favours certain types of rainforest while decreased rainfall favours woodlands and open forests. Regional precipitation patterns and topographic limitations suggest that such new habitat would be very far removed from that occurring in the Wet Tropics (Hilbert et al. 2004).

While global warming is the key factor causing more and rapid extreme events worldwide, such as hurricanes, drought, higher precipitation, flooding, and forest fires, the impacts of extreme events on biodiversity are not easy to predict (Easterling et al. 2000). The impacts include both direct effects on the biology of species and indirect significant alterations on the relationships between species and their habitats (Rouault et al. 2006), which may result in changing community associations (Chase 2007; Knapp et al. 2008), thereby limiting their adaptive capacity (Lavergne et al. 2010). Extreme drought and heat waves change the biological parameters of each individual by way of their reproductive status, specific physiological constrains such as homeostasis (Easterling et al. 2000), productivity of terrestrial ecosystems (Archaux and Wolters 2006; Ciais et al. 2005), and threat to invertebrate (Rouault et al. 2006) and vertebrate herbivores (Garel et al. 2004). For example, Arctic foxes face habitat loss, competition, and predation from red foxes, along with modifications in their prey's population cycles (IUCN 2009). Severe hurricanes result in the decline of forests, coral reefs, and mangroves (Krockenberger et al. 2004). Hurricanes have the potential to alter the food-web dynamics of insular ecosystems (Spiller and Schoener 2007), the successional ability of forests, and cause the greatest damage to early pioneer species of primary successional communities (Dallmeier and Comiskey 1998) which creates late successional hardwoods (Arévalo et al. 2000). Extreme floods also lead to species-specific mortality, changes in population dynamics, and the reorganisation of communities (Thibault and Brown 2008). Storms and floods threaten low-lying islands, coral reefs, mangroves, coastal marshes, and fisheries. According to Norkko et al. (2002), clay deposition increases with severe flooding which affects estuarine macrobenthic communities. Forest fires are intrinsically linked with droughts (Ellis et al. 2004), and extreme forest fires modify biodiversity, change the distribution of dominant plant species (Stocks et al. 1998), and increase the rate of vegetation transformation (Everham and Brokaw 1996).

The rising global sea level is perhaps the most obvious effect of global warming (Nicholls et al. 2011). For the past two to three million years, the global average sea level was about 120 m lower than it is today (Church et al. 2008). Sea-level rise affects coastal ecosystems in three ways, namely salt water intrusion and extensive inundation, especially in estuarine ecosystems such as mangroves and salt marshes; coastal erosion; and storm surge flooding (Nicholls et al. 2011). Intrusions of salt water and extensive flooding cause ecosystems to move inland, known as 'coastal squeeze', resulting in habitat loss for mangroves and salt marsh ecosystems, which serve as nursery habitats for fish, crustaceans, and insects (Bloomfield and Gillanders 2005; Legra et al. 2008). Longer flooding durations and higher salinity levels can result in plant deaths on the margins of mangrove habitat on the seaward side (He et al. 2007), as well as changes in species composition (Gilman et al. 2008) and productivity reduction (Castañeda-Moya et al. 2013). Fringing mangroves are vulnerable to species and sediment deposition differences (Sasmito et al. 2016), while increased submersion times also reduces salt marsh production and organic sedimentation (Nyman et al. 1994). The disappearance of these habitats combined with more frequent and intense storm surges can be quite detrimental to shore-nesting birds. Salt water intrusions cause large freshwater ecosystems to change to saline environments over time, leading to the extinction of freshwater species (Kennish 2002); loss of habitat, for example, wildfowl and wader species (Galbraith et al. 2002; Hughes 2004); loss of nesting sites, such as terns and Laysan finch (*Telespiza cantans*) (Baker et al. 2006); and loss of nursery and feeding sites, such as Laysan Albatross nests (Reynolds et al. 2015).

12.2 Islands and Biodiversity Hotspots

Islands, due to their discrete and isolated nature, provide rich genetic, species, and ecosystem biodiversity (Myers et al. 2000); therefore, they are considered 'biodiversity hotspots' from a global biodiversity perspective. Their contribution to global biodiversity is beyond the proportion to their land area, comprising some of the wealthiest centres of flora and fauna on Earth. The 180,000 islands around the world account for over 20% of the world's biodiversity (Kier et al. 2009). They harbour a variety of species and habitats, including endemic as well as endangered, and provide great opportunities to understand the origin, diversification, and extinction of terrestrial biotas (Reid 1998).

The United Nations Biodiversity A-Z has recognised 35 biodiversity hotspots (Mittermeier et al. 2004; Myers et al. 2000). These are North and Central America: California Floristic Province, Madrean pine-oak woodlands, Mesoamerica; the Caribbean: Caribbean Islands; South America: Atlantic Forest, Cerrado, Chilean Winter Rainfall-Valdivian Forests, Tumbes-Chocó-Magdalena, Tropical Andes; Europe: Mediterranean Basin; Africa: Cape Floristic Region, Coastal Forests of Eastern Africa, Eastern Afromontane, Guinean Forests of West Africa, Horn of Africa, Madagascar and the Indian Ocean Islands, Maputaland-Pondoland-Albany, Succulent Karoo; Central Asia: Mountains of Central Asia; South Asia: Eastern Himalaya, Nepal; Indo-Burma, India, Myanmar; Western Ghats, India; Sri Lanka; Southeast Asia and Asia Pacific: East Melanesian Islands, New Caledonia, New Zealand, Philippines, Polynesia-Micronesia, Southwest Australia, Sundaland, Wallacea; East Asia: Japan; Mountains of Southwest China; and West Asia: Caucasus Irano-Anatolian. In fact, nearly a third of all biodiversity hotspots are islands with a very high number of species (Fig. 12.1) (Gillespie and Roderick

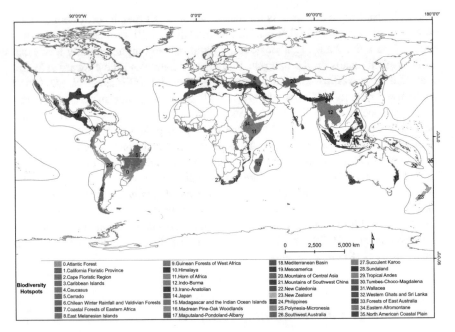

Fig. 12.1 The 35 biodiversity hotspot areas (with outer limits shown as red lines) in the world as recognised by the United Nations. (Adapted from Conservation Synthesis, Centre for Applied Biodiversity Science at Conservation International, https://databasin.org/datase/23fb5da15861411 09faf8d45de0a260 accessed on February 03, 2019)

2002). Islands support mainly intact natural ecosystems where native species and communities are well represented (Kier et al. 2009).

According to a study by Kier et al. (2009), of the top 20 regions in terms of endemism richness per standard area, 50% were island regions. All island regions except Japan ranked in the top 33% of the 90 regions. Similar patterns of endemic richness were also found for terrestrial vertebrates across the 90 regions. Considerably greater endemic richness of vertebrates was found on island regions (8.1 times higher than that for mainland regions), with island regions containing 23.2% of the range equivalents of this group (Table 12.1).

Although there are a wide variety of island types, they all have one common characteristic, that is, they are isolated and possess distinct borders in geographically well-defined regions (Paulay 1994). Islands are home to many endemic species, and the number of endemics rises with increasing isolation, island size, and topographical variation; as an example, more than 90% of the species on Hawaii are endemic (Nurse et al. 2001). Approximately 50% of all higher plants, mammals, birds, reptiles, and amphibians in Mauritius are endemic, while Seychelles has the highest rate of amphibian endemism in the world. Cuba's island has 18 endemic mammals, while nearby Guatemala and Honduras on the mainland have only 3. Madagascar has over 8000 endemic species, making it the country with the highest amount of endemic species in sub-Saharan Africa (Kier et al. 2009). The Hawaiian

Table 12.1 Summary of the number of range equivalents (RE) for vascular and terrestrial vertebrates in mainland and island regions

	Islands				Mainlands		
	Total RE	RE	%	RE/10⁴ km²	RE	%	RE/10⁴ km²
Vascular plants	315,903	82,546	26.1	172.3	233,357	73.9	18.2
Amphibians	4792	986	20.6	2.1	3806	79.4	0.3
Reptiles	7506	1952	26.0	4.1	5553	74.0	0.4
Birds	9585	2227	23.2	4.7	7358	76.8	0.6
Mammals	4703	1013	21.5	2.1	3690	78.5	0.3
Terrestrial vertebrates	26,586	6178	23.2	12.9	20,407	76.8	1.6

Source: Data from proceedings of the National Academy of Sciences, retrieved from https://doi.org/10.1073/pnas.0810306106

archipelago's indigenous flora is estimated to be comprised of 2000 species of angiosperms, 94–98% of which are endemic. Moreover, 76% of the plants of New Caledonia are endemic, as are 50% of Cuba, and 36% of Hispaniola. Large numbers of endemic species in terrestrial habitat islands may develop over long periods. This may be a reason for the incredible species richness of tropical rainforests. Since endemic species have geographically limited island distributions, they have often been shielded from diseases, predators, and rivals. However, they tend to be highly vulnerable to extinction as a result of climate change and other anthropogenic activities (Mimura 1999). Approximately three-quarters of the world's extinct species are island species. For example, of the 128 species of extinct birds, 122 (95.3%) are island extinctions (Loehle and Eschenbach 2012). Nearly 97% of Lord Howe Island's endemic plants are either extinct or threatened, as are 96% of Rodrigues and Norfolk Islands, 91% of Ascension Islands, and 81% of Juan Fernandez and Seychelles Islands (Sjöstedt and Povitkina 2015). Anthropogenic activities such as deforestation and the introduction of exotic species have a major impact on the species diversity of islands.

12.3   Vulnerability of Biodiversity on Small Islands to Climate Change

Island species are particularly at risk of extinction since islands have a small geographical area. They are restricted to the island or a specific area of the island, and they generally have low population numbers. Such factors increase their risk of extinction to natural variables such as normal population fluctuations, disease dissemination, and exposure to extremely destructive natural disasters. Small islands are particularly fragile and vulnerable to the effects of climate change for various reasons, but especially because of their geographical circumstances, that is, being comparatively small in size, low elevation, unconsolidated sediments as lithologic origins, ecological uniqueness and fragility, and proximity to oceans (Kumar and

Tehrany 2017). Recently, islands have garnered considerable focus due to concerns over their known sensitivity to climate change (IPCC 2013). Islands are influenced by climate change and ocean drivers such as temperature differences, rainfall, sea-level rise, wind and direction, and especially extreme events such as tropical cyclones, drought, and storm surges (Lo-Yat et al. 2011). Compared to land masses, the impact of climate change on small islands will be far larger considering their vulnerability and lower capacity to adapt (FAO 2008; IPCC 2007). Smaller islands and their biota are notably more vulnerable than larger land masses to environmental change due to their biogeography.

Rising sea levels and sea-surface temperatures associated with climate change have the most significant impacts on islands since most small islands have low altitudes and high coastal exposure compared to land masses. In addition, high population concentrations in coastal areas and the related anthropogenic activities make islands extremely vulnerable to sea-level rise. Projected climate changes for 2025 and 2100 predict that average sea level could rise by as much as 21 and 66 cm, respectively (www.biodiversity.be; accessed on 5 January 2019). This may lead to increased salinity due to sea encroachment, inundation, and storm surge or coastline erosion with the potential to negatively impact on island biodiversity. Another effect will be seawater intrusions into freshwater lenses, contributing to impacts on freshwater biodiversity. The upsurge in sea temperature causes coral bleaching, which negatively affects fishes and other sea creatures.

The survival strategies of island species are based on interdependence, coevolution, and mutualism. They have less dispersal capacity and compete with relatively few species, instead of defending systems against a wide spectrum of predators and competitors (Sax et al. 2002). Islands have a disproportional amount of biodiversity and species extinctions when compared to continental land masses. Narrow endemics found on islands are likely to have evolved traits such as reduced or absent inherent capabilities, including loss of flight in birds and insects and a loss of self-protective abilities with lower genetic variation (Gillespie and Roderick 2002; Whittaker and Fernández-Palacios 2007). These traits make them more vulnerable to climate change-related impacts and the resulting habitat fragmentation, thereby increasing their susceptibility to extinction (Wetzel and Likens 2013). The ability of island species to cope with climate change may be restricted due to a number of factors, including smaller geographical ranges, narrow genetic variation, small colonising populations, decreased species richness, and weak adaptations to prevent predation (Cox et al. 2016). Endemic species with highly restricted distributions are more threatened by rising temperatures (Taylor and Kumar 2016). Increasing temperatures may lead to the loss of cold climate zones in higher mountains or isolated mountainous regions which support cloud forests that harbour high levels of local endemism in the Pacific (Takeuchi 2003; Tuiwawa 2005). In the Balearic Islands, seagrass mortality was found to increase with increasing temperature (Marba and Duarte 2010). In the Hawaiian islands, a change in ocean and mean air temperature by 2–2.5 °C could cause the temperature tolerance zone of native species in montane cloud forest to shift upwards by 360–450 m (Loope and Giambelluca 1998).

Island reefs are renowned for their extraordinary hotspots of biodiversity and high levels of endemism (Allen 2008). Thermal stress and other factors such as increasing CO_2 levels and ocean pollution are anticipated to affect the functioning and viability of living reefs. Increasing temperature has been reported as the major reason for coral bleaching (De'ath et al. 2009; Jokiel and Coles 1990) as mass coral reef bleaching events have occurred in areas where ocean surface temperatures exceeded long-term averages by more than 1 °C during the preceding warm season (Goreau and Hayes 1994). Spawning of adult reef species has been negatively affected by high temperatures (Edmunds et al. 2001; Munday et al. 2009). A 100% coral mortality in the Kanton Atoll lagoon and 62% mortality on the outward lee-ward slope of the island and elsewhere across the Kiribati Phoenix islands in the central Pacific Ocean were reported due to excessive water temperatures over 6 months (Alling et al. 2007). In 2005, a mass coral bleaching event was discovered in the Caribbean Barbados island caused by accumulated thermal stress which persisted for many months after the temperature cooled down (Oxenford et al. 2008). This reveals the vulnerability of small islands and their limited defences against elevated seawater temperatures. Negative impacts on biodiversity, ecological functions, and services will significantly increase with increased greenhouse gas emissions, particularly if global warming exceeds the low emissions scenario.

12.4 Case Study: Climate Change Effects on Threatened Terrestrial Vertebrates of the Pacific Islands

12.4.1 Introduction

This case study reports on the potential impacts of climate change on terrestrial vertebrate species found in 26 Pacific Island countries. The aim was to investigate the distribution of terrestrial vertebrate species across these countries and identify those species that were most at risk of extinction due to them being present on only one or a few islands that had previously been classified as being most susceptible to climatic change (Kumar et al. 2018). Information on species vulnerability was obtained from the IUCN database and only the vulnerable, endangered, and critically endangered terrestrial vertebrate species were used in the analysis. The distribution of the species was combined with individual island susceptibility classes (very low, low, moderate, high, and very high) that were calculated from physical island attributes such as island lithology, area, shape (circularity), and elevation. Sea-level rise and significant wave height information across the Pacific were then combined with species vulnerability and island susceptibility to identify those species that occurred on islands that were most susceptible, those that occurred on the fewest islands, as well as islands that hosted the most species in the highest vulnerability classes. Parts of this work have been reported in more detail in Kumar and Tehrany (2017).

12.4.2   Methodology

The study region was limited to 30°06′56′N and 27°02′57′S latitudes and 126°06′56.938′E and 119°17′53.012′W longitudes (Fig. 12.2). All countries in this region, except Australia, New Zealand, East Timor, and the Indonesian Islands, were included. The study area was approximately 85 million square kilometres and has more than 2000 islands. A database of all islands in this region had been created as part of a previous study (Kumar et al. 2018; Nunn et al. 2015). The database consisted of 1779 islands that make up the 26 countries; however, only islands that were greater than one hectare in area were included since there are many thousands of much smaller island outcrops of which there are no records.

The database was used to develop an index of island susceptibility. Four island characteristics were used, namely area, shape (roundness), maximum elevation, and lithology of the islands. These four variables were either available for all islands or could be readily calculated from information available from other sources such as satellite imagery. There were other possibly better variables that could have been used; however, data on these variables were not available for all islands in the Pacific. One such variable was median elevation instead of maximum elevation, but median elevation for all islands was not available and could not be readily calculated since decent quality elevation data at appropriate scales were not available for all islands. Chapter 4 and Kumar et al. (2018) describe the calculation process in

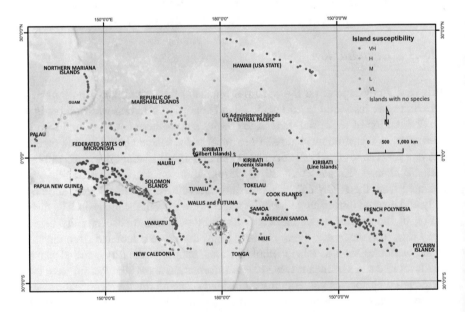

Fig. 12.2 Islands hosting at least one vulnerable, endangered, or critically endangered terrestrial vertebrate species with islands susceptibility rankings (VH, very high; H, high; M, moderate; L, low; and VL, very low susceptibility to climate change). Islands that do not host any of these species are shown as a blank circle

more detail. The result of this exercise was that we created an index of island susceptibility for each of the 1779 islands in the database. This was a relative measure but sufficient to rank islands from the lowest to highest susceptibility and use this to identify how species were distributed.

The IUCN Red List (IUCN 2016) database was searched for terrestrial vertebrate species present in the study area. The IUCN site lists the vulnerability of species mainly on a five-point scale: Least Concern (LC), Near Threatened (NT), Vulnerable (VU), Endangered (EN), and Critically Endangered (CR). There are other categories such as Extinct and Data Deficient, but these had no purpose in this study. For this study, we investigated 150 terrestrial vertebrate species classified as vulnerable, endangered, or critically endangered (Table 12.2). All threatened species belonging to the taxonomic groups of amphibians, mammals, and reptiles (terrestrial vertebrates) were downloaded. Marine species were excluded from this study since the island susceptibility data was for land area only. The list of threatened species included in the analysis is provided in Table 12.3. IUCN data are not island specific but stored as polygons identifying the region where each of the species is found; thus, the downloaded data consisted of 150 polygons. Each of these polygons (data for each species) were then overlaid on the full Pacific Islands vector file map consisting of island boundaries for each of the 1779 islands. This enabled us to identify all islands that fell within each of the species polygons and so were potentially home to these species. This resulted in a database of 1779 islands showing the number, names, and vulnerability class (vulnerable, endangered, and critically endangered) of all terrestrial vertebrate species for each of the islands.

This species database was then combined with the island susceptibility index database, providing spatial information on species vulnerability, island name, and island susceptibility (as a GIS map layer). The combined layer of island susceptibility and species information was overlaid with regional sea-level rise and projected mean significant wave height (H_s) difference data. The idea behind this analysis was to identify endemic and critically endangered species that were hosted by only one or a few islands that had a high ranking of susceptibility and would be affected by a relatively higher sea level and significant wave heights. Since the island susceptibility index incorporated only the physical characteristics of the individual islands with no variable referring to its location in the Pacific, and hence the impacts of different sea level rises and tropical cyclone activity, a separate spatial layer showing significant wave heights for the study area was generated.

Table 12.2 Vulnerable, endangered, and critically endangered species found in the study area

	Vulnerable	Endangered	Critically endangered	Total
Amphibians	10	0	1	11
Mammals	21	28	18	67
Reptiles	20	33	19	72
Total	51	61	38	150

Table 12.3 List of threatened species that were included in the analysis

Scientific name	Class	Red list category
Aproteles bulmerae[a]	Mammal	Critically endangered
Austrochaperina novaebritanniae	Amphibian	Vulnerable
Bavayia exsuccida[a]	Reptile	Endangered
Bavayia goroensis[a]	Reptile	Endangered
Bavayia ornata[a]	Reptile	Endangered
Brachylophus bulabula	Reptile	Endangered
Brachylophus fasciatus	Reptile	Endangered
Brachylophus vitiensis	Reptile	Critically endangered
Caledoniscincus auratus[a]	Reptile	Endangered
Caledoniscincus chazeaui[a]	Reptile	Endangered
Caledoniscincus orestes[a]	Reptile	Endangered
Caledoniscincus renevieri[a]	Reptile	Endangered
Caledoniscincus terma[a]	Reptile	Vulnerable
Celatiscincus euryotis[a]	Reptile	Endangered
Celatiscincus similis[a]	Reptile	Endangered
Chaerephon bregullae	Mammal	Endangered
Chalinolobus neocaledonicus[a]	Mammal	Endangered
Choerophryne siegfriedi	Amphibian	Critically endangered
Cophixalus nubicola	Amphibian	Vulnerable
Copiula minor	Amphibian	Vulnerable
Cornufer akarithymus	Amphibian	Vulnerable
Cornufer parkeri	Amphibian	Vulnerable
Dactylopsila tatei[a]	Mammal	Endangered
Dendrolagus dorianus[a]	Mammal	Vulnerable
Dendrolagus goodfellowi[a]	Mammal	Endangered
Dendrolagus inustus	Mammal	Vulnerable
Dendrolagus matschiei	Mammal	Endangered
Dendrolagus notatus[a]	Mammal	Endangered
Dendrolagus pulcherrimus	Mammal	Critically endangered
Dendrolagus scottae	Mammal	Critically endangered
Dendrolagus stellarum[a]	Mammal	Vulnerable
Dierogekko inexpectatus[a]	Reptile	Critically endangered
Dierogekko kaalaensis[a]	Reptile	Critically endangered
Dierogekko koniambo[a]	Reptile	Critically endangered
Dierogekko nehoueensis[a]	Reptile	Critically endangered
Dierogekko poumensis[a]	Reptile	Critically endangered
Dierogekko thomaswhitei[a]	Reptile	Critically endangered
Dierogekko validiclavis[a]	Reptile	Endangered
Dorcopsis atrata[a]	Mammal	Critically endangered
Dorcopsis luctuosa	Mammal	Vulnerable
Echymipera davidi[a]	Mammal	Endangered
Emballonura semicaudata	Mammal	Endangered

(continued)

Table 12.3   (continued)

Scientific name	Class	Red list category
Emoia adspersa	Reptile	Endangered
Emoia aneityumensis	Reptile	Endangered
Emoia boettgeri[a]	Reptile	Endangered
Emoia campbelli	Reptile	Endangered
Emoia erronan[a]	Reptile	Vulnerable
Emoia lawesi	Reptile	Endangered
Emoia loyaltiensis[a]	Reptile	Vulnerable
Emoia mokosariniveikau[a]	Reptile	Endangered
Emoia parkeri[a]	Reptile	Vulnerable
Emoia samoensis	Reptile	Endangered
Emoia slevini	Reptile	Critically endangered
Emoia trossula	Reptile	Endangered
Emoia tuitarere[a]	Reptile	Vulnerable
Eurydactylodes occidentalis[a]	Reptile	Critically endangered
Eurydactylodes symmetricus	Reptile	Endangered
Geoscincus haraldmeieri[a]	Reptile	Critically endangered
Graciliscincus shonae[a]	Reptile	Vulnerable
Hipposideros demissus	Mammal	Vulnerable
Hylarana waliesa	Amphibian	Vulnerable
Kanakysaurus viviparous[a]	Reptile	Endangered
Kanakysaurus zebratus[a]	Reptile	Endangered
Lacertoides pardalis[a]	Reptile	Vulnerable
Leiolopisma alazon	Reptile	Critically endangered
Lepidodactylus euaensis[a]	Reptile	Critically endangered
Lepidodactylus manni[a]	Reptile	Vulnerable
Lioscincus maruia[a]	Reptile	Endangered
Lioscincus steindachneri[a]	Reptile	Endangered
Lioscincus vivae[a]	Reptile	Critically endangered
Litoria becki	Amphibian	Vulnerable
Litoria lutea	Amphibian	Vulnerable
Loveridgelaps elapoides[a]	Reptile	Vulnerable
Marmorosphax boulinda[a]	Reptile	Vulnerable
Marmorosphax kaala[a]	Reptile	Critically endangered
Marmorosphax montana[a]	Reptile	Vulnerable
Marmorosphax taom[a]	Reptile	Critically endangered
Melomys matambuai[a]	Mammal	Endangered
Miniopterus robustior	Mammal	Endangered
Mirimiri acrodonta	Mammal	Critically endangered
Murexia rothschildi[a]	Mammal	Vulnerable
Nannoscincus exos[a]	Reptile	Critically endangered
Nannoscincus garrulus[a]	Reptile	Endangered
Nannoscincus gracilis[a]	Reptile	Vulnerable

(continued)

Table 12.3 (continued)

Scientific name	Class	Red list category
Nannoscincus greeri[a]	Reptile	Endangered
Nannoscincus hanchisteus[a]	Reptile	Critically endangered
Nannoscincus humectus[a]	Reptile	Endangered
Nannoscincus manautei[a]	Reptile	Critically endangered
Nannoscincus mariei[a]	Reptile	Vulnerable
Nannoscincus rankini[a]	Reptile	Vulnerable
Nannoscincus slevini[a]	Reptile	Endangered
Notopteris macdonaldi	Mammal	Vulnerable
Notopteris neocaledonica	Mammal	Vulnerable
Nyctimystes avocalis	Amphibian	Vulnerable
Nyctophilus nebulosus[a]	Mammal	Critically endangered
Oedodera marmorata[a]	Reptile	Critically endangered
Ogmodon vitianus[a]	Reptile	Endangered
Palmatorappia solomonis	Amphibian	Vulnerable
Paraleptomys rufilatus	Mammal	Endangered
Paramelomys gressitti	Mammal	Endangered
Perochirus ateles	Reptile	Endangered
Peroryctes broadbenti[a]	Mammal	Endangered
Petaurus abidi	Mammal	Critically endangered
Phalanger lullulae	Mammal	Endangered
Phalanger matanim	Mammal	Critically endangered
Pharotis imogene	Mammal	Critically endangered
Phoboscincus bocourti[a]	Reptile	Endangered
Pogonomys fergussoniensis[a]	Mammal	Endangered
Pteralopex anceps	Mammal	Endangered
Pteralopex atrata	Mammal	Endangered
Pteralopex flanneryi	Mammal	Critically endangered
Pteralopex pulchra	Mammal	Critically endangered
Pteralopex taki	Mammal	Endangered
Pteropus anetianus[a]	Mammal	Vulnerable
Pteropus cognatus[a]	Mammal	Endangered
Pteropus fundatus	Mammal	Endangered
Pteropus insularis	Mammal	Critically endangered
Pteropus mahaganus	Mammal	Vulnerable
Pteropus mariannus	Mammal	Endangered
Pteropus molossinus[a]	Mammal	VU
Pteropus nitendiensis	Mammal	Endangered
Pteropus ornatus[a]	Mammal	Vulnerable
Pteropus rennelli[a]	Mammal	Vulnerable
Pteropus tuberculatus	Mammal	Critically endangered
Pteropus ualanus	Mammal	Vulnerable
Pteropus vetulus[a]	Mammal	Vulnerable

(continued)

Table 12.3 (continued)

Scientific name	Class	Red list category
Pteropus woodfordi[a]	Mammal	Vulnerable
Pteropus yapensis[a]	Mammal	Vulnerable
Rattus vandeuseni	Mammal	Endangered
Rhacodactylus chahoua[a]	Reptile	Vulnerable
Rhacodactylus ciliatus[a]	Reptile	Vulnerable
Rhacodactylus sarasinorum[a]	Reptile	Vulnerable
Rhacodactylus trachyrhynchus[a]	Reptile	Endangered
Sigaloseps ruficauda[a]	Reptile	Vulnerable
Simiscincus aurantiacus[a]	Reptile	Vulnerable
Solomys ponceleti	Mammal	Critically endangered
Solomys salebrosus	Mammal	Endangered
Solomys sapientis[a]	Mammal	Endangered
Spilocuscus rufoniger	Mammal	Critically endangered
Syconycteris hobbit	Mammal	Vulnerable
Thylogale browni	Mammal	Vulnerable
Thylogale brunii	Mammal	Vulnerable
Thylogale calabyi	Mammal	Endangered
Thylogale lanatus[a]	Mammal	Endangered
Tropidoscincus aubrianus[a]	Reptile	Vulnerable
Uromys imperator	Mammal	Critically endangered
Uromys porculus	Mammal	Critically endangered
Uromys rex[a]	Mammal	Endangered
Xeromys myoides	Mammal	Vulnerable
Zaglossus bartoni	Mammal	Critically endangered

Species in bold are found on both very high- and high-susceptibility islands
[a]Endemic species

H_s was modelled on a monthly basis using two representative concentration pathways (RCPs) under the CMIP5 model by SOPAC (http://wacop.gsd.spc.int/). The RCPs used were RCP4.5 (a medium low emission scenario) and RCP8.5 ('business as usual case' or high emission scenario), while the climate models used were CNRM-CM5, HadGEM2-ES, INMCM4, and ACCESS1.0. The average ensemble H_s values were obtained for each month, and then one overall H_s layer for 2081–2100 was created by identifying the maximum H_s value for each location (pixel) across the 12 months. Similar values created for the 1986–2005 (historical) period were subtracted from the 2081–2100 layer to obtain the H_s difference layer, showing how much variation was expected in H_s over the 2081–2100 period compared to the 1986–2005 period.

12.4.3   Results and Discussion

Of the 150 terrestrial vertebrate species classified as vulnerable, endangered, or critically endangered in the IUCN database, 51 were vulnerable, 61 endangered, and 38 critically endangered. Overall, 11 were amphibians, 67 mammals, and 72 reptiles (Table 12.1). The results showed that 1105 of the 1779 islands did not host any of these 150 species. The other 674 islands hosted at least one species from the 150 threatened species. Of these 674 islands, 114 islands were ranked as having very low susceptibility, 171 as low susceptibility, 152 as moderate susceptibility, 178 as high susceptibility, and the remaining 59 islands as having very high susceptibility.

Table 12.4 shows the islands that hosted six or more of the vulnerable, endangered, or critically endangered species; there were 15 such islands. Of the other 659 islands, 3 had 5 species, 34 had 4 species, 143 had 3 species, 207 had 2 species, and 272 islands hosted only 1 species. The 15 islands that had six or more species were generally the larger islands, making up a combined area of 84% of the total area of the 1779 islands. The islands of New Guinea in Papua New Guinea and Viti Levu in Fiji were the largest two of this group. These 15 islands were of either continental (3), composite high (7), or volcanic high (5) origin, indicating that islands hosting a greater number of species were generally larger with higher elevations.

Islands that hosted at least one of the vulnerable, endangered, or critically endangered species are shown in Fig. 12.2, together with the susceptibility rankings for each island. Islands that did not host any threatened species from the list are shown as open circles. Figure 12.2 shows that islands on the eastern half of the study region did not host any of the vulnerable, endangered, or critically endangered terrestrial vertebrate species. Some of the countries in this group are Pitcairn Islands, French Polynesia, American Samoa, Kiribati, and the Cook Islands. The countries that hosted most of these species were Fiji, Vanuatu, Tonga, Samoa, Republic of Marshall Islands, Federated States of Micronesia, Palau, Guam, New Caledonia, and Papua New Guinea. These countries comprise the larger islands of the Pacific region: islands such as Viti Levu and Vanua Levu from the Fiji group, New Guinea and Bougainville from Papua New Guinea, Espiritu Santo and Malekula from Vanuatu, and La Grande Terre from New Caledonia. These islands have relatively larger areas than the other islands with threatened species and are of mostly volcanic or high-calcareous lithology. Guam and the Islands of Northern Mariana are exceptions to this.

Islands with a very low-susceptibility ranking hosted 74 vulnerable, 82 endangered, and 26 critically endangered species; with a low-susceptibility ranking hosted 84 vulnerable, 103 endangered, and 46 critically endangered species; with a moderate-susceptibility ranking hosted 100 vulnerable, 122 endangered, and 8 critically endangered species; with a high-susceptibility ranking hosted 71 vulnerable, 155 endangered, and 18 critically endangered species; and with a very high-susceptibility ranking hosted 5 vulnerable, 58 endangered, and 2 critically endangered species.

Table 12.4 Islands that host six or more species

	Islands	Lithology	Area (km²)	Island susceptibility ranking	Number of species
1	Bougainville	Composite high island	9318	L	9
2	Buka	Composite high island	936	VL	7
3	Choiseul	Composite high island	3400	L	8
4	Guadalcanal	Composite high island	5302	VL	8
5	Ile des Pins (Kunie)	Continental island	152	VL	8
6	Kadavu	Volcanic high island	408	VL	6
7	Kotomo (Koutoumo)	Continental island	16	L	6
8	New Caledonia (La Grande Terre)	Continental island	1628	L	53
9	New Guinea	Composite high island	786000	VL	31
10	Ovalau	Volcanic high island	103	VL	7
11	Santa Isabel	Composite high island	3490	L	6
12	Savai'i	Volcanic high island	1694	VL	6
13	Taveuni	Volcanic high island	435	VL	8
14	Vanua Levu	Volcanic high island	5534	VL	6
15	Viti Levu	Composite high island	10388	L	10

L and VL indicate low and very low susceptibility to climate change, respectively

While available resources and island size are important criteria for determining the risk of extinction, species distribution is also important, especially in light of extreme events and island susceptibility. In general, species found on only a few islands are at a greater risk of extinction compared to those found on multiple islands that are well distributed. Figure 12.3 shows the number of threatened species (of the 150 used in this study) hosted by each island. Many islands in Fiji hosted multiple species, as did New Caledonia and the Solomon Islands. Most of the islands in the Federated States of Micronesia and the Marshall Islands hosted one or two species. Table 12.5 presents a list of critically endangered and endangered species occurring on one island only. Out of the 150 species used in this study, 38 were critically endangered, and of these, 30 occurred on one island only. Similarly, out of the 61 endangered species, 31 occurred on one island only. However, most of these

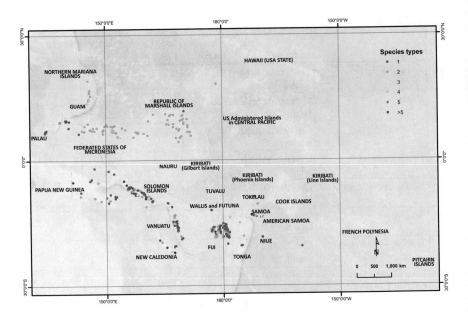

Fig. 12.3 Species types hosted by each of the islands

islands ranked in the low- and very low-susceptibility classes, while none of these islands ranked in the high- or very high-susceptibility classes.

H_s projections under RCP4.5 and RCP8.5 (Figs. 12.4 and 12.5) showed significant variability across the Pacific Ocean, with H_s differences of up to 0.4 m occurring by 2081–2100. Larger projected differences in H_s were seen in the north-west, around Palau, Guam, and Marshall Islands, and in the south around New Caledonia, Fiji, Tonga, and French Polynesia. The projected differences were generally higher under RCP8.5, especially in the southern Pacific Ocean around American Samoa, Niue, and Tonga. Figures 12.4 and 12.5 show that many of the critically endangered species were found in regions where the projected H_s differences were higher. This included some of the islands of Guam, Solomon Islands, Fiji, and Tonga. The situation was more critical under RCP8.5 (Fig. 12.5a), as these islands and the critically endangered species on them were projected to experience even higher H_s differences. When projected H_s values were compared with the island susceptibility rankings, we found that most islands from Tonga, Tokelau, Marshall Islands, and the Federated States of Micronesia, with high- and very high-susceptibility rankings, occurred in regions with moderate to high projected H_s. Thus, species that occur on these islands face a greater risk of extinction in the future and could be placed in the higher vulnerability ranking classes if climate change impacts are considered. More details on species distributions in relation to projected sea level changes are provided in Kumar and Tehrany (2017).

Of the 150 species used in this analysis, 84 were endemic to this region, meaning they do not occur anywhere else outside this region. If they become extinct in this

Table 12.5 List of critically endangered and endangered species found on one island only

Critically endangered species	Island susceptibility	Endangered species	Island susceptibility
*Aproteles bulmerae** (Bulmer's fruit bat)	VL	*Bavayia exsuccida** (Sclerophyll Bavayia)	L
Choerophryne siegfriedi (species of frog)	VL	*Bavayia goroensis** (species of lizard)	L
Dendrolagus pulcherrimus (Golden-mantled tree kangaroo)	VL	*Bavayia ornata** (ornate bavayia)	L
Dendrolagus scottae (Tenkile)	VL	*Caledoniscincus auratus** (Koumac Litter Skink)	L
*Dierogekko inexpectatus** (Key New Caledonian gecko)	L	*Caledoniscincus chazeaui** (Chazeau's Litter Skink)	L
*Dierogekko kaalaensis** (Kaala striped gecko)	L	*Caledoniscincus orestes** (Panié Litter Skink)	L
*Dierogekko koniambo** (Koniambo striped gecko)	L	*Caledoniscincus renevieri** (Renevier's Litter Skink)	L
*Dierogekko nehoueensis** (striped gecko)	L	*Celatiscincus similis*∗ (Northern Pale-hipped skink)	L
*Dierogekko poumensis** (Poum striped gecko)	L	*Dactylopsila tatei** (Tate's triok)	VL
*Dierogekko thomaswhitei** (Taom striped gecko)	L	*Dendrolagus goodfellowi** (Goodfellow's tree kangaroo)	VL
*Dorcopsis atrata** (black dorcopsis)	VL	*Dendrolagus notatus*∗ (Ifola tree-kangaroo)	VL
*Eurydactylodes occidentalis** (species of lizard)	L	*Dierogekko validiclavis** (bold-striped gecko)	L
*Geoscincus haraldmeieri** (Meier's skink)	L	*Echymipera davidi** (David's echymipera)	M
Leiolopisma alazon# (Ono-i-Lau ground skink)	L	*Emoia aneityumensis* (Anatom Emo skink)	VL
*Lioscincus vivae** (species of lizard)	L	*Emoia campbelli* (Vitilevu mountain tree skink)	L
*Marmorosphax kaala** (species of lizard)	L	*Eurydactylodes symmetricus* (large-scaled chameleon gecko)	L
*Marmorosphax taom** (species of lizard)	L	*Kanakysaurus viviparous** (species of lizard)	L
Mirimiri acrodonta (Fijian monkey-faced bat)	VL	*Kanakysaurus zebratus** (species of lizard)	L
*Nannoscincus exos** (northern dwarf skink)	L	*Lioscincus maruia** (Maruia Maquis skink)	L
*Nannoscincus hanchisteus** (Pindai dwarf skink)	L	*Lioscincus steindachneri** (White-lipped forest skink)	L
*Nannoscincus manautei**	L	*Melomys matambuai**	L

*Endemic species

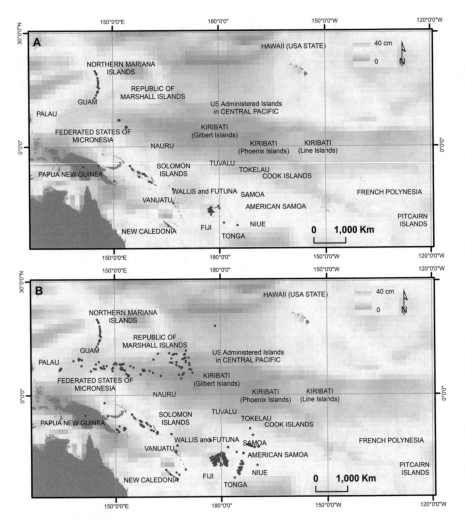

Fig. 12.4 Islands with critically endangered (**a**) and endangered (**b**) terrestrial vertebrate species in the study region in relation to projected significant wave height (H_s) differences (2081–2100 in relation to 1986–2005) under RCP4.5 (i.e. projected wave height under a medium low CO_2 emission scenario)

region, then it will mean global extinction. Of the 84 endemic species, 54 occurred on one island only, 11 occurred on two islands, and the other 19 occurred on three or more islands. Of the 54 endemic species occurring on one island only, 18 were critically endangered; however, many of these were found on the larger island of La Grande Terre in New Caledonia.

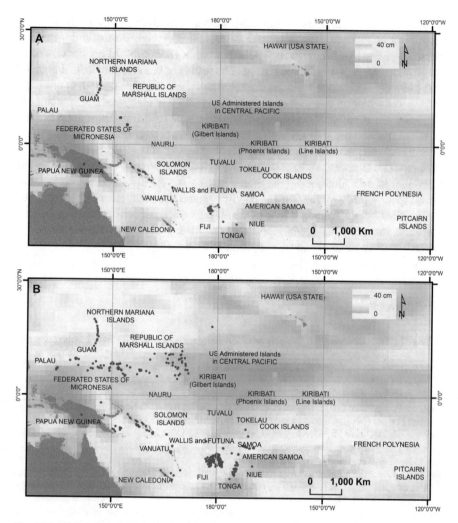

Fig. 12.5 Islands with critically endangered (**a**) and endangered (**b**) terrestrial vertebrate species found in the study region in relation to projected significant wave height (H_s) differences (2081–2100 in relation to 1986–2005) under RCP8.5 (i.e. projected H_s under 'business as usual case' or a high CO_2 emission scenario)

12.4.4 Conclusions

The islands in the Pacific, while small in size and spread over a vast region, are rich in biodiversity and host many species that are categorised as vulnerable, endangered, or critically endangered by the IUCN. Many of the species in the Pacific are also endemic to this region. The isolation of the islands, their generally small sizes, low elevations, as well as being primarily of coral and sandy origin make the islands in the Pacific vulnerable to climate change impacts. Analysis has shown that a large proportion of these islands fall in the high- and very high-susceptibility ranking classes. Species that occur on such islands are at a greater risk of extinction due to the projected impacts of climate change. The results of this study indicate that many of the terrestrial vertebrate species classified as vulnerable, endangered, or critically endangered occur on one or a few islands only and that many endemic species are hosted by one island only. The analysis could be used to identify those species most at risk, and action could be taken to afford them extra protection. It is also important to exercise care when implementing the findings of this study at a local scale. Further work is necessary to ascertain the actual presence of each species on the individual islands identified. This is important since the IUCN database does not record species presence data on an island-by-island basis but rather as polygon files on the distribution extent. While some islands may fall within these presence-polygons, they may not actually host the identified species; thus, the results of the analysis could potentially be more optimistic than reality.

References

Allen GR (2008) Conservation hotspots of biodiversity and endemism for indo-Pacific coral reef fishes. Aquat Conserv Mar Freshw Ecosyst 18(5):541–556. https://doi.org/10.1002/aqc.880

Alling A, Doherty O, Logan H, Feldman L, Dustan P (2007) Catastrophic coral mortality in the remote central pacific ocean: Kirabati Phoenix Islands. Atoll Res Bull 551:21. https://reposi-tory.si.edu/bitstream/handle/10088/4882/00551.pdf.

Archaux F, Wolters V (2006) Impact of summer drought on forest biodiversity: what do we know? Ann For Sci 63(6):645–652. https://doi.org/10.1051/forest:2006041

Arévalo JR, DeCoster JK, McAlister SD, Palmer MW (2000) Changes in two Minnesota forests during 14 years following catastrophic windthrow. J Veg Sci 11(6):833–840. https://onlineli-brary.wiley.com/doi/pdf/10.2307/3236553

Baker JD, Littnan CL, Johnston DW (2006) Potential effects of sea level rise on the terrestrial habi-tats of endangered and endemic megafauna in the northwestern Hawaiian islands. Endanger Species Res 2:21–30

Bellard C, Bertelsmeier C, Leadley P, Thuiller W, Courchamp F (2012) Impacts of climate change on the future of biodiversity. Ecol Lett 15(4):365–377. http://org.10.1111/geb.12093

Bloomfield A, Gillanders B (2005) Fish and invertebrate assemblages in seagrass, mangrove, salt-marsh, and nonvegetated habitats. Estuaries 28(1):63–77

Castañeda-Moya E, Twilley RR, Rivera-Monroy VH (2013) Allocation of biomass and net primary productivity of mangrove forests along environmental gradients in the Florida Coastal Everglades, USA. Forest Ecol Manage 307:226–241. https://doi.org/10.1016/j.foreco.2013.07.011

Chapin I, Zavaleta ES, Eviner VT, Naylor RL, Vitousek PM, Reynolds HL et al (2000) Consequences of changing biodiversity. Nature 405(6783):234–242. https://www.mrgscience.com/uploads/2/0/7/9/20796234/consequence_of_change_article.pdf

Chase JM (2007) Drought mediates the importance of stochastic community assembly. Proc Natl Acad Sci 104(44):17430–17434. https://doi.org/10.1073/pnas.0704350104

Church JA, White NJ, Aarup T, Wilson WS, Woodworth PL, Domingues CM et al (2008) Understanding global sea levels: past, present and future. Sustain Sci 3(1):9–22

Ciais P, Reichstein M, Viovy N, Granier A, Ogée J, Allard V et al (2005) Europe-wide reduction in primary productivity caused by the heat and drought in 2003. Nature 437(7058):529. https://doi.org/10.1038/nature03972

Cleland E, Chuine I, Menzel A, Mooney H, Schwartz M (2007) Shifting plant phenology in response to global change. Trends Ecol Evol 22(7):357–365. http://org.10.1016/j.tree.2007.04.003

Cox CB, Moore PD, Ladle R (2016) Biogeography: an ecological and evolutionary approach. Wiley, Hoboken, NJ

Currano E, Wilf P, Wing S, Labandeira C, Lovelock E, Royer D (2008) Sharply increased insect herbivory during the Paleocene–Eocene thermal maximum. Proc Natl Acad Sci 105(6):1960–1964. https://doi.org/10.1073/pnas.0708646105

Dallmeier F, Comiskey JA (1998) Forest biodiversity research, monitoring and modeling: conceptual background and old world case studies. Biodivers Conserv 12:2255–2277. https://link.springer.com/content/pdf/10.1023/A:1024593414624.pdf.

Davidar P, Arjunan M, Puyravaud J (2008) Why do local households harvest forest products? A case study from the southern Western Ghats, India. Biol Conserv 141(7):1876–1884. https://doi.org/10.1016/j.biocon.2008.05.004

De'ath G, Lough JM, Fabricius KE (2009) Declining coral calcification on the great barrier reef. Science 323(5910):116–119. https://science.sciencemag.org/content/sci/323/5910/116.full.pdf

Di Falco S, Chavas J (2008) Rainfall shocks, resilience, and the effects of crop biodiversity on agroecosystem productivity. Land Econ 84(1):83–96. http://le.uwpress.org/content/84/1/83.full.pdf

Drielsma MJ, Foster E, Ellis M, Gill RA, Prior J, Kumar L et al (2016) Assessing collaborative, privately managed biodiversity conservation derived from an offsets program: lessons from the southern Mallee of New South Wales, Australia. Land Use Policy 59:59–70. https://doi.org/10.1016/j.landusepol.2016.08.005

Easterling DR, Meehl GA, Parmesan C, Changnon SA, Karl TR, Mearns LO (2000) Climate extremes: observations, modeling and impacts. Science 289(5487):2068–2074. https://science.sciencemag.org/content/sci/289/5487/2068.full.pdf

Edmunds P, Gates R, Gleason D (2001) The biology of larvae from the reef coral Porites astreoides, and their response to temperature disturbances. Mar Biol 139(5):981–989. http://org.10.1007/s002270100634

Ellis S, Kanowski P, Whelan R (2004) National Inquiry into bushfire mitigation and management. Council of Australian Governments, Commonwealth of Australia

Everham EM, Brokaw NV (1996) Forest damage and recovery from catastrophic wind. Bot Rev 62(2):113–185. https://www.jstor.org/stable/4354268

FAO (2008) Agriculture organization 2003. State of the World's forests. FAO, Rome

FAO (2016) FAO yearbook. Fishery and aquaculture statistics 2016/FAO annuaire. Food and Agriculture Organization, Rome, p 108

Folke C, Holling CS, Perrings C (1996) Biological diversity, ecosystems and the human scale. Ecol Appl 6(4):1018–1024. https://org.10.2307/2269584

Galbraith H, Jones R, Park R, Clough J, Herrod-Julius S, Harrington B, Page G (2002) Global climate change and sea level rise: potential losses of intertidal habitat for shorebirds. Waterbirds 25(2):173–184. https://doi.org/10.1675/1524-4695(2002)025[0173:GCCASL]2.0.CO;2

Garel M, Loison A, Gaillard J-M, Cugnasse J-M, Maillard D (2004) The effects of a severe drought on mouflon lamb survival. Proc R Soc Lond B Biol Sci 271(Suppl 6):S471–S473. http://org.10.1098/rsbl.2004.0219

Gillespie RG, Roderick GK (2002) Arthropods on islands: colonization, speciation, and conservation. Annu Rev Entomol 47(1):595–632. https://doi.org/10.1146/annurev.ento.47.091201.145244

Gilman EL, Ellison J, Duke NC, Field C (2008) Threats to mangroves from climate change and adaptation options: a review. Aquat Bot 89(2):237–250. http://org.10.1016/j.aquabot.2007.12.009

Goreau TJ, Hayes RL (1994) Coral bleaching and ocean "hot spots". Ambio J Hum Environ Res Manage 23(3):176–180

Haines A, Kovats R, Campbell-Lendrum D, Corvalán C (2006) Climate change and human health: impacts, vulnerability and public health. Public Health 120(7):585–596. https://doi.org/10.1016/j.puhe.2006.01.002

He B, Lai T, Fan H, Wang W, Zheng H (2007) Comparison of flooding-tolerance in four mangrove species in a diurnal tidal zone in the Beibu gulf. Estuar Coast Shelf Sci 74(1–2):254–262. https://doi.org/10.1016/j.ecss.2007.04.018

Heller NE, Zavaleta ES (2009) Biodiversity management in the face of climate change: a review of 22 years of recommendations. Biol Conserv 142(1):14–32. http://org.10.1.1.700.5700&rep =rep1&type=pdf

Hilbert DW, Bradford M, Parker T, Westcott D (2004) Golden bowerbird (Prionodura newtonia) habitat in past, present and future climates: predicted extinction of a vertebrate in tropical highlands due to global warming. Biol Conserv 116(3):367–377. https://doi.org/10.1016/S0006-3207(03)00230-1

Hill R, Williams K, Pert P, Robinson C, Dale A, Westcott D, O'MALLEY T (2010) Adaptive community-based biodiversity conservation in Australia's tropical rainforests. Environ Conserv 37(1):73–82. https://doi.org/10.1017/S0376892910000330

Hughes R (2004) Climate change and loss of saltmarshes: consequences for birds. Ibis 146:21–28. https://doi.org/10.1111/j.1474-919X.2004.00324.x

IPCC (2007) The Physical Science Basis. Contribution of Working Group I to the Fourth Assessment Report. International Panel on Climate Change, 4, 2007

IPCC IWG (2013) Climate change 2013-the physical science basis. Intergovernmental Panel on Climate Change, Summary for Policymakers

IUCN (2009) IUCN Red List of Threatened Species (ver. 2009.1). www.iucnredlist.org. Accessed 28 Jan 2019

IUCN (2016) Red list spatial data, https://www.iucnredlist.org/technical-documents/red-list-training/iucnspatialresources. Spatial Data Download. https://www.iucnredlist.org/technical-documents/red-list-training/iucnspatialresources

Jokiel P, Coles S (1990) Response of Hawaiian and other indo-Pacific reef corals to elevated temperature. Coral Reefs 8(4):155–162

Kennish MJ (2002) Environmental threats and environmental future of estuaries. Environ Conserv 29(1):78–107

Kier G, Kreft H, Lee TM, Jetz W, Ibisch PL, Nowicki C et al (2009) A global assessment of endemism and species richness across island and mainland regions. Proc Natl Acad Sci 106(23):9322–9327

Knapp AK, Beier C, Briske DD, Classen AT, Luo Y, Reichstein M et al (2008) Consequences of more extreme precipitation regimes for terrestrial ecosystems. AIBS Bull 58(9):811–821. https://doi.org/10.1641/B580908

Kobayashi C, Umeda H, Nomoto K i, Tominaga N, Ohkubo T (2006) Galactic chemical evolution: carbon through zinc. Astrophys J Lett 1(653):1145. https://arxiv.org/pdf/astro-ph/0608688.pdf

Kogo BK, Kumar L, Koech R (2019) Forest cover dynamics and underlying driving forces affecting ecosystem services in western Kenya. Remote Sens Appl: Soc Environ 14:75–83. https://doi.org/10.1016/j.rsase.2019.02.007

Krockenberger A, Kitching R, Turton S (2004) Environmental crisis: climate change and terrestrial biodiversity in Queensland. Rainforest CRC, Australian Cooperative Research Centres Program, Australia

Kumar L, Tehrany S (2017) Climate change impacts on the threatened terrestrial vertebrates of the Pacific Islands. Sci Rep 7(1):5030. https://doi.org/10.1038/s41598-017-05034-4

Kumar L, Eliot I, Nunn PD, Stul T, McLean R (2018) An indicative index of physical susceptibility of small islands to coastal erosion induced by climate change: an application to the Pacific islands. Geomat Nat Haz Risk 9(1):691–702. https://doi.org/10.1080/19475705.2018.1455749

Lamsal P, Kumar L, Aryal A, Atreya K (2018) Invasive alien plant species dynamics in the Himalayan region under climate change. Ambio 47(6):697–710

Lavergne S, Mouquet N, Thuiller W, Ronce O (2010) Biodiversity and climate change: integrating evolutionary and ecological responses of species and communities. Annu Rev Ecol Evol Syst 41:321–350. https://doi.org/10.1146/annurev-ecolsys-102209-144628

Legra L, Li X, Peterson AT (2008) Biodiversity consequences of sea level rise in New Guinea. Pac Conserv Biol 14(3):191–199

Loehle C, Eschenbach W (2012) Historical bird and terrestrial mammal extinction rates and causes. Divers Distrib 18(1):84–91. http://org.10.1111/j.1472-4642.2011.00856.x

Loope LL, Giambelluca TW (1998) Vulnerability of island tropical montane cloud forests to climate change, with special reference to east Maui, Hawaii. Climate Change 39(2–3):503–517

Lo-Yat A, Simpson SD, Meekan M, Lecchini D, Martinez E, Galzin R (2011) Extreme climatic events reduce ocean productivity and larval supply in a tropical reef ecosystem. Glob Chang Biol 17(4):1695–1702. https://doi.org/10.1111/j.1365-2486.2010.02355.x

Magnusdottir R, Wilson B, Hersteinsson P (2008) Dispersal and the influence of rainfall on a population of the carnivorous marsupial swamp antechinus (Antechinus minimus maritimus). Wildl Res 35(5):446–454. https://doi.org/10.1071/WR06156

Marba N, Duarte CM (2010) Mediterranean warming triggers seagrass (Posidonia oceanica) shoot mortality.GlobChangBiol16(8):2366–2375.https://doi.org/10.1111/j.1365-2486.2009.02130.x

Memmott J, Craze P, Waser N, Price M (2007) Global warming and the disruption of plant–pollinator interactions. Ecol Lett 10(8):710–717. https://doi.org/10.1111/j.1461-0248.2007.01061.x

Mimura N (1999) Vulnerability of island countries in the South Pacific to sea level rise and climate change. Clim Res 12(2–3):137–143. https://www.int-res.com/articles/cr1999/12/c012p137.pdf

Mittermeier R, Robles Gil P, Hoffman M, Pilgrim J, Brooks T, Mittermeier C et al (2004) Hotspots revisited: earth's biologically richest and most endangered terrestrial Eco regions. CEMEX, S.A., Agrupación Sierra Madre, S.C, Mexico City

Munday P, Leis J, Lough J, Paris C, Kingsford M, Berumen M, Lambrechts J (2009) Climate change and coral reef connectivity. Coral Reefs 28(2):379–395. https://doi.org/10.1007/s00338-008-0461-9

Myers N, Mittermeier RA, Mittermeier CG, Da Fonseca GA, Kent J (2000) Biodiversity hotspots for conservation priorities. Nature 403(6772):853. http://wedocs.unep.org/bitstream/handle/20.500.11822/18446/Biodiversity_hotspots_for_conservation_priorit.pdf?sequence=1&isAllowed=y

Nicholls RJ, Marinova N, Lowe JA, Brown S, Vellinga P, De Gusmao D et al (2011) Sea-level rise and its possible impacts given a 'beyond 4 C world' in the twenty-first century. Philos Trans Royal Soc London A: Math Phys Eng Sci 369(1934):161–181. https://doi.org/10.1098/rsta.2010.0291

Norkko A, Thrush SF, Hewitt JE, Cummings VJ, Norkko J, Ellis JI et al (2002) Smothering of estuarine sandflats by terrigenous clay: the role of wind-wave disturbance and bioturbation in site-dependent macrofaunal recovery. Mar Ecol Prog Ser 234:23–42. https://www.int-res.com/articles/meps2002/234/m234p023.pdf.

Nunn P, Kumar L, Eliot I, McLean RF (2015) Regional coastal susceptibility assessment for the Pacific Islands: Technical Report. Australian Government and Australian Aid, Canberra, 123

Nurse LA, Sem G, Hay JE, Suarez AG, Wong PP, Briguglio L, Ragoonaden S (2001) Small Island states. In: Climate change. Cambridge University Press, Cambridge, pp 843–875

Nyman JA, Carloss M, DeLaune R, Patrick W Jr (1994) Erosion rather than plant dieback as the mechanism of marsh loss in an estuarine marsh. Earth Surf Process Landf 19(1):69–84

Oxenford HA, Roach R, Brathwaite A, Nurse L, Goodridge R, Hinds F et al (2008) Quantitative observations of a major coral bleaching event in Barbados, southeastern Caribbean. Clim Chang 87(3–4):435. https://doi.org/10.1007/s10584-007-9311-y

Parmesan C, Gaines S, Gonzalez L, Kaufman DM, Kingsolver J, Townsend Peterson A, Sagarin R (2005) Empirical perspectives on species borders: from traditional biogeography to global change. Oikos 108(1):58–75

Paulay G (1994) Biodiversity on oceanic islands: its origin and extinction. Am Zool 34(1):134–144

Pounds JA, Crump ML (1994) Amphibian declines and climate disturbance: the case of the golden toad and the harlequin frog. Conserv Biol 8(1):72–85. https://www.jstor.org/stable/2386722.

Rametsteiner E, Whiteman A (2014) State of the world's forests; enhancing the socio-economic benefits from forests enhancing the socioeconomic benefits from forests. Food and Agriculture Organization of the United Nations, Rome, p 133

Reid WV (1998) Biodiversity hotspots. Trends Ecol Evol 13(7):275–280

Reynolds MH, Courtot KN, Berkowitz P, Storlazzi CD, Moore J, Flint E (2015) Will the effects of sea-level rise create ecological traps for Pacific island seabirds. PloSone 10(9):e0136773. https://doi.org/10.1371/journal.pone.0136773

Rouault G, Candau J-N, Lieutier F, Nageleisen L-M, Martin J-C, Warzée N (2006) Effects of drought and heat on forest insect populations in relation to the 2003 drought in Western Europe. Ann For Sci 63(6):613–624. https://doi.org/10.1051/forest:2006044

Sasmito SD, Murdiyarso D, Friess DA, Kurnianto S (2016) Can mangroves keep pace with contemporary sea level rise? A global data review. Wetl Ecol Manag 24(2):263–278. https://doi.org/10.1007/s11273-015-9466-7

Sax DF, Gaines SD, Brown JH (2002) Species invasions exceed extinctions on islands worldwide: a comparative study of plants and birds. Am Nat 160(6):766–783. https://www.jstor.org/stable/10.1086/343877

Sjöstedt M, Povitkina M (2015) Vulnerability of Small Island development states. Does good governance help? QoG Working Paper Series 2015:12. University of Gothenburg

Spiller DA, Schoener TW (2007) Alteration of island food-web dynamics following major disturbance by hurricanes. Ecology 88(1):37–41. https://esajournals.onlinelibrary.wiley.com/doi/pdf/10.1890/0012-9658%282007%2988%5B37%3AAOIFDF%5D2.0.CO%3B2.

Staudinger MD, Carter SL, Cross MS, Dubois NS, Duffy JE, Enquist C et al (2013) Biodiversity in a changing climate: a synthesis of current and projected trends in the US. Front Ecol Environ 11(9):465–473. https://doi.org/10.1890/120272

Stocks B, Fosberg M, Lynham T, Mearns L, Wotton B, Yang Q et al (1998) Climate change and forest fire potential in Russian and Canadian boreal forests. Clim Chang 38(1):1–13. https://doi.org/10.1023/A:1005306001055

Takeuchi W (2003) Plant discoveries from PABITRA-related exploration in Papua New Guinea. Organ Divers Evol 3(2):77–84. http://www.senckenberg.de/odes/03-03.htm

Taylor S, Kumar L (2016) Global climate change impacts on pacific islands terrestrial biodiversity: a review. Trop Conserv Sci 9(1):203–223

Thibault KM, Brown JH (2008) Impact of an extreme climatic event on community assembly. Proc Natl Acad Sci 105(9):3410–3415. https://www.pnas.org/content/pnas/105/9/3410.full.pdf.

Tuiwawa M (2005) Recent changes in the upland watershed forest of Monasavu, a cloud forest site along the PABITRA gateway transect on Viti Levu, Fiji. Pac Sci 59(2):159–164. https://doi.org/10.1353/psc.2005.0028

Van Vliet E, Heijenbrok-Kal M, Hunink MG, Kuipers E, Siersema P (2008) Staging investigations for oesophageal cancer: a meta-analysis. Br J Cancer 98(3):547. https://doi.org/10.1038/sj.bjc.6604200

Welbergen JA, Davies NB (2011) A parasite in wolf's clothing: hawk mimicry reduces mobbing of cuckoos by hosts. Behav Ecol 22(3):574–579. https://doi.org/10.1093/beheco/arr008

Wetzel RG, Likens GE (2013) Limnological analysis. Springer Science & Business Media, New York

Whittaker RJ, Fernández-Palacios JM (2007) Island biogeography: ecology, evolution, and conservation. Oxford University Press, Oxford

Willis K, Bhagwat S (2009) Biodiversity and climate change. Science 326(5954):806–807. https://doi.org/10.1126/science.1181937

Chapter 13
Economic Impacts and Implications of Climate Change in the Pacific

Satish Chand

13.1 Introduction

At the start of writing of this chapter on 12 February 2018, Tropical Cyclone (TC) Gita was passing through the south-eastern parts of the Fiji Island group, gaining ferocity with wind gusts of up to 200 km/h, and headed for the small island nation of Tonga. The news the following morning was of widespread destruction with nearly every building on Tongatapu damaged and the century-old national parliament house and the weather office both reduced to rubble. While no deaths were reported, some 80% of the population were adversely affected by this category 4 tropical cyclone (TC).[1] The economic loss arising from this natural disaster is estimated at US$164.3 of millions[2]—a large sum for a nation of 106,000 people with a per capita GDP of US$5000. Samoa and the American territory of the same name were hit earlier by TC Gita, but the damage done was smaller as the cyclone was still gaining strength as it travelled over the warm waters of the Pacific Ocean. Fiji escaped the wrath of TC Gita as it slipped south of the most populated parts of the nation, affecting communities in the smallest islands only. A single cyclone within a single sweep has the potential to provide the 'knock-out' punch to several island economies, and TC Gita is a timely reminder of this fact.

Pacific Islands are highly prone to natural disasters. The sea rises by 12 mm a year within the western Pacific, swallowing eight islands in the Federated States of

---

[1] Tropical cyclones are categorised by wind speeds; category 1 has wind gusts of up to 125 km/h while category 5 has wind gusts exceeding 280 km/h.

[2] Damage assessment from the Asian Development Bank.

Cyclone Gita Recovery Project (RRP TON 52129) report, accessed online on 2 June 2019 at https://www.adb.org/sites/default/files/linked-documents/52129-001-ssa.pdf.

S. Chand (✉)
School of Business, Australian Defence Force Academy, UNSW Canberra,
Canberra, ACT, Australia
e-mail: Satish.Chand@unsw.edu.au

© Springer Nature Switzerland AG 2020
L. Kumar (ed.), *Climate Change and Impacts in the Pacific*, Springer Climate,
https://doi.org/10.1007/978-3-030-32878-8_13

Micronesia where the 'apocalyptic consequences of climate change have become a reality' (*The Economist*, 6th March 2018).[3] Kiribati and Tuvalu, two small island nations located within the central Pacific, are often used as the poster children for activists campaigning against the effects of emissions of greenhouse gases (GHG) into the atmosphere. These two nations amongst many other atoll island states face frequent storm surges, resulting from a rise in the sea-level and atmospheric temperature due to human activity. Indeed, the sequel to the movie '*An Inconvenient Truth*' depicts storm surges hounding homes in Tarawa, the capital of Kiribati. Pictures do not lie as the saying goes, but behind the movie clips are innocent victims least able to cope with the effects of climate change. Their homes will become inhabitable long before being inundated. This is because the encroaching salinity is inundating freshwater aquifers that lie on the shores, and the rising temperatures are killing coral that has served as a protective wall and the reefs therein, which have been a source of food for the people. Besides, the desertification of the little arable land available for agriculture is adversely affecting food production. It is possible to ship in food and water from abroad, but both require income which is lacking in the islands given their weak economies. Alternatively, the residents may be moved to higher ground, possibly abroad, but only if the option to emigrate were available. And even then, difficult as it may be, people loathe leaving homes they have occupied for generations.

To the extent that human activity is contributing to climate change, Pacific Islanders have contributed the least towards the emission of greenhouse gases, the anthropogenic cause of climate change. The average citizen of Kiribati used energy that amounted to some 98 kilograms of oil in 1990, the fourth lowest on the planet after Comoros (at 43), the Gambia (67), Guinea-Bissau (73), and Cape Verde (85). Per capita annual energy consumption for the Pacific Islands as a group at 334 kilograms of oil equivalent falls within the bottom sixth of the 206 countries for which this data is published by the World Bank.[4] At the other extreme in terms of energy consumption are Qatar (at 13,698), United Arab Emirates (at 19,979), and Bahrain (10,555); the corresponding figures for the USA, Canada, and Australia are 7672, 7602, and 5061, respectively; while the populous nations of China (609) and India (351) are ramping up their figures which already exceed those for the Pacific Islands. Cumulatively, the USA and EU countries contributed more than half of the total global emissions of greenhouse gases (GHG) from 1900 to 2005, while newly industrialising Asia is adding increasingly to the stock more recently (Stern 2007).

The economic effects of climate change are evident for many Pacific Islanders. The islanders have lived off their environment for millennia, and their fragile environments have sustained communities only because they have learned to live in

---

[3] The Economist, 2018 "Why climate migrants do not have refugee status"; accessed online at https://www.economist.com/blogs/economist-explains/2018/03/economist-explains-3?cid1=cust/ddnew/email/n/n/2018036n/owned/n/n/ddnew/n/n/n/nAP/Daily_Dispatch/email&etear=dailydispatch on 7 March 2018.

[4] This data is from the World Development Indicators (online) database, accessed on 1 February 2018.

harmony with nature. While there may be a short-term trade-off between economic prosperity and a healthy climate for industrialised communities, island residents do not have such luxury. Residents of small atoll states such as Kiribati, the Marshall Islands, and Tuvalu rely on aquifers for supplies of freshwater, the coral reefs for food and protection from tidal surges, and coastal land for agriculture. Most of their waste is also discharged into the oceans and the atmosphere, and thus the environment provides recycling services. A healthy environment is critical to the sustenance of these communities. And whenever a natural disaster strikes, as in the recent case of TC Gita, it is the poor who suffer the most as they live in the most disaster-prone areas, have poor housing, and have little (if any) savings to mitigate the adverse effects of extreme weather events such as storms, floods, and cyclones.

The sky and the planet that we inhabit is a global common. The Pacific Ocean a century ago was seen in much the same light as outer space now: it was the frontier for exploration and exploitation by industrialised Europe. It was believed up to the 1980s that human activity had a negligible impact on the environment but that the continued release of carbon dioxide (CO_2) through the burning of fossil fuel would impact on the climate (Nordhaus 1982). Common property creates the peculiar incentive for over-exploitation resulting in what economists call 'the tragedy of the commons'. The Pacific Ocean occupies a third of the planet's surface, is home to biodiversity that has enriched humanity, and continues to absorb carbon dioxide and thus keep the planet cool for all to benefit. As common property, the services provided by the Pacific Ocean are not valued by the marketplace. The warming of the Pacific Ocean, however, is contributing to devastating cyclones, many of which frequent several Pacific Island states. But these small island states have little leverage within the international community to draw attention to the price they are paying for climate change. Tropical Cyclone Pam, the strongest recorded in Vanuatu, struck on the 13th of March 2015 with winds of around 250 km/h and gusts peaking at 320 km/h, affecting more than half the total population of the country. The cyclone left in its aftermath a trail of torn buildings, displaced populations, and contaminated freshwater and food supplies. Less than a year later, TC Winston struck Fiji on the 19th to 20th of February, packing winds of up to 270 kilometres an hour, leaving behind 19 dead and hundreds of millions of dollars in damages to buildings, public infrastructure, and crops. Donors came to their assistance with relief supplies, but the islanders have in the main been left to their own devices to regroup and rebuild from the devastation caused by these natural disasters.

Some Pacific Islanders are already seeking means to mitigate the adverse effects of climate change. The people of Kiribati and Tuvalu, for example, purchased land in Fiji in 2016 to provide refuge if the sea rises to claim their homeland. However, globally coordinated effort at averting the worst effects of climate change and collective action to mitigate the impact on the poor are lacking. Stern (2007) opined that '[t]he costs of strong and urgent action to avoid serious impacts from climate change are substantially less than the damages thereby avoided'. That is, the costs of averting climate change—for example, through reductions in the stock of greenhouse gases in the atmosphere—are likely to be less than the costs of adapting to climate change. These costs are unlikely to be evenly spread across space: residents of atolls will end up paying a higher price than their rich industrial country cousins, at least in the short term. Should climate change leave the planet unhabitable, then surely residents of atolls will be the first casualties.

This chapter considers the peculiar economic characteristics of small island states and the interaction of these features with climate change. It considers the costs and the benefits of the release of greenhouse gasses and the distribution of these across populations of the islands versus the rest of humanity. Islanders suffer the consequences of climate change, but they also reap many of the rewards from industrialisation and international commerce. A devastating cyclone, for example, can destroy all local supplies and diminish domestic opportunities for income generation, leaving the community reliant on goods and services sourced from abroad and remitted funds to pay for them. Populations of small states now depend on international trade to access the full complement of goods and services for consumption, which in turn are paid for using the proceeds of their exports and remittances. The global market provides the opportunity for small states to diversify their risks from domestic production and local income generation. International commerce consequently may be the saviour for small states from climate change too.

Next, I present evidence on the impact and implications of climate change on the island states of the Pacific. My primary focus is on 21 Pacific Island Countries and Territories (PICTs henceforth) together with Australia and New Zealand (ANZ henceforth) that span much of the Southern Pacific Ocean. I argue that the adverse effects of climate change are felt disproportionately by the inhabitants of the small island states and territories and particularly the poor. I use their plight to argue for policy options in alleviating the problems arising from climate change. The rest of this chapter is structured as follows. Section 13.2 provides details on the economic settings of the island economies of the Pacific, Sect. 13.3 presents a succinct discussion on the economic effects of climate change, Sect. 13.4 presents options for policymakers in mitigating the consequences of climate-related disasters, and conclusion follows.

13.2 Economic Settings of Pacific Islands

There is a long and distinguished history of Pacific Islanders being portrayed as being laid back, carefree, and happy people living in comfort on their natural resources. Professor Edward Fisk first introduced the term 'subsistence affluence' in 1974 in the context of Papua New Guinea, where he observed that the level of productive capacity of residents within a subsistence unit exceeded that necessary to satisfy the localised demands for consumption of the unit. This latent excess capacity Fisk (1964) argued could be utilised for development 'if the necessary incentives were provided' (page 156). This view has retained currency since and is true in the few communities with abundant natural resources. However, population growth and climate change collectively are increasing the pressure on the natural resources, meaning that subsistence affluence may be an illusion for many islanders.

There is considerable heterogeneity in economic performance across the individual nations and territories of the Pacific. Table 13.1 provides data on land area, total population, average per capita income, aid receipts on a per capita basis, and life expectancy for three groups of nations: the 23 nations and foreign territories within the Pacific; Australia and New Zealand; and three island economies, namely, Maldives, Malta, and Mauritius (the 3Ms henceforth) that I use for comparisons with the PICTs. As shown in the table, there is far greater diversity on all of the above-listed indicators within the PICTs than across the three country groupings.[5] The 14 sovereign states (PICs henceforth) and another 9 foreign territories of Australia (Norfolk Islands), Britain (Pitcairn Is), France (New Caledonia, French Polynesia, Wallis and Futuna), New Zealand (Tokelau), and the USA (American Samoa, Guam, and Northern Mariana Islands) span most of the Pacific Ocean. These 23 Pacific Island Countries and Territories (PICTs) are each different in terms of size and economic well-being. Tokelau is the smallest with a land area of 12 square kilometres, while Papua New Guinea is the largest with some 452,860 square kilometres; on population, Tokelau has some 1000 people while PNG at the other extreme has a population of close to 8 million; on GDP per capita, Palau ranks topmost with a purchasing power parity (PPP) value at 2011 prices of US$14,982; and on social indicators, life expectancy averages 80 years in Wallis and Futuna compared to 65 years in Papua New Guinea. The PICTs are also amongst the largest recipients of official development assistance (i.e. aid): on per capita terms. The residents of Tuvalu receive US$4513 (figures for 2015) compared to the corresponding figure for PNG of US$74. There are large differences amongst the PICTs with respect to land area, population, per capita gross national income adjusted for differences in purchasing power across nations, receipts of official development assistance per person, and average life expectancy. Also provided in the same table are comparable statistics for Australia and New Zealand (ANZ henceforth) and a further three island nations, two from the Indian Ocean and one from the Mediterranean. A word of caution is in order here: data on the PICTs is of questionable quality. Thus, the figures reported must be interpreted with care. Income data on overseas territories are sparse and often dated; thus, these figures are only indicative of the general trends.

In terms of land area and population, Papua New Guinea is the largest amongst the PICTs and approximately twice the size of New Zealand. Generalisations are hard to make with regard to socio-economic indicators of the PICTs except to note that those with preferential access to industrial country labour markets rank high in terms of per capita income when compared to those lacking such access (shown in Fig. 13.1). Cook Islanders are citizens of New Zealand, those of the American Samoa are US citizens while citizens of Palau have unfretted access to the labour market of the USA through a 'Compact of Free Association'. But such access is not a sufficient explanation for high per capita income in the PICTs: Tokelauans are

---

[5] Data for several indicators for the PICTs are either unavailable or of poor quality. Data from the one consistent source, namely, the World Development Indicators database (WDI) has been used wherever possible. The *World Fact Book* has been used to fill data gaps left by the WDI.

Table 13.1 Basic statistics on Pacific Island countries and territories, Maldives, Malta, and Mauritius

Country/group	Land area (km²)	Population (000s)	Population density (per km²)	GDP per capita USSPPP (2011)	Net ODA received per capita (US$)	Life expectancy (years)	Air pollution (PM2.5)	Highest elevation (m)
PICTs								
American Samoa (a)	199	51	256	13,000	–	73	–	964
Cook Islands (a)	236	11	47	12,300	–	76	–	652
Fiji	18,270	892	49	8478	115	70	0	1324
French Polynesia	3660	278	76	–	–	77	–	2241
Guam	540	162	300	–	–	79	0	406
Kiribati	810	112	139	1967	578	66	0	81
Marshall Islands	180	53	294	3665	1077	..	0	14
Micronesia, Fed. Sts.	700	104	149	3271	779	69	0	782
Nauru	20	12	624	12,270	2505	–	–	70
New Caledonia	18,280	273	15	–	–	78	–	1628
Niue (a), (b)	260	2	8	5800	..	..	–	80
Papua New Guinea	452,860	7920	17	3867	74	65	65	4509
Pitcairn Is (a)	47	0.054	1	–	–	–	–	347
Northern Mariana Islands	460	55	119	–	–	–	100	965
Palau	460	21	46	14,982	654	–	–	242
Samoa	2830	193,759	68	5559	484	75	0	1857
Solomon Islands	27,990	587	21	2053	323	70	0	2335
Tokelau (a), (c)	12	1	83	1000	–	–	–	5
Tonga	720	106	148	5190	643	73	0	1046
Tuvalu	30	11	367	3324	4513	–	–	5
Vanuatu	12,190	265	22	2807	705	72	0	1877

Wallis and Futuna (a), (d)	142	15	106	3800	.-	80	.-	522
Pacific Island small states	*64,200*	*2359*	*37*	*5260*	*398*	*71*	*0*	
ANZ								
Australia	7,682,300	23,789	3	43,832	–	82	0	2228
New Zealand	263,310	4596	17	34,949	–	81	–	3724
The 3Ms								
Malta	320	432	1350	34,273	.-	82	100	253
Maldives	300	409	1364	14,015	66	77	100	5
Mauritius	2030	1263	622	18,864	61	74	100	828

Note: All data sourced from the World Development Indicators and for the year 2015, the most recent available, except for those with subscript 'a' where the data is from CIA World Fact Book; 'b' denotes that per capita GDP data is for the year 2003, 'c' is the same for the year 1993, and 'd' denotes that this data is for the year 2004; all data accessed online on 31 January 2018

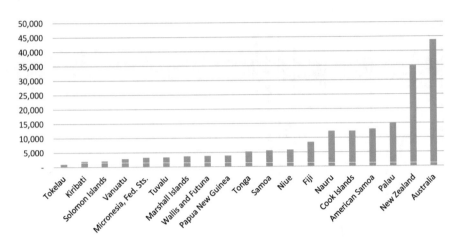

Fig. 13.1 Per capita gross national income (PPP, US$ at 2011 prices)

citizens of New Zealand, citizens of the Federated States of Micronesia and Marshall Islands also have a 'Compact of Free Association' with the USA and thus free access to the US labour market, and residents of Wallis and Futuna are citizens of France. Yet the per capita income in the islands is a fraction of the incomes in the metropolis with which they share the labour market. Solomon Islands and Papua New Guinea rank particularly low on per capita income, and neither has privileged access to industrial country markets for its labour. In terms of levels of per capita income, the PICTs as a group fall well short of Australia and New Zealand (ANZ). PNG, as an example, has per capita income that is less than one-tenth that of New Zealand and an average life expectancy that falls short by 16 years.

The main sources of income within the PICTs are agriculture, tourism, minerals, and fishing. The contribution of these industries to GDP differs across the individual members of the PICT. Subsistence agriculture provides the livelihoods for some 80 plus percent of the population in Papua New Guinea, the Solomon Islands, and Vanuatu; that is, subsistence agriculture is the main source of livelihood for the bulk of the population in three of the four Melanesian nations, but these nations have volcanic origins and therefore are rich in agricultural land. Tourism is the main export for Cook Islands, Fiji, Palau, and Vanuatu. License fees paid by foreign vessels who fish in the Exclusive Economic Zones of the PICs account for the bulk of foreign exchange earnings of Nauru, Kiribati, and Tuvalu. Mineral exports comprising gold, copper, cobalt, crude oil, and liquified natural gas accounted for close to 80% of the total commodity exports of K25 billion for Papua New Guinea in 2016.[6] Production is impacted upon heavily by climatic conditions and drought particularly. Most of the agriculture in the PICTs is rainfed, meaning that droughts affect food production, while cyclones and floods usually destroy a lot more. Major impediments to growth of commodity exports from the PICTs include the high freight costs given their isolation from industrial country markets. Papua New

---

[6] One Kina was worth US$0.31 on 18 February 2018.

Table 13.2 Visitor arrivals by year, 2009–2014

Country/territories	2009	2010	2011	2012	2013	2014
American Samoa	30,705	29,060	28,403	22,579	20,846	21,603
Cook Islands	101,229	104,265	113,114	122,384	121,115	121,458p
Federated States of Micronesia	24,473	24,422	–	38,263	42,109	35,440
Fiji Islands	605,478	689,502	733,712	740,593	767,249	780,271
Guam	1,052,871	1,197,408	1,159,778	1,308,035	1,334,497	1,343,092
Kiribati	3915	3490	5000	4907	4981	–
Marshall Islands	4923	4563	4559	4600	4333	4776
Nauru	3151	2120	2387	1753	1581	–
Niue	4662	6214	6094	5048	5129	7661
Northern Mariana Islands (CNMI)	353,956	–	–	401,219	438,908	459,240
Nouvelle-Calédonie	230,610	281,807	347,559	390,145	493,678	528,823
Palau	77,454	92,500	118,055	124,286	111,145	146,867
Papua New Guinea	124,199	147,000	163,000	224,182	245,844	118,418
Pitcairn	nc	nc	nc	nc	nc	nc
Polynésie française	160,447	153,919	162,776	168,978	164,393	180,602
Samoa	129,305	129,500	134,690	172,720	162,877	173,655
Solomon Islands	20,269	22,531	24,952	23,925	24,053	20,070
Tokelau	nc	nc	nc	nc	nc	nc
Tonga	74,669	65,005	68,373	57,230	59,665	64,219
Tuvalu	1604	1657	1232	1019	1032	1416
Vanuatu	225,493	237,648	248,898	321,404	357,405	329,013
Wallis et Futuna	nc	nc	nc	nc	nc	nc

Note: nc means data have not been compiled and means data may have been compiled but not made available to SPC. Source: Secretariat of the Pacific Community (http://prism.spc.int/regional-data-and-tools/economic-statistics), accessed 13 February 2018

Guinea, given its size, has a widespread of industries ranging from highly capital-intensive mines on the mainland to boutique tourism based largely around diving on the outer islands.

Tourism makes a significant contribution to GDP in several PICTs and has the potential to do more. The isolation of several PICTs from the major industrial centres has left the natural environment in many islands free of industrial pollutants.[7] The growth of tourism and its contribution to the economy has varied over time, but there remains considerable potential for further expansion. Table 13.2 provides data on visitor arrivals for 2009–2014, the most recent year for which this statistic is available for the PICTs. Guam leads: with a resident population of 162,000, this American territory received 1.3 million visitors in 2014. Fiji hosts some 800,000 tourists annually with the total earnings from the industry for 2016 amounting to

---

[7] Radioactive waste left after nuclear testing in the Northern Pacific is an exception, however.

Table 13.3 Value of tuna caught within national waters, 2011–2016, US$ million

FFA members	2011	2012	2013	2014	2015	2016
Australia	22	17	14	17	23	22
Cook Islands	54	132	46	51	51	46
Fiji	35	33	23	37	47	46
FSM	347	453	455	269	264	330
Kiribati	484	1349	696	1242	936	702
Marshall Islands	82	116	126	168	69	166
Nauru	181	112	344	268	86	195
New Zealand	30	30	37	23	13	13
Niue	–	–	1	1	1	0
PNG	1189	1318	1245	562	272	496
Palau	30	35	28	25	10	25
Samoa	10	12	7	5	6	9
Solomon Islands	389	291	325	264	312	326
Tokelau	37	50	34	39	70	25
Tonga	2	8	11	5	9	13
Tuvalu	117	165	124	153	107	184
Vanuatu	42	31	31	27	24	30
Subtotal	**3052**	**4153**	**3546**	**3156**	**2302**	**2628**

Source: Table 1, Pacific Islands Forum Fisheries Agency (FFA) (https://www.ffa.int/node/425), accessed online on 13 February 2018

FJD 1.6 billion or equal to 16% of GDP.[8] Fiji received some 110,000 tourists in 1970 but, by 2016, had increased this figure sevenfold but still behind Mauritius, a lot smaller nation with far fewer natural attractions, which received some 1.4 million tourists in 2016. Fiji and Mauritius are very similar in many respects including targeting the high-value end of the market, but Fiji has the advantages of being larger in physical size; better geographic positioning in being close to Australia, New Zealand, and Asia; and a lot more natural attractions. Papua New Guinea, a much larger country with far more natural and historical attractions, attracted only some 118,000 tourists in 2014. Growth of tourism in PNG has been hampered by the poor law and order image of the nation. The *Economist Intelligence Unit* ranks Port Moresby 137 out of 140 countries in terms of 'livability', warning visitors that the rate of crime in PNG is amongst the highest in the world.

Fisheries has been a sustainable source of income for many Pacific Islanders and that of protein for humanity. The Western and Central Pacific Ocean supplied 2.7 million tons (equal to 57%) of the total global catch of 4.7 million tons of tuna in 2015, and the Pacific Ocean remains the most fertile source of fisheries for the global population (FFA 2017, p. 1). The value of catch of albacore (*Thunnus alalunga*), bigeye (*Thunnus obesus*), skipjack (*Katsuwonus pelamis*), and yellowfin (*Thunnus albacares*) tuna within the 200-mile Exclusive Economic Zone of Pacific Island nations had passed USD2.6 billion annually by 2016 (see Table 13.3). There

---

[8] The data is from Fiji Islands Bureau of Statistics (http://www.statsfiji.gov.fj/statistics), accessed online on 13 February 2018.

is potential for increased harvest given that the assessment of stocks by the Scientific Community of the *Western and Central Pacific Fisheries Commission* (WCFPC) is that yields for skipjack and albacore tuna are below maximum sustainable harvest. However, changing weather patterns are affecting both the rate of catch and stock of tuna fisheries (Harley et al. 2015, p. 3).

Pacific Islanders are, in general, big and athletic people, therefore in themselves are a resource for the PICTs. However, and unfortunately, several Pacific Island states rank high in terms of the proportion of obese adults, defined as those with a body mass index (BMI) greater than 30. Of the 91 nations for whom this data is published by the World Health Organization, Pacific Island states take the top six positions: Nauru ranks first with a 78.5% (data for 1994, the most recent available), and American Samoa comes a close second at 74.6% (data for 2007), followed by Tokelau (at 63.4), Tonga (56.0), Kiribati (50.6), and French Polynesia (40.9) with Fiji, the only other PICT on the database, at 18th place with 23.9% of the adult population being obese.[9] Sports, rugby in particular, is a major preoccupation of many in the islands. Pacific Islanders have an international reputation in rugby and, in the 7-a-side version, have held the top ranks since the game commenced with Fiji having won the maiden gold medal in the sport in the 2016 Rio Olympic Games.

PICT economies overall have performed poorly relative to their potential. Two pairs of country comparisons are used to support this claim. Mauritius and Fiji have many similarities including their common colonial heritage, close to identical structure of production, and parallel historical dependence on access to preferential markets for their sugar and garment exports. Both are island economies isolated from their major markets, both have a heavy dependence on tourism, and both have been wrestling with the loss of significant export revenues as world trade for sugar and garments was liberalised. Mauritius gained its independence in 1968 and Fiji followed 2 years later, and from the same colonial power; thus, the two have very similar institutions. Fiji, however, has the advantage of having a land area that is some nine times larger and a population that is a third smaller than that of Mauritius. Population density in Fiji, consequently, is nearly one-thirteenth the corresponding figure for Mauritius. Fiji also had a 19% head start in terms of per capita GDP in 1980. By 2015, Mauritius had a per capita GDP more than double that of Fiji (see Table 13.1). Like Fiji and Mauritius, Maldives and Marshall Islands have many similarities, both being small atoll states heavily dependent on tourism and located some distance away from their major tourist markets. Marshall Islands, however, has the advantage of being in a *Compact of Free Association* with the USA. Compact funds and associated grants account for more than two-thirds of total government revenues. The compact, as explained earlier, allows for unfettered access for the Marshallese to the US labour market.[10] Despite these advantages, per capita income in Marshall Islands as of 2015 was roughly one-quarter that of Maldives.

---

[9] These are the only PICTs listed on the database and the data is for the most recent year available.

[10] The PICTs as a group receive high levels of aid on a per capita basis (see data in Table 13.1).

Generalisations drawn from cross-country comparisons are fraught with problems, but the pairwise comparisons made above are indicative of the poor performance of the PICTs vis-à-vis their potential. While the reasons for the poor performance of the PICTs compared to their counterparts from elsewhere are context-specific, a common feature is the poor governance in several PICTs. Nauru and Tuvalu, two similarly sized economies, have had very different experiences in managing their savings. While Tuvalu has been successful in growing its trust funds, Nauru has nearly depleted it (Graham 2005). Graham (2005), for example, notes that the Nauru Trust Fund had 'rampant systematic mismanagement' (page 47) and 'broke almost every rule in the book' (page 48).

Pressure on natural resources is uneven across the PICTs but on the whole is rising. Consequently, the opportunities to tap into idle capacity as first proposed by Professor Fisk have either disappeared or are rapidly disappearing. The difference in access to land across the individual PICTs illustrates this well. Population densities within the PICTs range from one person per square kilometre in Pitcairn Island to 624 on Nauru, but these figures on the whole compare favourably to the 3Ms (see Table 13.1). The country aggregates, however, mask the large variations in access to land within the PICTs. Urban slums have mushroomed in several of the PICTs, and overcrowding is a serious issue in many of the urban centres. Jenrok, a crowded settlement on Majuro, which is home to some 200 households or some 2000 individuals all housed on 6.5 hectares of land is a good case in point. The above figures equate to a population density of 28,000 people per square kilometre (ADB 2005).

Increases in infectious and lifestyle diseases across the PICTs are eroding the human resource base that is critical for income growth. People incapacitated by these preventable ailments raise demand for health outlays while depleting the stock of workers, both directly as the sick withdraw from the workforce and indirectly by needing carers. Nearly half of the population of Nauru has type II diabetes. Obesity, as noted earlier, is a serious problem in nearly every PICT. HIV/AIDS has already reached epidemic proportions in Papua New Guinea and is spreading to the rest of the PICTs. The disease affects every facet of life; the economic and social costs of an HIV epidemic could be far more catastrophic than any of the previous disasters encountered in the region.[11] Data on poverty rates are unavailable for most of the PICTs. But for those with this data, poverty rates are high. Some 35% of the households in Fiji in 2008 had income that fell short of the national poverty line while the corresponding figure for Papua New Guinea was 28%. More alarmingly, more than half of the households in the Republic of Marshall Islands were in poverty in 2002.[12]

This section has presented the economic setting of the Pacific Islands and Territories. It, in the main, has shown that many PICTs have weak economies and

---

[11] The experience of Botswana where some 40% of the population has been infected with HIV would be the worst nightmare for PNG. Interestingly, PNG had the first case of HIV about the same time as Botswana but did not experience a similar increase over time.

[12] The figures reported are drawn from the Pacific Regional Information System of the Secretariat of the Pacific Community, accessed online on 24 August 2015.

therefore short of the capacity to mitigate the adverse consequences of climate change. Next, I discuss the interaction between the economy and the environment to assess its impact on economic welfare.

13.3   Economic Effects of Climate Change

Economist have only recently turned their attention to the economic effects of climate change. Historically, economists have been taught the value of decentralised exchange through the market to maximise social welfare. This trust in the value of decentralised exchange through the market is premised on three specific assumptions, namely, that (1) preference orderings of consumers and the production functions of producers are independent of each other, (2) consumers maximise their individual welfare subject to their budget constraints, and (3) producers maximise profits for given market prices (Ayres and Kneese 1969, p. 283). The theoretical predictions follow from the above enumerated assumptions. The presence of externalities, either from consumption or production as is the case for GHG emissions, does not alter the theoretical prediction provided that property rights are assigned to the externality, and transactions take place with full information and without cost. Thus, clear rights to property, as Coase Theorem posits, would still lead to the maximisation of social welfare through decentralised exchange (Coase 1988). So at least at first glance, decentralised exchange in the marketplace already accounts for the effects of climate change. This conclusion, however, is flawed in the absence of property rights to commons and/or when transaction costs prevent the market from yielding efficient bargains.

Property rights to global commons such as the atmosphere and the international waters have been lacking for much of human history. Consequently, the services provided by the atmosphere and the oceans have been taken as being 'free for all'. This 'free for all' leads to the 'tragedy of the commons' wherein individuals and firms dump waste into commonly held property without compensation. An example may help illustrate a case of the tragedy of the commons. The coast, which is a source of protein such as mud crabs, is common property for most of the residents living close by. Consequently, anyone with the motivation is free to hunt for crabs on the coast. Each individual crab hunter has the incentive to catch whatever they can since leaving even a juvenile crab behind was the prize for others following close behind. In other words, it is in the interest of every hunter to catch as many crabs and as quickly as possible, leaving the coast bereft of the resource. Much of the coast is over-crabbed, overfished, and overrun—a typical case of the tragedy of the commons. The solution would be in assigning property rights and thus shifting the balance from destructive overuse of common property to sustainable use of private property. Private rights of property to the coast would provide the incentive for the new owner to sustainably harvest the resource. Regulation could have been an alternative solution but this requires enforcement. In some cases, traditional communities place bans on harvesting whenever, and wherever, necessary. Communities have through such practices lived through millennia on the produce of their environment without facing problems of overharvesting or over-exploitation of wildlife (Smith 1981).

On climate change specifically, economists first wrote about the potentially harmful effects of the emissions of GHG on the environment roughly half a century ago: well after a century of unfretted release of industrial pollutants into the atmosphere and the oceans. Ayres and Kneese (1969) were amongst the first to posit that 'the continued combustion of fossil fuels at a high rate could produce externalities affecting the entire world' (page 286). The costs of externalities moreover were not priced by the market given the fact that GHG released into the atmosphere spread across the globe, remaining there for centuries. This is not too different from the extant practice of sending rockets and associated debris into outer space. Most rich nations of the present benefitted in the process as they did not pay for the waste from industrialisation dumped into the atmosphere and the oceans, the effects of which on the climate are being felt now. Consequently, emissions of GHG have an externality across space and over time. On the former, islanders face the brunt of sea-level rise while on the latter, past generations have passed the costs of mitigation to the present.

While most scientists agree that climate change is taking place, and that this is the result of human activity, there is less than universal agreement on either of these claims. A full coverage of the science of climate change is beyond the scope of this chapter, but I accept the findings from the Intergovernmental Panel on Climate Change (IPCC) regarding climate change. At its core, the impact of the climate change is analysed by computing the balance of heat produced through incoming and the bouncing back of solar radiation from the planet's surface. Furthermore, the reflected radiation is believed to be blocked by a blanket of greenhouse gasses, carbon dioxide in the main produced through combustion of fossil fuel, that then raises the temperature of the planet above what it would be otherwise (Barnett and Campbell 2010; Le Treut et al. 2007). The released CO_2 diffuses through the oceans and the atmosphere, altering acidity, rainfall, and wind patterns on the planet's surface, thereby impacting on the biosphere in form and quantity yet to be fully understood. There is, however, agreement amongst scientists that climate change will have the biggest impact on agriculture and the coastlines (Nordhaus 1982).

So how much greenhouse gasses have the residents of the PICTs released? Table 13.4 provides data for 2014, the most recent available, on per capita emissions of greenhouse gasses and the ratio of the above to the average for the global citizen. The data presented is for five high-income industrialised economies of Australia, Canada, New Zealand, United Kingdom, and the USA; China and India, the two most populous nations of Asia; 11 Pacific Island nations; and another six island nations from elsewhere. The first column shows GHG emissions inclusive of the effects of changes in land use and forestry while the second excludes the latter two. The figures in the bottom most row show that average per capita emissions of GHG for the world was 6.73 tons of carbon dioxide equivalent (tCO_2e) when the effects of land use changes and forestry are included and 6.29 tCO_2e when not.

Table 13.4 Per capita emissions and ratios of greenhouse gas emissions (GHG), 2014

	Including land use change and forestry (tCO$_2$e)	Excluding land use change and forestry (tCO$_2$e)	Ratio to world average (figures from column 2)	Ratio to world average (figures from column 3)
HIIC				
Australia	22.30	25.14	3.32	3.99
Canada	24.39	20.96	3.63	3.33
New Zealand	13.38	16.80	1.99	2.67
UK	7.64	7.83	1.14	1.24
USA	19.84	20.00	2.95	3.18
Asia				
China	8.50	8.73	1.26	1.39
India	2.48	2.38	0.37	0.38
PICs				
Fiji	−0.87	2.51	−0.13	0.40
Kiribati	0.72	0.77	0.11	0.12
Marshall Islands	2.64	2.64	0.39	0.42
Micronesia	1.45	1.73	0.22	0.28
Nauru	4.52	4.52	0.67	0.72
Palau	20.55	20.55	3.05	3.27
PNG	9.80	2.16	1.46	0.34
Samoa	2.45	2.45	0.36	0.39
Solomon Islands	4.43	0.99	0.66	0.16
Tonga	3.15	3.15	0.47	0.50
Vanuatu	2.76	2.82	0.41	0.45
OSS				
Maldives	3.55	3.55	0.53	0.56
Malta	6.95	6.95	1.03	1.10
Mauritius	4.62	4.63	0.69	0.74
Saint Kitts and Nevis	7.27	7.13	1.08	1.13
Saint Lucia	6.64	6.45	0.99	1.03
Saint Vincent and Grenadines	2.68	2.51	0.40	0.40
World	6.73	6.29	1.00	1.00

Note: tCO$_2$e denotes tons of carbon dioxide equivalent, HIIC denotes high-income industrialised countries, PICs denotes Pacific Island Countries, and OSS denotes other small states. Data source: *CAIT Climate Data Explorer.* 2015. Washington, DC: World Resources Institute. Available online at: http://cait.wri.org, accessed online on 22 February 2018

The last two columns show the ratios of per capita emissions of GHG emissions for each nation relative to the average for the world. Australia ranks top with per capita emissions of GHG at 22.3 tons of CO$_2$e when changes in land and forestry are included, and 25.14 tons otherwise; these are multiples of 3.32 and 3.99 of the average emissions for the global citizen. China has per capita emissions inclusive of those from changes in land use and forestry 1.26 times the world average, while the corresponding figure for India is 0.37. Focussing just on the multiples of CO_2e inclusive of changes in land use and forestry, the figures for the Pacific Islands range from 3.05 for Palau to –0.13 for Fiji; the figure for Fiji implies that changes in land use and forestry in the nation exceed emissions of CO_2e by 0.87 tons per capita. GHG emissions for the other island nations are lower than those for the industrialised nations and closer to the global average but overall higher than those for the PICs. With the sole exception of Palau, the Pacific Islands as a group have low per capita emissions of GHG when compared to the industrialised nations and other island economies. At the very bottom is Kiribati where per capita GHG emissions are 11% that of the global average. These differences in emissions are for a single year, but they have persisted for centuries, meaning that the contribution to the overall stock of GHG emissions would be similar.

While the people of the PICTs have contributed minimally to the emissions of greenhouse gasses, they remain at the forefront in terms of facing the adverse effects of climate change. This is because the earth absorbs the most heat around the equator, where nearly every PICT is located, with the broad weather and climatic patterns determined by the flow of heat from the equator to the poles (Barnett and Campbell 2010, p. 8). The impact of climate change is assessed based on the balance of the quantity of heat and freshwater for the planet, and the contribution of the above to the salinity of the oceans, which in turn affects biological activity. While the effects of climate change on marine organisms are poorly understood, there is evidence that the distribution of tuna has been affected by changing salinity and temperature of the Central and Western Pacific Ocean. There is also evidence of a warming of the sub-tropical North and South Pacific Oceans, leading to changes in atmospheric and oceanic circulations (Bindoff et al. 2007b, p. 399). The IPCC has provided evidence of a statistically significant increase in and an eastward shift of the distribution of temperature with 'a human fingerprint on greenhouse gasses' as being responsible for changes in weather patterns.

Pacific Islanders are already facing extreme weather events that may be linked to climate change. Bindoff et al. (2007a), for example, note that '[t]here is evidence of an increase in the occurrence of extreme high water worldwide related storm surges, and variations in extremes during this period are related to the rise in mean sea level and variations in regional climate' (p. 387). The warming of the sea is linked to a rise in sea levels (from thermal expansion) leading to more damaging tropical cyclones, heavier downpours, coral bleaching, and prolonged droughts. Inhabitants of small island states and particularly those living on flood plains and close to the shorelines are most vulnerable. This includes many communities of the Pacific who may have to leave their homes permanently due to storm surges and a rise in sea levels.

The economic effects of climate change on the people of the PICTs are likely to be large as well. This is because the PICTs, as a group, are rich in and the economies depend heavily on marine resources (Hannesson 2008). Consequently, a change in climate will affect supplies of agricultural land, freshwater, and food including fisheries. Rising salinity is likely to lead to loss of coastal land that is currently used for agriculture, a rise in sea levels will result in the inundation of sources of freshwater, and coral bleaching will expose the coastline to storm surges. Some communities are completely dependent on rainwater, meaning that prolonged droughts will force them to move residence. Habitation within the PICTs is concentrated on coastal flats such as the Guadalcanal Plains in the Solomon Islands, meaning that small rise in sea levels and encroachment of agricultural land is likely to have devastating consequences on food production and settlement. The loss of reefs from coral bleaching and tropical cyclones will expose the resident population to the effects of storm surges and likely affect marine food supplies. The warming of the climate could also lead to the spread of mosquitoes and with them debilitating diseases such as malaria and dengue fever. Tourism will be affected through loss of beaches and tourism infrastructure such as hotels, which are nearly exclusively located on the shoreline.

The effects of climate change will reverberate through the PICTs to the rest of humanity. The Pacific Ocean spans one-third the surface area of the planet. It is the major carbon sink and a significant marine park for humanity. The United Nations Convention on the Law of the Sea provides for an Exclusive Economic Zone (EEZ) that extends some 200 miles (i.e. 370 km) from the coastline of a nation state wherein the coastal State has 'sovereign rights for the purpose of exploring and exploiting, conserving and managing the natural resources, whether living or non-living, of the waters superjacent to the seabed and of the seabed and its subsoil, and with regard to other activities for the economic exploitation and exploration of the zone, such as the production of energy from the water, currents and winds' (UN Convention on the Law of the Sea; Article 56(a)). Much of the Southern Pacific Ocean falls within the EEZ of Pacific Island Countries, and these nations have a responsibility to their own people and that of the globe to manage the resources therein sustainably. This responsibility arises for many reasons, including the fact that more than half of the total supplies of tuna for the global market is sourced from the Pacific Ocean (Gillett 2003).

13.4 Policy Options

So what are the policy options available to the people of the PICTs to minimise the impact of climate change? They cannot leave the remedies to the invisible hand of free markets. Room for them to reduce greenhouse gas emissions are limited, possibly with the exception of Palau, given the low base to begin with. Furthermore, any reductions in greenhouse gas emissions by the PICTs are unlikely to have noticeable impact on the aggregate stocks given their small population. The World Resources Institute, however, notes that '[s]tabilizing the global climate is the greatest challenge of the twenty-first century', noting that 'temperatures have exceeded global annual averages for 38 consecutive years' with the 'impacts being felt all around the world' (WRI, 2018).[13] Pacific Islanders have been at the forefront in

---

[13] See http://www.wri.org/our-work/topics/climate; accessed on 1 February 2018.

terms of facing the consequences of climate change, and there is every likelihood that they will remain so for the foreseeable future. They have two responses to climate change: (a) migration and (b) adaptation or a combination of the two. I look at these options separately before considering the role the international community may play in helping Pacific Islanders live with climate change.

(a) Migration

The reality is that not everyone affected by climate change will be able to emigrate. Therefore, deterministic models assigning the number of people who would move for a given rise in the sea level, or other adverse climatic shock, are fraught with difficulties (Tacoli 2009). Temporary movements to escape adverse climatic events with the people returning home after the event has passed are more likely. Populations residing on customary land have the least mobility since the rights to use and live on the land are limited to members of the family group. Such land tenure arrangements are prevalent in most of the PICTs. Some 97% of the total land in Papua New Guinea, for example, is held by traditional clans under customary title. The transfer of land across groups and even the leasing of land held under customary title is extremely difficult. Past attempts to permanently move communities from their lands have had mixed success. Curry and Koczberski (1999) report on a specific case from PNG where those resettled and their descendants faced risks of eviction decades later, while their prospects for returning home being eroded with time through loss of linkages to those who remained behind. Customary land tenure therefore impedes mobility of people, even on a temporary basis.

Customary land tenure may allow for movement across short distances and for short periods of time on the understanding that people will return to their lands once conditions have improved. This means that 'short-distance and short-term movements will probably increase, with the very poor and vulnerable in many cases unable to move' (Tacoli 2009, p. 523). The implication of this situation is that policies would have to be targeted at accommodating 'multiple and interactive causes' of emigration of which climate change is just one factor (Castles et al. 2014, p. 213). Sea-level rise within low-lying islands, however, may be sufficient reason for mass evacuation, as has been experienced by the people of Banaba and planned for through purchase of land by the people of Kiribati.

The people of Banaba, a small island located in the central Pacific, were relocated to Fiji in December 1945 after the ecological destruction of their homeland through mining of phosphate by the British colonisers that began in 1900. While the people were resettled half a century ago with many of the original settlers having passed away, their descendants continue to reminisce about their island home and what may have been the case in the absence of exploitative mining that left the island inhabitable. Reverend Tebuke Rotan, one of the representatives of his people, argued in his petition to the United Nations in 1974 that 'seventy-five years ago the Banabans lived in peace on Banaba, an independent people with our own skills as fishermen, our own culture and traditions.

We were not connected with any other island or peoples. Our parents were content' (quoted from Hermann 2005, p. 279).

If migration from locations adversely, and temporarily, affected by climate change is going to be an intermittent feature, this may fit well with notions of circular migration. Obviously, those permanently resettled such as the people of Banaba lack this option, but others likely to be affected by adverse climate events may plan for temporary escape with a view to returning home once conditions improve. The economic challenge is that of income diversification for groups of people vulnerable to climate change. Migration can be a means to the above, and if so then facilitated through training of the workforce for overseas employment and supported by governments through cooperative arrangements for international labour mobility. Many of the PICTs send members of the household abroad specifically to earn income and have this remitted. Tuvalu and Kiribati, for example, train seafarers at home for employment on foreign ships. Remittances sent in cash are a significant source of foreign exchange for Tonga (amounting to 39% of GDP), Samoa (26%), and Kiribati (15%), but it could be larger since transfers through private channels and of consumer durables and construction materials are not captured in the official statistics.[14] McKenzie (2006) uses survey data to reveal that the non-monetary transfers can amount to anywhere between 25 and 40% of the value of cash remitted and that reverse flows of goods are also common. Furthermore, Gibson and McKenzie (2014) provide compelling evidence of the benefits of seasonal employment of Pacific Islanders in New Zealand. Governments of both sending and receiving nations may negotiate access in advance of natural disasters, provide financial resources in the aftermath, and lend social support to people escaping adverse climatic events. Those partaking in such an activity could be assured of their political and economic rights in both places of abode, where flows of remittances between the source and destination countries are supported, and qualifications and pensions are portable across locations with such an approach likely to be more successful in keeping the two communities connected (De Haas 2005, p. 1273). This is in sharp contrast to existing practice of a prohibition on holding dual citizenships in some PICs and imposition of large fees to emigrants requesting reinstatement of their lost citizenships. PNG, for example, amended its Constitution in 2016 to allow former citizens to reacquire their citizenship but with an application fee of K5,000 (US$1550) for those above 19 years of age. Thus, immigration policies could be used to encourage freer circulation between source and destination to the benefit of both societies.

Finally, the moral case for allowing islanders from atolls facing seal-level rise to settle in industrial nations is compelling. Nancy Birdsall put their case to the United Nations in September 2012, noting that 'climate change is the biggest and most glaring example of a global problem that hits the poor people and

---

[14] Data on remittances is for 2005, the most recent available, from Browne and Mineshima (2007); Table 13.1.

countries hardest. By an unfortunate twist of fate, tropical countries that contributed least to the accumulation of gases are likely to suffer the worst declines
in agricultural productivity, in precisely the sector where the poor within
countries are heavily concentrated'.[15]

(b) Adaptation

The United Nations International Strategy for Disaster Reduction defines
adaptation as '[t]he adjustment in natural or human systems in response to
actual or expected climatic stimuli or their effects, which moderates harm or
exploits beneficial opportunities'. Accordingly, internal as well and international migration may be considered as an adaptation strategy to climate change.
Migration has already been discussed; thus, I will discuss other adaptation strategies. Small atoll states may invest in infrastructure including the construction
of artificial reefs and seawalls to cope with adverse weather events such as
storm surges. The costs of these investments are likely to be substantial and
possibly prohibitive given the weak economies of the PICTs. Therefore, they
will depend on international support including in the form of concessional
loans. The *Caribbean Planning for Adaptation to Global Climate Change* project coordinated by the *Organisation of American States* and funded through the
Global Environment Facility provides a model that may be emulated within the
Pacific Islands. Insurance schemes may be devised to underwrite the risks of
damage arising from adverse weather events. The World Bank has established a
disaster risk-pooling fund for the Pacific Islands highly prone to climate-related
hazards. These initiatives are intended to increase access to finance at the
national and sub-national levels to insulate island communities from the adverse
effects of climate change. Access to such finding will also provide certainty to
investors engaged in rebuilding in the aftermath of natural disasters. Both of the
above are fresh initiatives whose efficacy will be tested over time.[16] Finally,
climate change may provide benefits to the PICTs, but none are obvious at this
stage.

(c) International Assistance

There is a case for the PICTs to receive international assistance for adaptation to climate change. But as explained in Sect. 13.2, some of the PICTs
already receive large amounts of official development assistance. Tuvalu, for
example, received US$4513 per capita in official development assistance in
2015 while Nauru came second at US$2505 (see Table 13.1). The large receipts
of official development assistance is partly a reflection of the lack of development in the recipients. Small states in general suffer from the diseconomies of
providing government services; thus, per capita costs of providing basic services are higher than for larger states. Donors also have an incentive to provide
aid to small states to maximise leverage in international fora from the transfers.

---

[15] See http://www.cgdev.org/content/article/detail/1426491/?utm_&&&.

[16] On the *Caribbean Planning for Adaptation to Global Climate Change* project, see http://www.
cpacc.org/index.htm, and see http://www.worldbank.org/en/programs/disaster-risk-financing-and-
insurance-program on the World Bank's *Disaster Risk Financing and Insurance Program*.

As an example, Australia provided AUD479 million in official development assistance to PNG for the financial year ending 30 June 2017.[17] This transfer amounted to AUD60 for every man, women, and child in PNG. The corresponding figure for Tuvalu was AUD8.7 million which translates to a transfer of AUD791 per capita; that is 13 times the per capita sum for PNG. Australian aid in this instance has a larger impact on the average citizen of Tuvalu compared to their counterpart in PNG.

The PICTs have worked both at the national and regional levels to assist their people cope with the adverse effects of climate change. Relatively large island nations with volcanic origins such as Fiji, Papua New Guinea, and the Solomon Islands have assisted people adversely affected by climatic shocks. The national government is often the first to warn people of approaching cyclones and to assist those badly affected in the immediate aftermath. Regional governments also provide support following large natural disasters. TC Gita, which caused considerable damage to infrastructure and food supplies in Tonga, saw the community band together to support those worst affected; the Tongan government deployed staff to assess damage and direct support, while both Australia and New Zealand governments assisted in the recovery efforts. More can be done to prepare communities better for natural disasters. International assistance in early warning systems would provide additional time for communities to prepare for and/or move out of the path of a tropical cyclone. Similar warning systems for approaching droughts and tidal surges could prove valuable in minimising the effects of climate change. Financial assistance for investments in infrastructure such as seawalls and artificial reefs could also help minimise the effects of adverse weather events on the people.

The international community can contribute most by stabilising emissions of greenhouse gasses. The prospects for agreement on abatement targets for GHG emissions are slim given that the beneficiaries of GHG emissions are large industrial nations with the leadership of some being climate sceptics themselves. There is an alternative view of employing 'geoengineering' to directly control temperature increases, but the technology to achieve this is yet to be developed. The suggestion is to 'overshoot' levels of greenhouse gases to contain temperature increases now and thus remove the costs of mitigation from the present generation in wait of the development of the requisite technologies to allow stabilisation in future (Huntingford and Lowe 2007; Lemoine and Rudik 2017). Many islanders, however, may not have the time to wait, and their lives may be sacrificed in the interim. The risk of an 'overshoot strategy' moreover is that climate change may reach a point of no return, affecting all of humanity. The Australian government has acknowledged the potentially damaging effects of climate change on small island states within its neighbourhood (GoA 2017, p. 84). Issues of equity demands that rich nations who have contributed the most

---

[17] Data extracted from the Australian Department of Foreign Affairs and Trade, http://dfat.gov.au/about-us/publications/Documents/aid-fact-sheet-papua-new-guinea.pdf, accessed online on 14 February 2018.

to the prevailing levels of greenhouse gasses in the atmosphere make the largest sacrifice. Stern (2007) opines that holding the level below 550 ppm will reduce temperature rise from 2020 onwards by one to 1.25 °C which the 2015 Paris Agreement endorses. Such reductions in global temperatures are projected to reduce the likelihood of extreme rainfall events, floods, and droughts that will benefit the people of the islands at the coalface of climate change.

13.5 Conclusions

This chapter has provided an assessment of the economic effects of climate change, drawing on the experiences of 23 Pacific Island Countries and Territories (PICTs). The majority of these are located on or around the equator, away from the major industrial centres of the worlds, and face tropical cyclones and storm surges on a regular basis. At the time of writing in February 2018, Tonga and Fiji were both suffering the ravages of Tropical Cyclone (TC) Gita. If on cue, TC Gita provided a timely reminder of the havoc a single adverse weather event can cause island communities. Initial reports from Nukualofa, the capital of the tiny kingdom of Tonga, are that every building had been affected, with the century-old national parliament reduced to rubble. Recovery from the cyclone will take decades, and there is no guarantee that another similar or worse weather event will not strike before the community is back on its feet.

Climate change is real. I also accept international scientific evidence that this change is the result of human activity. While recognising the fact that there are sceptics, the evidence presented by the Intergovernmental Panel on Climate Change (IPCC) is overwhelming. Closer to home, extreme weather events in terms of prolonged droughts (including in Australia), devastating floods, and destructive tropical cyclones all point to a change in the climate for the worse. If Anthropocene is real, then the victims of extreme climate events are those least able to insulate themselves against these events. The poor and elderly without the means to escape the worst effects of tropical cyclones and storm surges in isolated atolls are cases in point.

Residents of several Pacific Island states and territories face the brunt of climate change on a regular basis. Atoll states such as Maldives (from the Indian Ocean), Tokelau, and Tuvalu have maximum elevation of five metres above sea level. Escaping the effects of climate change for their residents may mean emigration, but having such access requires international cooperation which is lacking. In many of the remaining PICTs where internal migration to evade climatic shocks is possible, most of the population lives close to the shore, freshwater is drawn from aquifers located close to the shoreline, and urban infrastructure and settlement is concentrated on the shoreline. The shoreline also provides most of the agricultural land while the coral reefs form a protective wall against sea surges and constitute a consistent source of food for the bulk of the population. Commercial fisheries and tuna in the main provide export and government revenues. Climate change risks throw a spanner into each of the above-mentioned and therefore affect the very survival of many island communities.

Are Pacific Islanders capable of withstanding the adverse effects of climate change? Their economies are weak, per capita incomes in many of the island states are low, while poverty and competition for natural resources are on the rise. Many islanders are amongst the largest recipients of official development assistance. Clearly, the economic potential to allow the islanders to adapt to climate change is variable across the PICTs but overall weak. While at the coalface of climate change, the islanders and their ancestors have contributed the least to emissions of greenhouse gasses into the atmosphere. The average Pacific Islander still lives on subsistence agriculture and releases a small fraction of greenhouse gasses compared to the level for his or her rich country counterpart. An islander facing the fury of a tropical cyclone has no time to shift blame for his/her predicament but to do everything possible to save life and property. I ask in this chapter if there is the responsibility of the residents of rich industrialised nations who have contributed the most to Anthropocene to assist the victims of climate change. Climate change can be an existential threat to atoll nation states such as Tuvalu and the Republic of Marshall Islands where a large tropical cyclone or a major tsunami can wipe out the entire population.

So what can be done? The answers are in allowing islanders to migrate both on a permanent basis as well as temporarily, depending on the circumstance. The preponderance of customary land tenure systems in the PICTs means that migration for many would have to be short term and possibly circular. But Pacific Islanders would have to be prepared to move following adverse weather events, and both source and host nations must support such mobility. Circular migration would allow the islanders to live in their homes when conditions permit and depart, hopefully temporarily, when conditions demand. Adaptation to climate change would also have to be brought into the mix. This could include investments in infrastructure including artificial reefs and seawalls to reduce the impact of storm surges and the like.

The international community has an obligation to work collaboratively towards reducing the emissions of greenhouse gasses as this is in the immediate interest of the islanders who are most vulnerable to adverse climatic shocks and in the long-term interest of humanity. Waiting for technology to develop to contain temperature increases to reduce the costs of mitigation on present generations could be a dangerous proposition if climate change reaches the point of no return. The Pacific Islanders, therefore, may be the canary in the coal mine signalling the plight of humanity collectively to the effects of climate change.

References

ADB (2005) Juumemmej: Republic of the Marshall Islands Social and Economic Report. Asian Development Bank, Manila

Ayres RU, Kneese AV (1969) Production, consumption, and externalities. Am Econ Rev 59(3):282–297

Barnett J, Campbell J (2010) Climate change and small island states: power, knowledge, and the South Pacific. Earthscan, London

Bindoff NL, Willebrand J, Artale V, Cazenave J, Gregory S, Gulev K et al (2007a) Historical overview of climate change. In: Solomon S (ed) Climate change 2007: The physical science basis. Contribution of Working Group I to the Fourth Assessment Report of the Intergovernmental Panel on Climate Change. Cambridge University Press, New York

Bindoff NL, Willebrand J, Artale V, Cazenave A, Gregory J, Gulev S et al (2007b) Observations: oceanic climate change and sea level. In: Solomon S (ed) Climate change 2007: the physical science basis. Contribution of Working Group I to the Fourth Assessment Report of the Intergovernmental Panel on Climate Change. Cambridge University Press, Cambridge and New York, NY

Browne C, Mineshima A (2007) Remittances in the Pacific Region. World Bank, Washington, DC

Castles S, De Haas H, Miller MJ (2014) The age of migration: international population movements in the modern world. The Guliford Press, New York

Coase R (1988) The firm, the market, and the law. University of Chicago Press, Chicago, IL

Curry G, Koczberski G (1999) The risks and uncertainties of migration: an exploration of recent trends amongst the Wosera Abelam of Papua New Guinea. Oceania 70(2):130–145

De Haas H (2005) International migration, remittances and development: myths and facts. Third World Q 26(8):1269–1284

FFA (2017) Economic and development indicators and statistics: Tuna Fisheries of the Western and Central Pacific Ocean 2016. Forum Fisheries Agency, Honiara

Fisk EK (1964) Planning in a primitive economy: from pure subsistence to the production of a market surplus. Economic Record 40(90):156–174

Gibson J, McKenzie D (2014) The development impact of a best practice seasonal worker policy. Rev Econ Stat 96(2):229–243

Gillett R (2003) Small island developing states of the Southwest Pacific. Review of the state of world marine capture fisheries management: Pacific Ocean, Food and Agricultural Organization of the United Nations, Rome, pp 121–140

Government of Australia (2017) Foreign Policy White Paper. Commonwealth Government of Australia, Canberra

Graham B (2005) Trust funds in the Pacific: their role and future. Asian Development Bank, Manila

Hannesson R (2008) The exclusive economic zone and economic development in the Pacific island countries. Mar Policy 32(6):886–897

Harley S, Williams P, Nicol S, Hampton J (2015) The Western and Central Pacific Tuna Fishery: 2013 Overview and status of stocks. Accessed https://spccfpstore1.blob. core.windows.net/digitallibrary-docs/files/8e/8e262ba47f426138d121b711c1736a43. pdf?sv=2015-12-11&sr=b&sig=l2gmsF4ttS05rjIF7HquTNo8p0kY78ui6JBLDvU1 hHY%3D&se=2018-08-12T05%3A46%3A33Z&sp=r&rscc=public%2C%20max-age%3D864000%2C%20max-stale%3D86400&rsct=application%2Fpdf&rscd=inline%3B%20 filename%3D%22Harley_15_Western_Tuna_2013_overview.pdf%22.

Hermann E (2005) Emotions and the relevance of the past: historicity and ethnicity among the Banabans of Fiji. Hist Anthropol 16(3):275–291

Huntingford C, Lowe J (2007) "Overshoot" scenarios and climate change. Science 316(5826):829–829

Le Treut H, Somerville R, Cubasch U, Ding Y, Mauritzen C, Mokssit A et al (2007) Historical overview of climate change. In: Solomon S (ed) Contribution of Working Group I to the Fourth Assessment Report of the Intergovernmental Panel on Climate Change. Cambridge University Press, Cambridge and New York, NY

Lemoine D, Rudik I (2017) Steering the climate system: using inertia to lower the cost of policy. Am Econ Rev 107(10):2947–2957

McKenzie DJ (2006) Remittances in the Pacific, Werner Sichel Lecture on "Immigrants and their International Money Flows" delivered on February 15, 2006, Washington, DC

Nordhaus W (1982) How fast should we graze the global commons? Am Econ Rev 72(2):242–246

Smith RJ (1981) Resolving the tragedy of the commons by creating private property rights in wildlife. Cato J 1:439

Stern NH (2007) The economics of climate change: the Stern review. Cambridge University Press, Cambridge

Tacoli C (2009) Crisis or adaptation? Migration and climate change in a context of high mobility. Environ Urban 21(2):513–525

Chapter 14
Adaptation to Climate Change: Contemporary Challenges and Perspectives

Patrick D. Nunn, Roger McLean, Annika Dean, Teddy Fong, Viliamu Iese, Manasa Katonivualiku, Carola Klöck, Isoa Korovulavula, Roselyn Kumar, and Tammy Tabe

14.1 Introduction

The need to adapt the ways we live to future climate change has become an almost universal truth. Yet it is widely viewed as something novel, which people have had to consider only recently as a response to unprecedentedly rapid changes in the

Manasa Katonivualiku was deceased at the time of publication.

P. D. Nunn (✉) · R. Kumar
University of the Sunshine Coast, Maroochydore, QLD, Australia
e-mail: pnunn@usc.edu.au; rnunn1@usc.edu.au

R. McLean
School of Physical, Environmental and Mathematical Sciences, University of New South Wales at the Australian Defence Force Academy, Canberra, ACT, Australia
e-mail: r.mclean@adfa.edu.au

A. Dean
University of New South Wales, Sydney, NSW, Australia
e-mail: annika.rose.dean@gmail.com

T. Fong · I. Korovulavula · T. Tabe
University of the South Pacific, Suva, Fiji
e-mail: teddy.fong@usp.ac.fj; isoa.korovulavula@usp.ac.fj; tammy.tabe@usp.ac.fj

V. Iese
Pacific Centre for Environment and Sustainable Development, University of the South Pacific, Suva, Fiji
e-mail: viliamu.iese@usp.ac.fj

M. Katonivualiku (Deceased)
United Nations Economic and Social Commission for Asia and the Pacific, Bangkok, Thailand

C. Klöck
SciencePo, Paris, France
e-mail: carola.kloeck@sciencespo.fr

© Springer Nature Switzerland AG 2020
L. Kumar (ed.), *Climate Change and Impacts in the Pacific*, Springer Climate,
https://doi.org/10.1007/978-3-030-32878-8_14

earth's climate. As many people in the Pacific Islands region know well, this is not true. Islands in the middle of oceans are places where life is uncommonly exposed to environmental adversity. Compared to larger landmasses, there are fewer places where adverse conditions can be avoided; there are fewer options for livelihoods compared to many larger landmasses (Nunn and Kumar 2018).

For people living on such islands, adaptation was necessary in the past to overcome such adversity. And given that people have been living on islands in the western Pacific, thousands of cross-ocean kilometres from continents, for more than three millennia, it is easy to understand how adaptation became an integral part of Pacific societies. Adaptation was the key to survival in the past, just as it will become for many Pacific Islanders today in the future (Hay and Mimura 2013; Robinson 2017).

This chapter is a snapshot of climate-change adaptation on Pacific Islands in 2018–2019 with emphasis on the degree to which strategies that are being planned or implemented acknowledge Pacific Islanders' adaptive capacity and coping histories.

14.2 Adaptation on Pacific Islands

The following eight case studies represent the views of scientists who have an intimate knowledge of issues concerning climate-change adaptation in Pacific Island countries, largely in rural communities where its impacts are likely to be greatest. Some of the case studies are reasonably positive, outlining promising ways forward for embedding effective and sustainable adaptation strategies in such communities. Other case studies are more negative, focused on failed interventions and the reasons why these persist. All case studies agree that future interventions need to be better aligned with the nature of island environments and the communities that occupy them and which will have to bear the brunt of future climate-change impacts.

Case study 1 discusses seawalls and questions why these remain so popular outside iconic urban locations in the Pacific when they manifestly cause more problems than they solve. Case study 2 looks at relocation, one of the most tortuous challenges for Pacific Island peoples, and the importance of learning from past analogues. Case study 3 discusses peripherality as a measure of the diversity of community coping capacity in the Pacific, asking whether it is better to be peripheral or not. Case study 4 tackles the issue of climate finance, how its effects can be optimised in Pacific Island contexts, as illustrated by revision of the development planning process in Tonga. Case study 5 focuses on how traditional island food systems can be made more resilient in the face of climate change with an emphasis on science-informed solutions that are acceptable to rural communities. Case study 6 takes issue with the nature and process of many interventions for climate-change adaptation over the past few decades, demonstrating that their results have not met their expected goals. Case study 7 recounts how forward planning has been achieved for communities throughout Kadavu Island (Fiji) and the benefits it is already accruing for rural dwellers that are expected to sustain their livelihoods into the future.

Case study 8 tracks the resettlement schemes that saw Gilbertese (I-Kiribati) people shunted around the Pacific during the twentieth century, explaining how their experience should inform similar resettlement in the future.

All the case studies rely heavily on 'grey' (unpublished) literature and the (as-yet) unreported results of particular projects. As such, they represent not just the 'snapshot' in time of the current state of climate-change adaptation in the rural Pacific but also a more complete picture of its successes and failures and challenges than anyone might be able to glean from the published literature. In particular, failure is something rarely talked about in publications, yet it is clearly something glaringly apparent to many practitioners in the region. The issue of the lack of meaningful (rather than token) community participation is another theme that comes out in many of the case studies. Finally, the absence of external (donor) understanding of the region and the challenges its average people face from climate change is also clear. Hopefully, this chapter helps bring interested parties together to develop workable, effective and sustainable solutions to these challenges in the future.

14.3 Case Study 1: Seawalls as Intuitive Yet Maladaptive Responses to Shoreline Erosion in Rural Island Contexts

Sea-level rise as a result of climate change is having impacts on all Pacific Island countries (Nurse et al. 2014). Along island coasts throughout the Pacific (and beyond), shoreline erosion is widespread, together with the common associated effects of spreading groundwater salinisation and increasing lowland flooding. Yet while climate change and sea-level rise contribute to shoreline erosion, this process also has many other drivers including dredging, coastal infrastructure construction, mangrove clearance and sand mining, for example. All put pressure on coastal ecosystems and may initiate or exacerbate shoreline erosion (Yamamoto and Esteban 2013; Connell 2013).

Shoreline erosion per se is unproblematic: beaches are dynamic systems and subject to natural variation (Cooper and McKenna 2008). Yet shoreline erosion becomes problematic when it affects settlements, infrastructure or agriculture. This is the case in the Pacific Islands today, where most settlements and human activities are concentrated along the shore and are thus threatened by shoreline erosion, saltwater intrusion and coastal inundation. This contrasts with the situation in the past, at least a few hundred years ago when high-island settlements on Pacific Islands were mostly located inland, away from the threats of shoreline erosion and extreme waves (Nunn 2007; Siméoni and Ballu 2012).

Today's Pacific Islanders can respond to the climate-driven threats of coastal change in three broad ways: protection, accommodation or retreat (Williams et al. 2018). Protection, also referred to as defence, seeks to hold the existing shoreline in place, generally through engineering solutions; accommodation aims to increase the flexibility of both human activities and infrastructure allowing them to absorb the effects of sea-level rise; retreat means leaving the shoreline to change naturally and invariably requires the relocation of vulnerable people and activities (Wong et al. 2014). On Pacific Islands, protection—what has also been termed 'resistance'—has been the most common response (Cooper and Pilkey 2012).

Defence measures, such as seawalls, often appear intuitively the correct response at first sight: they are relatively easily built, visible and seemingly solid and safe. Seawalls are therefore a popular 'solution' to coastal erosion and inundation across Pacific Island countries (Donner and Webber 2014; Monnereau and Abraham 2013). Yet in practice, most seawalls, particularly along rural island coasts, fail to protect coastal communities from erosion and inundation in the way they anticipated. In fact, given the way in which seawalls are constructed along many Pacific shorelines, these structures often increase rather than reduce vulnerability (Fig. 14.1).

Fig. 14.1 Some issues with seawalls in rural Pacific Island communities. (**a**) Shows the coastline at Navunievu village (Vanua Levu, Fiji) where clearance of the mangrove fringe in the 1950s exposed the shore to wave erosion, to which local residents responded by building a (now-degraded) seawall in the 1970s and another (now crumbling) in the 1990s (photo by Patrick Nunn). (**b**) Shows the seawall, funded by USAID, along the front of Karoko village (Vanua Levu, Fiji) that does not extend beyond the village limits, places where shoreline erosion is visibly occurring (photo courtesy of Annah Piggott-McKellar). (**c**) Shows a house newly built on reclaimed land, bordered by a seawall made from uncemented reef rock, at Kabangani (Abemama, Kiribati). (Photo courtesy of Virginie Duvat)

To construct a seawall, local people usually use concrete or other locally available material to build a vertical structure at the top of the beach. Without the necessary financial and technical resources, villages do not evaluate how the seawall may affect local conditions such as currents and sedimentation patterns prior to construction, nor do they generally have the resources to regularly maintain and repair the seawall after construction. In many cases, seawall construction leads to scouring of the foreshore and *increases* in erosion at the front of and/or at the extremities of the seawall (Cooper and Pilkey 2012). Often 18–24 months after construction, many seawalls start collapsing, leaving the population behind more vulnerable to inundation than before since they now have an eroded beach that is less able to dissipate waves and at the same time fewer resources to expend on alternative measures (Yamamoto and Esteban 2013; Donner and Webber 2014; Nunn 2009).

Why do seawalls then remain a popular 'solution'? To a large extent, this has to do with the apparent success of seawalls in core areas. In iconic locations such as capital city shorelines, expensive modern seawalls are built, mostly with external assistance. The design of these showcase seawalls takes into consideration local coastal dynamics, such as waves or currents, and resources for regular maintenance are available. Under such circumstances—adequate design and regular maintenance—seawalls and other hard engineering structures can effectively protect the infrastructure behind although this comes mostly at the expense of the beach in front (Cooper and Pilkey 2012). Yet, most seawalls in the Pacific are neither adequately designed nor regularly maintained. Rather, local communities emulate what they see in core areas, the apparent 'one-size-fits-all' solution that seems to work elsewhere (Nunn 2009, 2013).

Unfortunately, there is no one-size-fits-all solution to coastal erosion and inundation. Armouring shorelines in a manner that is appropriate and effective requires not only substantial and ongoing levels of funding, but also technical expertise and local data. Most of these requirements are lacking in Pacific Island countries. While local communities may be able to attract funding, either from donors or from the central government, such funding is invariably time-limited so that it is impossible to adequately maintain seawalls. Additionally, there is an absence of data on coastal dynamics in rural locations and, often more importantly, on the potential effects of a seawall on coastal processes. Among rural (community-level) decision makers in countries like Fiji that are contemplating their response to undesirable coastal change, there is generally little discussion or evaluation of the drivers of shoreline erosion in this particular location, or of alternative responses (other than seawalls) and their respective advantages and problems, or indeed of the efficacy of past response measures (Nunn 2009; Betzold and Mohamed 2017). Without such information, it is hard to identify the most suitable and long-term solution for a specific context—and once a maladaptive solution like an inadequate seawall is in place, other, more sustainable, measures may subsequently become even more costly to implement.

To effectively address coastal erosion and inundation, we should not uncritically transfer a specific solution but use knowledge and technology—both traditional and modern—to identify viable solutions in association with the vulnerable population.

In some instances, this may be some form of shoreline fortification, but it is more likely that other measures may be better able to address coastal erosion and inundation. The importance of finding effective and sustainable solutions is becoming ever more pressing as the rate of sea-level rise in the western Pacific accelerates (Chen et al. 2017).

14.4 Case Study 2: Planned Relocation as an Adaptation Strategy—Examples from the Fiji Islands

According to Fiji's latest census report (2017), Fiji's total population stands at 884,887, 56% of which resides in urban areas and the rest in rural areas: a level of urbanisation that is unusually high for Pacific Island countries. Together with most revenue-generating activities, most of Fiji's rural and urban communities are situated along the coast where climate-change impacts are likely to be most severe. The Fiji government has set ambitious development objectives to improve the welfare of all Fijians, but the achievement of these objectives is likely to be rendered more challenging by future climate change. Fiji's 2017 Vulnerability Assessment Report provides significant new analysis of Fiji's vulnerability to climate change, with projections outlining potential impacts for Fiji over the coming decades, covering the impact of climate change on Fiji's economy, livelihoods and poverty levels, health and food security together with key industries including agriculture and tourism, as well as potential impacts of sea-level rise on coastal areas and low-lying islands (Government of Fiji et al. 2017).

Satellite data indicate sea level has risen in Fiji by about 6 mm each year since 1993, a rate much larger than the global average of 2.8–3.6 mm per year. This higher rate of rise may be partly related to natural fluctuations caused by phenomena such as decadal oscillations of Pacific climate. There is very high confidence that mean sea level will continue to rise over the course of the twenty-first century. It is projected that this rise will be in the range of 8–18 cm by 2030 and 41–88 cm by 2090 under a high-emissions scenario (medium confidence on this range) (Church et al. 2013). Current climate models also indicate that the frequency and intensity of extreme rainfall events will increase in Fiji (with high confidence). The current 1-in-20-year daily rainfall event will become, on average, a 1-in-9-year event under a very low-emissions scenario and a 1-in-4-year event under a very high-emissions scenario by 2090 (Power et al. 2017).

The above data highlight the vulnerability of coastal communities to the impacts of sea-level rise, flooding and inundation, saltwater intrusion into freshwater (subterranean) aquifers, and the salinisation of gardens and cash crops. Inundation incidents are increasing as sea-level rates have accelerated since the 1900s. More regular future coastal inundation will intensify coastal erosion; reduce crop yields (and recovery times between events); impact on other economic activities, such as fishing and lead to more frequent flooding of shoreline villages and coastal infrastructure. Coral reefs, which support marine ecosystems and protect coastlines from the impacts of storm surges, will also be impacted by increased ocean acidification, coral bleaching and irreparable damage to fish stocks attributable in part to ocean-surface deoxygenation. Soil degradation and soil erosion as a result of prolonged dry seasons and extreme temperatures are also predicted to increase (BOM 2014; Walsh et al. 2012).

The high vulnerability of Fiji's coastal communities to sea-level rise and inundation has seen drastic measures being taken in the form of relocation in a few places. Examples of climate-associated relocation in Fiji (largely because of coastal inundation and coastal erosion) are Vunidogoloa village in Vanua Levu (first community to relocate in 2014), closely followed by Denimanu village on Yadua Island (McNamara and Des Combes 2015; Charan et al. 2017). The Vunidogoloa and Denimanu communities moved with the support of the Fiji government *within* their own land boundaries and with the cooperation and support of the affected people.

Relocation is a complex process so in situ adaptation alternatives (accommodation and protection; see above) should be exhausted first before the option of relocation (retreat) is considered. Any community move must avoid maladaptive impacts and have the full, informed and participatory consent of all affected communities and individuals. Studies elsewhere have emphasised that the availability of land is key to the success of any relocation (see below), and this is especially so in the Pacific, where people and land have a very special, intuitive relationship and where most of the land is under customary land ownership (Campbell 2010).

The availability of adequate finance to underwrite relocation costs is crucial. In order to avoid the further impoverishment of relocated communities, substantial monetary support and training provision are required to re-establish communities and to restore sustainable livelihoods (De Sherbinin et al. 2011). The three elements of land, financial resources and ongoing support are vital to the sustainability and long-term success of any planned relocation.

Fiji plans to move more than 40 villages to higher ground to escape expected coastal inundation and is also working on ways to help resettle future migrants from other Pacific Island nations as sea-level rises over the next few decades (Hermann and Kempf 2017). In anticipation of sharply increased internal climate-related relocation, the Fiji government has undertaken to develop a national relocation guideline. This is currently undergoing stakeholder consultations and incorporates lessons from past relocations including those instigated by non-climatic causes.

14.5 Case Study 3: Peripherality—A Blessing or a Curse for Climate-Change Adaptation in Pacific Island Communities?

Pacific Island communities are manifestly more vulnerable to the effects of climate change than most in continental nations. Rising ocean temperature, the primary manifestation of climate change, is the main driver of rising sea level. As discussed above, sea-level rise threatens coastal settlements, infrastructure and livelihoods in the Pacific Islands region; community-level planners in particular face a daunting challenge to sustain livelihoods in such contexts as future sea-level rise continues at above world average rates (Nunn et al. 2014; Nunn and Kumar 2018). Some Pacific Island communities will need to undertake transformational adaptation, particularly islands in archipelagic groups like Fiji (discussed in this case study) that are experiencing sea-level rise 2–3 times the global average.

As island societies feel the effects of climate change, sustainable community-specific solutions to address these challenges in the Pacific are needed to ensure that islands continue to remain habitable for humans in the future. To achieve effective and sustainable adaptation, it is important to choose solutions that are tailored for individual communities to optimise their chances of success. Not all solutions are suitable for every community in such archipelagic countries. Peri-urban (or near-core) communities require differently designed strategies to those faced by communities in more remote (peripheral) locations. This case study evaluates whether peripherality influences climate-change adaptive capacity among non-urban communities in Fiji based on a recent project funded by the Asia Pacific Network for Global Change Research (CRRP2015-FP02).

The study of communities in Bua Province on Vanua Levu Island in Fiji (Fig. 14.2) and in Serua on neighbouring Viti Levu Island gathered data that allowed the development of indices to measure 'peripherality'. A few examples are discussed below to illustrate the point and demonstrate how peripherality can be used by national planners and their donor partners to tailor interventions for climate-change adaptation in particular communities that are likely to be both effective and sustainable.

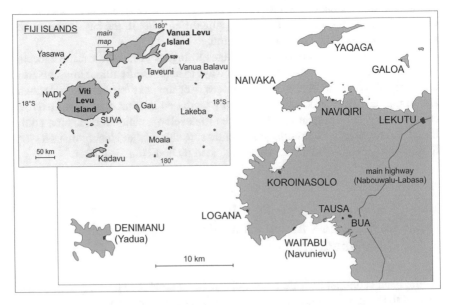

Fig. 14.2 Communities studied in Bua Province, Fiji, showing the locations of Logana and Tausa

In Bua Province, already peripheral in terms of Fiji's developmental and urban cores, two villages (Logana and Tausa) are deemed peripheral, given the long travel times to a full-service hospital, recognised as a key measure of peripherality.

With no land connection to the main highway on Vanua Levu Island, the people of Logana need to travel by boat to Bua village before they are able to catch a bus to reach hospital. The inhabitants of Tausa have to walk two kilometres to reach the bus route. Neither community has mains electricity, relying instead on generators for energy. There are no government offices within thirty minutes' walk of either village. No one in either is in full-time employment. Drinking water comes from springs or wells; rainwater is used mostly for washing. The people of these two villages remain connected to the world through battery-operated radio, television (signals are generally poor) and mobile phones, although reception can often be obtained only in certain spots; Facebook is the most common social media platform used. People in both communities are deeply religious and commonly rely on traditional remedies rather than Western medicine, except for complex complaints such as cardiac arrests, strokes and cancer. Both communities are entirely self-sufficient, obtaining food from land and sea.

Both communities have experienced the effects of climate change, ranging from rising sea level, felt more at Logana than Tausa given its proximity to the sea, prolonged periods of drought, increasingly strong hurricanes and shifts in the growing/fruiting seasons of important food crops. Yet both communities have a wealth of traditional environmental knowledge that enables them to anticipate imminent disasters and also prepares them for their impacts. For example, people routinely grow, preserve and stockpile surplus foods before the start of the cyclone season to ensure they will not starve if disaster strikes. More than this, both communities have good understanding of the need to conserve vegetation, especially mangroves that protect shorelines from erosion and reduce extreme-wave impacts (see below). It is clear that the people of Logana and Tausa have considerable culturally grounded coping ability for withstanding the effects of most natural disasters, something developed over generations of intimacy with local environments and their food-production potential. It seems self-evident that climate-change adaptation strategies targeting communities like Logana and Tausa should not sideline their inherent coping ability, but rather use this as a starting point to build resilience. Yet to date, most interventions in such communities have ignored baseline resilience and attempted to develop completely new adaptive strategies that invariably fail. It is suggested that measuring a community's peripherality may be a useful proxy for inherent resilience.

When community memory and coping ability are not considered—or are devalued—climate-change impacts can be unnecessarily amplified, a point illustrated by the less-peripheral community of Togoru in Serua Province on Viti Levu Island. Within 20-min drive of the main Suva-Nadi Highway from which Suva, Fiji's capital city, can be reached within half an hour, Togoru is not strictly peripheral. The village has been experiencing comparatively rapid shoreline erosion for decades (Fig. 14.3) that has not only forced them to move iteratively inland, almost to the edge of their landholdings, but has also reduced the degree to which they can grow

Fig. 14.3 At Togoru on the south coast of Viti Levu Island, Fiji, the community has been forced to adapt to rapid shoreline recession over the past few decades. Its former graveyard is now surrounded by the sea (**a**), representing about 100 m of shoreline movement in 43 years. Erosion of the shoreline continues, as shown by the undermining of the roots of coconut palms (**b**). (Photos by Patrick Nunn)

their own food and obtain potable water from wells, largely a result of seawater intrusion into groundwater. Knowing well what was happening, the people have discussed relocation (retreat) and consider it their preferred long-term option, but outside bodies have favoured in situ interventions, notably an elaborate mangrove planting scheme that seemed doomed to fail in such a high-energy wave environment. Togoru provides another example of how community priorities, informed by intimacy with the local environment over many generations, were ignored in favour of interventions that failed.

The measurement of community peripherality as a proxy for community coping capacity and local environmental knowledge in archipelagic countries like Fiji would appear to be viable—and potentially of great value to outside bodies that seek to implement effective and sustainable climate-change adaptation. In some cases, it is clearly a curse to be peripherally located in a 'developing' country away from the places where globally informed change is most rapid. Yet it can also be seen as a blessing to be peripheral, for you to have the opportunity to conserve and utilise your traditional knowledge—and inform yourselves through observing the errors made by less-peripheral communities as they adapt to climate change (Maru et al. 2014).

14.6 Case Study 4: Making Climate Finance Work—The Experience of Tonga

A number of key challenges face Pacific Island countries in accessing and utilising climate finance effectively. One major issue in relation to accessing climate finance is that Pacific Island countries often do not have the capacity to navigate the complex requirements of different climate funds because of very small (limited capacity) government administrations. Another major issue in terms of effective management and utilisation of climate finance is lack of mainstreaming of climate-change adaptation and disaster risk management into development plans, policies and budgets both horizontally (across sectors) and vertically (from the national to the community and household level).

Since climate-change adaptation and disaster risk management are cross-cutting, cross-sectoral and multi-scalar, their lack of integration compromises the effective utilisation of climate finance. Lack of horizontal integration has meant that many past climate-change projects in Pacific Island countries have been perceived, governed and implemented in a stand-alone fashion, parallel to development processes (Schipper and Pelling 2006; Dean 2017). Lack of vertical integration has meant that climate finance has frequently circulated at the national level, with priorities at the community level—especially in rural areas—being neglected or addressed only in an ad hoc manner on a project-by-project basis (Barnett and Campbell 2010).

Supporting mainstreaming of climate-change adaptation and disaster risk management into development processes is one of the current objectives of several global financing institutions such as the Green Climate Fund (GCF), the European Union's Global Climate Change Alliance (GCCA), the Asia Development Bank (ADB), and the German Agency for International Cooperation (GIZ).

Over recent years, many Pacific Island countries have recognised the benefits of integrating institutional and governance arrangements for climate-change adaptation and disaster risk management and have created, or are in the process of creating, Joint National Action Plans.[1] Several Pacific Island countries have also made steps forward to horizontally integrate climate-change adaptation and disaster risk management into sustainable development processes, policies and plans across various sectors including health, agriculture, water and sanitation. The regional Framework for Resilient Development in the Pacific (FRDP), which Pacific Island leaders

---

[1] Joint National Action Plans are being implemented in the Pacific to fulfill the requirement of the UNFCCC Cancun Adaptation Framework to develop National Adaptation Plans (NAPs). Joint National Action Plans have been created in Tonga, the Cook Islands, Fiji and Niue. Joint National Action Plans are underway in several other Pacific Island countries.

endorsed in 2016, echoes the need to mainstream climate-change adaptation and disaster risk management into sustainable development processes in order to make development resilient.

While progress has been made in many Pacific Island countries to integrate climate change and disaster risk management into horizontal sustainable development, what is lacking are clear vertical development processes with similar intent. Without this, communities have few ways of communicating their needs to the national government level (see below) or accessing finance to address risks and vulnerabilities either from the national budget or from external funding (Nunn et al. 2014). The example of Tonga, which is in the process of developing a risk-integrated vertical development approach, illustrates how risk-integration and bottom-up vertical development processes might help climate finance reach the subnational scale.

In Tonga, the non-government organisation Mainstreaming of Rural Development Innovation (MORDI) Tonga Trust, with the support of the Ministry of Internal Affairs, the International Fund for Agricultural Development (IFAD) and other partners, has been working to develop a strong, bottom-up vertical development process over the past 10 years. Enabling the voices of communities to be heard at the national government level, while still empowering communities to address their own needs, has been the vision driving this work. The foundation of this approach is the community development plan. Since 2014, climate change and disaster risk management have been integrated into the community development planning process with support from partnerships with the Pacific Risk Resilience Program (PRRP) and the University of the South Pacific (USP). The community development plan has become a central tool for mobilising funding from multiple sources to address self-determined community risks and vulnerabilities, including but not limited to funding that could be categorised as climate finance. The simplicity and efficiency of the process has enabled its widespread uptake within Tonga. Community development plans have already been completed in almost 150 communities, and the MORDI Tonga Trust and the Ministry of Internal Affairs anticipate completing plans with all rural communities by 2022.

Community development planning is based on the tools of Participatory Learning and Action (PLA), emphasising inclusiveness and empowerment. The process is based on three core steps: (1) enabling communities to identify their problems, (2) enabling communities to identify the possible causes of these problems, and (3) facilitating identification of solutions to these problems. Problems are first brainstormed, ranked and prioritised by separate groups of men, women and youth. After identifying causes and potential solutions, problems are merged to form the community development plan. If any problems appear in all groups they become top-rank priorities, then problems are prioritised in the following order: those common to women and youth, then men and women, then men and youth, and then women, youth and men. In this way, women and youth priorities are elevated. Common priorities across community development plans are then combined to form district development plans. The process of creating district development plans is also highly participatory, with town officers using cards and butchers' paper to actively merge community development plans (Fig. 14.4).

Fig. 14.4 Town officers and district officers merge priorities from community development plans to form a district development plan in Vava'u, Tonga. (Credit: MORDI Tonga Trust)

The priorities from district development plans are included in island development plans, which also include recognition of strategic development opportunities (e.g. in tourism, agriculture or forestry). Priorities within island development plans are aligned under sectors, which makes them easy to link to national governance processes. Sectoral priorities from island development plans are then collated to form sector plans. Although only a few sector plans have been created so far in Tonga, the intention is that these will be used to inform the annual corporate plans and budgets of government ministries and also feed into the overarching National Development Plan.

Following this pathway (Fig. 14.5), community priorities are elevated to the island level where they are linked to national budget processes. This means that national budget funding including climate finance in the form of national budget support is more likely to reach island, district and community levels where it is demonstrably needed for agreed purposes. At the same time, communities are trained how to mobilise funds from multiple sources to implement the priorities in their risk-integrated community development plans. The Ministry of Meteorology, Energy, Information, Disaster Management, Environment, Climate Change and Communications (MEIDECC) recently launched the Tonga Climate Change Trust

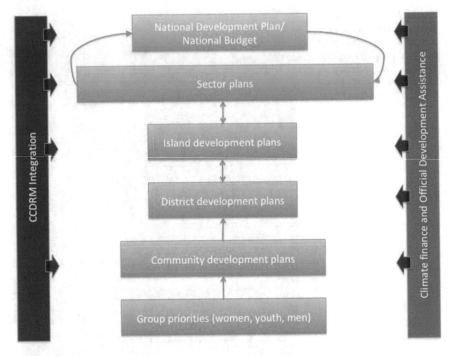

Fig. 14.5 Visual depiction of risk-integrated planning process in Tonga

Fund with the objective of providing community grants for climate-change projects. Having risk-integrated community development plans makes it easier for communities to access grants from this fund.

In Pacific Island countries, risk-integrated vertical development processes might contribute to building resilience to climate change and natural hazards in several ways. First, climate change, climate variability and hazards are considered in the process of brainstorming and prioritising problems in community development plans. Climate change and disaster risk management are also factored into island development plans, sectoral plans and ultimately National Development Plans. Second, when projects are developed and implemented, they are climate-proofed, meaning they are based on risk profiles that incorporate future climate-change projections. Third, the community development planning approach is ultimately focused on empowering communities to address their own needs. Building community capacity to cope with present risks, including natural hazards and climate variability, is widely acknowledged in the scientific literature as an effective way of reducing vulnerability to the effects of future climate change (Ayres and Forsyth 2009; Van Aalst et al. 2008).

This approach also encourages communities to use their own assets to address small problems that they face and to utilise their networks to fundraise as required. Problems that are common to many communities or districts can be pitched to development partners, who invariably value the rigour of the process and appreciate how efficiencies can be achieved (in terms of time and resources) by utilising economies of scale and not having to consult communities directly. Community development plans are expected to attract pooled finance from multiple sources.

In Tonga, the MORDI Tonga Trust, the Ministry of Internal Affairs and MEIDECC are partnering with the University of the South Pacific to strengthen the risk integration and to formalise guidelines for climate-proofed implementation. This work will further finesse the approach described above. What is blatantly clear now though is that integrating risk considerations into a strong development 'spine', as is being done in Tonga, is critical for enabling effective utilisation of climate finance at the subnational scale. Risk-integrated community development plans also have the great benefit of reducing consultation fatigue in communities.

14.7 Case Study 5: Adapting Island Food Systems to Climate Change

Food security adaptation interventions in Pacific Island countries are diverse (Campbell 2015; McGregor et al. 2009). They range from broadly targeting sustainable development (to make it more resilient) to targeting individual components of the agriculture system. Multiple approaches exist including the replacement of traditional (varieties of) crops with more resilient varieties, collecting and sharing traditional farming knowledge and distributing appropriate farming technologies across the Pacific Islands.

The traditional food system in Pacific Island countries is designed such that traditional staple crops and trees provide an all-year round supply of food (Thaman 1982; FAO 2008; Allen 2015). For example, breadfruit and wild nuts provide food when yam and *taro* are in transition between the planting and harvesting seasons. Yet this traditional food system is especially vulnerable to climatic and environmental changes. The high dependence on rainfall of some traditional crops makes communities unable to plant them when the wet season shifts, or there is a prolonged drought. In addition, normal fruiting seasons for some important food security trees are during the cyclone or 'bad weather' season in the Pacific Islands; recent and future changes in west Pacific cyclonicity increase the likelihood that such fruits will become damaged and be wasted (Power et al. 2017).

More generally, the more frequent occurrence and increased intensity of sudden-onset climatic events such as floods, droughts, cyclones and storm surges are reducing the availability and accessibility of food in Pacific Island communities. Slow-onset events such as sea-level rise and rising temperatures are also reducing agricultural production by reducing yield, affecting transportation of food and subsequently driving up prices. These factors reduce the availability and accessibility of foods at certain times of the year. Pests and diseases also impact upon traditional staple crops widening this food security 'gap', an issue that needs to be addressed in order to achieve food security in Pacific Island countries (FAO 2008; Campbell 2015).

In an effort to close this gap, Pacific Island countries are learning to cope better with sudden-onset extreme events and adapt sustainably to slow-onset impacts of climate change. Food system adaptation approaches range from community-focused interventions targeting subsistence farmers to large-scale programs targeting commercial farmers as well as driving changes to agricultural legislation and policies intended to climate-proof future production.

This case study on adapting island food systems to climate change in the Pacific draws on examples from Pacific Island countries where many partners and organisations are working together to adapt Pacific food systems to climate change and achieve food security from the community scale to the national and regional scales. The objectives of these adaptation interventions are threefold: to improve availability of food by reducing impacts and increasing survivability of crops, to improve accessibility by increasing production and improving access to locally produced food and to increase the nutritional value of food and reduce food-borne diseases.

In low-lying (atoll) island countries with comparatively poor soils and high exposure to saltwater intrusion, the Development of Sustainable Agriculture in the Pacific (DSAP) project, funded by the European Union, worked to increase sustainable food production. People in the Republic of the Marshall Islands (RMI) who participated in this project decided to plant sweet potato because it protects the soil from moisture loss due to its tendency to grow wild quickly. Sweet potato is now becoming a popular staple crop (together with breadfruit) for communities in Majuro, RMI's capital. In neighbouring Kiribati, DSAP facilitated a competition between communities to plant crops including sweet potato to improve the health of rural communities. Sweet potato is now a popular crop on Marakei Island in Kiribati and elsewhere. In Nauru, DSAP focused on planting sweet potato as well as *taro* and other vegetables to increase food production and boost farmers' income. The increase in local food production in such places is helping buffer them against the high cost of imported food. For all of these projects, sweet potato varieties were transferred from the 'climate ready' collection developed in the Centre for Pacific Crops and Trees (CePaCT) at the Pacific Community (SPC) in Suva.

Another major regional project focusing on food security was the Pacific Adaptation to Climate Change (PACC) Program, which covered 14 Pacific Island countries. Under PACC, four countries (Fiji, Palau, Papua New Guinea, Solomon Islands) employed a 'no-regrets' approach to adaptation for food security and agriculture. For two decades, Palau had experienced saltwater intrusion (from sea-level rise) and loss of soil nutrients (from heavy rain events) that affected *taro* production. The PACC project helped farmers in Ngatpang state test varieties of *taro* more tolerant of saline groundwater. The program also facilitated the construction of dikes to reduce saltwater intrusion. In 2013, three varieties of *Colocasia taro* were identified as salt tolerant: *dungersuul*, *dirrubong* and *kirang*. Suckers from these were distributed to farmers throughout Palau.

Across the Fiji Islands, flooding and sea-level rise are affecting coastal communities and farmers' livelihoods in low-lying areas. Working closely with the Ministry of Agriculture, UNDP's PACC project helped two communities (Nakelo in Tailevu and Deuba in Serua) to develop better drainage systems and to test crop varieties more tolerant of waterlogging and salty groundwater. In 2013, three varieties of climate-resilient *taro*, cassava and sweet potato were identified and successfully planted in the two communities. Creeks were dredged to remove sediment while drainage systems were redesigned to optimise water supply and runoff.

In Vanuatu, a community-based food security project, funded by the EU-GCCA, was implemented by the University of the South Pacific (USP). On Tanna Island, fishponds were constructed and *tilapia* introduced, with management training provided for community members. A livestock expert also offered training on chicken farming. Fish from the pilot ponds and chickens have been harvested already, and the success of this intervention has encouraged other farmers to engage in these activities. At Tassiriki on Moso Island, improved (allegedly climate-proofed) farming systems were introduced to communities with the assistance of local Forestry and Agriculture Officers. In the improved farming systems, several plots were set up where proven drought-resistant varieties of sweet potato, *taro*, cassava and cabbage (*Abelmoschus manihot*) were intercropped with sandalwood, *namamau* (*Seruniga flexuosa*, timber tree), gliricidia (*Gliricidia sepium*, nitrogen-fixing tree) and a variety of citrus species. It is anticipated that gliricidia will continually fix nitrogen into the soil for the crops and help raise soil fertility, reducing fallow periods. The intention is for farmers to rely on drought-resilient crops in the short term and harvest citrus fruits, timber and sandalwood trees in the longer term.

In all the adaptation interventions mentioned above, practitioners and communities have reported increased production, increased food availability and increased incomes from the sale of excess produce. Both practitioners and communities have also reported increased consumption of nutritious locally produced foods resulting in better individual health.

Yet despite the evident success of the adaptation interventions described above, important questions remain. How efficient and sustainable are food system adaptation interventions? How effective are these in reducing the yield losses against the specific climatic hazards to which a specific community is vulnerable? With the increased frequency and intensity of both sudden- and slow-onset climatic events, will food system adaptation interventions be capable of minimising agricultural losses?

An unpublished study by the University of the South Pacific and UNESCO on community-based climate-change loss and damage in the agriculture sector revealed that about 90% of household agriculture adaptation interventions in Cook Islands, Fiji, Samoa, Solomon Islands and Timor Leste did not completely prevent (climate-change) loss and damage to agriculture. This is both because the increased intensity and frequency of hazards is forcing communities to be forever in recovery mode and because sociocultural barriers persist in planning and distribution of effective agricultural technologies. It is time to make our food system adaptation focused, specific and targeted to both current and future hazards. Good decision support tools as well as innovative traditional and modern food system technologies will help Pacific Island countries not only to bridge the gap in food systems but also to understand the factors affecting food security now and in the future.

14.8   Case Study 6: Challenges of Aligning Community Priorities for Climate-Change Adaptation with Donor Agendas

Experiences with climate-change adaptation projects in over 80 communities throughout nine Pacific Island countries reveal a disquieting trend, namely that donors and their 'for-profit' contractors have agendas that seldom align with community needs or priorities. This is commonly because these contractors prefer the goals of their clients, usually donor partners of Pacific Island governments, rather than those of the communities in whose interests they are purportedly working. Such approaches often leave communities more reliant than resilient, and the interventions unsustained and eventually forgotten.

The tension between outside contractors and in-country non-government organisations (NGOs) has become increasingly acute over the past decade. On the one hand, this is because the impacts of climate change on rural Pacific Island communities are becoming increasingly manifest, which exposes the impotence of many outside 'solutions'. On the other hand, it is becoming clear that the majority of external funding for climate-change adaptation is going to such contractors—the so-called boomerang aid—rather than to NGOs that are often better able (if adequately funded) to design, implement and sustain solutions to assist communities adapt to climate change (Iati 2008; Smith and McNamara 2015). In addition, it is becoming increasingly apparent that many of the 'solutions' being proposed, developed and implemented for rural Pacific Island communities are far too complex, too alien and ignorant of the recipients' cultural contexts to stand any chance of being accepted and sustained. A good example is the recent development of a risk-mapping toolkit that cannibalised selected parts of existing Vulnerability and Adaptation Assessments to produce an impressive risk map intended to form the basis for community planning for future climate change, particularly sea-level rise. The fact that this toolkit ignored community governance, cultural sensitivities, land-use patterns and livelihoods makes it irrelevant for subnational planning in such contexts (Robinson 2019). Toolkits appeal to donors because they identify readily implementable solutions and have milestones that are contractually measurable. Yet they have little regard for existing priorities, community development plans or innovative local solutions using local materials. Such approaches exemplify the criticisms that have been directed towards the misfit between external interventions and community needs in the Pacific over recent years (Nunn 2009; Nunn et al. 2014; McNamara 2013; Lebel 2013).

Examples of 'failed' interventions for climate-change adaptation in the Pacific are well known to practitioners but generally not formally reported. The examples of seawalls (and other 'hard' shoreline-protection structures) are discussed above. The authors worked on one shoreline-stabilisation project in Samoa that was hailed as innovative because it used sand-filled geotextile containers; to save money, the weight of these containers was reduced, so it came as no surprise that they were dislodged by the first storm surge to hit this particular shoreline. The whole project rendered useless in consequence. Another project on which one of the authors worked was a rainwater-harvesting system in Kiribati that failed because the donor was not satisfied with the standard of sand and gravel available in the country, requiring it to be imported from Fiji; clearly, such an expensive arrangement could not be sustained (Piggott-McKellar et al. 2019a).

The authors have also participated in a project that saw the construction of evacuation centres for rural communities in Fiji, Tuvalu and Vanuatu. The metal frames of the buildings came from New Zealand and the calcium-fibre cement board walls from China. The buildings are said to be weatherproof and (Category 5) cyclone resistant but were never formally certified as such. It seemed clear to the authors that the donor prioritised the act of intervention and its completion rather than its effectiveness and sustainability, issues that have been repeatedly recognised as characterising much climate-change adaptation aid in Pacific Island nations and others in recent years (Nunn and Kumar 2018; Weiler et al. 2018).

Within a year of completion, one of these evacuation centres had water seeping through its walls, which had rendered its toilets unusable, but no one has answered such community concerns. This underscores the point that post-intervention monitoring is really at the whim of the implementing agencies and the donors who pay them. Recipient communities have no voice and are passive participants in the fiction that such interventions are actually helping to meet the challenges that Pacific Island communities face from climate change.

14.9 Case Study 7: Engaging Communities for Meaningful Forward Planning for Climate-Change Adaptation— Examples from Kadavu Island, Fiji

Future climate change will have a multiplicity of impacts on people and resources in the Pacific Islands. To avoid the least desirable impacts, informed forward planning is essential (see above). In countries like Fiji, where long-term systematic planning is not a deeply embedded cultural trait, it is more common for communities to be reactive rather than proactive. To address this, some success has been achieved by assisting communities to consider the future and plan in advance to limit the impacts of likely changes, typically those affecting livelihoods that include food security and settlement relocation (see above).

This case study relates the experience of implementing planning for future climate change among rural *iTaukei* communities using Integrated Coastal Management (ICM) on the Fiji Island of Kadavu, where a largely traditional (non-Westernised) way of life is maintained. The introduction of ICM to Kadavu proved an eye-opener for chiefs and commoners alike, not least because it showed how collaborative planning could strengthen unity and cohesiveness. The preliminary to establishing the Kadavu ICM plan was to engage the community at both village and district level, acknowledging traditional hierarchies and decision-making structures.

The Kadavu Integrated Coastal Management planning process proved a prolonged exercise because it needed the effective participation of community members of different socioeconomic and cultural traditional status. It also required the specific involvement of women, youth and those with disabilities. This is key to ensuring a clear sense of ownership of the entire planning process among community members, a process involving the formulation of visions, goals and objectives of the plan through cooperation and participation.

Owing to the cultural norms and sensitivities, this process involved dividing the community members into different groups. There was a group for the elders, mainly clan heads including the chief. Another group consisted of women and another just young people. Each group was then able to share thoughts openly on environmental, social, economic and climate-change issues that affect them. The issues and strategies to address these issues were then presented in a plenary session and the strategies and actions ranked and prioritised. The top four issues for each community were then identified for action, the foremost being those that were practical and doable within an agreed timeframe.

To maintain momentum in the community engagement process, one issue was identified that could be addressed with simple, efficient and fast-acting management interventions. An example is the management action that the Lomanikoro, Nakaunakoro and Nakoronawa villages in western Kadavu implemented to address the flooding events that they regularly experience during heavy rainfall. Collectively, they agreed to improve village drainage systems and to carry out soil conservation measures on hillsides, both by planting vetiver grass around their *kava* gardens and by revegetation of creek banks. Such 'low-hanging fruits' provide short-term wins and therefore encourage continued momentum to tackle some of the harder or more complex climate-change challenges identified through the ICM process.

The best example of a hard challenge is the relocation of entire villages along the western and southern coasts of Kadavu that will be needed to avoid the ongoing disruption attributable to sea-level rise. Such upslope relocation is considered impractical given the immense resources needed and also the lack of community willingness to consider such a drastic option. Such larger and complex challenges will be explored in the future through community dialogues, consultations and a consensus-building process. On a practical note, a number of village headmen in the island's west indicated that dismantling and rebuilding wooden residences would be much easier than concrete ones. The challenge for the community and the authorities are the communal concrete buildings such as the community village hall, dispensary and the village church which invariably also serves as the main disaster evacuation centre.

Engaging communities in forward planning also creates a platform for the integration of traditional ecological knowledge and modern science in developing practical solutions to potential climate-change impacts. For instance, by applying traditional environmental knowledge, the appearance of an overwhelming abundance of *Acanthurus triostegus* (local name is *jivikea*) is a warning of the onset of a prolonged period of dry weather. In addition, farming of more drought-resistant root crops such as the wild yam (locally *tivoli*) and cocoyam (*dalo-ni-tana*) would improve the abundance of food stock during drought periods.

Indiscriminate burning and intensive agricultural practices have been identified as exacerbating negative climate-change impacts. Burning in particular has been responsible for the degradation of arable land in most parts of Kadavu. One of the achievements of the community engagement process has been the unequivocal agreement of all the island's villages to stop indiscriminate burning. Intensive farming cultivation practices, specifically in the districts of Ravitaki and Nabukelevu, have also contributed to land degradation; as a result of the community engagement process, communities in these districts have agreed to undertake reforestation in affected areas.

Most Kadavu coastal communities understand that shoreline erosion can be attributed to climate change. Through the community engagement process described above, villages like Dravuwalu, Lawaki, Namara and Soso have proclaimed the values of mangroves protecting their villages from wave erosion, especially during storms. Most other coastal villages do not have mangrove protection so are interested in ways of acquiring this. An interim solution is to identify coastal plants, such as *dilo* (*Calophyllum sp*), *dabi* (*Xylocarpus sp*) and the coastal almond, *tavola* (*Terminallia sp*), that can stabilise eroding coasts.

Engaging communities in forward planning for climate-change adaptation is a critical process to ensure that they are optimally prepared for the various negative impacts of future climate change. The ICM planning process allows collaborative planning and identification of practical solutions for addressing these impacts as well as identifying effective and sustainable adaptation options.

14.10   Case Study 8: Relocation as Climate-Change Adaptation—Precedents from Solomon Islands

In the late 1930s, groups of Gilbertese[2] families were relocated from the southern Gilbert Islands to the Phoenix Islands by the Gilbert and Ellice Islands Colony (GEIC) Administration as a result of what it perceived as overpopulation, scarcity of land and poverty in their home islands (Maude 1968). The Phoenix Islands Resettlement Scheme was conducted in the hope of providing the Gilbertese with an opportunity to access adequate land and improve livelihoods, while simultaneously reducing the population pressure on limited land and resources in the Gilbert Islands—now part of the sovereign nation of Kiribati.

The Phoenix Islands proved unsuitable as a home for the Gilbertese. The islands turned out to be vulnerable to long periods of drought which created more problems for the GEIC due to the isolation of these islands and the high cost of transporting food and water there for the settlers. In 1963, the Phoenix Islands Resettlement

---

[2] Today referred to as I-Kiribati.

Scheme was finally ended and the people relocated to the British Solomon Islands Protectorate (BSIP), today known as Solomon Islands (Donner 2015).

The relocation to Solomon Islands was conducted by the GEIC and BSIP in two stages. The first relocation took place in the 1950s where families from Manra in the Phoenix Islands were relocated to Titiana on Gizo Island in the Western Province. The second wave of relocation was carried out in 1963–1964, whereby the remaining settlers in the Phoenix Islands were relocated to Wagina Island, in Choiseul Province. The BSIP had declared Wagina a 'wasteland', thus Crown Land, since it had been unoccupied for many years. This illustrates one challenge of contemporary climate-change forced relocation between islands in the Pacific, namely that the place/s to which a community is proposed to be relocated should ideally be unoccupied and lacking any substantive claim to its land title; an example of from Fiji was recently discussed (Hermann and Kempf 2017).

The Gilbertese settlers on Wagina faced many challenges, not least its isolation, the lack of food in the densely forested landscape—a huge contrast to the atoll environments in the low Gilbert and Phoenix islands to which they were accustomed—and the novel presence of malaria. The inevitable cultural conflicts arising from emplacing a particular cultural group within a group of islands occupied by another, quite distinctive, group were amplified when Solomon Islands became independent in 1978. The Gilbertese settlers on Wagina believe they were given the land in exchange for that they occupied in Kiribati whereas groups of Solomon Islanders have since laid claim to Wagina. These issues illustrate other challenges associated with proposed climate-forced migration in the Pacific, namely that the relocated people must feel secure in their new homes and able to pursue their traditional ways of life and maintain their traditional culture, issues that go straight to the heart of identity. In response to stories like those of the Gilbertese on Wagina, the former President of Kiribati, Anote Tong, declared that his people wished for 'migration with dignity', the only option he saw for sustainably successful migration (Dreher and Voyer 2015). Recent work has focused on the key ingredients for relocation in Pacific rural contexts (Piggott-McKellar et al. 2019b).

14.11 Conclusions and Future Priorities

Numerous insights can be drawn from the case studies enunciated in this chapter. Among the most common is that the voices of people living in rural communities, which is a huge majority in the Pacific Islands region, are not being effectively heard in most cases by the people who make decisions about the ways they live—and the ways they aspire to live in the future. If this situation does not change, then it may become a recipe for disaster, where rural dwellers are disproportionately impacted by the progressive effects of climate change during the twenty-first century while their urban counterparts are largely shielded from these effects. There is clearly an onus on Pacific Island governments and their donor partners and international bodies to effectively overcome the challenges of engaging with rural communities, an issue that applies to rural communities in 'developing' countries elsewhere (Betzold 2015; Barbier 2015). The participatory approaches outlined in case studies 4 and 7 hold great promise for the future but, to be most effective, should ideally be driven by persons with considerable in-depth knowledge of Pacific Island communities and the ways they function (Remling and Veitayaki 2016). It is clear that uninformed approaches to Pacific Island adaptation can produce negative impacts, even maladaptation.

A second major conclusion is that while much community-level decision-making is not science-informed, as shown in case study 1 and others, it is also clear that most external interventions in rural communities for climate-change adaptation fail to (adequately) acknowledge their traditional, culturally grounded coping ability, as shown in case studies 3, 6 and 7. Barriers have two sides. The barriers that outsiders often identify as hindering climate-change adaptation in rural Pacific communities are viewed quite differently from within those communities. Mutual respect and recognition are needed to inform effective and sustained adaptation (McNaught et al. 2014).

A final point is that much money has been squandered on climate-change adaptation in the Pacific Islands over the past three decades, largely because of the misalignment between the goals of donors and those of recipient communities. There is little point in dredging up the past as the need to address future challenges becomes ever more exigent. As discussed in case studies 2 and 8, the need for the future relocation of vulnerable communities to places where they are less vulnerable is inescapable. We must learn from the mistakes of the past and persuade funders to invest in long-term goals (such as relocation) in order that future climate-change impacts on Pacific Island people are minimal.

References

Allen MG (2015) Framing food security in the Pacific Islands: empirical evidence from an island in the Western Pacific. Reg Environ Chang 15:1341–1353

Ayres J, Forsyth T (2009) Community-based adaptation to climate change: strengthening resilience through development. Environ 55:23–31

Barbier EB (2015) Climate change impacts on rural poverty in low-elevation coastal zones. Estuar Coast Shelf Sci 165:A1–A13

Barnett J, Campbell J (2010) Climate Change and small island states: Power, knowledge and the south pacific. London, Earthscan

Betzold C (2015) Adapting to climate change in Small Island developing states. Clim Change 133:481–489

Betzold C, Mohamed I (2017) Seawalls as a response to coastal erosion and flooding: a case study from Grande Comore, Comoros (West Indian Ocean). Reg Environ Chang 17:1077–1087

BOM (2014) Climate variability, extremes and change in the Western tropical Pacific: new science and updated country reports. In: Pacific-Australia climate change science and adaptation planning program technical report. Australian Bureau of Meteorology (BOM) and Commonwealth Scientific and Industrial Research Organisation (CSIRO), Melbourne

Campbell JR (2010) Climate-induced community relocation in the Pacific: the meaning and importance of land. In: MCADAM J (ed) Climate change and displacement: multidisciplinary perspectives. Hart, Oxford

Campbell JR (2015) Development, global change and traditional food security in Pacific Island countries. Reg Environ Chang 15:1313–1324

Charan D, Kaur M, Singh P (2017) Customary land and climate change induced relocation—a case study of Vunidogoloa Village, Vanua Levu, Fiji. In: Filho WL (ed) Climate change adaptation in Pacific countries. Springer International Publishing, Cham

Chen X, Zhang X, Church JA, Watson CS, King MA, Monselesan D, Legresy B, Harig C (2017) The increasing rate of global mean sea-level rise during 1993–2014. Nat Clim Chang 7:492–495

Church JA, Clark PU, Cazenave A, Gregory JM, Jevrejeva S, Levermann A, Merrifield MA, Milne GA, Nerem RS, Nunn PD, Payne AJ, Pfeffer WT, Stammer D, Unnikrishnan AS (2013) Sea level change. In: Stocker TF, Qin D, Plattner G-K, Tignor M, Allen SK, Boschung J, Nauels A, Xia Y, Bex V, Midgley PM (eds) Climate change 2013: the physical science basis. Working Group I Contribution to the Fifth Assessment Report of the Intergovernmental Panel on Climate Change. Cambridge University Press, Cambridge

Connell J (2013) Islands at risk? Environments, economies and contemporary change. Edward Elgar, Cheltenham

Cooper JAG, McKenna J (2008) Working with natural processes: the challenge for coastal protection strategies. Geogr J 174:315–331

Cooper JAG, Pilkey OH (2012) Introduction. In: Cooper JAG, Pilkey OH (eds) Pitfalls of shoreline stabilization: selected case studies. Springer, Dordrecht

De Sherbinin A, Castro M, Gemenne F, Cernea MM, Adamo S, Fearnside PM, Krieger G, Lahmani S, Oliver-Smith A, Pankhurst A, Scudder T, Singer B, Tan Y, Wannier G, Boncour P, Ehrhart C, Hugo G, Pandey B, Shi G (2011) Preparing for resettlement associated with climate change. Science 334:456–457

Dean A (2017) Funding and financing for disaster risk reduction including climate change adaptation. In: Kelman I, Mercer J, Gaillard JC (eds) The Routledge handbook of hazards and disaster risk reduction. Taylor and Francis, London

Donner SD (2015) The legacy of migration in response to climate stress: learning from the Gilbertese resettlement in the Solomon Islands. Nat Res Forum 39:191–201

Donner SD, Webber S (2014) Obstacles to climate change adaptation decisions: a case study of sea-level rise and coastal protection measures in Kiribati. Sustain Sci 9:331–345

Dreher T, Voyer M (2015) Climate refugees or migrants? Contesting media frames on climate justice in the Pacific. Environ Commun—J Nat Culture 9:58–76

FAO (2008) Climate change and food security in Pacific Island countries. Food and Agriculture Organisation of the United Nations, Rome

Government of Fiji, World Bank, Recovery, G. F. F. D. R. A (2017) Fiji 2017: climate vulnerability assessment—making Fiji climate resilient. World Bank, Washington, DC

Hay JE, Mimura N (2013) Vulnerability, risk and adaptation assessment methods in the Pacific Islands region: past approaches, and considerations for the future. Sustain Sci 8:391–405

Hermann E, Kempf W (2017) Climate change and the imagining of migration: emerging discourses on Kiribati's land purchase in Fiji. Contemp Pac 29:231–263

Iati I (2008) The potential of civil society in climate change adaptation strategies. Pol Sci 60:19–30

Lebel L (2013) Local knowledge and adaptation to climate change in natural resource-based societies of the Asia-Pacific. Mitig Adapt Strat Glob Chang 18:1057–1076

Maru YT, Smith MS, Sparrow A, Pinho PF, Dube OP (2014) A linked vulnerability and resilience framework for adaptation pathways in remote disadvantaged communities. Global Environ Change-Hum Policy Dimen 28:337–350

Maude HE (1968) Of islands and men: studies in Pacific history. Oxford University Press, Melbourne

McGregor A, Bourke RM, Manley M, Tubuna S, Deo R (2009) Pacific island food security: situation, challenges and opportunities. Pacific Econ Bull 24:24–42

McNamara KE (2013) Taking stock of community-based climate-change adaptation projects in the Pacific. Asia Pac Viewp 54:398–405

McNamara KE, Des Combes HJ (2015) Planning for community relocations due to climate change in Fiji. Int J Disaster Risk Sci 6:315–319

McNaught R, Warrick O, Cooper A (2014) Communicating climate change for adaptation in rural communities: a Pacific study. Reg Environ Chang 14:1491–1503

Monnereau I, Abraham S (2013) Limits to autonomous adaptation in response to coastal erosion in Kosrae, Micronesia. Int J Global Warm 5:416–432

Nunn PD (2007) Climate, environment and society in the Pacific during the last millennium. Elsevier, Amsterdam

Nunn PD (2009) Responding to the challenges of climate change in the Pacific Islands: management and technological imperatives. Climate Res 40:211–231

Nunn PD (2013) The end of the Pacific? Effects of sea level rise on Pacific Island livelihoods. Singap J Trop Geogr 34:143–171

Nunn PD, Aalbersberg W, Lata S, Gwilliam M (2014) Beyond the core: community governance for climate-change adaptation in peripheral parts of Pacific Island countries. Reg Environ Chang 14:221–235

Nunn PD, Kumar R (2018) Understanding climate-human interactions in Small Island developing states (SIDS): implications for future livelihood sustainability. Int J Clim Change Strategies Manage 10:245–271

Nurse LA, Mclean RF, Agard J, Briguglio LP, Duvat-Magnan V, Pelesikoti N, Tompkins E, Webb A (2014) Small Islands. In: Barros VR, Field CB, Dokken DJ, Mastrandrea MD, Mach KJ, Bilir TE, Chatterjee M, Ebi KL, Estrada YO, Genova RC, Girma B, Kissel ES, Levy AN, MacCracken S, Mastrandrea PR, White LL (eds) Climate change 2014: impacts, adaptation, and vulnerability. Part B: Regional aspects. Contribution of Working Group II to the Fifth Assessment Report of the Intergovernmental Panel on Climate Change. Cambridge University Press, Cambridge and New York

Piggott-McKellar A, McNamara K, Nunn PD, Watson J (2019a) What are the barriers to successful community-based climate change adaptation? A review of grey literature. Local Environ 24(4):374–390

Piggott-McKellar A, McNamara KE, Nunn PD, Sekinini S (2019b) Moving people in a changing climate: lessons from two case studies in Fiji. Soc Sci 8(5):133

Power SB, Delage FPD, Chung CTY, Ye H, Murphy BF (2017) Humans have already increased the risk of major disruptions to Pacific rainfall. Nat Commun 8:14368

Remling E, Veitayaki J (2016) Community-based action in Fiji's Gau Island: a model for the Pacific? Int J Clim Change Strategies Manage 8:375–398

Robinson SA (2019) Mainstreaming climate change adaptation in small island developing states. Clim Dev 11:47–59

Robinson S-A (2017) Climate change adaptation trends in small island developing states. Mitig Adapt Strat Glob Chang 22:669–691

Schipper L, Pelling M (2006) Disaster risk, climate change and international development: scope for, and challenges to, integration. Disasters 30:19–38

Siméoni P, Ballu V (2012) Le mythe des premiers réfugiés climatiques: mouvements de populations et changements environnementaux aux îles Torrès (Vanouatou, Mélanésie). Annales de Géogr 3:219–241

Smith R, McNamara KE (2015) Future migrations from Tuvalu and Kiribati: exploring government, civil society and donor perceptions. Clim Dev 7:47–59

Thaman RR (1982) Deterioration of traditional food systems, increasing malnutrition and food dependency in the Pacific Islands. J Food Nutr 39:109–125

Van Aalst MK, Cannon T, Burton I (2008) Community level adaptation to climate change: the potential role of participatory community risk assessment. Glob Environ Chang 18:165–179

Walsh KJE, Mcinnes KL, Mcbride JL (2012) Climate change impacts on tropical cyclones and extreme sea levels in the South Pacific—a regional assessment. Global Planet Change 80-81:149–164

Weiler F, Klöck C, Dornan M (2018) Vulnerability, good governance, or donor interests? The allocation of aid for climate change adaptation. World Dev 104:65–77

Williams AT, Rangel-Buitrago N, Pranzini E, Anfuso G (2018) The management of coastal erosion. Ocean Coast Manag 156:4–20

Wong PP, Losada IJ, Gattuso J-P, Hinkel J, Khattabi A, Mcinnes KL, Saito Y, Sallenger A (2014) Coastal systems and low-lying areas. In: Field CB, Barros VR, Dokken DJ, Mach KJ, Mastrandrea MD, Bilir TE, Chatterjee M, Ebi KL, Estrada YO, Genova RC, Girma B, Kissel ES, Levy AN, MacCracken S, Mastrandrea PR, White LL (eds) Climate change 2014: impacts, adaptation, and vulnerability. Part A: Global and sectoral aspects. Contribution of Working Group II to the Fifth Assessment Report of the Intergovernmental Panel on Climate Change. Cambridge University Press, Cambridge and New York, NY

Yamamoto L, Esteban M (2013) Atoll Island states and international law: climate change displacement and sovereignty. Springer, Heidelberg

Index

A
Actual evapotranspiration, 416
Adaptation, 3, 4, 18, 22–24
  agricultural, 333
  CBA, 381, 382
  climate change, 329, 336
  economic development, 385, 386
  food security, 382–383
  government revenue, 385, 386
  livelihoods, 384–385
  and mitigation, 330, 351, 352
  national and community-based, 335
  and risk reduction, 324
  USAID food security project, 337–339
Adaptation-related finance, 330
Adaptive capacity, 22–24
Agricultural Production Systems Simulator
  (APSIM) model, 334
Agricultural resilience, 325, 329, 335, 339,
  342–344, 352
Agriculture-focused NGOs, 324
Agriculture sector
  adaptation actions, 324
  annual production rates, 325
  arable lands, 324
  atoll soil health project, 339–341
  climate change, 324
  commercial crops, 323
  development assistance, 330
  extreme events and rising sea levels,
  331–334
  farmers, 325
  farming, 324
  GDP, 326, 327
  hazards, 325

  household level, 328
  invasive seaweed, 341, 342
  land areas, 323
  livelihood and food, 323
  mucuna, 342–343
  national and community-based
    adaptation, 335
  private sector, 323
  rain fed, 324
  recovery mode, 325
  resourcing, 329, 330
  risk reduction, 324
  sea level, 325
  smaller and low-lying islands, 324
  social cultural activities, 328, 329
  strategic directions (*see* Strategic
    directions)
  TASP, 335–337
  temperatures, 325
  USAID Food Security Adaptation Project,
    337–339
Agroforestry farming, 348
Annual average significant wave
  height (H_s), 209
Antarctica's ice sheets, 3
Anthropogenic CO_2, 16
Anthropogenic impacts, 370
Anxiety, 23
APEC Climate Center (APCC), 336
Aquaculture
  economic development, 362–364
  food security, 362–364
  livelihoods, 362–364, 384, 385
  marine resources, 379, 380
Arable lands, 324

© Springer Nature Switzerland AG 2020
L. Kumar (ed.), *Climate Change and Impacts in the Pacific*, Springer Climate,
https://doi.org/10.1007/978-3-030-32878-8

Archetypal Pacific islands, 42
Areas of islands, 43, 44
Asian Development Bank, 19, 509
A2 storyline, 184
Atmosphere-ocean general circulation models
    (AOGCMs), 183
Atmospheric stabilisation, 295
Atoll/reef islands, 439
Atoll soil health project, 339–341
Australia and New Zealand (ANZ), 482
Australian Centre for International
        Agricultural Research
        (ACIAR), 340

B
B1 and B2 storyline, 184
Bigeye tuna, 375
Bilateral commitments, 329
Biodiversity
    atmospheric CO_2, 451
    climate change, 449
    climatic envelopes, 450
    crops *vs.* pollinators, 451
    drylands, 450
    extinction, 450, 451, 453, 455–457, 465,
        466, 468, 470
    extreme drought, 452
    extreme floods, 452
    food production/agriculture, 449
    forest fires, 452
    fringing mangroves, 453
    global warming, 452
    heat waves change, 452
    human welfare, 449
    IPCC models, 451
    life support system, 449
    livelihoods, 449
    Polar regions, 450
    rainfall, 451, 452
    salt water intrusions, 453
    sea-level rise, 452
    terrestrial, 450
    vulnerability, 455–457
Biodiversity hotspots
    islands, 453–455
    mainland and island regions, 455
    United Nations Biodiversity, 453, 454
Bonriki freshwater lens, 437
Boomerang aid, 516
British Solomon Islands Protectorate
        (BSIP), 520
Building capacity, 385
Business-as-usual scenario, 19

C
Carbon dioxide (CO_2), 2, 477
Cassava varieties, 334
Centre for Pacific Crops and Trees (CePaCT),
        338, 514
Chemical fertilizers, 339
Ciguatera, 18
Circularity, 205, 229
Circular migration, 497
Climate-change adaptation (CCA), 329, 336
    community-level decision-making, 521
    community priorities, 516–517
    finance work, 509–513
    forward planning, 517–519
    maladaptation, 521
    Pacific Islands, 500, 501, 521
    peripherality, 505–509
    planned relocation, 504–505
    resettlement schemes, 519–520
    seawalls, 501–504
    traditional food systems, 513–515
Climate change projections, 429, 430
Climate changes
    accelerating pace, 19
    adaptation, 4
    CCA (*see* Climate-change adaptation
        (CCA))
    climate descriptors, 1
    climate variables, 1
    concept, 1
    displacement, 21, 22
    environmental, physical and socio-
        economic aspects, 1
    global warming (*see* Global warming)
    human health risks, 6, 7
    human-induced, 2
    and impacts (*see* Climate impacts)
    implications, 1
    migration, 21, 22
    mitigation, 4
    Pacific island countries (*see* Pacific island
        countries)
    perspectives, 1
    pressure on water and food, 6
    projections (*see* Climate projections)
    public involvement, 1
    system's components, 1
    variability (*see* Climate variability)
    weather patterns and extreme events, 5, 6
    wildlife and ecosystems, 7
Climate finance work
    bottom-up vertical development
        process, 510
    communities/districts, 512

community development planning, 510, 512
community priorities, 511
complex requirements, 509
disaster risk management, 509
district development plans, 511
GCF, 509
horizontal integration, 509
management and utilisation, 509
MEIDECC, 511
MORDI, 510
PICs, 510, 512
PRRP and USP, 510
risk-integrated vertical development
    approach, 510
Climate impacts
annual rainfall, 14
anthropogenic CO$_2$, 16
changing weather patterns and extreme
    events, 5, 6
direct and indirect impacts, 18
ENSO characteristics, 14
foodborne disease, 18
global warming, 4–5
human health, 6, 7, 18
IPCC report, 12, 13
mean significant wave height (H_s) data,
    16, 17
mitigation policies, 14
observed and projected
    rainfall, 14
    temperature, 13
observed time series, 16
pressure on water and food, 6
rainfall and temperature data, 13, 15
sea-level rise, 13
seawater pH level, 17
Suva, 14
temperatures, 13, 14
tropical cyclones, 13
wildlife and ecosystems, 7
Climate-influenced hazards, 339
Climate Model Intercomparison Project Phase
    3 (CMIP3), 252
Climate modelling experiments
change and adaptation pathways, 267
CMIP5 experiments, 267
coarse horizontal resolutions, 265
frequency, 265
GCMs, 265
multi-model study, 265
RCP, 267
Climate projections
climate models
    AOGCMs, 183

CMIP, 184
    evaluations, 187
    GCMs, 184
    primary tools, 183
    RCMs, 183
climate trends, 183
emission and pathways, 183–185
ENSO, 187, 188
extreme rainfall, 190, 192
model evaluations, 185–187
multiple linear and non-linear processes,
    183
rainfall, 189–191
sea-level rise, 192–194
SPCZ, 188, 189
Climate risk exposure, 292
Climate smart agriculture, 348, 349
Climate variability, 331–333
adaptation strategies, 171
average positions, 173
ENSO, 174, 175, 178
EOF, 176
human-induced global warming, 172
IPO, 175–177
observed (*see* Observed climate variability)
particular emphasis, 172
PDO, 175–177
SPCZ, 172–174
Climatological characteristics
formation, 254
motion, 254
peak activity, 254
SPCZ, 255
CMIP3 models, 184–186
CMIP5 models, 185, 186
Coarse-scale satellite imagery, 38
Coastal communities, 378–379
Coastal fisheries
direct effects, 373, 374
indirect effects, 373
Coastal flooding, 324
Coastal fringe, 229
Coastal habitats, 383
Coastal protection structure, 20
Coastal regions, 297
Coastal resources, 386, 387
Coastal sharks, 377
Coastal squeeze, 452–453
Coastal vulnerability, 229, 275, 276, 279, 286,
    292
Coastline data, 281
CO$_2$ emission, 183
Commonwealth Scientific and Industrial
        Research Organisation (CSIRO), 178

Community-based adaptation (CBA), 381, 382
Community-based methods, 338
Community development planning, 510, 511
Community-focused approach, 337
Community priorities
    approaches, 516
    evacuation centres construction, 517
    'failed' interventions, 516
    'for-profit' contractors, 516
    NGOs, 516
    rainwater-harvesting system, 517
    risk-mapping toolkit, 516
Community readiness approach, 337
Community resilience, 379, 381, 382
Compact of Free Association, 479
Composite high islands, 40
Composite low islands, 40
Composite water-level, 209
Continental islands, 40
Convention on International Trade in
        Endangered Species (CITES), 376
Cook Islands, 303, 328
Coral reefs, 8, 18, 368–370
Coupled Model Intercomparison Project
        (CMIP), 184, 187
Coupled Model Intercomparison Project Phase
        5 (CMIP5), 16
Crop-based approach, 337
Crop modeling simulation, 332
Crop models, 333–334
Crop nurseries, 338
Crop production
    foundation, 339
Crown-of-thorn starfish (COTS), 342
Current climate variability
    droughts, 422, 423
    floods, 423, 424
    groundwater inundation, 425
    landslides, 425
    salinity profiles, 426
    storms, 423, 425
    tropical cyclones, 423–425
    wave overwash, 425

D
DAC Creditor Reporting System (CRS)
        database, 329
Decision Support System for Agrotechnology
        Transfer (DSSAT version 4.5)
        model package, 333
Decision support tools, 337
Demography, 360
Demonstration farms, 339

Desalination, 408, 409
Desiree potato variety, 334
Development Assistance Committee
        (DAC), 329
Development of Sustainable Agriculture in the
        Pacific (DSAP), 514
Disappearing islands, 22
Disaster risk management, 325
Disaster risk reduction (DRR), 336, 352
Disastrous events, 286
District development plans, 511
Diverse farming systems, 348
Diversification of crops, 339
Donor-funded projects, 335
Donor funds, 329
Downscaling
    coastal systems, 227
    continental shelf coasts, 227
    definition, 227
    geologic and morphologic features, 227
    hierarchical framework, 227, 228
    islands coverage, 249
    sparse data availability, 249
Droughts, 276, 422, 423

E
Earthquakes, 276
Earth system models (ESMs), 183
Economic development, 380, 385, 386
    aquaculture, 362–364
    fisheries, 362–364
Economic effects
    assumptions, 487
    atmosphere/international waters
            lacking, 487
    climate change, 487
    Coase Theorem posits, 487
    EEZ, 491
    GHG, 488
    habitation, 491
    IPCC, 488, 490
    Pacific Islanders, 490
    PICs, 490
    PICTs, 488, 490
    private rights, 487
    tCO_2e, 488
Economic infrastructure, 277
Economic settings
    GDP, 482
    heterogeneity, 479
    labour markets, 479
    PPP, 479
    production, 482

questionable quality, 479
subsistence affluence, 478
Tokelau, 479
Ecosystem-based adaption (EBA), 381
Ecosystems, 7
Education systems, 361
El Niño, 174, 175
El Niño drought, 422
El Niño Modoki, 175, 257
El Niño-Southern Oscillation (ENSO), 14,
174–179, 181, 182, 185–190, 255,
292, 334, 365, 366, 374, 385,
422, 429
Empirical orthogonal function (EOF), 176
Energy, 278
ENSO impact
basin-wide cooling, 256
cyclone intensity, 257
data homogeneity, 256
equatorward shift, 257
extratropical transition, 258
genesis location and frequency, 257
genesis regimes, 257
intraseasonal variability, 260
non-traditional type, 257
over climatological characteristics, 257
poleward shift, 257
SLP, 256
SST, 256
systematic shifts, 256
track clusters, 257
Environmental stressors, 34
Ethnic and cultural diversity, 360
European centres, 33
Evapotranspiration, 415–417
Exclusive economic zones (EEZs), 8, 280,
359, 360, 362, 365, 374, 375, 380,
385, 386, 491
Exposure index
annual average significant wave heights,
209, 211–213
climate and oceanic factors, 208
composite water-level parameter, 209,
210, 220
cut-off values, 213
ENSO, 209
Pacific islands, 214, 215
parameters, 208, 209
process sensitivity spatial distribution, 214
sea-level rise, 220
tidal range, 209
tropical cyclone frequency, 211–213, 220
tropical cyclone influence, 212
variables, 210

water-level range, 213
wave height parameter, 210
Extreme events, 5, 6, 403, 438, 439
Extreme events and rising sea levels on
agriculture
climate change
and climate variability, 331–333
crops in PICs, 333, 334
climatic hazards, 331

F
Farmers, 325
Farming, 324
Federated States of Micronesia (FSM), 43,
280, 305, 324
Fisheries
coastal, 373, 374
economic development, 362–364
food security, 362–364
livelihoods, 362–364, 384, 385
oceanic (tuna), 374–376
sharks and rays, 376, 377
Fish habitats, 20
Flood events, 369
Flooding, 331
Food production, 6
Food security
adaptation options
filling the gap, 383
minimising the gap, 383
agricultural production, 349
aquaculture, 362–364
fisheries, 362–364
freshwater species, 379
marine resources, 361
in Pacific Island region, 382
PICTs, 389
project, 337–339
vulnerability and risk assessments, 350
Food webs, 367, 368
Forward planning
coastal plants, 519
communal concrete buildings, 518
community engagement process, 518
cultural norms/sensitivities, 518
engaging communities, 518, 519
importance, 517
intensive farming cultivation practices, 519
iTaukei communities, 517
Kadavu Integrated Coastal
Management, 518
low-hanging fruits, 518
upslope relocation, 518

The Foundation of the South Pacific
        International (FSPI), 343
Fourth Assessment Report (AR4), 192
Freshwater, 332
Freshwater availability
    Pacific Islands (*see* Pacific Islands)
    types (*see* Water resources)
Freshwater lakes, 406
Freshwater lens, 332, 407
Freshwater zone, 407, 408

G
GDP, 19, 20, 24, 326, 327
Geographic Information System (GIS), 279
Geomorphic sensitivity index
    coarse assessment, 220
    distributions, 216–219, 222
    exposure and indicative sensitivity
        indices, 221
    indicative susceptibility and process-based
        indices, 214, 215
    islands, 215, 216, 220, 221
    modal geomorphic sensitivity, 217, 222
    Palau, 222
    parameters, 220, 221
    regional scale, 222
    value, 223
Geomorphic susceptibility, 203
Giant swamp taro, 341
Gilbert and Ellice Islands Colony
        (GEIC), 519
Gilbertese settlers, 520
GIS shapefile, 37
Global climate models (GCMs), 13–17,
        183–186
Global community, 34
Global rainfall patterns, 5
Global Self-Consistent, Hierarchical and
        High-Resolution Geography
        Database (GSHHG), 302
Global warming, 452
    1.5 °C global warming, 4
    CO_2 emitted, 20
    impacts, 3
    natural cycles, 2
    and sea-level rise, 4, 5
    scientific community, 3
    scientific evidence, 3
Google Earth imagery, 248
Government revenue, 380, 385, 386
Green Climate Fund (GCF), 330, 509
Green farming technology, 343
Greenhouse effect, 2

Greenhouse gases (GHGs), 2
    agreement, 495
    global emissions, 476, 488
    Kiribati and Tuvalu, 476
    per capita emissions, 488–490
Greenland, 3
Green manure, 339
Groundwater, 406, 408, 412, 423
    management strategy, 437
    in Pacific Islands, 332
    reserves, 333
    salinity, 333
    salinization, 333
    sea level rise, 332

H
Health services, 277
High-elevation ecosystems, 18
Highest astronomical tide (HAT), 209
Households, 328
Human activity, 171
Human health risks, 6, 7
Human mobility, 21

I
ICT infrastructure, 24
Impacts of climate change
    coastal infrastructure (*see* Infrastructure)
Imported bottled water, 409
Improved drinking water, 420
Indicative susceptibility index, 203
    breakdown, 206, 207
    circularity, 205
    distribution, 205, 206
    islands, 203–205, 208
    Kiribati, 208
    lithology, 204, 205
    maximum elevation, 204
    modal indicative susceptibility, 207
    physical variables, 208
    Tokelau, 208
Indicative vulnerability
    definition, 228
    island scale, 244
    matrix, 232, 233
    pacific region, 228
    rankings, 245–247
    susceptibility rankings, 232, 233
    value, 228
    weightings, 233, 234, 244
Industrial revolution, 2
Inequities, 34

Information communication infrastructure,
22–24
Infrastructure
adaptation strategies, 286
assessments, 281, 286
built infrastructure
counts, 282, 283
cumulative values, 283, 284
individual bands, 282, 283
replacement value, 284, 285
characteristics, 279, 281
coast, 281
coastal erosion, 276
coastline data, 281
commercial, 284
components, 277
Cook Islands, 280
development and sustainable livelihood,
278, 279
distribution
Funafuti, Tuvalu, 286, 290, 292
Koror, Palau, 286, 287, 292
Rarotonga, Cook Islands, 286, 288, 292
Tongatapu, Tonga, 286, 289, 292
Upolu, Samoa, 286, 291, 292
economic, 277
energy, 278
GIS, 279
health services, 277
ICT costs, 278
individual countries, 282
industrial, 284
multitude of ways, 277
number and replacement value, 282
PacRIS, 279
PCRAFI, 280, 281
in PICs, 275–277
policies and institutional management, 278
population, 286
public, 284
and public facilities, 278
reef islands, 280
risk assessment, 276
road, 282
socio-economic activities, 278, 279
soil types, 282
South Pacific region, 279
study region, 280
transport, 277
urban centres, 286
weaknesses, 278
Insularity, 230
Integrated Coastal Management (ICM), 517
Integrated vulnerability method, 339

Interdecadal Pacific Oscillation (IPO),
175–177, 179, 182, 185, 264,
365, 430
Intergovernmental Panel on Climate Change
(IPCC), 184, 193, 194, 201, 366,
488, 496
International assistance, 494, 495
International community, 34, 497
International Fund for Agriculture
Development (IFAD), 337
Internet, 23
access, 24
sources, 37
Inundation, 324
Invasive seaweed, 341, 342
Irrigation, 340
Island areas, 37, 38
Island characteristics
areas, 43, 44
distribution, 41–43, 46–48
maximum elevations, 45, 46
Island classification
areas, 37, 38
database, 37
elevations, 38
lithologies, 38, 39
locations and shapes, 37
names, 38
physical and natural attributes, 37
Island database, 35, 36, 49–167
Island distribution, 41–43, 46–48
Island elevations, 38
Island groups, 297
Island lithologies, 38, 39
Island locations and shapes, 37
Island names, 38
Island settings, 35
Island species, 455
Island susceptibility index, 459
Island types
composite high/low islands, 40
continental islands, 40
limestone high/low islands, 39, 40
reef islands, 40
volcanic high/low islands, 39
Island vulnerability assessment
apparent limitations, 249
bias, 249
coastal fringe, 244
coastal landforms skirting, 226
coastal management, 225
criteria, 248
downscaling, 226, 248
elevations, 248

Island vulnerability assessment (*cont.*)
   fine-scale coastal, 248
   geology, 226
   indicative, 228, 232
   instability, 226, 230, 232
   management purposes, 244
   metocean processes, 248
   regional-scale analysis, 226
   regional-scale maps, 244
   suites, 248
   susceptibility, 226, 229–230
   transition, 226

K
Kiribati's agriculture sector, 276, 305,
    316, 326
Kyoto Protocol, 3

L
Labor force employment, 328
Land ownership, 428
Land use survey, 338
La Niña, 174, 175, 422
Larger Melanesian countries, 324
Last Glacial Maximum (LGM), 43
Leakage, 412
LiDAR-generated digital elevation
    models, 316
Lifetime maximum intensity (LMI), 264
Limestone high islands, 39
Limestone islands, 406
Limestone low islands, 40
Lithology, 204
Litres per person per day (Lpd), 411
Livelihoods
   aquaculture, 362–364, 384, 385
   fisheries, 362–364, 384, 385
Livestock Sector Policy, 329
Local early action planning (LEAP), 381, 382
Lowest astronomical tide (LAT), 209
Low-lying coastal areas, 21
Low-lying coral islands, 406

M
Madden-Julian Oscillation (MJO) impact, 255
   El Niño events, 260
   environmental factor, modulates, 260
   lower-frequency mode variability, 261
   propagating disturbance, 260
Mainstreaming of Rural Development
    Innovation (MORDI), 510

Mainstreaming of Rural Development
    Innovations Tonga Trust (MORDI
    TT), 337
Mandarin trees, 332
Mangroves, 8, 371–373
Marginalization, 34
Marine climate, 365
Marine habitats
   coral reef, 368–370
   food webs, 367, 368
   mangroves, 371–373
   oceanic habitats, 367, 368
   seagrass, 370, 371
Marine plants, 342
Marine protected areas (MPAs), 384, 387
Marine resources
   aquaculture, 362–364, 379, 380
   coastal resources, 386, 387
   communities and culture, 378–379
   cultural and social importance, 361
   demography, 360
   economic development, 380
   ethnic and cultural diversity, 360
   fisheries, 362–364
   government revenue, 380
   physical and biological features, 359, 360
   tourism, 362, 365, 379, 380
   vulnerability, 365–367
Marshall Islands, 306, 307, 316
Maximum elevations of islands, 45, 46
Mean annual rainfall contours, 413
Mean rainfall and evaporation changes
   Australian Bureau of Meteorology and
    CSIRO, 431
   CMIP5 projections, 430
   drought frequency and duration, 431
   rainfall stations, 432
   rainfall trends, 432
   rainfall variability, 431
   water resources availability, 431
Melanesia, 9
Meteorology, Energy, Information, Disaster
    Management, Environment, Climate
    Change and Communications
    (MEIDECC), 511
Methane (CH_4), 2
Metocean, 229
Micronekton, 368
Micronesia, 9
Midterm, 229
Migration, 21, 22
Ministry for Agriculture, Forestry, Food and
    Fisheries (MAFFF), 329
Mitigation, 3, 4, 14

Mitigation-related finance, 330
Mobile phone connectivity, 23
Mobile technology, 23
Model ensembles, 186
Modoki-type, 188
Mucuna, 342–343
Multi-agro-ecological sites study, 343

N
NASA's Gravity Recovery and Climate
    Experiment, 3
Nationally Determined Contributions
    (NDCs), 349
Natural climate variability
    ENSO impact, 256–260
    MJO impact, 260–262
Natural disasters, 300, 361
Natural hazards, 276, 336
Nauru, 307, 308, 316
Nitrogen leaching, 334
Nitrous oxide (N₂O), 2
Niue, 308, 309
Non-climate-related natural hazards, 426, 427
Noumea strategy, 384
Nutrient omission trials, 340
Nutrient-rich waters, 367
Nutritional value of crops, 343

O
Observed climate variability
    communities, 178
    greenhouse gases, 178
    high-quality data, 179
    human activities, 178
    natural indicators, 178
    rainfall, 179, 181
    sea-level rise, 181, 182
    temperatures, 179, 180
    tropical cyclone, 180
Ocean acidification, 374
Oceanic (tuna) fisheries, 374–376
Oceanic habitats, 367, 368
Oceanic islands, 34
    diversity, 34
    genesis, 34
Oceans, 20, 369
Ocean waters, 3
OECD-DAC database, 330
Organic composting, 339
Organic fertilizer, 341, 342
Orographic effect, 414
Outgoing longwave radiation (OLR), 261

P
Pacific Adaptation to Climate Change (PACC)
    Program, 514
Pacific-Australia Climate Change Science
    Adaptation Planning (PACCSAP),
    172, 178, 195
Pacific Basin, 35, 36
Pacific Catastrophe Risk Assessment and
    Financing Initiative (PCRAFI),
    280, 302
Pacific Climate Change Science Program
    (PCCSP), 172, 178, 180, 190, 195
Pacific Climate Futures version 2.0, 333
Pacific Community (SPC), 337–340
Pacific Decadal Oscillation (PDO), 175, 177
Pacific Equatorial Divergence, 368
Pacific island countries (PICs), 405
    agriculture (see Agriculture sector)
    challenges, 275, 324
    coastal populations, 301
    community relocation, 320
    coral/sandy formations, 276
    critically endangered species, 465,
        467–469
    database, 458, 459
    endangered species, 465, 467–469
    ESRI's base map, 302
    geographical isolations, 275
    governments, 278
    H_s values, 466
    islands hosting, 458, 464, 465
    IUCN site lists, 459
    low economic stability, 275
    lower accessibility, 278
    marine species, 459
    natural hazards, 276
    PCRAFI, 302
    population distribution (see Population
        distribution)
    RCPs, 463
    residential buildings, 302
    resources and island size, 465
    species vulnerability, 457
    terrestrial vertebrate species, 457, 459
    threatened species, 459–463, 465, 466
Pacific Island Countries and Territories
    (PICTs), 359, 360, 362–364
    climate change, 478, 496
    customary land tenure systems, 497
    economic effects (see Economic effects)
    economic settings, 478–487
    policy options, 491–496
    TC Gita, 496
Pacific Islanders, 497, 501

Pacific islands
    adaptation, 22–24
    adaptive capacity, 22–24
    apocalyptic consequences, 475
    biodiversity, 18
    biogeochemical cycles, 8
    characteristics, 10, 11
    climate change, 297
    coral reef formation, 10
    current climate variability (*see* Current
        climate variability)
    demographic indicators, 297–299
    economic impacts, 19–20
    EEZ, 8
    El Niño, 296
    emissions, 10–12
    energy consumption, 476
    environmental events, 300–301
    equator divides, 7
    ethno-geographic and cultural lines, 9
    geological origin, 9, 10
    global climate, 8
    governments, 520
    high-elevation ecosystems, 18
    human activities, 427–429
    indigenous communities, 19
    information communication infrastructure,
        22–24
    international community, 477
    lack of information, 22–24
    Melanesia/Micronesia, 9
    natural disasters, 300
    non-climate-related natural hazards,
        426, 427
    northeastern quadrant, 8
    PICs, 300
    Polynesia, 9
    population density, 297
    population distribution, 10
    problems, 300
    rainfall characteristics, 415
    and resources, 8
    safe water, 420, 421
    sea-level rise, 10, 295
    shoreline ratio, 320
    and territories, 21
    topography, 8
    urban populations, 420, 421
    volcanic islands, 10
    water security, 419
Pacific Risk Information System (PacRIS), 279
Pacific Risk Resilience Program (PRRP), 510
Pacific Small Island Developing States
    (PSIDS), 341, 342

Palaeoclimatologists, 2
Palau, 309, 310
Paleotempestology, 263
Palintest SKW 500 Quick Soil Test Kit, 340
Papua New Guinea (PNG), 323
Participatory Learning and Action (PLA), 510
Participatory productivity index, 340
Participatory rural appraisal, 338
Particularly cyclones, 361
PCRAFI database, 281
Pelagic sharks, 377
Peripherality
    communities, 506, 507
    coping ability, 507
    effective and sustainable adaptation, 506
    Logana and Tausa, 507
    measurement, 508, 509
    media, 507
    Pacific Island communities, 506
    rising ocean temperature, 505
    traditional environmental knowledge, 507
Phoenix Islands Resettlement Scheme, 519
Physical character, 229
Phytoplankton, 367, 368
Piped water supply systems, 427
Planned relocation
    coastal vulnerability, 504
    coral reefs, 505
    Denimanu communities, 505
    Fiji government, 504
    finance, 505
    high-emission scenario, 504
    in situ adaptation alternatives, 505
    natural fluctuations, 504
    rural areas, 504
    sea-level rise, 505
Plant-derived fertilizers, 342
Poleward migration, 372
Policy options
    adaptation, 494
    GHG reduction, 491
    international assistance, 494, 495
    migration, 492–494
    responses, 492
Polynesia, 9
Polynesian countries, 323
Poor water governance and management, 428
Population distribution
    coastal communities, 297
    Cook Islands, 303
    FSM, 305
    Kiribati, 305, 306
    Marshall Islands, 306, 307
    Nauru, 307, 308

Niue, 308, 309
Palau, 309, 310
Samoa, 310, 311
Solomon Islands, 311, 312
Tonga, 313
Tuvalu, 313, 314
Vanuatu, 314, 315
Post-disaster needs assessments, 331
Potential evapotranspiration (PET), 336
Private sector, 323
Production index, 350
Projected mean sea level (MSL), 434–436
Proximity, 230
Pumping systems, 428
Purchasing power parity (PPP), 479

R
Rain fed, 324
Rainfall, 179, 181, 189–191, 408, 413, 414, 451, 452
Rainwater, 408
Raising awareness, 23
Reconstructed paleoclimate temperature records, 178
Recovery mode, 325
Reducing Emissions from Deforestation and Forest Degradation (REDD+), 330
Reefal lithology, 319
Reef islands, 40, 316
Regional climate models (RCMs), 183
Regional scale, 229
Relative instability, 229
Representative Concentration Pathways (RCPs), 185, 192, 194, 267, 366, 367, 463
Republic of the Marshall Islands (RMI), 324
Resettlement schemes
    BSIP, 520
    GEIC Administration, 519
    Gilbertese settlers, 520
    Phoenix Islands, 519
Residential buildings
    calculations, 302
    coastal hazards, 301
    database, 301
    exposure database, 302
    population distribution, 301
    projection, 302
Resilient crops, 339
Resilient Development in the Pacific (FRDP), 509
Resourcing agriculture, 329, 330
Reverse osmosis (RO), 408

Rio markers, 329
Rising temperatures, 276
Risk assessments, 350
Risk-integrated planning process, 512
Road infrastructure, 282
Rural water supply systems, 412

S
Safe drinking water, 420
Salinity
    crops, 333
    groundwater, 333
    intrusion, 332
    low-lying coastal areas, 332
    soil, 332, 333, 344
    water, 332
Saltwater intrusion, 324, 333
Samoa, 310, 311
Samoa Agriculture Sector Plan 2016–2020, 329
Samoan culture, 19
School of Agriculture and Food Technology, 343
Seabed/shoreface gradient, 230
Seagrass, 8, 370, 371
Sea-level pressure (SLP), 174, 256
Sea-level rises, 2, 4, 5, 10–13, 16, 18, 20, 22, 171, 181, 182, 192–194, 276, 292, 369, 452
    climate impacts, 295, 296
    El Niño event, 296
    extreme climate events, 296
    pacific community, 301
    Solomon Islands, 295
Seamounts, 8
SEAPODYM model, 387
Sea surface temperatures (SSTs), 172, 174, 175, 186, 187, 256, 366, 369
Seawalls
    advantages and problem, 503
    coastal dynamics, 503
    coastal erosion and inundation, 502, 503
    Pacific Islanders, 501
    shoreline erosion, 501
    vertical structure, 503
Seawater, 410
Seawater inundation, 438
Seaweed fertilizer, 342
Shallow-water organisms, 17
Sharks and rays, 376, 377, 388
Shoreline erosion, 501
Sinking islands, 22
Skipjack tuna, 374

Slow-onset events, 276
Slow-onset stressors, 331
Small Island Developing States (SIDS), 403
Social cultural activities, 328, 329
Socio-economic activities, 278, 279
Socio-economic benefits, 362
Socio-economic reality, 19
Soil fertility, 324, 326, 334, 338, 339, 342, 343
Soil-moisture balance maps, 336
Soil testing, 340
Soil types, 282
Solomon Islands, 295, 311, 312, 328, 520
Southern Oscillation, 174
Southern Oscillation index (SOI), 175
South Pacific Applied Geoscience
    Commission (SOPAC), 16
South Pacific Convergence Zone (SPCZ),
    172–174, 179, 185–190, 255
South Pacific Convergence Zone index
    (SPCZI), 173, 174
Southwest Pacific Enhanced Archive for
    Tropical Cyclones (SPEArTC), 253
Spatial ecosystem and population dynamics
    model (SEAPODYM), 374
Spider-web diagrams, 340
Stocks, 387, 388
Stormy weather, 20
Strategic directions, agriculture sector plans
    access to land, 344, 348
    applied and accessible research, 350–351
    climate smart agriculture, 348, 349
    diverse farming systems, 348
    and food security, 344–347, 349, 350
    genuine partnerships, 351
    healthy soils, 344, 348
    resourcing, 352, 353
    sustainable land management, 344, 348
    sustainable water supply, 348
Subsistence-based communities, 378
Subsistence-based economies, 361
Subsurface waters, 366
Sudden-onset climatic events, 513
Surface water, 406, 412
Susceptibility, 202, 203
    circularity, 229, 230
    criteria, 230–232
    insularity, 230
    proximity, 230
    seabed/shoreface gradient, 230
    whole-island, 229
Susceptibility index
    exposure index (*see* Exposure index)
    indicative susceptibility (*see* Indicative
        susceptibility index)

Sustainable water supply, 348
Suva, 14
Systems-oriented approach, 335

T
Taro production, 333, 334
Taro varieties, 340
Telecommunications, 23
Teleconnections, 175
Temperatures, 450
Terrestrial biodiversity, 450
Three-dimensional groundwater model, 436
Tonga, 313
Tonga Agriculture Sector Plan (TASP),
    335–337
Tons of carbon dioxide equivalent (tCO$_2$e), 488
Total dissolved salts (TDS), 423
Tourism, 20, 362, 365, 384, 385
    marine resources, 379, 380
Traditional communities, 378
Traditional food systems
    adaptation interventions, 515
    agriculture system, 513
    CePaCT, 514
    climate-change loss and damage, 515
    community-based food security project, 515
    community-focused interventions, 514
    dependence on rainfall, 513
    DSAP, 514
    namamau, 515
    PACC program, 514
    Pacific food systems, 514
    staple crops/trees, 513
    sudden-onset climatic events, 513
Traditional knowledge, 361
Transient islands, 36
Transportation, 284
Tripole index (TPI), 177
Tropical cyclones (TCs), 5, 13, 180, 276, 331,
    340, 366, 369, 423, 434
    classification, 253
    climate change
        anthropogenic activities, 262
        climate modelling experiments,
            265–267
        IPO, 264
        LMI, 264, 265
        long-term reanalysis products, 262
        paleotempestology, 262, 263
        SPEArTC, 264
        systematic assessments, 262
        warning centres, 263
    climate modelling, 252

climatological characteristics, 253–256
data records and homogeneity
    comprehensive compilation, 253
    drawbacks, 253
    Ontong Java, 252
    quantitative risk assessments, 252
    research, 253
    SPEArTC, 253
economic loss, 251
factors, 251
frequency, 211, 212
Gita, 475, 477, 496
natural climate variability, 256–262
Pacific, 251
Pam, 477
socio-economic impacts, 268
Tsunami, 276
Tuna fisheries, 387
Tuvalu, 313, 314
Typical community water supplies, 412
Typical per capita water, 412

U
United Nations Biodiversity A-Z, 453, 454
University of the South Pacific (USP), 510, 515
Urban areas, 328
Urban centres, 277
Urban water supply systems, 412
USAID Food Security Adaptation Project,
        337–339

V
Vanuatu, 314, 315
Vessel day scheme (VDS), 385
Volcanic high islands, 39
Volcanic islands, 10, 276
Volcanic low islands, 39
Vulnerability
    analytical tool, 202
    definition, 201, 202
    marine resources, local changes
        dissolved oxygen concentrations, 366
        ENSO, 366
        human interference, 365
        interannual variations, 365
        IPCC-AR, 366
        IPCC predictions, 366
        RCP, 366, 367
        SST, 366
        subsurface waters, 366
        superimposed, 366
        TC, 366

physical and biologic factors, 202
risk assessment, 202

W
Walker cell, 174
Walker circulation, 174
Warmer conditions, 2
Warmer temperatures, 6
Water balance
    characteristics, 417
    rainfall trends, 418
    rainfall *vs.* actual evapotranspiration, 417
    variation of rainfall, 417
Water-borne disease, 427
Water meters, 341
Water monitoring programs, 428
Water resources
    desalination, 408, 409
    extraction, water, 410
    freshwater lens, 407
    groundwater, 406, 408
    importation, 409, 410
    non-potable uses, 410
    Pacific islands, 405
    publications, 410
    rainwater, 408
    seawater, 410
    surface water, 406
Water security, 419
    actual sea level, 435
    atoll islands, 438, 439
    climate change projections, 429, 430
    ENSO effects, 435
    extreme rainfall frequency, 432, 433
    fresh groundwater, 435–437
    intensity changes, 432, 433
    mean and extreme temperature
            changes, 433
    mean rainfall and evaporation changes, 430
    MSL, 434, 435
    tropical cyclones, 434
    wave overwash, 437, 438
Water use
    agriculture, 412
    hydrological influences
        characteristics and rainfall, 413, 414, 418
        evapotranspiration, 415–417
        rainfall, 413, 414
    limited industry, 412
    mining, 412
    non-consumptive, 412
    water requirements, 411
    water supply, 411, 412

Wave height (H$_s$), 459, 463
Wave overwash, 425, 437, 438
Weighted and non-weighted estimates
    instability ranking, 234, 239–243
    island susceptibility ranking,
        234–238
Weightings
    island susceptibility, 233
    secondary assessment, 244
    secondary criteria, 244, 249
    variables, 230

Western and Central Pacific Commission
    (WCPFC), 386
Western and Central Pacific Ocean (WCPO),
    360, 374
Wildlife, 7
World Geodetic System 1984 (WGS84), 37
World Vector Shorelines (WVS) database, 37

Z
Zooplankton, 368

Printed in the United States
By Bookmasters